T0251689

GRAPH-BASED SOCIAL MEDIA ANALYSIS

Chapman & Hall/CRC
Data Mining and Knowledge Discovery Series

SERIES EDITOR

Vipin Kumar

University of Minnesota
Department of Computer Science and Engineering
Minneapolis, Minnesota, U.S.A.

AIMS AND SCOPE

This series aims to capture new developments and applications in data mining and knowledge discovery, while summarizing the computational tools and techniques useful in data analysis. This series encourages the integration of mathematical, statistical, and computational methods and techniques through the publication of a broad range of textbooks, reference works, and handbooks. The inclusion of concrete examples and applications is highly encouraged. The scope of the series includes, but is not limited to, titles in the areas of data mining and knowledge discovery methods and applications, modeling, algorithms, theory and foundations, data and knowledge visualization, data mining systems and tools, and privacy and security issues.

PUBLISHED TITLES

ACCELERATING DISCOVERY : MINING UNSTRUCTURED INFORMATION FOR HYPOTHESIS GENERATION
Scott Spangler

ADVANCES IN MACHINE LEARNING AND DATA MINING FOR ASTRONOMY
Michael J. Way, Jeffrey D. Scargle, Kamal M. Ali, and Ashok N. Srivastava

BIOLOGICAL DATA MINING
Jake Y. Chen and Stefano Lonardi

COMPUTATIONAL BUSINESS ANALYTICS
Subrata Das

COMPUTATIONAL INTELLIGENT DATA ANALYSIS FOR SUSTAINABLE DEVELOPMENT
Ting Yu, Nitesh V. Chawla, and Simeon Simoff

COMPUTATIONAL METHODS OF FEATURE SELECTION
Huan Liu and Hiroshi Motoda

CONSTRAINED CLUSTERING: ADVANCES IN ALGORITHMS, THEORY, AND APPLICATIONS
Sugato Basu, Ian Davidson, and Kiri L. Wagstaff

CONTRAST DATA MINING: CONCEPTS, ALGORITHMS, AND APPLICATIONS
Guozhu Dong and James Bailey

DATA CLASSIFICATION: ALGORITHMS AND APPLICATIONS
Charu C. Aggarawal

GRAPH-BASED SOCIAL MEDIA ANALYSIS

Edited by

Ioannis Pitas

Aristotle University of Thessaloniki

Greece

CRC Press

Taylor & Francis Group

Boca Raton   London   New York

CRC Press is an imprint of the
Taylor & Francis Group, an **informa** business

A CHAPMAN & HALL BOOK

CRC Press
Taylor & Francis Group
6000 Broken Sound Parkway NW, Suite 300
Boca Raton, FL 33487-2742

© 2016 by Taylor & Francis Group, LLC
CRC Press is an imprint of Taylor & Francis Group, an Informa business

No claim to original U.S. Government works

Printed on acid-free paper
Version Date: 20160223

International Standard Book Number-13: 978-1-4987-1904-9 (Hardback)

Visit the Taylor & Francis Web site at
http://www.taylorandfrancis.com

and the CRC Press Web site at
http://www.crcpress.com

Contents

Preface

The great recent changes in the World Wide Web (WWW) facilitated media information sharing among users, user-based content creation and remote collaboration. Nowadays, social networks allow users to post public profiles and share them with their friends, thus creating virtual communities and networks. Posting relevant media data (notably images, videos, music and text) in social media sites enabled a dramatic increase of multimedia data storage, communication and consumption. Social media are concentrated on the creation and exchange of user-generated content, allowing users to create, search, share, rate and access multimedia data, thus creating a totally new media experience. Furthermore, we now see a rapid convergence of classical broadcasted media and social media, where content seamlessly flows and is exchanged between these two worlds.

This edited book focuses on the use of graph analysis in the study of social media and digital media. It covers the following topics: graphs in social media, graph and hypergraph fundamentals, mathematical foundations coming from linear algebra, algebraic graph analysis, graph clustering, community detection, graph matching, web search based on ranking, label propagation and diffusion in social media, graph-based pattern recognition and machine learning, matrix and tensor decomposition factorization in multimedia recommendations, multimedia social search based on hypergraph learning, graph signal processing, big data approaches for social media analytics, evolving social data analysis and big graph storage, processing and visualization. Emphasis is on big data aspects, as social media address inherently strong big data issues related both to the size of the stored multimedia content and to the social graph size.

This book addresses an important scientific and technological challenge, namely the confluence of graph analysis and network theory with linear algebra, digital media, machine learning, big data analysis and signal processing. I firmly believe that this convergence can develop novel ground-breaking approaches in social media and digital media analysis, towards unleashing their full potential. To this end, novel lines of research can be followed in: a) graph-based approaches in media representation (e.g., graph signals), machine learning (e.g., graph embedding techniques), use of context (notably rich tags and associated graphs), b) big data analysis methods (distributed, approximate, sub-sampled or adaptive analysis) and c) incremental methods to address the dynamic/evolving nature of social media data.

It can be used as a textbook in undergraduate or graduate courses on digital media, social media or social networks. Its targeted audience is University students in computer science, electrical engineering, mathematics, information science, digital media or sociology departments that want good scientific and technical coverage of social media and digital media analysis. The number of relevant courses offered steadily increases with the popularity of social media. Furthermore, almost all major Universities offer courses on web topics, or on digital media/multimedia, which have special parts devoted to social media/networks. Students in the above mentioned fields can now have a complete reference textbook in graph-based social media analysis. Finally, scientists and engineers working in social media and in digital media production and distribution, as well as the general public with scientific interests, can obtain a thorough understanding of social media analysis.

The practical outcome for the audience can be the following:

1. Understanding the social media structure.

2. Understanding graph theory, particularly the algebraic description and analysis of graphs and their use in social media studies.

3. Helping scientists, artists and sociologists understand complex social media phenomena, like information diffusion, marketing and recommendation systems in social media and evolving systems.

4. Acquiring expertise to analyze the social and digital media markets.

5. Providing insight into processing, storing and visualizing big social media data and social graphs.

As an editor, I would like to thank the various chapter authors for their excellent work in compiling a very good overview of graph-based social media analysis. Furthermore, I am grateful to Mr. N. Tsapanos and Ms. F. Patrona for helping to prepare this edited book. Finally, I would also like to thank the following reviewers of the various book chapters: G. Arce, B. Baingana, N. Barbalios, N. Bassiou, A. Bors, T. Chen, A. Delopoulos, X. Dong, B. Huet, G. Giannakis, W. Hu, A. Iosifidis, G. Karypis, E. Kofidis, I. Kompatsiaris, C. Kotropoulos, O. Lezoray, A. Lykas, Y. Manolopoulos, S. Narang, J. Pokorny, D. Rafailidis, P. Simeonidis, A. Tefas, S. Theodoridis, N. Tsitsas, A. Vakali, P. Vandergheynst, N. Vretos and O. Zoidi.

Acknowledgements. Chapters 1 and 5 are related to the research done within the EU funded FP7 R & D projects IMPART and 3DVTS, respectively.

Prof. Ioannis Pitas
Department of Informatics
Aristotle University of Thessaloniki
Thessaloniki, Greece

Contributors

Brian Baingana
University of Minnesota
Minneapolis, U.S.A.

Nikoletta Bassiou
Aristotle University of Thessaloniki
Thessaloniki, Greece

Eftychia Fotiadou
Aristotle University of Thessaloniki
Thessaloniki, Greece

Georgios Giannakis
University of Minnesota
Minneapolis, U.S.A.

Alexandros Iosifidis
Aristotle University of Thessaloniki
Thessaloniki, Greece

Santosh Kabbur
University of Minnesota
Minneapolis, U.S.A.

George Karypis
University of Minnesota
Minneapolis, U.S.A.

Constantine Kotropoulos
Aristotle University of Thessaloniki
Thessaloniki, Greece

Gonzalo Mateos
University of Rochester
New York, U.S.A.

Sunil Narang
University of Southern California
Los Angeles, U.S.A.

Ioannis Pitas
Aristotle University of Thessaloniki
Thessaloniki, Greece

Jaroslav Pokorny
Charles University
Prague, Czech Republic

Vaclav Snasel
Charles University
Prague, Czech Republic

Panagiotis Symeonidis
Aristotle University of Thessaloniki
Thessaloniki, Greece

Andrea Tagarelli
University of Calabria
Arcavacata di Rende (CS), Italy

Anastasios Tefas
Aristotle University of Thessaloniki
Thessaloniki, Greece

Panagiotis Traganitis
University of Minnesota
Minneapolis, U.S.A.

Nikolaos Tsapanos
Aristotle University of Thessaloniki
Thessaloniki, Greece

Olga Zoidi
Aristotle University of Thessaloniki
Thessaloniki, Greece

Editor Biography

 **Prof. Ioannis Pitas** (IEEE fellow, IEEE Distinguished Lecturer, EURASIP fellow) received a diploma and PhD degree from the Department of Electrical Engineering, Aristotle University of Thessaloniki, Greece. Since 1994, he has been a Professor in the Department of Informatics of the same University. He has served as a Visiting Professor at several universities. His current interests are in the areas of intelligent digital media, image/video processing, machine learning and human centered computing. He has published over 800 papers, contributed to 44 books in his areas of interest and edited or (co-)authored another 10 books. He has also been a member of the program committee of many scientific conferences and workshops. In the past, he served as Associate Editor or co-Editor of eight international journals and was General or Technical Chair of four international conferences. He participated in 68 R&D projects, primarily funded by the European Union and is/was a principal investigator/researcher in 40 such projects. He has 20600+ citations to his work and an h-index of 67+ (2015).

Contact information:

Prof. Ioannis Pitas, Department of Informatics
Aristotle University of Thessaloniki,
Thessaloniki 54124, Greece
http://www.aiia.csd.auth.gr/
Email: pitas@aiia.csd.auth.gr

Chapter 1

Graphs in Social and Digital Media

Alexandros Iosifidis, Nikolaos Tsapanos and Ioannis Pitas

Aristotle University of Thessaloniki, Greece

1.1 Introduction

Over the past few years, great changes have occurred in the World Wide Web (WWW). Web evolution has introduced new applications, which facilitate the interactive exchange of information, user-based content creation, and remote collaboration. Contrary to the static web pages prevalent in the past, where users could only access information, current web applications allow users to communicate, upload, and modify content. *Social networks* are web-based services that allow their users to create a public profile, create a list of users with whom they share connections and material, and view such connections within the system [Ell08]. Such prominent social network applications, which have nowadays become part of everyday life and are definitely a very important factor in web development, continue growing, while many web services renovate to meet growing demand and provide new functionalities.

Naturally, these trends have dramatically increased data flow, especially that of multimedia content (images, videos, music, text), which is now more than ever "user generated." One of the most important revelations in this transition is undoubtedly the emergence of social media. *Social media* is the social interaction among people in which they create, share, or exchange information and ideas in virtual communities and networks [ABHH08]. Social media have also been defined as a group of Internet-based applications that build on the ideological and technological foundations of Web 2.0 and that allow the creation and exchange of *user-generated content* [KH10]. These applications allow users to join communities and create, find, share, rate, and access the multimedia data that is in the respective social media sites. They also allow people to find, connect, inform, and inspire others. It should be noted here that the term social media is used both for the above described tools, as well as for the multimedia data describing social activity of the users. In this book this term will be used interchangeably for describing both tools and media data.

Typically, traditional media, e.g., newspapers, magazines, TV, and radio, provide one-way information diffusion, since the audience can read or listen, but cannot share his/her opinion on a subject, but to a very limited extent. On the contrary, social media provide two-way information diffusion that allows the user to share his/her information with other users, access their information, and communicate with them.

This information communication and sharing can be simple, such as commenting on one's status on social networks, or rating a video in video sharing sites, or more complex such as movie recommendation, based on user profiling and clustering techniques. Note that social media are tools with which users can generate content, contrary to social networking websites, whose purpose is to connect users and allow interaction among them. Social media popularity has offered countless opportunities for creating not only personal, but also professional relationships. This is why webometrics, i.e., web statistics and analytics sites, like Alexa and Google Analytics, show that the most popular websites are Social Media-based pages [Ell08].

Some popular Social Media platforms and their purpose follow:

- Social networking sites: Facebook, Hi5, Last.FM, LinkedIn, MySpace.

- Video and image sharing sites: YouTube, Flickr, Picasa.

- Collaborative work sites: Wikipedia, Wikia, Wikitravel.

- Social bookmarking services: Del.icio.us, Digg, Blinklist, Simpy.

- Blogging and micro-blogging: Blogger, ExpressionEngine, LiveJournal, Twitter, Wordpress.

- Virtual worlds for social gaming: Active Worlds, Forterra Systems, Second Life.

Before proceeding to the presentation of popular social networks and their relevant data, we describe how such networks can be depicted by exploiting graphs [dSP65]. Let us consider a toy example social network of a small number of users. By exploiting social media data, each user i can be represented by a feature vector $\mathbf{v}_i$, as illustrated in Figure 1.1.1. Such a feature vector $\mathbf{v}_i$ is an abstraction of the media data (e.g., images, video, tags) that reside inside a user model i. It must be noted though that, typically, such data are highly unstructured and heterogeneous. As will be further described in the following sections, this social network can be represented by a graph $\mathcal{G}(\mathcal{V}, \mathcal{E}, \mathbf{W})$, where each graph node $\mathbf{v}_i \in \mathcal{V}$ represents a unique user and graph edges $e_{ij} \in \mathcal{E}$ between nodes exist, whenever a relation connects two users i and j. The matrix $\mathbf{W}$ contains the weight values corresponding to the graph edges e_{ij}. Edges e_{ij} can be established by using graph node relationships defined by the corresponding social networks. For example, an edge e_{ij} connects two users i and j if they have a "friendship" relationship in a social network. Alternatively, edges e_{ij} can be established by using similarity measures defined over the node representations of $\mathcal{G}$. For example, an edge e_{ij} will connect two users i and j, if the user representations $\mathbf{v}_i$ and $\mathbf{v}_j$ are "similar," according to a similarity measure $s(\mathbf{v}_i, \mathbf{v}_j)$. In the first case, the corresponding graph is usually unweighted, i.e., $w_{ij} = 1$ or 0, in the case where the users i and j are connected or unconnected, respectively. In the latter case, the corresponding graph is usually weighted, i.e., w_{ij} take values defined by the similarity measure $s(\mathbf{v}_i, \mathbf{v}_j)$. Typically, such similarities are scaled in the range $[0, 1]$ or $[-1, 1]$. We consider that the i-th and j-th graph nodes are connected, if e_{ij} exists or, equivalently, if $w_{ij} \neq 0$.

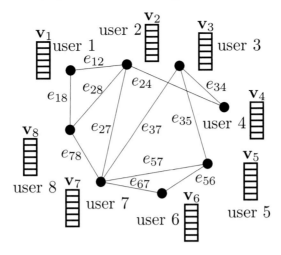

FIGURE 1.1.1: A toy example of a social graph.

1.2 Dominant social networking/media platforms

In this section, some of the most widely used social media/network platforms, namely YouTube, Facebook, Twitter, Instagram, Flickr and LinkedIn, are presented, along with the services and information they provide to users. These platforms are not only very popular, but they also cover a wide range of social networking services. Subsequently, Blogs and bibliographic databases are briefly described.

YouTube is a video sharing website, where users can upload, view, rate videos, and comment on them. It was established in 2005 as an independent video sharing service. In 2006, it became part of Google's services. Registration and use is free of charge, since its profits come from advertisements. According to YouTube statistics, 100 hours of video are uploaded every minute [You]. It is also estimated that nearly 2 billion videos are being viewed daily. Most videos in YouTube come from regular users and some come from businesses, organizations, etc. YouTube content is typically user-generated. Two other interesting aspects of YouTube are a) the ability to embed videos in other sites and b) the YouTube recommendation system. The latter is a very important and complex mechanism that employs information about user activities and interests, in order to recommend videos.

Founded in 2004, *Facebook* is a social networking service that connects people with friends, family, etc. As of March 2014, it had over one billion users, making it the largest social network on the web, in terms of user base and global reach. After registration, users are able to update their profile and add other users in their friend list. Each user is able to upload photographs and videos, exchange messages with other users, share his or her thoughts by updating the status of his/her profile, share external website URLs, play games and much more. Facebook users can decide whether they want their profile media content to be public or hidden. In addition, users are able to create web pages, called *groups*, which are devoted to common interests with other users, and advertisement web pages, where they can, e.g., promote their business. One of the most characteristic Facebook features is the *Like* button. Moreover, the users are able to comment on photos, videos and status of their friends. This allows users to express their opinion about a status, a photo, a brand etc, thus providing information for both content and user profile analysis.

Twitter was launched in 2006 as a social platform, whose main purpose was the creation of microblogs, i.e., posts (*tweets*) that are no more than 140 characters long. Today, with more than 280 million active users, it has become a popular platform of asynchronous communication, where people can exchange ideas, and opinions, and promote themselves or their companies. Compared to other social networking sites, like Facebook, it provides fewer networking services. It has been designed to provide each user with a profile containing some personal information, according to user's privacy settings. A user's profile is the 'place' where one's posts can be seen. The relationship established between two users is that of *followee-follower*. A user (follower), by choosing to follow another Twitter user (followee), has the ability to see all the followee's posts on his/her homepage. Thus, a user homepage turns into a sort of newspaper, where all relevant posts are gathered and can be viewed. *Hashtags*, i.e., one or more words collected into unspaced labels prefixed by the # sign, are one of the core features of Twitter, helping users discover content and other people relevant to their interests. Twitter is available on all sorts of portable devices, such as tablets, smartphones, etc.

Founded in 2010, *Instagram* is a photo/video sharing and social networking service. Its name is a combination of "instant camera" and "telegram." It is now owned by Facebook. As of November 2013, Instagram had over 200 million monthly active users, sharing more than 20 billion photos and videos. Instagram users can capture pictures or videos, edit them by using mobile application tools (crop, rotate, apply predefined filters and tilt-shift) and share them with their followers. Each photo or video can be associated with a certain Foursquare venue. This is made possible by using the GPS on the capture device and the search feature provided by Instagram. Regarding its social aspect, Instagram users can connect with others by using the *Follow* function. The service has adopted the same follower model also used in Twitter. If a user is characterized as a "public" person, he or she can be followed by anyone, whereas, if someone is characterized as a "private" person, he/she must approve each follow request separately. Users can like or comment on photos shared by public or private profiles that they follow. Similar to Twitter, hash tags are one of the core features of Instagram, helping users to discover content and other people.

Flickr is a website where people can upload and share pictures or videos. It was created in 2004 and became part of Yahoo! services. Today it has more than 87 million users. The platform offers two kinds of user accounts: Free and Pro. Users of Free accounts can upload up to 300 MB of information every month, while users of Pro accounts have unlimited space and bandwidth. The site automatically deletes users who are not active for more than 90 days. Besides uploading and downloading pictures and videos, users also have the ability to comment, describe, tag, categorize data and edit privacy settings, i.e., specify which users can view their posts and categorize their data. The users can also include data in third party sites, like Facebook, Twitter, several blogs, etc. Last, the services of Flickr are available on portable devices and can be also accessed via RSS feeds, emails and posts from other sites.

LinkedIn is a business-oriented social network. It was created in 2003 and today has more than 259 million users in more than 200 countries [Hem13]. The basic functionality of LinkedIn allows users to create profiles and *connections* with other users. Each user is able to update his/her professional profile, exchange messages with other users, share his/her thoughts by uploading the status of his/her profile, comment on the status of his/her connections, follow company accounts and seek for professional opportunities. In addition, users are able to create and follow Groups, which mainly concern professional and career issues. Groups keep their members informed through messages containing updates to the group.

Blogs (derived from the word weblog) are informational sites, in which content is uploaded in the form of *posts*, which mostly consist of text and, sometimes, images, embedded video clips or sound files. Each post is usually tagged with relevant labels that also serve as search terms. Newer posts are usually sorted before older ones, in reverse chronological order. Originally, each blog was the work of a single person and often covered a single subject. Recently, *multi-author blogs* (MABs) have been developed, where a large number of authors are allowed to write posts. Examples of MABs are blogs operated by newspapers and universities. The majority of blogs is interactive, allowing visitors to interact by commenting the posts. Thus, blogs can be considered to be social networking services. A typical blog includes text, images, videos and links to other web pages or blogs.

In scientific publishing, information appearing in bibliographic databases, like Thomson ISI's Web of Science and MEDLINE, has been exploited for the study of scientific activity and the importance of scientific articles, journals and scientists, typically by counting citations to each paper. Such information is studied by bibliometrics [DB09]. It has attracted much attention, since the realization that bibliographic data have a natural mathematical representation using graph structures as citations are essentially links between one paper and the papers to which it refers [dSP65]. Such structures have led to the creation of *author networks* and *citation networks*, describing connections between scientists and specific contributions, respectively [dSP65].

1.3 Collecting data from social media sites

Social media can provide *social metrics*, i.e., measurable data that can be accessed by users. Social metrics are very important, in that they can be used in social graph analysis to extract useful information, termed as *social analytics* [Sch96]. For example, the number of views of a YouTube video, or the number of "Likes" of a Facebook group post can be further analyzed in order to define trends or cascades. Collecting data from social media may also be handy, in order to visualize social graphs, their structure and their evolution. Such social graphs may be either static, describing a 'snapshot' of the social network in time instance $t = t_0$, or evolving over time, having the form $\mathcal{G}(\mathcal{V}(t), \mathcal{E}(t), \mathbf{W}(t))$. In the latter case, transitions of the graph structures from an initial state to future states can be evaluated and analyzed [PBV+07].

To satisfy the need of gathering data, social media sites provide *Application Programming Interfaces* (APIs) that allow users to interact with their database and have access to media content data, metadata, statistics, and other social metrics. In the following, we describe some of the functionalities and the data types that are available for download.

Being one of the largest video databases on the web, YouTube has one of the most powerful and complete APIs. The available YouTube data can be either *video-oriented* or *user-oriented* and can be seen in Tables 1.3.1 and 1.3.2, respectively. Some of the above data can only be accessed if their creator allows public access to them. In addition, the API allows a programmer to use the application/platform and its services. For instance one can search for videos, filter the results, upload videos, etc. A very important feature is that of YouTube recommended videos. While displaying search results, YouTube returns a list of, at most, 50 recommended videos to a specific user. Such a feature can be very useful, since it allows access to parts of the YouTube recommendation graph for further analysis [TPP14].

Facebook has a vast API that provides a high level of interaction with their database. The data types that can be accessed from Facebook are shown in Table 1.3.3. The Facebook API provides almost all data that appear in a user profile. However, the user needs the correct

TABLE 1.3.1: Video-Oriented data available in YouTube.

Type	Data
Entry	Unique video ID
	Video title
	Author name and address
	Video category (e.g., comedy)
	Date published
	Last update
	Video description
	Video keywords
	Video duration in seconds
	Minimum, Maximum, Average rating
	Number of users who rated
	Thumbnails
Content/User Statistics	Number of views
	Number of users having the video to their favorite list
	User's last access on YouTube
	Number of videos the user has viewed
	Number of users subscribed to this channel
	Total upload views on the user's channel
Comments	Number of comments
	Comments feed link

TABLE 1.3.2: User-Oriented data available in YouTube.

Data
ID (URI)
First name
Last name
Age
Gender
Location
Relationship
Occupation
School
List of hobbies
List of movies
List of music
List of books
Hometown
Various descriptions regarding the user

permissions and access token to access them. Data of particular interest for further analysis of the Facebook graph include the "Likes," expressing the user's appreciation of a status, a photo, a brand, etc., which can be used in combination with users' personal details to estimate trends, e.g., on new products. In addition, information related to tags appearing in

TABLE 1.3.3: Data available on Facebook.

Type	Data
Users	Person using Facebook (ID, name)
	List of user's friends
	Photo album
	Facebook message
	User's photos published in Facebook
	User's tag of the user at a place in an object
	Post published in Facebook
Object	Set of comments on a particular object
	Set of Likes on a particular object
	Shares of a particular object
	Usage metrics for several object types
App	Facebook app (ID)
	An app link object created by an app
	Review of a Facebook app
	User gaining a game achievement in an App
	Games achievement type created by an App
Others	Comment published on any other node
	Link shared on Facebook
	Facebook group
	Status message or post published to Facebook
	Message thread in Facebook Messages
	Event
	Video published in Facebook
	Web domain claimed within Facebook Insights
	Facebook Page
	Facebook Page milestone
	Offer published by a Page
	Details of a payment made via Facebook
	Test user created by a Facebook App
Thumbnails	Information related to shares, Open Graph
	and App Links about a URL

media data, like photos, can be exploited in order to create different types of social graphs, e.g., graphs of closely related Facebook friends, as detailed in Section 1.4.

Tweets are the fundamental ingredients of Twitter. They consist of 140 character long messages, along with additional meta-data. The data available on the Twitter platform are illustrated in Table 1.3.4.

Similar to YouTube, Flickr allows users to access vital information via its API. Specifically, it provides an XML description that contains the information appearing in Table 1.3.5.

Regarding author and citation networks, analysis of huge bibliographic datasets allows the collection of data that can be exploited for the analysis of scientific activity. Such data include the name of a scientist, his/her affiliation and research fields, a list of papers published in journals and conferences, a list of journals and conferences where his/her work has been published, the number of citations and a list of papers referring to his/her work, as well as the relationships between scientists, like *advisor*, *advisee* and *co-author*.

TABLE 1.3.4: Data available in Twitter.

Type	Data
Users	User's ID
	User's name
	Location
	Number of followers
	Number of favorite tweets
	Number of tweets the user has made
	Number of friends the user has
Followers	The followees that a person is following
	The users that follow a person
	Relationship between two users (e.g. friendship)
Tweet	A specific tweet (ID)
	100 most recent re-tweets of the tweet
	Up to 100 user id's who have re-tweeted the tweet (id)
	Hashtags appearing in the tweet
	Media data (e.g., images) appearing in the tweet
	Links appearing in the tweet
	Users mentioning the tweet
	Additional information about a tweet's content
Search	A collection of tweets that answer a certain question
Favorites	20 of the most recent tweets that were favored by the user
	Number of comments
Trends	Top 10 most popular topics for a particular WOEID
	(Yahoo! Where On Earth ID)

1.4 Social media graphs

After roughly describing some of the most popular social networks and the data that can be retrieved from them by using the corresponding APIs, we describe relevant graph structures that can be constructed and analyzed, as described in subsequent chapters. We shall focus our attention on graph structures that can be constructed by using Facebook and Twitter data. While the remaining social media are very popular, most of their users either do not give permission to access their data, or there are no user-item connections. For example, most YouTube users exploit the ability to view videos without creating an account.

1.4.1 Graphs from Facebook data

As previously described, friendship is the basic relationship between two Facebook users. By using this information, a *friendship graph* $\mathcal{G}(\mathcal{V}, \mathcal{E}, \mathbf{W})$ can be constructed, where each graph node $\mathbf{v}_i \in \mathbf{V}$ represents a unique user (ID) and graph edges $e_{ij} \in \mathcal{E}$ denote a friendship relationship between two users, i and j. Matrix $\mathbf{W}$ contains the weight values corresponding to the graph edges e_{ij}. Because the friendship relationship is undirected, the friendship graph is also undirected, i.e., $\mathbf{w}_{ij} = \mathbf{w}_{ji}$. Since the friends of a user typically also have their own friends, the friendship graph of the user can be expanded to include friends of friends. An

TABLE 1.3.5: Data available in Flickr.

Type	Data
User	Username
	User's name
	Location
	URL of user's photos
	URL of user's profile
	Date of first photo
	Total number of user photos
Photo	Title
	Description
	Photo owner
	Visibility, i.e., public or private photo
Dates	Date published
	Date the photo was taken
	Last update
	Editability, i.e., if a user can edit this picture or not
Statistics	Number of photo views
	Number of comments
	Number of user "favorites"

example of such a friendship graph depicting a friend and friend-of-friend relationship is shown in Figure 1.4.1. The weights of the edges appearing in this graph are equal to one, $w_{ij} = 1$. It can be observed that, in this graph three groups (person clusters) are formed indicating that the nodes (persons) belonging to each cluster are better connected to each other, when compared to nodes belonging to different clusters.

Using such graphs, one can estimate the number of friendship relationships needed, in order to connect two users. This is also known as the *number of hops* used to connect the two users. It has been estimated that the mean number of hops needed to connect two random nodes in the Facebook friendship graph is equal to 4.74 [UKBM11, BBR$^+$12]. Furthermore, research findings indicate that the number of users that can be accessed within three hops is relatively small, while each additional hop enlarges considerably the number of users that can be accessed. This is also known as the Six Degrees of Separation theory [Bar03].

Since it is obvious that the structure of such a dense friendship graph can represent only aspects of the real relationships between the users, data provided by the Facebook API can be exploited, in order to construct graph structures that better describe the relationships between users. For instance, one can exploit the photo tagging information provided by the users, when tagging themselves and their friends in their photos, in order to explore groups of friends that are also related in real life. For example, if two persons are both tagged in multiple photos, then there is high probability that they are strongly related in real life as well. By using this additional information, the second degree friendship graph in Figure 1.4.1 can be transformed to the one shown in Figure 1.4.2. Since the number of mutual photo tagging of two users may vary, the graph shown in Figure 1.4.2 is a weighted graph, where the weight of an edge connecting two graph nodes represents the total number of common photo tags for two connected users. By observing this graph, it can be seen that several small groups of closely related friends are created (also noted as *cliques*). This indicates that the users forming these groups are probably more strongly related in real life too. Therefore, it is manifested that real relationships between a person and other users

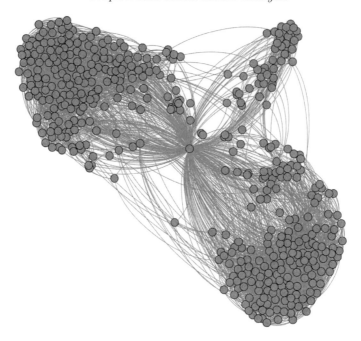

FIGURE 1.4.1: A graph describing the friendship relationships between a user and the friends of his/her friends on Facebook.

can be revealed by exploiting data available in social media, e.g., in Facebook. It should be noted here that, while tagging information used in this example should be provided by the users, ways for automatic metadata extraction, using computer vision techniques have also been explored, e.g., by employing face verification techniques [TYRW14].

Several other statistics related to users, like the relationship between the age of connected users, can also be revealed by analyzing the available Facebook data [UKBM11]. Users tend to connect to other users belonging to the same age category, while most of the users are likely to have multiple friends belonging to the age range of the 20s [UKBM11]. Another statistic related to the Facebook friendship graph refers to degree of each user node, i.e., the number of friends a user has [UKBM11].

Finally, an interesting graph structure can be constructed by using the available Facebook data to form the adjacency matrix $\mathbf{A}$ of the friendship graph constructed by taking into account the residence information of the Facebook users living in 54 countries [UKBM11]. The block structure of this adjacency matrix proves that most of the Facebook friendships are between users living in the same country. Specifically, 84.2% of the graph edges connect users living to the same country. In addition, this adjacency matrix can show friendships between users living in different countries.

1.4.2 Graphs from Twitter data

As previously described, the basic relationship between two Twitter users is that of following. By using this information, a follower graph $\mathcal{G}(\mathcal{V}, \mathcal{E}, \mathbf{W})$ can be constructed, where each node $\mathbf{v}_i \in \mathcal{V}$ of the graph represents a unique user (ID) and graph edges $e_{ij} \in \mathcal{E}$ between nodes exist whenever a followee-follower relationship connects two users, where user i follows user j. The matrix $\mathbf{W}$ contains the weight values corresponding to the graph edges e_{ij}. Because such a relationship is non-symmetric, the follower graph is a directed

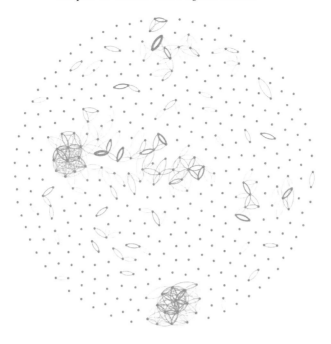

FIGURE 1.4.2: A graph describing the friendship relationships between a user and the friends of his/her friends on Facebook exploiting photo tagging information.

one, i.e., w_{ij} can be different from w_{ji}. Since the followees that a user follows can also follow other users, the follower graph of a user can be expanded, so as to include followers of the followers. An example of such a graph is shown in Figure 1.4.3. It should be noted here that the graphs in Figures 1.4.1 and 1.4.3 have been created using the profiles of the same person. Compared to the Facebook friendship graph, it can be seen that the Twitter graph is much bigger and denser, consisting of 90,000 nodes. This can be attributed to Twitter's inherent characteristics, where the objective is to connect with multiple users and exchange ideas, essentially forming an active newspaper, as has been described in Section 1.3. Most of this paragraph also applies to Instagram, as it is similar to Twitter in this regard. An example of an Instagram graph can be found in Figure 1.4.4.

By observing the graph in Figure 1.4.3, it can be seen that its analysis for determining meaningful clusters or cliques is difficult. However, by exploiting the data provided by the Twitter API, one can construct graph structures that better describe user interests. For instance, one can exploit the hashtag information provided by the users, when they express their opinion on a subject, in order to explore groups of users who share interests. By using this additional information, the enhanced first-degree follower graph of the same person can be seen in Figure 1.4.5. This graph is formed by the users followed by a particular user (the same as in Figures 1.4.1 and 1.4.3), where the edge weight connecting the followee and a follower is equal to 1, if they have not included any common hashtag in their last 200 tweets. The edge weight is increased by 1, for every common hashtag used by the two users. That is, the weight of the edge connecting the follower and his/her j-th followee is equal to $w_j = N + 1$, where N is the number of common hashags in their last 200 tweets. Similar to the Facebook case, it can be seen that by observing the second-degree followee-follower graph in Figure 1.4.5, users with common interests can be revealed. That kind of information can be exploited in various ways, e.g., it can be useful to advertisers on the Twitter platform who want to target the most suitable customers.

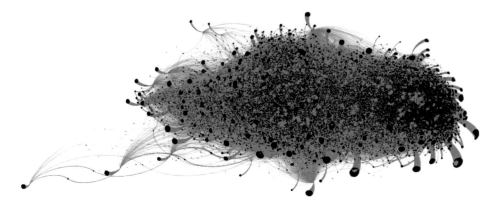

FIGURE 1.4.3: A followee-follower graph between a user and the followers of his/her followers on Twitter.

1.4.3   Graphs from bibliographic data

As previously described, bibliographic data can be exploited in order to describe relationships between scientists, or between their publications. A co-authorship graph describing the scientific network of an author is illustrated in Figure 1.4.6. As can be seen, author graphs are weighted, i.e. the weight of an edge connecting two authors depends on the number of papers co-authored by the two authors, corresponding to the two graph nodes. Furthermore, additional information describing author relationships, e.g. advisor-advisee relations, is described by using different node colors. Given that author networks describe information related to paper authorship, such networks are undirected. However, citation, as well as advisor-advisee networks are directed and unweighted. In particular, citation and

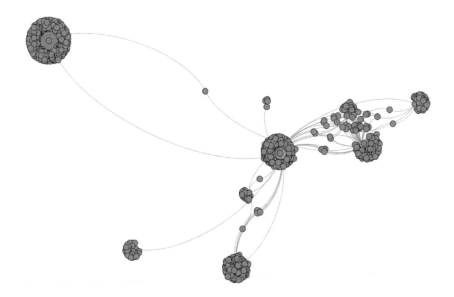

FIGURE 1.4.4: A followee-follower Instagram graph describing the relationships between a user and the followers of his/her followers.

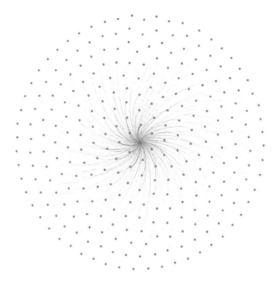

FIGURE 1.4.5: The first degree followee-follower Twitter graph of a user enhanced by using hashtag information.

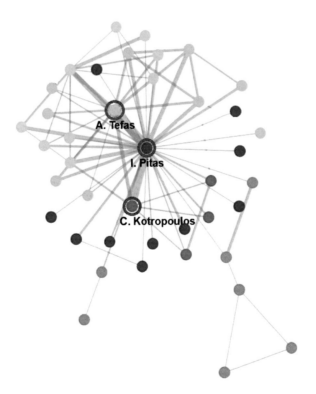

FIGURE 1.4.6: A co-authorship graph of the editor of this book (I. Pitas).

citation networks form the basics of *bibliometrics* [DB09]. Such a citation network is shown in Figure 1.4.7. In such networks, influential papers (like P_1 in Figure 1.4.7) describing new concepts that are frequently cited essentially form cascades.

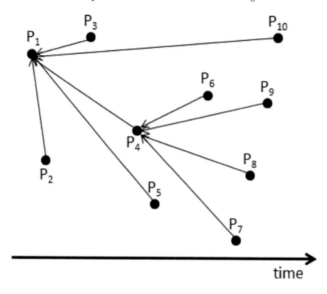

FIGURE 1.4.7: Citation network.

1.5   Graph storage formats and visualization

After constructing a social graph, one should be able to store it so that it can be used for future processing. To this end, several file formats have been designed. The most widely used ones are: Trivial Graph Format (TGF), DOT format, Graph Modelling Language (or Graph Meta Language) (GML) format, Graph eXchange Language (GXL) format, eXtensible Graph Markup and Modeling Language (XGMML) format, Directed Graph Markup Language (DGML) format, GraphML format and XCM format and XTM format.

Graph file formats can be categorized in plain text-based formats and XML-based formats. TGF, DOT and GML formats belong to the first category. They have been designed to have simple structure. A simple graph description can be made by using a list of nodes followed by a list of edges, which specify node pairs. The XML-based graph file formats use an XML-based syntax. Such a graph description contains a "graph" element that consists of "node" and "edge" elements. Attributes "ID," "source" and "target" are used in order to define graph nodes and edges. Finally, it should be noted that some of these maps have been designed specifically for the description of graphs defined from geographical map topologies, i.e., the XCM and XTM ones.

In addition, visualization applications, open source libraries and tools for graph manipulation have been developed. The most widely used graph visualization applications and softwares include Gephi, Graph Visualization Software (Graphviz), Cytoscape, and Tulip. Graph manipulation libraries and tools include igraph, Open Graph Drawing Framework (OGDF), and yEd.

1.6 Big data issues in social and digital media

An underlying problem of graph-based social media analysis methods is their computational complexity, which may scale cubically with respect to the number of data to be analyzed. For instance, for the analysis of a graph $\mathcal{G}(\mathcal{V}, \mathcal{E})$ consisting of N nodes $\mathbf{v}_i \in \mathcal{V}$, $i = 1, \ldots, N$ using algebraic graph-based dimensionality reduction and classification methods discussed in Chapter 6, one should compute, store and analyze the corresponding graph weight matrix $\mathbf{W} \in \mathbb{R}^{N \times N}$. The computation of $\mathbf{W}$ usually has a computational complexity equal to $O(DN^2)$, where D denotes the dimensionality of $\mathbf{v}_i$, while the application of algebraic graph analysis techniques on $\mathbf{W}$ usually requires the eigendecomposition or inversion of $\mathbf{W}$, having a computational complexity equal to $O(N^3)$ [PC99]. Taking into account the size of social media graphs, e.g. a graph describing the relationships between the members of a Facebook group may consist of hundreds of thousands of nodes N, while the number N of images appearing in the profiles of the members of a Facebook group may be in the order of millions. In such cases, the application of algebraic graph analysis methods in a single computer may be practically impossible.

At such size scales, a single computer is simply unable to handle the required processing tasks, as it does not have enough memory, or even disk space, to store all the required data. In order to overcome restrictions concerning the cardinality of social media data sets, three approaches can be followed: a) approximate solutions, b) incremental learning approaches and c) distributed computing approaches. Approximate solutions tackle the problem by solving an approximation of the original problem that requires less memory and computational cost. Common approaches to this end operate on a subset of the available data [AMS02, BW09], or approximate Linear Algebra methods, e.g. matrix multiplication and Singular Value Decomposition, and their application in analysis tasks [SS00, WS01, DM05]. Incremental learning approaches create an initial model by employing a (usually small) part of the data. Subsequently, they update the model by exploiting the remaining data, used in small batches. Detailed descriptions of several incremental learning methods will be given in Chapter 11.

In general, distributed computing may provide the only viable means to handle big (social) media processing or analysis tasks. Data can be distributed for storage in individual nodes of a computing cluster, or a computing cloud. Each node can contribute its processing cores and memory to the distributed system, bypassing the cost and hardware limits of having a single computer with the same amount of processing cores and memory. Clusters are also highly scalable, as nodes can be easily added, removed, or replaced. Chapter 10 is dedicated to the description of various big media data analysis tasks.

1.7 Distributed computing platforms

In distributed computing, different computers are connected through a network, while communication between them is performed through messages for coordination or data exchange purposes [CDK05]. Such machines form a computing *cluster*. The usual architecture of a computing cluster involves a *master node* coordinating several *worker nodes*. Its main advantage over super-computers with extraordinary large numbers of CPUs and amount of RAM is the scaling capability of a computing cluster, which may also use super-computers as worker nodes. The scalability of computing clusters allows them to handle Big Data problems, which may be too big for even the most powerful single machine.

Message Passing Interface (MPI) is a standardized, language-independent protocol for distributed computations that involves several processes running on several different computers [GLDS96]. It provides specifications for message-passing distributed processing library development, so that such libraries can have the same behavior and be compatible with each other. The current revision of the protocol is MPI-2, while MPI-3 is under development.

The *MapReduce* distributed programming model [DS08] was invented by Google specifically for handling extremely big datasets. It was inspired by the corresponding map and reduce primitives offered by functional programming languages, such as Lisp. In general, it consists of two steps. The first one applies a function on all the elements separately (*Map*) and the second one collects the results, using a commutative and associative operation (*Reduce*). An advantage of this model, over using a standard MPI system, is that the programmer does not have to handle the low-level details of an implementation, e.g., data distribution to the worker nodes, fault-tolerance, or load balancing, because it provides tools for writing high-level programs for clusters. While it may not provide a suitable solution for every possible problem, the MapReduce model particularly lends itself to problems that involve running simple operations on a large number of data elements on several worker nodes.

As the name implies, there are two major components in the MapReduce programming model. The Map command, when every worker applies a user defined function to each element of the dataset. Each worker can then return the results to the master node, thus computing that function output for the entire dataset. Additionally, using the Reduce command, a worker applies a commutative and associative operation to collect the data elements, or the results of a previously mapped function, into a single result. As the operation is commutative and associative, the results for each worker are independent from other workers and they can also be combined in the same way on the master node. A variation of the Reduce command is ReduceByKey [DS08], in which, given a distributed set of (key, value) pairs and a target operation, the operation is performed on the value parts for each key separately. If there were k total keys, then the output would be k (key, total) pairs, where each total is the result of performing the operation only on the value parts that are associated with the specific key. MapReduce has been successfully used in a number of big media data problems, e.g., distributed big graph mining [KF12] or machine learning [GKP+11].

There are several different implementations of the MapReduce framework besides Google's own, such as Apache *Hadoop* [SKRC10], which is also an open source project. MapReduce implementations often come with a distributed file system, like the *Google File System (GFS)* or the *Hadoop File System (HDFS)*, as such file systems synergize very well with and are, at most times, required by MapReduce tasks. Both GFS and HDFS are typically used for storing extremely big files. These file systems split a big file into blocks (or *chunks*) and distribute them to the computers serving as computing cluster nodes, allowing for multiple copies of a block to be stored in several different computers for fault tolerance. File blocks can also be redistributed, in order to balance the load. GFS includes a single Master node, which holds the metadata regarding the location of file chunks and it is not involved in any actual data transfers. It instead directs an application to the node that contains the requested chunk. When running Map or Reduce tasks, a node that already contains the dataset elements is selected to perform the task on them, instead of a random node, thus reducing the communication costs between nodes. However, such file systems are not well suited for frequent file reads and writes.

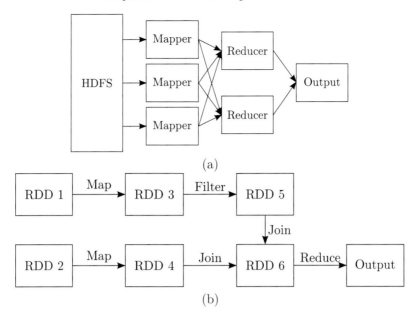

FIGURE 1.7.1: The difference between Hadoop and Spark DAG scheduling: a) Hadoop's two-stage MapReduce and b) a more complex Spark DAG schedule.

The Apache *Spark* cluster computing framework [ZCF$^+$10] builds upon Hadoop, in order to improve computation speed. It is also compatible with HDFS. Its advantages over Hadoop include a) its ability to create and operate on the more complex *Directed Acyclic Graph* (*DAG*) for scheduling tasks rather than Hadoop's two-stage MapReduce DAG scheduling, as illustrated in Figure 1.7.1, and b) its ability to cache data in the distributed memory. The data in memory are stored in collections of elements called *Resilient Distributed Datasets* (*RDDs*), which support Map and Reduce operations and, additionally, other operations, such as *join* and *filter*. It also allows data from RDDs to "spill" to the hard drive, in case the amount of memory is insufficient. All the above memory handling and DAG task scheduling is done by the framework, without input from the programmer. To form the cluster, one computer runs the master component. Computers join the cluster by connecting a worker component to the master. The cluster is very flexible: dedicated computers may always operate in the cluster, while other computers, e.g., workstations, can contribute any resources they can spare, while still allowing people to work on them.

Finally, it is possible for programmers to create their own DAG schedule of distributed tasks through Apache *Storm*, a distributed computation system. In order to perform a distributed task, a so-called *topology* has to be created. A topology can include *Spouts* and *Bolts*. The data flow is done through *streams*. Streams are generated by Spouts, while Bolts perform computations on streams they receive from Spouts or other Bolts and output a processed stream. Both of these entities can have various levels of parallelism. The user can create a network of Spouts and Bolts by defining directed connections between them. The resulting network is packaged into a topology, which can then be submitted for execution to a Storm cluster. Again, the user only needs to define the Spouts, Bolts, and the topology. Everything else is handled by the system. An alternative to Storm can be found in Microsoft's *Dryad* project.

1.8   Conclusions

In this chapter, an overview of the dominant social networks and social media platforms to date has been given. It has been shown that social networks can be represented by using graph structures. First-degree and second-degree graphs of the dominant social networks have been illustrated and discussed. Such graphs, created by using social media data available through the corresponding APIs, can be exploited for further analysis, as will be discussed in the following chapters. Issues concerning the computational complexity and memory requirements of graph-based social media analysis methods have been discussed. Finally, big data issues and distributed computing platforms that are useful in social media analysis are reviewed.

Bibliography

[ABHH08] T. Ahlqvist, A. Back, M. Halonen, and S. Heinonen. *Social media road maps exploring the futures triggered by social media.* VTT, 2008.

[AMS02]   D. Achilioptas, G. McSherry, and B. Scholkopf. Sampling techniques for kernel methods. In *Proc. Neural Information Processing Systems*, 2002.

[Bar03]   A. L. Barabási. *Linked: How Everything is Connected to Everything Else and What It Means for Business.* Plume, 2003.

[BBR+12] L. Backstrom, P. Boldi, M. Rosa, J. Ugander, and S. Vigna. Four degrees of separation. In *Proc. of the 4th Annual ACM Web Science Conference*, pages 33–42, 2012.

[BW09]   M. A. Belabbas and P. J. Wolfe. Spectral methods in machine learning and new strategies for very large datasets. In *Proceedings of the National Academy of Sciences*, volume 106, pages 369–374, 2009.

[CDK05]   G. Coulouris, J. Dollimore, and T. Kingberg. *Distributed Systems: Concepts and Design (International Computer Science).* Addison-Wesley Longman, 2005.

[DB09]   N. De Bellis. *Bibliometrics and citation analysis: from the science citation index to cybermetrics.* Scarecrow Press, 2009.

[DM05]   P. Drineas and M. W. Mahoney. On the Nystrom Method for Approximating a Gram Matrix for Improved Kernel-based Learning. *Journal of Machine Learning Research*, 6:2153–2275, 2005.

[DS08]   J. Dean and G. Sanjay. MapReduce: simplified data processing on large clusters. *ACM Communications Magazine*, 51(1):107–113, 2008.

[dSP65]   D. J. de Solla Price. The pattern of bibliographic references indicates the nature of scientific research front. *Science*, 149:510–515, 1965.

[Ell08]   N. Ellison. Social network sites: Definition, history, and scholarship. *Journal of Computer-Mediated Communication*, 13:21–23, 2008.

[GKP+11] A. Ghoting, B. Krishnamurhty, E. Pednault, B. Reinwald, V. Sindhwani, S. Takikonda, Y. Tian, and S. Vaithyanathan. SystemML: Declarative machine learning on MapReduce. In *Proc. IEEE International Conference on Data Engineering*, pages 231–242, 2011.

[GLDS96] W. Gropp, E. Lusk, N. Doss, and A. Skjellum. A high-performance, portable implementation of the MPI message passing interface standard. *Parallel computing*, 22(6):789–828, 1996.

[Hem13] J. Hempel. LinkedIn: How it's changing business. *Fortune*, pages 69–74, 2013.

[KF12] U. Kang and C. Faloutsos. Big graph mining: algorithms and discoveries. *ACM SIGKDD Explorations*, 14(2):29–36, 2012.

[KH10] A. M. Kaplan and M. Haenlein. Users of the world, unite! The challenges and opportunities of social media. *Business Horizons*, 53(1):59–68, 2010.

[PBV+07] G. Palla, A. Barabasi, T. Vicsek, Y. Chi, S. Zhu, X. Song, J. Tatemura, and B. L. Tseng. Quantifying social group evolution. *Nature*, 446:664–667, 2007.

[PC99] V. Y. Pan and Z. Q. Chen. The complexity of the matrix eigenproblem. In *Proc. of the 31st annual ACM Symposium on Theory of Computing*, pages 507–516, 1999.

[Sch96] L. H. Schmidt. Commonness across cultures. In *Cross-Cultural Conversation: Initiation*, pages 119–132. Oxford University Press, 1996.

[SKRC10] K. Shvachko, H. Kuang, S. Radia, and R. Chansler. The Hadoop distributed file system. In *Proc. IEEE Symposium on Mass Storage Systems and Technologies*, pages 1–10, 2010.

[SS00] J. Smola and B. Scholkopf. Sparse greedy matrix approximation for machine learning. In *Proc. International Conference on Machine Learning*, 2000.

[TPP14] I. Tsingalis, I. Pipilis, and I. Pitas. A statistical and clustering study on YouTube 2D and 3D video recommendation graph. In *Proc. International Symposium on Communications, Control and Signal Processing*, 2014.

[TYRW14] Y. Taigman, M. Yang, M. Ranzato, and L. Wolf. Deepface: Closing the gap to human-level performance in face verification. In *Proc. IEEE Conference on Computer Vision and Pattern Recognition*, 2014.

[UKBM11] J. Ugander, B. Karrer, L. Backstrom, and C. Marlow. The anatomy of the Facebook social graph. *Computing Research Repository*, 2011.

[WS01] C. K. I. Williams and M. Seeger. Using the Nystrom method to speed up kernel machines. In *Proc. Neural Information Processing Systems*, pages 682–688, 2001.

[You] http://www.youtube.com/yt/press/statistics.html.

[ZCF+10] M. Zaharia, M. Chowdhury, M. J. Franklin, S. Shenker, and I. Stoica. Spark: Cluster computing with working sets. In *USENIX Conference on Hot Topics in Cloud Computing*, 2010.

Chapter 2

Mathematical Preliminaries: Graphs and Matrices

Nikolaos Tsapanos, Alexandros Iosifidis and Ioannis Pitas

Aristotle University of Thessaloniki, Greece

2.1   Graph basics

Formally, a *graph* $G = (\mathcal{V}, \mathcal{E})$ is an ordered pair of a set of vertices $\mathcal{V} = \{v_i\}$ and a set of edges $\mathcal{E} \subseteq \mathcal{V} \times \mathcal{V}$ [BM76, Wes01]. Graphs are abstract constructs that can model relationships (*edges*) between entities (*vertices*). The edge (u_i, u_j) connecting vertices u_i and u_j is *incident* to u_i and u_j and signifies that vertex u_i is *adjacent* to u_j. The graph is called *undirected*, if $(v_i, v_j) \in \mathcal{E}$ implies $(v_j, v_i) \in \mathcal{E}$. If the order of the vertices in an edge (*source-sink*) is important, then the graph is called *directed* (or *digraph*).

If a graph has more than one distinct edge connecting the same vertices, then it is called a *multigraph* [GY99]. The number of edges that connect vertex v_i with other vertices is called *degree* d_i of the vertex. In directed graphs, one can distinguish between the *in-degree* and the *out-degree* of a vertex, which are the number of the incoming and the outgoing edges, respectively. Vertices with a degree equal to 0 have no connections and are characterized as *isolates*. The degree is one factor in determining the importance of a vertex in the graph. Vertices with a high degree are more likely to be considered important. If all graph vertices have the same number of connections, i.e., they have the same number of incident edges, the graph is a *regular* one. In the opposite case, the graph is characterized as *irregular*.

A graph where every vertex is connected to all other vertices is called a *complete graph*. The vertices that are connected with v_i form the *neighborhood* of v_i. It is possible that edges in both directed and undirected graphs have a *weight value*, in which case the degree of a vertex is determined by the sum of the weights of its incident edges. It is also possible that vertices, edges, or both, have either labels or attributes associated with them, resulting in a *labeled* or an *attributed graph*, respectively. Typically, in social media, graph vertices may refer to media content, e.g., images or video, which are vectorial data. In such cases, each graph vertex v_i can be described by a feature vector $\mathbf{x}_i \in \mathbb{R}^D$, where D is the dimensionality of the vector.

A *subgraph* S of G is a graph whose vertex and edge sets are a subset of the vertex and edge sets of G, respectively. A subgraph is called an *induced* subgraph of G, if vertices v_i and v_j being connected in G implies that they are also connected in S.

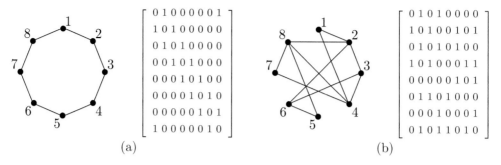

$$\begin{bmatrix} 0\,1\,0\,0\,0\,0\,0\,1 \\ 1\,0\,1\,0\,0\,0\,0\,0 \\ 0\,1\,0\,1\,0\,0\,0\,0 \\ 0\,0\,1\,0\,1\,0\,0\,0 \\ 0\,0\,0\,1\,0\,1\,0\,0 \\ 0\,0\,0\,0\,1\,0\,1\,0 \\ 0\,0\,0\,0\,0\,1\,0\,1 \\ 1\,0\,0\,0\,0\,0\,1\,0 \end{bmatrix}$$

$$\begin{bmatrix} 0\,1\,0\,1\,0\,0\,0\,0 \\ 1\,0\,1\,0\,0\,1\,0\,1 \\ 0\,1\,0\,1\,0\,1\,0\,0 \\ 1\,0\,1\,0\,0\,0\,1\,1 \\ 0\,0\,0\,0\,0\,1\,0\,1 \\ 0\,1\,1\,0\,1\,0\,0\,0 \\ 0\,0\,0\,1\,0\,0\,0\,1 \\ 0\,1\,0\,1\,1\,0\,1\,0 \end{bmatrix}$$

(a)   (b)

FIGURE 2.1.1: Example graphs and their respective adjacency matrices. a) Ring graph and b) arbitrary graph.

An ordered sequence of connected vertices $(v_i, \ldots, v_k, \ldots, v_j)$ that starts with vertex v_i and ends with vertex v_j forms a *path* between v_i and v_j. The graph is connected, iff there is a path from any vertex to any other vertex in the graph. The shortest path between two vertices is called *geodesic* and its length is called the *distance* of the two vertices. The longest distance between two vertices is called the *diameter* of the graph.

The *adjacency matrix* $\mathbf{A}$ of a graph $G = (\mathcal{V}, \mathcal{E})$ is a square $|\mathcal{V}| \times |\mathcal{V}|$ matrix such that:

$$a_{ij} = \mathbf{A}(i,j) = \begin{cases} 1, & \text{if } (v_i, v_j) \in \mathcal{E} \\ 0, & \text{otherwise,} \end{cases} \tag{2.1.1}$$

where $|\cdot|$ denotes the cardinality of a set. Some example graphs and their adjacency matrices can be seen in Figure 2.1.1. It is easy to prove that the elements of the k-th power $\mathbf{A}^k \triangleq \mathbf{A}\mathbf{A}\ldots\mathbf{A}$ (k times) of the adjacency matrix $\mathbf{A}$ provide the total number of possible paths between the corresponding vertices that have exactly length k [GY99]. If the graph is undirected, the adjacency matrix $\mathbf{A}$ is a *symmetric* matrix, i.e., $\mathbf{A} = \mathbf{A}^T$, where $\mathbf{A}^T$ denotes the transpose of $\mathbf{A}$. If the graph is weighted, the adjacency matrix can contain the numerical weight of each edge in the appropriate entry in the adjacency matrix, in which case it is also called the graph *weight matrix*, typically denoted by $\mathbf{W} = [w_{ij}]$. In the case where $\mathbf{W}$ denotes only connections between the graph vertices, i.e., $w_{ij} = \{0,1\}$, $\mathbf{W}$ is identical to the graph adjacency matrix $\mathbf{A}$.

An alternative matrix representation of a graph is provided by its *Laplacian* matrix $\mathbf{L} = \mathbf{D} - \mathbf{A}$ [Chu97], where $\mathbf{D}$ is a diagonal matrix, whose i-th diagonal entry is the degree d_i of vertex u_i. If the graph is weighted, the Laplacian matrix is similarly defined as $\mathbf{L} = \mathbf{D} - \mathbf{W}$, where $\mathbf{W}$ is the weight matrix and the entries of the degree matrix $\mathbf{D}$ are given by $d_i = \sum_j w_{ij}$. The normalized Laplacian matrix $\mathcal{L}$ is defined as:

$$\ell_{ij} = \mathcal{L}(i,j) = \begin{cases} 1, & \text{if } i = j \text{ and } d_i \neq 0 \\ -\frac{1}{\sqrt{d_i d_j}}, & \text{if } (v_i, v_j) \in \mathcal{E} \\ 0, & \text{otherwise} \end{cases} \tag{2.1.2}$$

or in a more compact form:

$$\mathcal{L} = \mathbf{I} - \mathbf{D}^{-\frac{1}{2}} \mathbf{A} \mathbf{D}^{-\frac{1}{2}}, \tag{2.1.3}$$

where $\mathbf{I} \in \mathbb{R}^{N \times N}$ is the *identity matrix* and $\mathbf{X} \triangleq \mathbf{Y}^{\frac{1}{2}}$, iff $\mathbf{X}^2 = \mathbf{Y}$, so that $\mathbf{D}^{-\frac{1}{2}}$ is essentially a diagonal matrix, whose elements are $\frac{1}{\sqrt{d_i}}$. If the graph is weighted, the corresponding normalized Laplacian matrix is defined as:

$$\mathcal{L} = \mathbf{I} - \mathbf{D}^{-\frac{1}{2}} \mathbf{W} \mathbf{D}^{-\frac{1}{2}}. \tag{2.1.4}$$

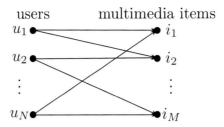

users   multimedia items

FIGURE 2.1.2: Bipartite graph describing user recommendations.

If the vertices of a graph can be partitioned into two disjoint sets $\mathcal{V}_1$ and $\mathcal{V}_2$, such that every edge $e \in \mathcal{E}$ connects a vertex from one set to a vertex in the other set, the graph is called *bipartite*. Such graphs arise in recommendation systems, where one vertex set can represent users and the other set can represent items. If a user is associated with an item, an edge connects their respective vertices, as shown in Figure 2.1.2.

Graphs are also often used to represent similarity between samples of a multimedia dataset. Let us assume that a set of digital media data, e.g., a set of N facial images, has been preprocessed so that each sample $i = 1, \ldots, N$ is represented by a D-dimensional vector $\mathbf{x}_i \in \mathbb{R}^D$, $i = 1, \ldots, N$. Let us also define a *similarity measure* $s_{ij} = s(\mathbf{x}_i, \mathbf{x}_j)$ that is used to measure the similarity between two vectors $\mathbf{x}_i$ and $\mathbf{x}_j$. $s_{ij} = s(\cdot, \cdot)$ may be any similarity measure providing non-negative values (usually $0 \leq s_{ij} \leq 1$). The most widely adopted choice is the *heat kernel* (also known as the *diffusion kernel*), defined by [KL02]:

$$s(\mathbf{x}_i, \mathbf{x}_j) = e^{-\frac{\|\mathbf{x}_i - \mathbf{x}_j\|_2^2}{2\sigma^2}}, \tag{2.1.5}$$

where $\| \cdot \|_2$ denotes the L_2 norm of a vector and σ is a scaling parameter for the Euclidean distance between $\mathbf{x}_i$ and $\mathbf{x}_j$. By using s_{ij}, $i = 1, \ldots, N$, $j = 1, \ldots, N$, we can form the graph weight matrix $\mathbf{W} \in \mathbb{R}^{N \times N}$, $\mathbf{W} = [s_{ij}]$. That is, we can assume that the vectors $\mathbf{x}_i$ are embedded in a graph $G = (\mathcal{V}, \mathcal{E}, \mathbf{W})$, where $\mathcal{V} = \{\mathbf{x}_i\}_{i=1}^N$ denotes the graph vertex set and $\mathcal{E}$ the set of edges connecting $\mathbf{x}_i$. In this context, $w_{ij} = s_{ij}$ denotes the weight value of the edge connecting the graph vertices $\mathbf{x}_i$ and $\mathbf{x}_j$.

It is also possible to use multiple similarity measures, in order to obtain a labeled multigraph. In this case, there can be multiple edges connecting the same vertices [BR12]. Edges with the same label can correspond to similarities between vertices according to the same measure, e.g., color similarity or motion similarity in the case of digital videos residing on graph vertices.

A *hypergraph* is a graph whose edges may connect more than two vertices [Ber89] and can be used to represent multiple relationships between vertices. A hypergraph $\mathcal{H}$ is defined as a pair $\mathcal{H} = (\mathcal{V}, \mathcal{E})$, where $\mathcal{V} = \{v_i\}$ is a finite set of vertices and $\mathcal{E} = \{e_j\}$ is a set of non-empty subsets of vertices, called *hyperedges*, representing relationships between vertices. If no hyperedge contains more than two vertices, the hypergraph becomes equivalent to a normal graph. An example of a hypergraph is shown in Figure 2.1.3, where the set of vertices is $\mathcal{V} = \{v_1, v_2, v_3, v_4, v_5, v_6, v_7, v_8\}$ and the set of hyperedges is $\mathcal{E} = \{e_1, e_2, e_3\}$, where $e_1 = \{v_1, v_2, v_5, v_7\}$, $e_2 = \{v_2, v_3, v_4, v_8\}$ and $e_3 = \{v_1, v_5, v_6\}$. In a weighted hypergraph, denoted by $\mathcal{H} = (\mathcal{V}, \mathcal{E}, w)$, each hyperedge e_j is assigned a weight $w(e_j)$. If $v_i \in e_j$, the hyperedge $e_j \in \mathcal{E}$ is *incident* to the vertex $v_i \in \mathcal{V}$. A hypergraph can be represented by the *incidence matrix* $\mathbf{H} \in \mathbb{R}^{|\mathcal{V}| \times |\mathcal{E}|}$, where:

$$H(i, j) = \begin{cases} 1, & \text{if } v_i \in e_j \\ 0, & \text{if } v_i \notin e_j. \end{cases} \tag{2.1.6}$$

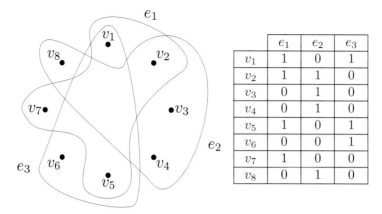

	e_1	e_2	e_3
v_1	1	0	1
v_2	1	1	0
v_3	0	1	0
v_4	0	1	0
v_5	1	0	1
v_6	0	0	1
v_7	1	0	0
v_8	0	1	0

FIGURE 2.1.3: Example of a hypergraph.

The degree of a vertex $v_i \in \mathcal{V}$ is equal to the number of its incident edges. In weighted hypergraphs, the degree of a vertex v_i is defined as the sum of the weights associated with its incident hyperedges $d(v_i) = \sum_{e_j \in \mathcal{E}} w(e_j) H(i, j)$. In a similar way, the degree of a hyperedge $e_j \in \mathcal{E}$ is equal to its cardinality $\delta(e_j) = \sum_{v_i \in \mathcal{V}} H(i, j)$. The *hypergraph adjacency matrix* $\mathbf{A}$, as defined by Zhou et al. in [ZHS07] according to the random walk model, is calculated by:

$$\mathbf{A} = \mathbf{H}\mathbf{W}\mathbf{H}^T - \mathbf{D}_v, \tag{2.1.7}$$

where $\mathbf{W}$ and $\mathbf{D}_v$ are the diagonal matrices of the hyperedge and vertex weights, respectively. The normalized Laplacian matrix of a hypergraph is defined accordingly:

$$\mathcal{L} = \mathbf{I} - \frac{1}{2}\mathbf{D}_v^{-1/2}\mathbf{H}\mathbf{W}\mathbf{H}^T\mathbf{D}_v^{-1/2} = \frac{1}{2}\left(\mathbf{I} - \mathbf{D}_v^{-1/2}\mathbf{A}\mathbf{D}_v^{-1/2}\right). \tag{2.1.8}$$

2.2   Linear algebra tools

Linear algebra [Str88] plays an important role in graph analysis. Therefore, its tools are reviewed in this section.

There are various products defined between two matrices. The *matrix product* $\mathbf{Z} \triangleq \mathbf{X}\mathbf{Y}$ of $N \times P$ matrix $\mathbf{X}$ and $P \times M$ matrix $\mathbf{Y}$ is the $N \times M$ matrix $\mathbf{Z}$, whose elements are given by:

$$z_{ij} = \sum_{k=1}^{P} x_{ik} y_{kj}. \tag{2.2.1}$$

The *Hadamard product* $\mathbf{Z} \triangleq \mathbf{X} \circ \mathbf{Y}$ of $N \times M$ matrices $\mathbf{X}$ and $\mathbf{Y}$ is the $N \times M$ matrix $\mathbf{Z}$, whose elements are given by:

$$z_{ij} = x_{ij} y_{ij}. \tag{2.2.2}$$

The *Kronecker product* $\mathbf{Z} \triangleq \mathbf{X} \otimes \mathbf{Y}$ of $N \times M$ matrix $\mathbf{X}$ and $P \times Q$ matrix $\mathbf{Y}$ is the $NP \times MQ$

matrix $\mathbf{Z}$, whose elements are given by:

$$\mathbf{Z} = \begin{bmatrix} X(1,1)\mathbf{Y} & X(1,2)\mathbf{Y} & \ldots & X(1,n)\mathbf{Y} \\ X(2,1)\mathbf{Y} & X(2,2)\mathbf{Y} & \ldots & X(2,n)\mathbf{Y} \\ \vdots & \vdots & \ddots & \vdots \\ X(n,1)\mathbf{Y} & X(n,2)\mathbf{Y} & \ldots & X(n,n)\mathbf{Y} \end{bmatrix}. \tag{2.2.3}$$

The *Khatri-Rao product* $\mathbf{Z} \triangleq \mathbf{X} \odot \mathbf{Y}$ of two likewise partitioned block matrices, namely $N \times M$ matrix $\mathbf{X}$ and $P \times Q$ matrix $\mathbf{Y}$, is the likewise partitioned block matrix $\mathbf{Z}$, whose blocks are given by:

$$\mathbf{Z}_{ij} = \mathbf{X}_{ij} \otimes \mathbf{Y}_{ij}. \tag{2.2.4}$$

The *rank* of a matrix $\mathbf{A} \in \mathbb{R}^{N \times M}$ is defined to be the number of its linearly independent column (or row) vectors. In general, $\text{rank}(\mathbf{A}) \leq \min(N, M)$. Matrix $\mathbf{A}$ is a *full rank* matrix, iff $\text{rank}(\mathbf{A}) = \min(N, M)$, otherwise $\mathbf{A}$ is *rank deficient*. There are two important subspaces associated with the $N \times M$ matrix $\mathbf{A}$. The *range* $\mathcal{R}(\mathbf{A})$ of $\mathbf{A}$ is a subspace of $\mathbb{R}^N$ defined as follows:

$$\mathcal{R}(\mathbf{A}) = \{\mathbf{y} \in \mathbb{R}^N : \text{ there exists at least one } \mathbf{x} \in \mathbb{R}^M, \text{ such that } \mathbf{A}\mathbf{x} = \mathbf{y}\}. \tag{2.2.5}$$

The *null space* $\mathcal{N}(\mathbf{A})$ of $\mathbf{A}$ is a subspace of $\mathbb{R}^M$ defined by:

$$\mathcal{N}(\mathbf{A}) = \{\mathbf{x} \in \mathbb{R}^M : \mathbf{A}\mathbf{x} = \mathbf{0}_N\}, \tag{2.2.6}$$

where $\mathbf{0}_N \in \mathbb{R}^N$ is a vector having all its elements equal to zero. By using the above notation, the following relation holds:

$$\text{rank}(\mathbf{A}) + \dim(\mathcal{N}(\mathbf{A})) = \min(N, M), \tag{2.2.7}$$

where $\dim(\cdot)$ denotes the vector space dimension.

The *Frobenius norm* $\|\mathbf{A}\|_F$ of a matrix is given by:

$$\|\mathbf{A}\|_F = \sqrt{\sum_{i=1}^{N} \sum_{j=1}^{M} a_{ij}{}^2}, \text{ or } \|\mathbf{A}\|_F = \sqrt{tr(\mathbf{A}^T \mathbf{A})}, \tag{2.2.8}$$

where $tr(\cdot)$ is the trace of the matrix, i.e., the sum of the elements of its diagonal.

A square matrix $\mathbf{A} \in \mathbb{R}^{N \times N}$ is called *invertible* (or *non-singular*), if there is a matrix $\mathbf{B} \in \mathbb{R}^{N \times N}$ such that:

$$\mathbf{A}\mathbf{B} = \mathbf{I} \quad \text{and} \quad \mathbf{B}\mathbf{A} = \mathbf{I}. \tag{2.2.9}$$

In this case, matrix $\mathbf{B} = \mathbf{A}^{-1}$ is called the *inverse* of $\mathbf{A}$ and vice versa. A square matrix that is not invertible is called *singular*. A square matrix $\mathbf{A}$ is singular, if its rank is less than N. A symmetric matrix $\mathbf{A}$ is *positive-definite*, iff $\forall \mathbf{x} \in \mathbb{R}^N, \mathbf{x}^T \mathbf{A} \mathbf{x} > 0$.

An *eigenvector* of a real-valued, square matrix $\mathbf{A} \in \mathbb{R}^{N \times N}$ is a non-zero vector $\mathbf{v} \in \mathbb{R}^N$ that satisfies the following equation:

$$\mathbf{A}\mathbf{v} = \lambda \mathbf{v}, \tag{2.2.10}$$

meaning that the vector $\mathbf{A}\mathbf{v}$ follows the direction of $\mathbf{v}$. λ is the *eigenvalue* of $\mathbf{A}$ corresponding to the eigenvector $\mathbf{v}$. The eigenvalues are solutions of the equation:

$$\det(\mathbf{A} - \lambda \mathbf{I}) = 0, \tag{2.2.11}$$

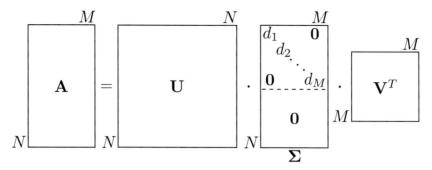

FIGURE 2.2.1: Singular Value Decomposition of matrix $\mathbf{A}$.

where $\det(\cdot)$ denotes the *determinant* of a matrix. Any symmetric matrix $\mathbf{A}$ has real eigenvalues and can always be written in the form:

$$\mathbf{A} = \mathbf{V}\mathbf{\Lambda}\mathbf{V}^T, \qquad (2.2.12)$$

where $\mathbf{\Lambda}$ is a diagonal $N \times N$ matrix, whose diagonal values are the distinct matrix eigenvalues λ_i, $i = 1, \ldots, N$ and matrix $\mathbf{V} = [\mathbf{v}_1, \mathbf{v}_2, \ldots, \mathbf{v}_N]$ is an *orthogonal* matrix, i.e., $\mathbf{V}\mathbf{V}^T = \mathbf{V}^T\mathbf{V} = \mathbf{I}$, formed by the eigenvectors $\mathbf{v}_i$, $i = 1, \ldots, N$. Equation (2.2.12) provides the *eigendecomposition* of $\mathbf{A}$.

Example 2.2.1. *Matrix eigenanalysis.*

Matrix $\mathbf{A} = \begin{bmatrix} 1 & 2 & 3 \\ 2 & 5 & 4 \\ 3 & 4 & 2 \end{bmatrix}$ *has the following eigenvalues and eigenvectors:*

$$\lambda_1 = -1.8, \mathbf{v}_1 = [0.6, 0.3, -0.7]^T, \lambda_2 = 2.6, \mathbf{v}_2 = [0.7, -0.6, 0.3]^T,$$
$$\lambda_3 = 9.2, \mathbf{v}_3 = [0.4, 0.7, 0.6]^T.$$

For simplicity, the eigenvalues and eigenvector entries have been rounded to the first digit.

The *Singular Value Decomposition (SVD)* [GVL96] of a $N \times M$ matrix $\mathbf{A}$ is given by:

$$\mathbf{A} = \mathbf{U}\mathbf{\Sigma}\mathbf{V}^T, \qquad (2.2.13)$$

where $\mathbf{\Sigma}$ is a $N \times M$ matrix, whose diagonal elements are the $\min(N, M)$ *singular values* $d_1 \geq d_2 \geq \ldots \geq d_{\min(N,M)} \geq 0$ of $\mathbf{A}$, as shown in Figure 2.2.1. It can be proven that the singular values of $\mathbf{A}$ are the square roots of the eigenvalues of matrix $\mathbf{A}^T\mathbf{A}$. The rank of a rectangular $N \times M$ matrix $\mathbf{A}$ is equal to the number of the non-zero singular values d_i.

The columns of $\mathbf{U}$, $\mathbf{V}$ are the left and right *singular vectors* of $\mathbf{A}$, i.e., the eigenvectors of the $N \times N$ $\mathbf{A}\mathbf{A}^T$ and $M \times M$ $\mathbf{A}^T\mathbf{A}$ matrices, respectively. Matrices $\mathbf{U}$ and $\mathbf{V}$ are orthogonal, i.e., $\mathbf{U}^T\mathbf{U} = \mathbf{U}\mathbf{U}^T = \mathbf{I}$ and $\mathbf{V}^T\mathbf{V} = \mathbf{V}\mathbf{V}^T = \mathbf{I}$. They are not uniquely defined, as, e.g., $-\mathbf{U}$ and $-\mathbf{V}$ can also be used towards the same result. The columns of matrix $\mathbf{U}$ corresponding to the non-zero singular values span the range of matrix $\mathbf{A}$, i.e., they provide an orthonormal basis for that space. The columns of matrix $\mathbf{V}$ corresponding to the zero singular values span the null space of matrix $\mathbf{A}$. If matrix $\mathbf{A}$ is square, symmetric and positive definite, then its SVD coincides with its eigendecomposition.

Example 2.2.2. *Matrix SVD.*

$$Matrix\ \mathbf{A} = \begin{bmatrix} 5 & 2 & 7 & 3 \\ 2 & 9 & 4 & 9 \\ 0 & 6 & 1 & 5 \end{bmatrix}$$

has the following SVD:

$$\mathbf{U} = \begin{bmatrix} -0.4 & 0.9 & 0.2 \\ -0.8 & -0.2 & -0.6 \\ -0.4 & -0.4 & 0.8 \end{bmatrix}, \mathbf{\Sigma} = \begin{bmatrix} 16.8 & 0 & 0 & 0 \\ 0 & 7 & 0 & 0 \\ 0 & 0 & 0.4 & 0 \end{bmatrix},$$

$$\mathbf{V} = \begin{bmatrix} -0.2 & 0.6 & -0.1 & -0.8 \\ -0.6 & -0.4 & 0.6 & -0.2 \\ -0.4 & 0.7 & 0.2 & 0.6 \\ -0.6 & -0.2 & -0.7 & 0.1 \end{bmatrix}.$$

Therefore, it has three singular values: $d_1 = 16.8$, $d_2 = 7$ and $d_3 = 0.4$. For simplicity, the singular values and singular vector entries have been rounded to the first digit.

A square $N \times N$ matrix $\mathbf{A}$ is non-singular, i.e., it can be inverted, iff all its singular values are non-zero $d_1 \geq d_2 \geq \ldots \geq d_{\min(N,M)} > 0$. The inverse of a non-singular $N \times N$ matrix $\mathbf{A}$ can be written as:

$$\mathbf{A}^{-1} = \mathbf{V}\mathbf{\Sigma}^{-1}\mathbf{U}^T. \tag{2.2.14}$$

The Frobenius norm $\|\mathbf{A}\|_F$ of a matrix can be found from its singular values:

$$\|\mathbf{A}\|_F = \sqrt{\sum_i d_i^2}. \tag{2.2.15}$$

A *homogeneous system* of N linear equations and M unknowns described by a $N \times M$ matrix $\mathbf{A}$ [GVL96]:

$$\mathbf{A}\mathbf{x} = \mathbf{0}, \tag{2.2.16}$$

satisfying $N \geq M - 1$ and rank $(\mathbf{A}) = M - 1$, has two solutions: a) the trivial solution $\mathbf{x} = \mathbf{0}$ and b) a non-trivial solution up to a scale factor, provided by the SVD decomposition of matrix $\mathbf{A} = \mathbf{U}\mathbf{\Sigma}\mathbf{V}^T$. Namely, it is equal to the scaled column of matrix $\mathbf{V}$, $\mathbf{x} = \lambda\mathbf{v}_i$, whose corresponding singular value is zero $d_i = 0$. Since the rank of matrix $\mathbf{A}$ is rank$(\mathbf{A}) = M - 1$, all other singular values are non zero (positive). In practice, due to numerical errors, we choose the singular value d_i that has the lowest absolute value (close to zero).

A system of N linear equations with N unknowns residing in vector $\mathbf{x}$ described by a $N \times N$ matrix $\mathbf{A}$:

$$\mathbf{A}\mathbf{x} = \mathbf{b} \tag{2.2.17}$$

has the following solution:

$$\mathbf{x} = \mathbf{A}^{-1}\mathbf{b}, \tag{2.2.18}$$

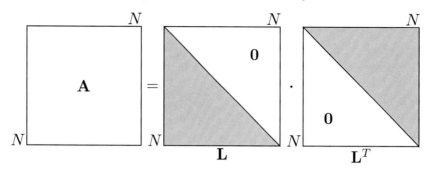

FIGURE 2.3.1: Cholesky decomposition of matrix **A**.

provided that matrix **A** is non-singular. Numerous methods exist for the numerical solution of linear equation systems [Str88, Mey00], whose presentation is beyond the overview provided by this section.

If $N > M$, the linear system of N equations and M unknowns $\mathbf{Ax} = \mathbf{b}$ is over-determined, i.e., it has more equations than unknowns. In this case, no exact solution may exist, if **b** does not belong to the space spanned by the columns of **A**. Such a system has the following *least squares solution* [LH74]:

$$\mathbf{x} = (\mathbf{A}^T\mathbf{A})^{-1}\mathbf{A}^T\mathbf{b}. \tag{2.2.19}$$

The matrix $\mathbf{A}^\dagger \triangleq (\mathbf{A}^T\mathbf{A})^{-1}\mathbf{A}^T$ is called the (left) *generalized inverse* (or *Moore-Penrose pseudoinverse*) of matrix **A** [Jam78].

2.3   Matrix decompositions

Cholesky decomposition (or *Cholesky triangle*, or *Cholesky factorization*) decomposes an $N \times N$ symmetric, positive-definite matrix (*SPD*) **A** into the product of a lower triangular matrix with positive diagonal elements **L** and its transpose $\mathbf{L}^T$ [Wil88]:

$$\mathbf{A} = \mathbf{LL}^T, \tag{2.3.1}$$

as shown in Figure 2.3.1. Equation (2.3.1) can be written as:

$$\mathbf{A} = \begin{pmatrix} a_{11} & \mathbf{a}_{21}^T \\ \mathbf{a}_{21} & \mathbf{A}_{22} \end{pmatrix} = \begin{pmatrix} l_{11} & \mathbf{0}^T \\ \mathbf{l}_{21} & \mathbf{L}_{22} \end{pmatrix} \begin{pmatrix} l_{11} & \mathbf{l}_{21}^T \\ \mathbf{0} & \mathbf{L}_{22}^T \end{pmatrix} =$$

$$= \begin{pmatrix} l_{11}^2 & l_{11}\mathbf{l}_{21}^T \\ l_{11}\mathbf{l}_{21} & \mathbf{l}_{21}\mathbf{l}_{21}^T + \mathbf{L}_{22}\mathbf{L}_{22}^T \end{pmatrix}. \tag{2.3.2}$$

It is obvious that $l_{11} = \sqrt{a_{11}}$ and $\mathbf{l}_{21} = \frac{1}{l_{11}}\mathbf{a}_{21}$. Then, the computation proceeds to recursively compute $\mathbf{L}_{22}$ through the equation:

$$\mathbf{A}_{22} - \mathbf{l}_{21}\mathbf{l}_{21}^T = \mathbf{L}_{22}\mathbf{L}_{22}^T. \tag{2.3.3}$$

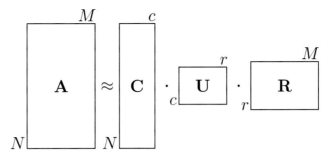

FIGURE 2.3.2: CUR approximation of matrix $\mathbf{A}$.

Example 2.3.1. *Cholesky decomposition.*

Matrix $\mathbf{A} = \begin{bmatrix} 9 & 3 & 1 \\ 3 & 9 & 3 \\ 1 & 3 & 9 \end{bmatrix}$ *has the following Cholesky decomposition:*

$$\mathbf{L} = \begin{bmatrix} 3 & 0 & 0 \\ 1 & 2.8 & 0 \\ 0.3 & 0.9 & 2.8 \end{bmatrix}, \mathbf{L}^T = \begin{bmatrix} 3 & 1 & 0.3 \\ 0 & 2.8 & 0.9 \\ 0 & 0 & 2.8 \end{bmatrix}.$$

For simplicity, the matrix entries have been rounded to the first digit.

Cholesky decomposition is mainly used for the numerical solution of linear equations $\mathbf{A}\mathbf{x} = \mathbf{b}$, when $\mathbf{A}$ is SPD. We first compute the matrix $\mathbf{L}$ as described above. Then, we solve the equation $\mathbf{L}\mathbf{y} = \mathbf{b}$, where $\mathbf{y} = \mathbf{L}^T\mathbf{x}$, using forward substitution and, finally, we solve $\mathbf{L}^T\mathbf{x} = \mathbf{b}$ using back substitution. This approach, when applicable, is preferable to other decompositions, in terms of both efficiency and numerical stability.

CUR approximation is described by three matrices $\mathbf{C} \in \mathbb{R}^{N \times c}, \mathbf{U} \in \mathbb{R}^{c \times r}, \mathbf{R} \in \mathbb{R}^{r \times M}$ that, when multiplied, closely approximate a given matrix $\mathbf{A} \in \mathbb{R}^{N \times M}$, i.e., $\mathbf{A} \approx \mathbf{CUR}$. In other words, given the matrix $\mathbf{A}$ we have to find the three matrices $\mathbf{C}, \mathbf{U}, \mathbf{R}$ such that the approximation error $\|\mathbf{A} - \mathbf{CUR}\|_F$ is small [DKML04]. Figure 2.3.2 illustrates CUR approximation.

Matrix $\mathbf{C}$ contains c columns of $\mathbf{A}$, while $\mathbf{R}$ contains r rows of $\mathbf{A}$. Given specific instances of matrices $\mathbf{C}$ and $\mathbf{R}$, then $\mathbf{U}$ is a $c \times r$ matrix provided by $\mathbf{U} = \mathbf{C}^\dagger \mathbf{A} \mathbf{R}^\dagger$, where $\mathbf{C}^\dagger$ is the Moore-Penrose pseudoinverse of $\mathbf{C}$ (respectively, $\mathbf{R}$) [MD09]. The most important issue is how to select these columns and rows [DKML04]. Two methods are more popular for this problem subspace, namely sampling with replacement, which allows a row or column to be selected multiple times, and sampling without replacement, which does not. It is necessary to estimate the absolute $\|\mathbf{A} - \mathbf{CUR}\|_F$ and the relative error bound, in order to decide which one is better [DMM08]. Thus, it is obvious that CUR matrix approximation is not unique.

Example 2.3.2. *Matrix CUR approximation.*

$$\text{Matrix } \mathbf{A} = \begin{bmatrix} 5 & 2 & 7 & 3 \\ 2 & 9 & 4 & 9 \\ 0 & 6 & 1 & 5 \\ 6 & 8 & 0 & 2 \end{bmatrix} \text{ can be approximated with the following matrices:}$$

$$\mathbf{C} = \begin{bmatrix} 5 & 3 \\ 2 & 9 \\ 0 & 5 \\ 6 & 2 \end{bmatrix}, \mathbf{U} = \begin{bmatrix} 0.2 & -0.3 \\ 0.1 & 0.1 \end{bmatrix}, \mathbf{R} = \begin{bmatrix} 2 & 9 & 4 & 9 \\ 0 & 6 & 1 & 5 \end{bmatrix}.$$

For simplicity, the entries of matrix $\mathbf{U}$ *have been rounded to the first digit. The error of this approximation is* $\|\mathbf{A} - \mathbf{CUR}\|_F = 9.1796$.

CUR approximation is less accurate than SVD, but has certain advantages. CUR, unlike SVD, maintains sparsity. If $\mathbf{A}$ is big and sparse, then matrices $\mathbf{C}$ and $\mathbf{R}$ are also big and sparse, while matrices $\mathbf{U}$ and $\mathbf{V}$ of the SVD would be dense. Moreover, since the matrices $\mathbf{C}$ and $\mathbf{R}$ are made of a subset of data samples and variables respectively, it is much easier to interpret them, than the SVD left and right singular vectors, which represent the data in a transformed space.

Another matrix decomposition technique is the *Non-negative Matrix Factorization* (*NMF*), whose purpose is to factorize a non-negative matrix $\mathbf{X}$ into two matrices $\mathbf{H}, \mathbf{F}$, which are also non-negative [PT94]. NMF has important applications in cases, where the data contained in matrix $\mathbf{X}$ are non-negative by definition, e.g., when vectors $\mathbf{x}_i, i = 1, \dots, N$ represent vectorized images. Since the problem can not be solved exactly nor is there a unique solution, it is commonly approximated numerically [LS01]:

$$\mathbf{X} \approx \mathbf{HF}. \tag{2.3.4}$$

If $\mathbf{X}$ is an $N \times M$ matrix, then $\mathbf{H}$ and $\mathbf{F}$ are $N \times P$ and $P \times M$ matrices, respectively. Usually, P is smaller than N and M. Typically, matrix $\mathbf{X}$ contains M feature data vectors $\mathbf{X} = [\mathbf{x}_1, \mathbf{x}_2, \dots, \mathbf{x}_M]$. In that case, equation (2.3.4) can be written as: $\mathbf{x}_i \approx \mathbf{Hf}_i, i = 1, \dots, M$. This means that each data vector $\mathbf{x}_i$ is approximated by a linear combination of the columns of $\mathbf{H}$ (called *basis vectors*) weighted by the components of $\mathbf{f}_i$ [SD05]. In the case of images, the columns of $\mathbf{H}$ represent *basis images*.

There are different types of non-negative matrix factorizations, each using different cost functions for measuring the divergence between $\mathbf{X}$ and $\mathbf{HF}$ and possibly regularizing $\mathbf{H}$ and/or $\mathbf{F}$ matrices [LS01]. The cost functions quantify the approximation error in (2.3.4). The Frobenius norm (i.e., the *Euclidean distance*) between matrices $\mathbf{K}$ and $\mathbf{L}$ is lower bounded by zero and clearly vanishes, iff $\mathbf{K} = \mathbf{L}$. The *Kullback-Leibler divergence* [KL51]:

$$D(\mathbf{K}\|\mathbf{L}) = \sum_{ij} (k_{ij} \log \frac{k_{ij}}{l_{ij}} - k_{ij} + l_{ij}) \tag{2.3.5}$$

is also lower bounded by zero and vanishes iff $\mathbf{K} = \mathbf{L}$. Thus, two NMF variants arise from the following minimization problems:

- Minimize $\|\mathbf{X} - \mathbf{HF}\|^2$, s.t. $\mathbf{H}, \mathbf{F} \geq 0$. (2.3.6)
- Minimize $D(\mathbf{X} \| \mathbf{HF})$, s.t. $\mathbf{H}, \mathbf{F} \geq 0$. (2.3.7)

Neither the Frobenius norm nor Kullback-Leibler divergence are convex in $\mathbf{H}$ and $\mathbf{F}$ simultaneously. Gradient descent can be applied for minimizing either (2.3.6) or (2.3.7) [Lin07], but has slow convergence and is sensitive to the choice of the step size. The so-called *multiplicative update rules* [LS01] can be used for iteratively solving the above mentioned minimization problems. They take the following forms for the Frobenius norm:

$$\mathbf{F}_{(t+1)} = \mathbf{F}_{(t)} \frac{(\mathbf{H}^T \mathbf{X})_{(t)}}{(\mathbf{H}^T \mathbf{HF})_{(t)}}, \mathbf{H}_{(t+1)} = \mathbf{H}_{(t)} \frac{(\mathbf{XF}^T)_{(t)}}{(\mathbf{HFF}^T)_{(t)}} \quad (2.3.8)$$

and KL divergence:

$$\mathbf{F}_{(t+1)} = \mathbf{F}_{(t)} \frac{\sum_i \mathbf{H}_{(t)} \mathbf{X}_{(t)} / (\mathbf{HF})_{(t)}}{\sum_k \mathbf{H}_{(t)}}, \mathbf{H}_{(t+1)} = \mathbf{H}_{(t)} \frac{\sum_m \mathbf{F}_{(t)} \mathbf{X}_{(t)} / (\mathbf{HF})_{(t)}}{\sum_n \mathbf{F}_{(t)}}, \quad (2.3.9)$$

respectively, where t is an iteration index.

2.4 Vector and matrix derivatives

In the following, we provide some frequently used vector and matrix derivatives [Bel70]:

$$\nabla_{\mathbf{x}} \left(\mathbf{x}^T \mathbf{a} \right) = \nabla_{\mathbf{x}} \left(\mathbf{a}^T \mathbf{x} \right) = \mathbf{a} \quad (2.4.1)$$

$$\nabla_{\mathbf{X}} \left(\mathbf{a}^T \mathbf{X} \mathbf{b} \right) = \mathbf{a} \mathbf{b}^T \quad (2.4.2)$$

$$\nabla_{\mathbf{X}} \left(\mathbf{a}^T \mathbf{X}^T \mathbf{b} \right) = \mathbf{b} \mathbf{a}^T \quad (2.4.3)$$

$$\nabla_{\mathbf{x}} \left(\mathbf{x}^T \mathbf{A} \mathbf{x} \right) = \left(\mathbf{A} + \mathbf{A}^T \right) \mathbf{x} \quad (2.4.4)$$

$$\nabla_{\mathbf{X}} \left(\mathbf{b}^T \mathbf{X}^T \mathbf{A} \mathbf{X} \mathbf{c} \right) = \mathbf{A}^T \mathbf{X} \mathbf{b} \mathbf{c}^T + \mathbf{D} \mathbf{X} \mathbf{c} \mathbf{b}^T \quad (2.4.5)$$

$$\nabla_{\mathbf{X}} \left((\mathbf{X} \mathbf{b} + \mathbf{c})^T \mathbf{A} (\mathbf{X} \mathbf{b} + \mathbf{c}) \right) = \left(\mathbf{A} + \mathbf{A}^T \right) (\mathbf{X} \mathbf{b} + \mathbf{c}) \mathbf{b}^T \quad (2.4.6)$$

$$\nabla_{\mathbf{X}} \, tr \, (\mathbf{X}) = \mathbf{I} \quad (2.4.7)$$

$$\nabla_{\mathbf{X}} \, tr \, (\mathbf{X} \mathbf{A}) = \mathbf{A}^T \quad (2.4.8)$$

$$\nabla_{\mathbf{X}} \, tr \, (\mathbf{A} \mathbf{X} \mathbf{B}) = \mathbf{A}^T \mathbf{B}^T \quad (2.4.9)$$

$$\nabla_{\mathbf{X}} \, tr \, (\mathbf{A} \mathbf{X}^T \mathbf{B}) = \mathbf{B} \mathbf{A} \quad (2.4.10)$$

$$\nabla_{\mathbf{X}} \, tr \, (\mathbf{A} \mathbf{X} \mathbf{B} \mathbf{X}^T \mathbf{C}) = \mathbf{A}^T \mathbf{C}^T \mathbf{X} \mathbf{B}^T + \mathbf{C} \mathbf{A} \mathbf{X} \mathbf{B}. \quad (2.4.11)$$

Bibliography

[Bel70] R. Bellman. *Introduction to matrix analysis*. McGraw-Hill, 2nd edition, 1970.

[Ber89] C. Berge. *Hypergraphs: combinatorics of finite sets*, volume 45. North Holland, 1989.

[BM76]   J. A. Bondy and U. S. R. Murty. *Graph theory with applications*, volume 6. Macmillan London, 1976.

[BR12]   R. Balakrishnan and K. Ranganathan. *A textbook of graph theory.* Springer Science & Business Media, 2012.

[Chu97]   F. R. K. Chung. *Spectral Graph Theory.* American Mathematical Society, 1997.

[DKML04]   P. Drineas, R. Kannan, M. W. Mahoney, and A. Let. Fast Monte Carlo algorithms for matrices III: Computing a compressed approximate matrix decomposition. *SIAM Journal on Computing*, 36, 2004.

[DMM08]   P. Drineas, M. W. Mahoney, and S. Muthukrishnan. Relative-error CUR matrix decompositions. *SIAM Journal on Matrix Analysis and Applications*, 30(2):844–881, 2008.

[GVL96]   G. H. Golub and C. F. Van Loan. *Matrix Computations.* JHU Press, 3rd edition, 1996.

[GY99]   J. L. Gross and J. Yellen. *Graph theory and its applications.* CRC Press, 1999.

[Jam78]   M. James. The generalised inverse. *The Mathematical Gazette*, pages 109–114, 1978.

[KL51]   S. Kullback and R. A. Leibler. On information and sufficiency. *The Annals of Mathematical Statistics*, pages 79–86, 1951.

[KL02]   R. I. Kondor and J. D. Lafferty. Diffusion kernels on graphs and other discrete input spaces. *International Conference on Machine Learning*, 2002.

[LH74]   C. L. Lawson and R. J. Hanson. *Solving least squares problems*, volume 161. SIAM, 1974.

[Lin07]   C.-J. Lin. Projected gradient methods for nonnegative matrix factorization. *Neural computation*, 19(10):2756–2779, 2007.

[LS01]   D. D. Lee and H. S. Seung. Algorithms for Non-negative Matrix Factorization. In T. K. Leen, T. G. Dietterich, and V. Tresp, editors, *Advances in Neural Information Processing Systems 13*, pages 556–562. MIT Press, 2001.

[MD09]   M. W. Mahoney and P. Drineas. CUR matrix decompositions for improved data analysis. *Proceedings of the National Academy of Sciences*, 106(3):697–702, 2009.

[Mey00]   C. D. Meyer. *Matrix Analysis and Applied Linear Algebra.* SIAM Press, 2000.

[PT94]   P. Paatero and U. Tapper. Positive matrix factorization: A non-negative factor model with optimal utilization of error estimates of data values. *Environmetrics*, 5(2):111–126, 1994.

[SD05]   S. Sra and I. S. Dhillon. Generalized nonnegative matrix approximations with bregman divergences. In *Advances in neural information processing systems*, pages 283–290, 2005.

[Str88]   G. Strang. *Linear Algebra and Its Applications.* Brooks Cole, 1988.

[Wes01]   D. B. West. *Introduction to graph theory*, volume 2. Prentice Hall, 2001.

[Wil88]   J. H. Wilkinson. *The Algebraic Eigenvalue Problem*. Clarendon Press, 1988.

[ZHS07]   D. Zhou, J. Huang, and B. Schölkopf. Learning with hypergraphs: Clustering, classification, and embedding. In *Proc. Advances in Neural Information Processing Systems*, volume 19, page 1601, 2007.

Chapter 3

Algebraic Graph Analysis

Nikolaos Tsapanos, Anastasios Tefas and Ioannis Pitas

Aristotle University of Thessaloniki, Greece

3.1 Introduction

Graphs provide an intuitive way of representing connected or interacting entities [GY99]. One can very easily model and study various real-life and scientific structures using graphs, e.g., web pages linking to each other [BV03], friendship in social networks, the relative location of image features [DPZ01], transportation networks, electric circuits and computer network topologies. Mathematicians have been studying graph theory since the 18th century, starting with a published paper on the Seven Bridges of Konigsberg by Euler [Big93].

Many solutions to classical graph problems involve combinatorics, either with respect to vertex/edge selection or to vertex/edge ordering. As such, the best available algorithms are super-polynomial in complexity [GJ90], making their solution intractable, when large graphs are involved. While understanding and studying graphs seems deceptively easy, especially when the size of a graph is relatively small, the nature of graphs makes them very difficult to represent in vector form in order to handle them with algebraic methods directly. Most difficulties stem from the fact that the same graph can have radically dissimilar interconnectivity under different vertex permutations. While there are several graph invariant properties, most of them are scalars and, as such, are not particularly useful in meaningfully representing a graph. The graph spectrum is an invariant that provides a much richer representation of the graph and is the subject of spectral graph theory [Chu97].

This chapter is meant as an introduction to the intersection of graph theory and algebraic or statistical approaches. It is organized as follows: Section 3.2 provides a quick introduction to spectral graph theory. Applications of the various graph analysis tasks are summarized in Section 3.3. The chapter continues by surveying techniques for several subfields, starting with random graph generation in Section 3.4 and continuing with graph clustering and community detection in Section 3.5, graph matching techniques in Section 3.6, random walks in Section 3.7 and anomaly detection in Section 3.8. Finally, Section 3.9 concludes this chapter.

3.2 Spectral graph theory

3.2.1 Adjacency and Laplacian matrix

The most well-known representation of a graph with n vertices is the adjacency matrix $\mathbf{A} \in \{0,1\}^{n \times n}$ of the graph. An adjacency matrix is a means of representing which vertices (or nodes) of a graph are adjacent to each other. That is, if the vertex i is adjacent (connected/related) to the vertex j then $A(i,j) = 1$. The adjacency matrix of an undirected simple graph is symmetric (i.e., $A(i,j) = A(j,i)$), and, therefore, has a complete set of real eigenvalues and an orthogonal eigenvector basis. Spectral graph theory studies the relationships between the set of eigenvalues and eigenvectors of a graph matrix representation and various properties of the graph. In general, the *eigenvalues* of its adjacency matrix $\mathbf{A}$ are given by solving the following eigenanalysis problem:

$$\mathbf{A}\mathbf{u}_i = \lambda_i \mathbf{u}_i, \; i = 0 \ldots n - 1 \tag{3.2.1}$$

and are related to various graph properties. The *graph spectrum* is defined to be the set of these eigenvalues. The vector $\mathbf{u}_i$ that can satisfy (3.2.1) for a given λ_i is the corresponding *eigenvector* of $\mathbf{A}$. Alternatively, the eigenvalues of a matrix are defined as the roots of the characteristic polynomial of the matrix:

$$|\mathbf{A} - \lambda \mathbf{I}| = 0, \tag{3.2.2}$$

where $|\cdot|$ denotes the matrix determinant. The characteristic polynomial of the adjacency matrix $\mathbf{A}$ is of n-th degree and, as such, has n roots (eigenvalues). Some eigenvalues may coincide, in which case the algebraic *multiplicity* of an eigenvalue λ_i is the number of times λ_i appears as a solution to (3.2.2). Finally, the eigenvalues and eigenvectors of an adjacency matrix $\mathbf{A}$ leads to its *eigendecomposition* [Fra00]:

$$\mathbf{A} = \mathbf{U}\mathbf{\Lambda}\mathbf{U}^T, \tag{3.2.3}$$

where $\mathbf{\Lambda}$ is a diagonal matrix (i.e., $[\mathbf{\Lambda}]_{ii} = \lambda_i$) containing the n eigenvalues and the columns of matrix $\mathbf{U} = [\mathbf{u}_1, \mathbf{u}_2, \ldots, \mathbf{u}_n]$ contain the corresponding eigenvectors.

The graph spectrum can be calculated using the adjacency matrix $\mathbf{A}$ or other related matrices, notably the normalized *Laplacian* one. The Laplacian matrix $\mathcal{L}$ is a related matrix representation of a graph that is given by:

$$\mathcal{L} = \mathbf{D} - \mathbf{A}, \tag{3.2.4}$$

where $\mathbf{D}$ is the degree matrix of the graph. The degree matrix is a diagonal matrix containing information about the degree of each vertex (i.e., the number of edges attached to each vertex $[\mathbf{D}]_{ii} = \sum_j [\mathbf{A}]_{ij}$). For undirected graphs the Laplacian matrix is symmetric positive semi-definite (i.e., $\lambda_i \geq 0 \ \forall i$). The symmetric normalized Laplacian is defined as

$$\mathcal{L}_{sym} := \mathbf{D}^{-1/2} \mathcal{L} \mathbf{D}^{-1/2} = \mathbf{I} - \mathbf{D}^{-1/2} \mathbf{A} \mathbf{D}^{-1/2} \tag{3.2.5}$$

and the random walk normalized Laplacian is defined as

$$\mathcal{L}_{rw} := \mathbf{D}^{-1} \mathcal{L} = \mathbf{I} - \mathbf{D}^{-1} \mathbf{A}. \tag{3.2.6}$$

The normalized Laplacians are used in various graph analysis tasks, such as graph clustering and random walks on graphs.

The eigenvalues λ_i of any graph matrix representation are typically sorted in non-decreasing order $\lambda_0 \leq \lambda_1 \leq \cdots \leq \lambda_{n-1}$. The graph spectrum calculated using different graph matrices are not the same and, depending on which matrix was used to calculate the spectrum, the eigenvalues can be in different value ranges and relate to graph properties in different ways. The graph spectrum coming from any such matrix is invariant to graph isomorphisms, i.e., the calculated eigenvalues are invariant to vertex permutations [Chu97]. Unfortunately, it is possible that two non-isomorphic graphs are co-spectral, i.e., have the same eigenvalues [Chu97]. Thus, comparing the two graph spectra is not a completely reliable way to test graph isomorphisms.

We will now provide an overview of some bounds and interesting properties of graph spectra calculated from the normalized Laplacian matrices. The interested reader may refer to [Chu97] for more in-depth discussion and proofs. The normalized Laplacian eigenvalues of a graph G with n vertices satisfy:

$$0 = \lambda_0 \leq \lambda_1 \leq \cdots \leq \lambda_{n-1} \leq 2. \tag{3.2.7}$$

The smallest eigenvalue, λ_0 is always zero, because a vector $\mathbf{1}$ having all entries equal to 1, is an eigenvector for $\lambda = 0$ for any Laplacian matrix. The equality $\lambda_{n-1} = 2$ applies, if and only if G is bipartite. The sum of the eigenvalues is always equal to $\sum_{i=0}^{n-1} \lambda_i = n$. The second smallest eigenvalue λ_1 is the so-called *algebraic connectivity* of the graph. If $\lambda_1 > 0$ then G is connected. Otherwise, the multiplicity of eigenvalue 0 is equal to the number of connected components in G [Chu97]. This follows directly from the fact that the spectrum of the union of two disjoint graphs is the union of their respective eigenvalue sets. This property is very useful in designing spectral graph clustering algorithms, as it will be described in Section 3.5.

3.2.2 Similarity matrix and nearest neighbor graph

In many cases, the graph that connects the data is not available. In order to use spectral graph analysis techniques, a graph has to be constructed over the data [vL07]. The data samples $\mathbf{x}_i \in \mathbb{R}^n$ are usually in vectorial form and they form the data matrix $\mathbf{X} = [\mathbf{x}_1, \mathbf{x}_2, \ldots, \mathbf{x}_n]$. In order to use graph analysis techniques on such a dataset, we form a weighted graph, where the data samples reside on the graph nodes and the edges between the nodes are constructed based on various strategies. The adjacency matrix of such a graph is typically a *similarity matrix*, containing scores that measure the similarity between two data samples. The value $[\mathbf{S}]_{ij}$ in the matrix is usually calculated using the heat kernel

$e^{-||\mathbf{x}_i-\mathbf{x}_j||^2/2\sigma^2}$ [AR13]. It is evident that the similarity matrix represents a complete graph, whose edge weights are equal to the similarity between the corresponding graph vertices that are connected by each edge. Alternatively, one can construct sparse adjacency matrices based on the similarity matrix, as described below.

The *nearest neighbor graph* is constructed using the similarity matrix of the data and a specific method to prune it to produce the adjacency graph. A well-known approach is to produce symmetric nearest neighbor adjacency matrices as follows:

$$A(i,j) = \begin{cases} S(i,j) \text{ or } 1, & i \in \mathcal{N}(j) \text{ or } j \in \mathcal{N}(i) \\ 0, & \text{otherwise,} \end{cases} \tag{3.2.8}$$

where $\mathcal{N}(j)$ denotes the set containing the indices of the k-nearest neighbors of sample j. Alternatively, one can use a threshold ϵ on the similarity, in order to find the neighbors of the vertex i. In that case the graph is called ϵ-neighborhood graph [vL07]. Obviously, the adjacency matrix in (3.2.5) and (3.2.6) can be replaced by $\mathbf{S}$ to produce the corresponding Laplacian matrices. The way the adjacency graph is constructed affects spectral graph analysis [MM13]. All the above matrices capture data geometry and can be used in various tasks, ranging from clustering and classification to label propagation and dimensionality reduction.

3.3   Applications of graph analysis

Graphs provide a flexible, powerful, and useful tool for the representation of a wide variety of entities. As such, the applications of the graph related tasks surveyed in this Chapter are numerous. A concise summary of applications can be found in Table 3.3.1.

Random graph generation is useful for providing realistic synthetic graphs, so that graph-based algorithms can be tested for performance and scaling capabilities. As it is often the case, real graph data are limited in size, or completely unavailable. Furthermore, the determination of the ground truth may be almost impossible. In such cases, randomly generated graphs provide the only means for algorithms to be tested. The temporal evolution of real graphs can also be studied and predicted by appropriate random graph generation models. It is also possible to generate random graphs for scheduling simulations on distributed systems [CMP+10].

Graph clustering is a very useful tool for the analysis of graph-based data. Graph connectivity is easier to analyze when the graph is clustered and interesting substructures become easier to identify [MGSZ02]. The computational load can also be split for various graph analysis algorithms, as, depending on the task, clusters can be processed separately [Ide04]. It can be employed to improve the performance of tag recommendation methods in social media [PKV10], i.e., for the vast quantities of user generated content (photographs, video and audio clips) that are available online. This is mostly accomplished by finding additional appropriate tags for a media item that the original user/owner may have omitted, in order to find more items that are relevant to user searches. It can also help graph compression algorithms to lower the bits per link required to encode the graph. In database systems, graph clustering can improve the speed with which relative data are accessed through more appropriate paging [DRSS96]. Graph compression algorithms are also more efficient, if a good graph clustering can be attained [BRSV11].

Graph matching has a wide variety of applications in pattern recognition and computer vision [DPZ01]. There has been an abundance of graph-based works in 2D [EF86] and 3D

[SF83] image analysis [CFSV04]. Biometric identification problems, such as face recognition, can be expressed in terms of graph matching, more specifically elastic graph matching [TKP01]. In this framework, vertices are arranged in a square grid (graph) and then overlaid on a training image. The image features (for example, 2D Gabor filter bank output) at the image location of every vertex are computed and stored. In order to classify a test image, the trained grid vertices are placed on a test image and are displaced, trying to find a location where the image features of the training grid vertex most closely match the features of the test image. This procedure deforms the grid (graph). The strain on the graph edges cause by this deformation, and the graph dissimilarity to the local test image features, can be used to measure the similarity between the training and test image. The same technique, with some appropriate modifications, can also be used for facial expression recognition [KP07]. Several image registration [PF97] and retrieval [HH98] techniques are based on graph matching [CCLS07]. A graph can be used to capture the relative positions of various image features, with respect to each other. This provides a more abstract image representation that avoids the most common problems that plague appearance based image representation methods, such as illumination changes and geometric transformations. Reliable similarity measurements between two images can be obtained, provided that there are enough correctly established correspondences between the vertices of the two image representation graphs. Graph matching techniques for document processing, such as optical character recognition (OCR) [CL90], string matching [FGK95], and symbol recognition [LMV01] have also been developed, though they can be considered as subcases of image registration.

Random walks have found applications in local graph clustering, as a random walk starting from any vertex is more likely to remain in the vertex cluster than to move to another cluster. Another application of random walks is to estimate the number of elements in a set, or the volume of an object, whose exact computation would otherwise be intractable, through random sampling. Mobile agents often perform random walks, for example in wireless networks, and some web spiders crawl the web in this fashion. Therefore, random walks can be used to model their behavior [Ber09] and provide performance metrics for various algorithms, like how long it is expected to take for a mobile agent to reach a target node, or until two agents reach the same node [KKR06], under a proposed scheme. Random walks can even be used for image segmentation. By seeding some pixels with a label, e.g., through user initialization, the probability of pixels belonging to each label propagates through the image as a random walk that is less likely to cross image edges [Gra06]. Google's PageRank models the behavior of a web surfer largely as a random walk.

Anomaly detection methods can be used for fraud detection; as stolen credit cards and financial scandal detection, such an example that had a significant impact was the Enron scandal [SA05]. Using a measure of graph entropy, important nodes can be identified through the magnitude of the change in graph entropy caused by their removal from the graph. Network intrusions, attempts to gain unauthorized access to systems and online attacks can be classified as anomalies. A general way of detecting such attacks is by measuring the deviation of the network under attack when compared to its normal operation. Once detected, these attacks can be prevented or remedied [Ide04].

With the proliferation of social networks and social media in the last few years, there has been an increasing interest in their analysis, while businesses already viewed them as potential new markets to be exploited. Several, if not all, tasks related to social networks map well to graph analysis algorithms that are reviewed in this chapter. *Community detection*, i.e., finding groups of users that are densely connected with each other, is almost synonymous to vertex clustering in graphs. *Recommendation systems*, finding pictures, music, or video clips that are relevant to specific user interests can be provided with more and better suggestions based on social graph clustering according to user interests. It is sometimes

TABLE 3.3.1: Overview of graph analysis applications.

Graph analysis task	Application	Field
Graph generation	Algorithm testing	Software engineering
	Network evolution Algorithm simulation	Social network analysis
Graph clustering	Data storage Data compression	Database systems
	Popularity prediction Tag recommendation	Social network analysis
	Substructure indentification Network usage optimization	Computer networks
Graph matching	2D,3D Image analysis Face recognition Face verification Object registration/retrieval	Computer vision
	Document analysis	Language engineering
	Molecular structure study	Computational chemistry
Random walks	Enumeration	Multiple
	Volume computation	Computational geometry
	Mobile agent modelling	Distributed systems
	Web crawling	Internet computing
Anomaly detection	System intrusion detection Network attack detection	Computer security
	Financial fraud detection	Law enforcement
	Influential individual detection	Social network analysis

possible to *classify* communities with a label. *Centrality* is one way of identifying *important* vertices in a social network. Graph anomaly detection methods provide alternative interpretations on what is important, because vertex importance can have various meanings. For example, influential people, including but not limited to celebrities, can help to faster propagate ideas, news, and stories. People that pose security threats, such as criminals, can also be considered important and their identification would be of great interest. Information diffusion is another important issue that can be very well described by graphs. With the plethora of new ideas born daily and their rapid propagation provided by social networks, it is interesting to be able to predict which idea will catch on and where they will originate from, thus allowing people to identify, predict, or capitalize on new trends. This can be done by studying diffusion mechanisms over graphs.

3.4   Random graph generation

Being able to randomly generate graphs that have certain properties can be very useful in many applications [WS98]. For example, one may wish to generate graphs with predetermined vertex clusters, in order to test a clustering algorithm, since ground truth is difficult to define on real data sets. A generation method that models the growth of, e.g., a social network may well be used to study the future evolution of the said network and, thus, test the scalability of the relevant graph analysis algorithms.

3.4.1 Desirable random graph properties

There are several ways to evaluate the clustering results in graph vertex clustering. A measure to evaluate how densely connected the graph vertices are is provided by the average *clustering coefficient* [WS98]. A vertex clustering coefficient is defined as the fraction of edges in the induced subgraph of the vertex neighbors (excluding the vertex) over the total possible edges of that subgraph. The average graph clustering coefficient is obtained by averaging the clustering coefficient of all the graph vertices.

In various real life graphs, such as the ones corresponding to social networks, two major observations have been made: a) the vertex degrees follow a *power law*, i.e., they have a heavy-tailed probability distribution and b) the diameter of these graphs (i.e., the greatest distance between any pair of graph vertices) actually decreases as they grow larger over time, contrary to expectations [LKF05]. The small diameter of large social networks, otherwise known as the *small world phenomenon*, was studied as early as in the 1960s [Mil67]. It appears that almost everyone in the world can be connected to almost everyone else with a relatively small number of mutual acquaintances, typically about 6 or 7. Studies in web social networks indicate that this holds even as the network graph expands with new vertices [LKF05]. Therefore, any generation model for such graphs should produce graphs that retain these two properties, which will subsequently be described in detail.

A random variable x is formally said to follow a power law, if its probability distribution $p(x)$ has the form $p(x) \propto x^{-\alpha}$, where typically $2 < \alpha < 3$ [CSN09]. Usually, it is assumed that $x \geq x_{\min}$ to avoid indeterminacy (division by zero). Such probability distributions are called *heavy-tailed*, because they do not approach zero quite as fast as x increases. This, in effect, means that extreme values, such as very high vertex degrees, are not as unlikely to appear as other distributions would suggest (e.g., the Gaussian one). An interesting property of a power law is that it is scale invariant. Supposing that $p(x) = x^{-\alpha}$, one can easily see that:

$$p(cx) = (cx)^{-\alpha} = c^{-\alpha}x^{-\alpha} = c^{-\alpha}p(x). \qquad (3.4.1)$$

Graphs, whose vertex degrees follow a power law, are called *scale-free* graphs (or networks) [BB03]. Another interesting property of a power law distribution is that its logarithmic function $\ln p(x)$ vs. $\ln x$ (in the graph case, the logarithm of number of vertices having a given degree vs the logarithm of that vertex degree) plot is closely approximated by a straight line. This method of plotting is usually referred to as a log-log plot. An example of the power law in a graph constructed from video data obtained from YouTube [TPP14] can be seen in Figure 3.4.1, where the log-log plot appears to follow the power law within the degree range from 10 to 1000.

3.4.2 Random graph generation models

The *Erdös-Rényi* method is the simplest one for generating a graph with n vertices, by connecting any two vertices by an edge with probability p [Gil59]. The degree distribution of graphs generated using this method is a Poisson one. This is not a particularly good model, as the uniformly random edge placement can not guarantee the presence of dense clusters. Hence, it is rather unlikely that the generated graph will have the desirable properties observed in real networks. This generation model has been thoroughly studied in [ER60].

The *Watts-Strogatz* method [WS98] starts from a graph, whose vertices are placed on a circle and are ordered numerically. Each vertex is connected to its adjacent vertex

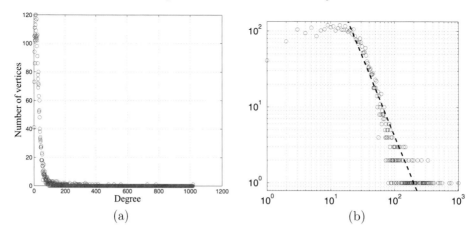

FIGURE 3.4.1: The power law in a graph constructed from data obtained from YouTube: a) The degree distribution and b) the corresponding log-log plot.

and every k_n-th vertex for some integers k_n. In a more formal way, an edge $(u_i, u_j) \in E$ is generated, iff $i \equiv j \mod k_n$. Then some edges are relocated, avoiding self loops and duplicate connections, with probability p. Depending on the value of p, the graph can retain most of its original structure for p values close to 0 and become completely random for values close to 1. For intermediate p values, the generated graphs have high clustering coefficients and low diameters. However, the probability distribution of the vertex degrees does not follow a power law.

In order to satisfy the power law distribution of vertex degrees, [RB02] employs a *preferential attachment* procedure, which favors increasing the degree of vertices already having high degrees ("the rich get richer" approach), in order to ensure that there will be sufficient vertices with much higher degrees than average and, thus, satisfy the power law distribution. Starting from an initial set of vertices with degree 1, vertices are incrementally added into the current graph. Any new vertex gets a set amount of new connections. However, the selection of its neighbors is not uniformly random. An existing vertex u_i has a $\frac{\deg(u_i)}{\sum_j \deg(u_j)}$ probability of being chosen for the new vertex neighborhood. This makes practical sense, as, for example, there are popular people in communities and anyone new in the community is very likely to encounter the popular people first. In the case of web pages, an informative web page has many incoming links, and new web pages on the same subject are more likely to link to it. The drawback of this method, however, is that the diameter tends to increase logarithmically with the number of vertices.

A random graph generation model that actually produces graphs that have a high clustering coefficient, a small diameter, and whose degree distribution observes a power law is presented in [LCKF05]. The method is based on the *Kronecker multiplication* $\mathbf{A}$ of the $n \times n$ matrix $\mathbf{A}_1$ and the $m \times m$ matrix $\mathbf{A}_2$:

$$\mathbf{A} = \begin{bmatrix} a_1(1,1)\mathbf{A}_2 & a_1(1,2)\mathbf{A}_2 & \dots & a_1(1,n)\mathbf{A}_2 \\ a_1(2,1)\mathbf{A}_2 & a_1(2,2)\mathbf{A}_2 & \dots & a_1(2,n)\mathbf{A}_2 \\ \vdots & \vdots & \ddots & \vdots \\ a_1(n,1)\mathbf{A}_2 & a_1(n,2)\mathbf{A}_2 & \dots & a_1(n,n)\mathbf{A}_2 \end{bmatrix}. \tag{3.4.2}$$

It is easy to see that the Kronecker product $\mathbf{A}$ is an $mn \times mn$ matrix. The Kronecker product of two graphs is described by the product of their adjacency matrices. The k-th Kronecker power of a graph G can be defined by the k Kronecker multiplications of its adjacency matrix by itself. Figure 3.4.2 presents an example of a graph Kronecker multiplication by itself. The main idea for its use in graph generation is raising a suitable seed graph G to the appropriate power, in order to obtain a graph of the required size. To include certain randomness in the generation procedure, instead of using the binary adjacency matrix $\mathbf{A}$ of G, it is suggested to replace the 1s and 0s with probabilities $p_{(1)}$ and $p_{(0)}$ ($0 \leq p_{(0)} < p_{(1)} \leq 1$), respectively, in the adjacency matrix and, thus, obtain a probability matrix. When raised to the k-th Kronecker power, the elements of the resulting matrix will be the product of l times $p_{(1)}$ and $(k - l)$ times $p_{(0)}$, $0 \leq j \leq k$. This Kronecker product provides a probability that the corresponding element in the final graph adjacency matrix will be 1; otherwise it will be zero. The adjacency matrix is determined accordingly. Graphs generated using this method are proven to maintain the non-increasing diameter and, with a careful choice of the seed graph G, have a power law probability distribution of the vertex degree [LCKF05].

$$
A_1 = \begin{bmatrix} 1 & 1 & 0 \\ 1 & 1 & 1 \\ 0 & 1 & 1 \end{bmatrix}, A_2 = A_1 \otimes A_1 =
\begin{array}{|c|c|c|}
\hline
A_1 & A_1 & 0 \\
\hline
A_1 & A_1 & A_1 \\
\hline
0 & A_1 & A_1 \\
\hline
\end{array}
$$

FIGURE 3.4.2: The second Kronecker power of an example graph G.

According to [MX07], the graphs generated by Kronecker multiplication are not searchable by a distributed greedy algorithm. In order to rectify this, a Kronecker-like operation is used in [BBHW10], in which a seed graph G is not Kronecker multiplied by itself but by suitable graphs H_i from a family of graphs $\mathcal{H}$. Depending on the definition and parameter selection of the graphs H_i, this model can generate graphs similar to the Watts-Strogatz method and the original Kronecker product method. The authors, however, suggest selecting the graphs H_i in such a way, so that, in the final probability matrix, the probability of vertices u_i and u_j being connected is determined by the Hamming distance $h(i, j)$ of their indices i and j:

$$
p(u_i, u_j) = \begin{cases} 1, & \text{if } h(i, j) \leq 1 \\ h(i, j)^{-\alpha}, & \text{otherwise.} \end{cases} \tag{3.4.3}
$$

The resulting graphs are essentially extensions of a n-dimensional hypercube. There has been evidence that this assumption is not far from reality [KPBV09]. It can be proven that such graphs are searchable [Kle00].

3.4.3 Spectral graph generation

A generative model for graphs that is based on spectral graph theory is presented in [XH06]. The method begins by embedding the graph into a heat kernel $\mathbf{H}_t$, which is achieved by exponentiating the elements of $\mathbf{\Lambda}$ of the Laplacian eigenspectrum $\mathcal{L} = \mathbf{U}\mathbf{\Lambda}\mathbf{U}^T$ and then performing a Young-Householder decomposition of the resulting matrix $\mathbf{H}_t$ to retrieve the coordinate matrix $\mathbf{Y}$ [Bro87].

$$
\mathbf{H}_t = \mathbf{U}e^{-\frac{1}{2}\mathbf{\Lambda}t}\mathbf{U}^T = \mathbf{Y}^T\mathbf{Y}. \tag{3.4.4}
$$

The columns of matrix $\mathbf{Y}$ are the coordinates of the mapping of the corresponding eigenvector to the heat kernel space. Given a set of graphs $\mathcal{T} = \{G_1, G_2, \ldots, G_N\}$ (not necessarily of the same size), the coordinate matrices are truncated to remove extraneous eigenvectors that the smallest graph does not have. Let $\hat{\mathbf{Y}}_i$ be the truncated coordinate matrix of graph G_i, after the eigenvectors of the least significant eigenvalues have been removed. The largest graph is selected as the reference graph for the generative model.

Since the proper order of the eigenvectors may not be the same for every sample graph, the correspondences between the retained eigenvectors of the reference graph and each of the rest of the sample graphs must be established. The Scott and Longuet-Higgins algorithm [SLh90] is used to find these correspondences, which are stored in the binary correspondence matrix $\mathbf{C}_i$ for each graph G_i. The mean and the covariance matrix of the distribution of graph embeddings over $\mathcal{T}$ are calculated as follows:

$$\hat{\mathbf{X}} = \frac{1}{N} \sum_{1}^{N} \mathbf{C}_i^T \hat{\mathbf{Y}}_i \tag{3.4.5}$$

$$\mathbf{\Sigma} = \frac{1}{N} \sum_{1}^{N} (\mathbf{C}_i^T \hat{\mathbf{Y}}_i - \hat{\mathbf{X}})(\mathbf{C}_i^T \hat{\mathbf{Y}}_i - \hat{\mathbf{X}})^T. \tag{3.4.6}$$

In a final preprocessing step, the covariance matrix $\mathbf{\Sigma}$ is eigendecomposed into:

$$\mathbf{\Sigma} = \mathbf{\Psi}\mathbf{\Gamma}\mathbf{\Psi}^T. \tag{3.4.7}$$

Where $\mathbf{\Gamma}$ denotes the diagonal matrix of eigenvalues and $\mathbf{\Psi}$ denotes the matrix, whose columns are formed by the eigenvectors, ordered by the corresponding eigenvalues in $\mathbf{\Gamma}$. An input observation map on the heat kernel space is fit into the model through a parameter vector $\mathbf{b}$, whose optimal value $\mathbf{b}^*$ is calculated as:

$$\mathbf{b}^* = \arg\min_{b} (\hat{\mathbf{Y}} - \hat{\mathbf{X}} - \mathbf{\Psi}\mathbf{b})^T(\hat{\mathbf{Y}} - \hat{\mathbf{X}} - \mathbf{\Psi}\mathbf{b}) = \mathbf{\Psi}^T(\hat{\mathbf{Y}} - \hat{\mathbf{X}}). \tag{3.4.8}$$

In order to reconstruct the adjacency matrix of a graph embedded in the heat kernel at point $\hat{\mathbf{Y}}$, the coordinate matrix $\hat{\mathbf{Y}}^*$ is reconstructed first:

$$\hat{\mathbf{Y}}^* = \hat{\mathbf{X}} + \mathbf{\Psi}\mathbf{\Psi}^T(\hat{\mathbf{Y}} - \hat{\mathbf{X}}). \tag{3.4.9}$$

Then the heat kernel and the Laplacian matrix are reconstructed as follows:

$$\hat{\mathbf{H}}_t^* = (\hat{\mathbf{Y}}^*)^T \hat{\mathbf{Y}}^* \tag{3.4.10}$$

$$\hat{\mathbf{\Lambda}}^* = -\frac{1}{N} \ln \hat{\mathbf{H}}_t^*. \tag{3.4.11}$$

Finally, the reconstructed graph adjacency matrix can be recovered from its reconstructed Laplacian:

$$\mathbf{A}^* = \mathbf{D} - \mathbf{D}^{\frac{1}{2}}\hat{\mathbf{\Lambda}}^*\mathbf{D}^{\frac{1}{2}}, \tag{3.4.12}$$

according to (3.2.4).

The incorporation of the eigenvalues into the long-vector eigenvector representation in (3.4.7) can lead to reconstruction issues [WW07]. Thus, it is suggested that the eigenvalues and the concatenated eigenvectors forming the aforementioned long-vector are split up. Let $\mathbf{e}$ be the eigenvalue vector and $\mathbf{z}$ be the eigenvector long-vector (concatenated eigenvectors). As before, the means $\bar{\mathbf{e}}$ and $\bar{\mathbf{z}}$ and covariance matrices $\mathbf{\Sigma}_e$ and $\mathbf{\Sigma}_z$ are calculated. The covariance matrices are eigendecomposed into $\mathbf{\Gamma}_e$, $\mathbf{\Psi}_e$ and $\mathbf{\Gamma}_z$, $\mathbf{\Psi}_z$, respectively. Parameter vectors $\mathbf{b}_e$ and $\mathbf{b}_z$ are used to fit the data into the generative model:

$$\hat{\mathbf{e}} = \bar{\mathbf{e}} + \mathbf{\Psi}_e\mathbf{b}_e, \quad \hat{\mathbf{z}} = \bar{\mathbf{z}} + \mathbf{\Psi}_z\mathbf{b}_z. \tag{3.4.13}$$

In order to generate a new graph, random parameter vectors $\mathbf{b}_e$ and $\mathbf{b}_z$ are drawn from a normal distribution and vectors $\hat{\mathbf{e}}$ and $\hat{\mathbf{z}}$ are converted into the eigenvalue matrix $\hat{\mathbf{\Lambda}}$ and eigenvector matrix $\hat{\mathbf{\Phi}}$, respectively. Since the parameter vectors were randomly generated, the eigenvectors may not form an orthogonal base. This is corrected by projecting the eigenvectors to form an orthogonal matrix $\hat{\mathbf{\Phi}}_O$:

$$\hat{\mathbf{\Phi}}_O = (\hat{\mathbf{\Phi}}\hat{\mathbf{\Phi}}^T)^{-\frac{1}{2}}\hat{\mathbf{\Phi}}. \tag{3.4.14}$$

The graph Laplacian is reconstructed as per equations (3.4.11), (3.4.12). However, due to the split between eigenvalues and eigenvectors and the orthogonalization step, the adjacency matrix contains negative entries. A final thresholding on its values is needed to acquire the adjacency matrix of the randomly generated graph that is similar to the original sample set.

When compared to non-spectral graph generation techniques, using the graph spectrum allows for generating variants of an already existing graph. It is, therefore, better suited for testing graph similarity algorithms. When the changes remain small-scale, the realism of the generated graph is not far from the original graph, in case real data are used for seeding the generator. However, there are no theoretical guarantees that the generated graphs will retain the power law distribution of vertex degrees, or the small diameter, particularly if the eigenvector displacement is significant.

3.5 Graph clustering

The graph clustering objective is to gather graph vertices into groups (clusters), so that the vertices in each cluster are closely related to each other. There is no formal definition of what a cluster is and exactly what properties it has. In general, vertices of the same cluster should be heavily connected to other vertices within the same cluster, while being sparsely connected to the rest of the graph. In general, interconnection can be measured by the density of the cluster induced subgraph, while connectivity outside the cluster can be measured by the size of the *graph cut* removing the cluster from the graph.

A thorough survey on graph clustering is provided in [Sch07]. There are two broad categories of clustering algorithms: global and local ones. As the name implies, global algorithms

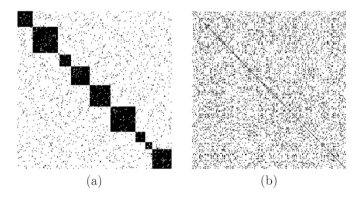

(a)   (b)

FIGURE 3.5.1: The adjacency matrices of the same graph: a) where vertices are ordered in such a way that they are grouped together, if they belong to the same cluster, b) under arbitrary vertex ordering.

require knowledge of the entire graph. Since many real graphs, such as the Internet, can number vertices in the billions, global algorithms become unsuitable for such cases. Global methods can be further subdivided into iterative, divisive, agglomerative and hierarchical ones. Local clustering algorithms use adjacency lists of a vertex and its neighbors.

3.5.1   Global clustering algorithms

Iterative methods, in general, go through all vertices and assign them to clusters. The decision can be final, but it can usually be revised, by going through the vertices again and changing vertex assignments, based on the optimization of a metric and on previous results. It is also possible to process one vertex at a time and gradually create and update clusters based on what has been encountered thus far. Such methods are called *online* ones.

Clusters need not be rigidly defined and subclusters can be contained within the same cluster. This is a reasonable hypothesis in real-life applications. For example, companies have departments and employees, who, while still closely interconnected with other people in the same company are even more interconnected with employees in the same department. This implies a hierarchical structure of clusters to which a vertex can belong. There are several clustering methods that accommodate this. An illustration of this concept can be found in Figure 3.5.2, which shows a graph vertex dendrogram, where vertices are grouped into increasingly larger clusters. At each dendrogram level, vertices are considered to belong to the same cluster if they have a common ancestor.

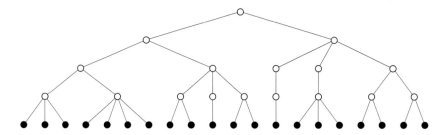

FIGURE 3.5.2: A dendrogram of graph vertices providing a hierarchical clustering.

An online hierarchical clustering method can be *divisive*, starting from the entire graph in one cluster, then recursively splitting clusters off in a top-down fashion. The majority of these methods boil down to selecting proper graph cuts at each step. The size of the cuts, i.e., the number of edges removed, should be minimal. It can be proven that the minimal cut can be acquired through a maximum-flow algorithm [CSRL01]. The issue that arises, however, is how to prevent the algorithm from choosing trivial cuts, like isolated vertices of 2,3-cliques, over bigger cuts that separate clusters. Imposing size restrictions on the cut renders the graph clustering problem NP-hard [Sch07].

Another class of divisive clustering methods is based on the removal of appropriate edges. The *betweenness* measure of an edge [Fre77] is defined as the number of closest paths between any two vertices that go through that edge. The idea is that edges with high betweenness are more likely to connect the vertices of a cluster to ones belonging to the rest of the graph, rather than vertices in the same cluster. By iteratively removing the edge with the highest current betweenness, and recomputing the measure for the remaining graph, it is expected that the clusters will eventually become disconnected from each other.

Agglomerative online hierarchical clustering methods are an alternative to the divisive approaches. They start from an empty set of clusters and then add clusters or assign new processed vertices to existing clusters. It is even possible to merge two clusters into a bigger one. This usually involves the optimization of some fitness measure [Sch07].

An issue with such approaches, however, is that the order in which the vertices are presented to the algorithm can significantly change the clustering output. If an actual cluster is presented in sequence to an algorithm, it may assign incoming vertices to different clusters, thus splitting the cluster up. Another possibility is that the first few vertices of a cluster get assigned to an already existing cluster and, thus, both the existing cluster and the new cluster will be sub-optimal choices. If the decisions that an algorithm makes are irreversible, then the clustering results can be quite inadequate. To rectify this, most algorithms can revise intermediate clustering results at later iterations guided by a global clustering suitability measure.

Taking a cue from physics, one can consider the graph edges as resistances in an electrical circuit, connected to the corresponding vertices. If a battery was connected to this circuit, the voltage would spread in all the vertices, as Figure 3.5.3 illustrates. Vertices belonging to the same cluster are expected to have similar voltage values, as they are densely connecting the cluster vertices. Alternatively, vertices can be classified as being closer to the source of electricity or closer to the sink of electricity, depending on their voltage. Thus, a graph can be split in two, with the process being recursively applied to the new subgraphs. The computational concerns of this approach can be efficiently tackled with circuit analysis tools. The issue with this approach is that there is no way to determine the best placement of the battery ends (source and sink).

A stochastic, global approach to graph clustering is presented in [TPN12]. As illustrated in Figure 3.5.1, when vertices belonging to the same cluster are placed in consecutive places in the ordering of the adjacency matrix, intense square blocks appear along the diagonal. In terms of *Discrete Cosine Transform* (DCT) [Pit00], this means that the sum of the diagonal elements of the DCT matrix are maximized, when the vertices are properly ordered. In a proof of the concept of the reverse, *Simulated Annealing* (SA) was used, in order to maximize the objective function, which is the sum of the k first diagonal elements of the DCT transform of the graph adjacency matrix. At each iteration, two randomly selected vertices switch places in the adjacency matrix order and the change in the sum is computed. If the objective function increases, the transition is automatically accepted. If it decreases, then the transition is accepted stochastically, as per standard SA operation. Results indicated that the method can be effective in graph clustering.

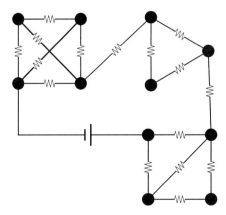

FIGURE 3.5.3: The graph as a circuit comprised of resistances.

3.5.2   Local clustering algorithms

As mentioned before, some graphs (such as the World Wide Web) are extremely large. Thus, any clustering algorithm with superlinear complexity is not a feasible option. Furthermore, storing the entire adjacency matrix in memory is impossible. Local clustering techniques are designed to operate with only a small part of the graph being visible at a time, usually through vertex adjacency lists or subgraphs. First, a seed vertex u is selected. Then the task is to find which other vertices belong to the same cluster with u, by accessing its neighborhood up to a small maximum distance. There are several heuristic or probabilistic local search methods that can expand the cluster of u, by optimizing a suitable criterion. Simulated annealing can also be used for local graph clustering, as it has the advantage that it can escape local optima by stochastically accepting worse solutions, in order to find even better ones later down the road. Local clustering algorithms can also be used in order to obtain a global graph clustering, e.g., by running a local clustering algorithm for every vertex and then combining the results through a voting scheme to determine the global clusters.

Random walks can be exploited to cluster graph vertices [vD00]. The probabilities of a random walk ending at a given vertex is closely related to the spectrum of the transition matrix. The basic idea is that, when a vertex is visited, then the random walk will most likely visit vertices belonging to the same cluster, before eventually proceeding to a vertex of a different cluster. Figure 3.5.4 shows an illustrative example of this idea for two clusters being connected through an edge. A random walk algorithm can only move from one cluster to another one through the edge with the highlighted vertices. Even when the walk is at those vertices, it is still more likely to remain within the cluster.

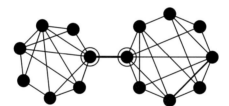

FIGURE 3.5.4: Random walk in a two-cluster graph.

3.5.3   Spectral clustering algorithms

Spectral graph clustering techniques were inspired by the observation that, when a graph is comprised of k disjoint cliques, the k smallest eigenvalues of the normalized Laplacian are 0 and the i-th corresponding eigenvector ($0 \leq i \leq k - 1$) has distinct, non-zero values for vertices that belong to the i-th clique [Sch07]. Adding edges to such a graph will gradually cause the eigenvalues to increase (except for the first, smallest eigenvalue, which will always be 0) and the corresponding eigenvectors to change slightly with each edge addition.

A large family of spectral vertex clustering techniques are based on *spectral bisection* [GM98]. These techniques use the so-called *Fiedler vector* [Fie73], i.e., the eigenvector $\mathbf{u}_1$ corresponding to the second smallest eigenvalue λ_1, $0 = \lambda_0 \leq \lambda_1 \leq \cdots \leq \lambda_{n-1} \leq 2$ of the Laplacian matrix of a graph. This eigenvalue is also called the *algebraic connectivity* of a graph. In the simplest case, in which there are two clusters, each of which has strong internal connectivity and are sparsely connected with each other, then the Fiedler vector entries corresponding to vertices of one cluster will be positive, while the entries corresponding to vertices of the other will be negative. This provides a bisection of the graph.

In general, there are two ways the Fiedler vector can be used to bisect a graph. For an edge-based bisection, after the Fiedler vector has been computed, the vertices are split into two groups, based on the relation of their relevant Fiedler vector entry to the median of all such entries. Finally, edges between these two vertex groups are cut. For vertex-based bisections, there are several options [HK92]: a) the sign of the entry is used to split the vertices, b) the largest gap in values is found and the vertices are split accordingly [OTNP14], or c) the graph is split at the value that provides the best cut quotient [GM98]. A special option uses the *normalized cut* measure and is also known as the *Ncut* approach [SM00]. The Fielder vector is also useful, when trying to find the central vertex of a neighborhood [QH04]. An application of such a graph clustering technique to facial image clustering is presented in [VSP11, OTNP14].

The most common algorithms for *spectral clustering* use one of the normalized Laplacians presented in Section 3.2 and perform eigenanalysis to extract the r eigenvectors that correspond to the smallest eigenvalues $\lambda_1 \leq \lambda_2 \leq \cdots \leq \lambda_r$ of the normalized Laplacian, excluding $\lambda_0 = 0$. The first r eigenvectors are stored in a $n \times r$ matrix $\mathbf{U}$. The rows of this matrix $\mathbf{U}$ are the new data representations. A standard clustering algorithm (e.g., k-means) can be used to cluster them. This new data representation has been generally found to offer better data clusterability [JB03, KVV04]. It is obvious that the construction of the similarity graph as well as the selection of the normalized Laplacian plays a crucial role in clustering performance [vL07, MM13]. Several variants of the basic algorithm have been proposed in the literature [vL07]. Some of them are discussed below.

In a weighted complete graph, the first r eigenvectors of matrix $\mathbf{D}^{-1/2}\mathbf{W}\mathbf{D}^{-1/2}$ are computed and stored in a $n \times r$ matrix $\mathbf{U}$. These eigenvectors can be used to compute a r-way normalized cut cost function [JB03]. The orthogonal projection operator defined by matrix $\mathbf{U}\mathbf{U}^T$ is compared to the orthogonal projection matrix $\mathbf{\Pi}_0 = \sum_{i=1}^{r} \mathbf{D}^{1/2}\mathbf{e}_i\mathbf{e}_i^T\mathbf{D}^{1/2}/(\mathbf{e}_i^T\mathbf{D}\mathbf{e}_i)$ of a proposed partition into clusters, as indicated by the binary cluster membership vectors $\mathbf{e}_i$. The comparison is performed using the Frobenius norm, thus implicitly comparing the subspaces spanned by the columns of the ideal cluster assignment matrix, represented by $\mathbf{U}\mathbf{U}^T$, and the proposed partition, represented by $\mathbf{\Pi}_0$ [GvL96]. This norm provides the cost function for the r-way cut-based clustering algorithm.

A method to split a graph into k clusters using k-way normalized cuts is presented in [YS03]. The task is formulated as a relaxed trace maximization problem:

$$\text{maximize } \frac{1}{k} tr(\mathbf{Z}^T\mathbf{W}\mathbf{Z}), \tag{3.5.1}$$

where $\mathbf{Z} = \mathbf{X}(\mathbf{X}^T\mathbf{D}\mathbf{X})^{-\frac{1}{2}}$, $\mathbf{D}$ is the diagonal of the graph weight matrix $\mathbf{W}$ and $\mathbf{X}$ is a $n \times k$ cluster assignment matrix. The solution to this maximization problem is found by forming matrix $\mathbf{Z}$ using the top k eigenvectors of the weight matrix $\mathbf{W}$. It is interesting that the weighted kernel k-means clustering algorithm can also be formulated as a similar trace maximization problem $tr(\mathbf{Y}^T\mathbf{W}^{\frac{1}{2}}\mathbf{K}\mathbf{W}^{\frac{1}{2}}\mathbf{Y})$, where $\mathbf{K}$ is the kernel matrix, $\mathbf{W}$ is the weight matrix and $\mathbf{Y}$ is an orthonormal matrix containing parameters related to the k centers in the kernel space [DGK04]. Thus, the k-way normalized cuts and kernel k-means are equivalent problems.

This is useful, as using the kernel k-means algorithm to perform k-way normalized cuts circumvents eigenvector computation. Computing the eigenvectors is a time and memory consuming task, more specifically it runs in practically $O(n^3)$ time and $O(n^2)$ space. This renders classic spectral approaches infeasible for a large number of graph nodes. On the other hand, kernel k-means can be implemented in a way that takes $O(n_z)$ time and memory, where n_z is the number of non-zero elements of the weight matrix. It must be noted that n_z can be equal to n^2 in the worst case. Alternatively, an algorithm to perform kernel k-means using only a small fraction of the rows of the kernel matrix, with minimal losses in the clustering performance, was proposed in [CJHJ11].

Spectral graph clustering techniques require less user input, in terms of determining parameters and are, therefore, more automated. They are also more elegant from a theoretical stand point. Detecting trivial clusters (isolated 2-, 3-cliques) is more easily avoided, when using spectral techniques. Generally speaking, it is more difficult for spectral techniques to result in very bad clustering. Since they require the Laplacian matrix of the graph, they cannot be employed for extremely large graphs, due to memory limitations. Even if a large graph could fit into memory, the eigenanalysis process would take a significant amount of time. In these cases, the kernel k-means algorithm provides a feasible alternative. The use of kernel k-means in clustering big datasets of facial images is presented in [TTNP15].

3.5.4 Overlapping community detection

While graph clustering techniques can also be used for community detection purposes, the exclusive assignment of a graph vertex to a specific cluster can be inaccurate, as there can be several overlapping communities that a vertex belongs to, as can be seen in Figure 3.5.5. For example, a person can be considered to belong to the community of their family, of their co-workers and of a hobby group, among many others. As such, several overlapping community detection techniques have been developed [For10, XKS13].

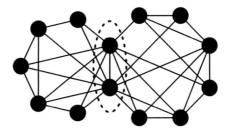

FIGURE 3.5.5: Two overlapping communities. Their common vertices are highlighted by the dotted ellipse.

A rather popular approach to overlapping community detection is the *Clique Percolation Method (CPM)* [PDFV05], based on k-cliques, which are fully connected subgraphs of k-nodes. Two k-cliques are considered to be adjacent if they have $k - 1$ vertices in common. Several adjacent k-cliques form a k-*clique chain*. A k-clique community is defined to be a maximal k-clique chain. This allows vertices to belong to overlapping communities, while also tolerating vertices in the same community that are not directly connected by an edge. The method begins by finding maximal cliques. Although this is an NP-complete problem, it was found that it can run in a reasonable amount of time on sparse graphs, such as those encountered in real problems. Let n_c be the number of maximal cliques found. A $n_c \times n_c$ clique-clique overlap matrix is then computed. The communities correspond to the connected components of this matrix [EB98]. An implementation of CPM by the original authors named CFinder[1] can be found in [APF+06]. The method can be accelerated by starting from an empty set of cliques. It continues by sequentially inserting a new edge and by updating the existing cliques. This extension is called *Sequential Clique Percolation (SCP)* [KKKS08] and can also be efficiently applied to weighted graphs, by processing the edges in descending weight order.

Some alternative overlapping community detection approaches are presented in

---

[1] http://cfinder.org

[BGK+05]. *Rank Removal* works by assigning a score to each vertex, corresponding to the importance of that vertex, using, e.g., betweenness, or even PageRank [BP98]. The most important vertices are then removed, until the graph only consists of disconnected components no larger than a user-defined cardinality. These components form the core of their respective community. The removed vertices are then added back to the graph and included in a community, if doing so increases a chosen metric, measuring the internal vs. external connectivity of that community. It is possible to add the same vertex to more than one community, thus allowing communities to overlap on such vertices. *Iterative Scan* begins with a community using a seed vertex, then adds and removes vertices, greedily maximizing an appropriate metric similar to Rank Removal, until no change can further improve that metric. Once such a point is reached, the currently included vertices form a community and a new seed vertex is selected. The process continues using new seeds, until the process fails to produce a community that has not already been detected a set number of times. It has been found that applying Iterative Scan on the community cores detected by Rank Removal provides the best results [BGMI05]. Finally, because it is possible for Iterative Scan to produce disconnected communities, a variant called *Connected Iterative Scan* checks each community for correctness after every insertion and deletion [Kel09].

It is also possible to work on graph edges, instead of graph vertices, for overlapping community detection. The main concept of such approaches is to work on the *line graph* of the original graph, i.e., the graph in which the original graph edges are the new graph vertices. Two vertices of the new graph are connected, iff the corresponding edges in the original graph are incident to the same vertex. Performing clustering on the line graph to obtain a non-overlapping vertex partition results in an edge partition in the original graph. Communities can then be formed, by assigning the vertices incident to every edge in a partition to the corresponding community. While edge partition is non-overlapping, a vertex can belong to every community its edges were clustered into, thus allowing for overlapping vertices. Approaches based on this concept utilize hierarchical clustering into a dendrogram of edge communities [ABL10] and random walks [EL09]. The construction of the line graph is not always necessary, as [KJ11] presents an extension of *Infomap* [RB08] to directly find edge communities.

A spectral community detection approach [WS05] is designed to find k communities and involves computing the top $k-1$ eigenvectors. These vectors are used to obtain an embedding of the graph in a Euclidean space and perform traditional clustering methods in that space, such as k-means. This, however, results in non-overlapping communities. An extension of this approach is presented in [ZWZ07] and can perform overlapping spectral community detection through fuzzy clustering. Fuzzy c-means is used to acquire a fuzzy cluster membership vector for each vertex, whose k entries correspond to the probability that the vertex belongs to each of the k clusters. These entries are non-negative and their sum is 1. Thresholding the cluster memberships can assign a vertex to multiple communities.

3.6 Graph matching

In graph matching, the basic task is to establish a correspondence between the vertices of two graphs, whose vertices can have additional features, such as those illustrated, with the assistance of a tool provided by [ZD13], in Figure 3.6.1. In general, finding the optimal correspondence can be reduced to solving the following *quadratic assignment problem*

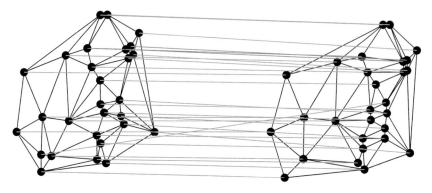

FIGURE 3.6.1: Graph matching between two graphs, whose vertices contain image features.

[Ans03]:

$$c^* = \arg\max_c \left(\sum_i \sum_j c_{ij} f(i,j) + \sum_i \sum_j \sum_k \sum_l c_{ik} c_{jl} g(i,j,k,l) \right), \qquad (3.6.1)$$

where $c_{ij} = 1$, iff element i (a vertex, in the case of graphs) is assigned to element j and 0 otherwise. Function $f()$ is a matching score function between two elements and function $g()$ is, respectively, a matching function of the pair wise assignment between elements i and j to elements k and l. Various restrictions can be further imposed, such as $\sum_j c_{ij} = 1$, $\forall\, i$, so that no double assignments to the same element are permitted. However, the quadratic assignment problem is NP-hard. An extensive survey of both exact (isomorphic) and inexact graph matching can be found in [CFSV04].

Continuous optimization methods have been adapted to suit the graph matching problem. By defining matching in terms of a continuous function, known optimization methods can be used to compute the function maximum and then use it to retrieve the respective graph matching. A technique based on relaxation labeling [FE73] employs a probability vector for every vertex of the first graph, representing the probability that the vertex is matched with each vertex of the other graph. In an iterative process, these probabilities are recalculated, based on the probabilities of neighboring vertices, until a termination criterion is met, e.g., when the optimization reaches a static point, or when the maximum number of iterations is reached. Insight on the probabilistic framework behind this approach can be found in [CKP95].

The problem of matching an input graph $G_1 = (V_1, E_1, \mathbf{A}_1)$ with a relational model graph $G_2 = (V_2, E_2, \mathbf{A}_2)$, where $\mathbf{A_1}$ and $\mathbf{A_2}$ contain data associated with each vertex of V_1 and V_2, respectively, is tackled in [WH97], by introducing a maximum a-posteriori (MAP) based consistency measure to rectify initialization errors and vertices that correspond to, essentially, noise, in order to improve matching performance [SH85]. The relational graph model G_2 is allowed to be appended with dummy vertices, so that the cardinality of its subgraph vertex set matches the cardinality of the vertex set of G_1. MAP criterion calculates the probability of a matching f given the data $\mathbf{A}_1$ and $\mathbf{A}_2$, using the probabilities of vertex v_1 of G_1 matching vertex v_2 of G_2, given their individual data in a Bayesian framework. The optimal matching is determined by finding the individual matching assignments that maximize this probability. Relational consistency is measured on a vertex neighborhood scale, using dictionaries of possible relations between the input graph and the model graph and a Hamming distance measure. In a post-processing step, unmatched vertices in G_1 are discarded.

A way to parameterize the matching function (3.6.1), so that it favors matching graphs the way a human would match them, by using manually annotated graph matches, is proposed in [CCLS07]. Given a pair of graphs G_1 and G_2, and a manually provided graph match c, the objective is to find the parameters of a predictor:

$$g_w(G_1, G_2) = \arg\max_c \langle w, \Phi(G_1, G_2, c) \rangle, \qquad (3.6.2)$$

i.e., the weights w of a linear discriminant function and the feature function Φ for which the error:

$$\frac{1}{N} \sum_{n=1}^{N} \Delta(g_w(G_1^n, G_2^n), c^n) + \lambda\Omega(w) \qquad (3.6.3)$$

is minimized, when the graph matching is c. This is essentially an inverse optimization problem [AO01] that the authors solve, by utilizing the large margin approach [TJHA05]. Once these parameters have been properly determined, the minimization of (3.6.3) provides better matches than the solutions to the original quadratic assignment problem.

3.6.1 Spectral graph matching

The isomorphism, or near isomorphism, of two graphs G_1 and G_2 with adjacency matrices $\mathbf{A}_1$ and $\mathbf{A}_2$, respectively, can be described as the task of finding a permutation matrix $\mathbf{C}$, such that $\mathbf{C}\mathbf{A}_1\mathbf{C}^T = \mathbf{A}_2$ [XK01]. This problem can then be reformulated as finding the matrix $\mathbf{C}^*$ minimizing the trace (sum of diagonal elements) of:

$$\mathbf{C}^* = \arg\min_{\mathbf{C}} tr\left[(\mathbf{C}\mathbf{A}_1\mathbf{C}^T - \mathbf{A}_2)(\mathbf{C}\mathbf{A}_1\mathbf{C}^T - \mathbf{A}_2)^T\right]. \qquad (3.6.4)$$

This can be approximated with a matrix $\mathbf{\Phi}$ that is orthogonal, but does not follow the additional restrictions that the permutation matrix $\mathbf{C}$ does. It can be proven that:

$$\min_{\mathbf{\Phi}} tr\left[(\mathbf{\Phi}\mathbf{A}_1\mathbf{\Phi}^T - \mathbf{A}_2)(\mathbf{\Phi}\mathbf{A}_1\mathbf{\Phi}^T - \mathbf{A}_2)^T\right] = \sum_i (\lambda_i - \mu_i)^2, \qquad (3.6.5)$$

where λ_i and μ_i are the ordered eigenvalues of $\mathbf{A}_1$ and $\mathbf{A}_2$, respectively, thus indicating that graphs can be compared through their spectra.

A spectral approach to graph matching can be found in [SB92], where the graph vertices represent feature points in an image. The task is to find correspondences between the feature points in two different images. Instead of an adjacency matrix, a proximity matrix $\mathbf{H}$ is constructed by measuring the Euclidean distance between any two features and using a Gaussian activation function to obtain the final elements of $\mathbf{H}$ for each image. The spectrum of $\mathbf{H}$ is then calculated to be used for graph matching and the eigenvectors are ordered in a decreasing eigenvalue order to become the column vectors of matrix $\mathbf{E}$. Since the number of features is not necessarily the same, the least important eigenvectors, i.e., the eigenvectors corresponding to the smallest eigenvalues of the largest graph, are discarded. If $\mathbf{E}_1$ and $\mathbf{E}_2$ are the matrices of the two images, the sign of the eigenvectors in $\mathbf{E}_2$ are determined in such a way that the orientation of the orthogonal axes of the eigenvectors agree with those of $\mathbf{E}_1$. The rows of $\mathbf{E}_1$, $\mathbf{E}_2$ form a set of feature vectors for the image features they correspond to. These feature vectors from both images are correlated and the resulting matrix $\mathbf{Z}$ indicates the dissimilarity score for each pair of features.

Building upon the work in [SB92] described above, in order to improve robustness to noise and outliers, the use of different activation functions on the elements of $\mathbf{H}$, such as the sigmoidal function, or an increasing weighting function or a Euclidean weighting function, is investigated in [CH03]. Additionally, instead of simply assigning a correspondence between features with the lowest dissimilarity score, an extension of the Expectation-Maximization (EM) algorithm [CH98] is used to determine correspondences. Finally, the fact that the eigenvalue order has been compromised, due to noise, is taken into account, by incorporating the probability that rows have been misplaced in the formulations used in the EM algorithm.

Spectral graph matching methods are more deterministic, though probabilities can still be incorporated in them. They also need less user input, in order to perform as well. The graph spectrum can also constitute a descriptor of a graph for feature extraction purposes.

3.6.2   Frequent subgraph mining

The goal of *Frequent subgraph mining (FSM)* algorithms is to find all the frequent subgraphs occurring in either a set of medium sized graphs (called "transactions" in the literature) or a single, very big graph. Since there are no spectral approaches in this field, we will cover it only briefly. For a more detailed survey, the interested reader may refer to [JCZ13]. There are two main components to FSM: isomorphism testing and candidate generation. As previously mentioned, subgraph isomorphism is NP-complete and there is no known polynomial algorithm for graph isomorphism. Two isomorphic graphs always have the same labeling when canonically labeled. However, the canonical labeling algorithm itself has an exponential worst-case running time. Candidate generation can either follow the *Breadth First Search (BFS)* approach, in which candidates of $k+1$ size are considered only after all candidates of size k have been checked, or the *Depth First Search (DFS)* approach. In both cases, the aim it to avoid generating duplicate candidates. *Frequent subtree mining* is a subfield of FSM that includes more specialized algorithms.

3.7   Random walks

Random walks, as the name suggests, are graph vertex traversals that randomly move from one vertex to a neighboring one. Most applications of random walks relate to sampling and approximation [Lov93], the most obvious being the uniform generation of random data. The following measures are of particular interest, when a random walk is concerned: access time, cover time, and mixing rate. The *access time* of vertex u_i from vertex u_j is the expected number of steps a random walk takes to reach the former vertex, starting from the latter one. The *cover time* is the expected number of steps the walk takes to reach every vertex at least once. The *mixing rate* measures how fast the walk converges to its limiting probability distribution. Supposing that the walk is currently at vertex u, the probability of each of its neighbors being selected for a visit is assumed to be uniform, i.e., $1/d(u)$, where $d(u)$ is the degree of vertex u. We can gather all the transition probabilities in a matrix $\mathbf{M}$ and model the random walk as a Markov chain following the rule:

$$\mathbf{p}^{(t+1)} = \mathbf{M}^T \mathbf{p}^{(t)}, \tag{3.7.1}$$

where $\mathbf{p}^{(t)}$ is the vector of the probabilities $p_i^{(t)}$ that the walk is at vertex u_i after t steps, given an initial probability vector $\mathbf{p}^{(0)}$. Alternatively, $\mathbf{p}^{(t)}$ can be connected to $\mathbf{p}^{(0)}$ by:

$$\mathbf{p}^{(t)} = (\mathbf{M}^T)^t \mathbf{p}^{(0)}. \tag{3.7.2}$$

If the walk is allowed to go on infinitely on a graph $G = (\mathcal{V}, \mathcal{E})$, the stationary probability distribution converges to:

$$\pi(u_i) = \frac{d(u_i)}{|\mathcal{E}|}, \tag{3.7.3}$$

assuming that G is not bipartite.

Suppose that there is a set whose elements have n characteristics, and the problem we want to solve is that of enumeration, i.e., to estimate N, the number of all the possible elements of the set [Bab79]. If N is exponential on n, then exhaustively enumerating every element becomes intractable. It is possible, however, to estimate the population of the set by randomly sampling the elements with a random walk traversal on the characteristics and studying how the number $N^{(t)}$ of known elements increases after t iterations. Some restrictions apply, e.g., that the initial set size is known, which is trivially 1, when starting from a random element of the set. It must also be possible to uniformly generate new random set elements in polynomial time. Finally, the ratio of consecutive population measurements $N^{(t+1)}/N^{(t)}$ must be polynomially bound. If these restrictions apply, then the total number can be extrapolated from the ratios $N^{(t+1)}/N^{(t)}$.

A similar application is finding the volume of a convex body that lacks an analytical mathematic expression for this task [DF88]. The exact computation of the volume of an n-dimensional polytope, i.e., a flat sided geometric shape generalized to n dimensions, is NP-hard [DF88]. The complexity of approximation algorithms that have a bounded error for this volume computation is also exponential [Ele86]. However, a randomized polynomial time algorithm, similar to the above approach, can provide estimates whose error is theoretically improbable to exceed any given positive number.

There are some applications, in which the desired stationary probability distribution of the random walk is not uniform, so that it may be used to generate random elements with a non-uniform distribution. For example, a stochastic optimization algorithm may prefer to make transition chains with vertices near optima rather than chains that randomly lead away from these optima. Random walks can be modified, so that their stationary distribution converges to any arbitrary probability distribution function $F()$. In an idea similar to the Monte Carlo approach, whenever a transition from vertex u_i to vertex u_j is decided and $F(u_j) < F(u_i)$, then the transition is performed with a further probability of $F(u_j)/F(u_i)$, otherwise the walk stays in u_i. It can be proven that the stationary distribution Q_F of this modified random walk is given by [MRR$^+$53]:

$$Q_F(u) = \frac{F(u)}{\sum_{v \in V} F(v)}. \tag{3.7.4}$$

There are many interesting connections between the spectral graph theory and random walk measures, as thoroughly described in [Chu97], [Lov93]. Access time can be expressed as a function of the graph eigenvectors. A function of either the second, or the second and last eigenvalue of the Laplacian matrix provides an upper bound for the relative pairwise graph vertex distance. In a vertex transitive graph, a random walk of total s steps converges to the uniform distribution in a number of steps bound by a function of the eigenvalues of the Laplacian matrix. The mixing rate on a non-bipartite graph is the absolute value of either the second or the last eigenvalue of the adjacency matrix, whichever has the greater absolute value.

The PageRank algorithm, developed by the founders of Google, models the behavior of a web surfer as a random walk on the graph of the World Wide Web [BP98]. This "random surfer" may click on any link in the current web page he is visiting with the same probability, while periodically being "bored" and jumping to another page randomly. The probability that this surfer will visit a web page constitutes its PageRank. This probability is given by the dominant eigenvector, i.e., the eigenvalue corresponding to the eigenvalue with the largest magnitude, of the graph transition probability matrix [BL06].

3.8   Graph anomaly detection

Graph anomalies can refer to basic graph properties, such as the presence or absence of a vertex or edge, or even the existence of "anomalous" subgraphs, such as the near-stars or near-cliques. *Near star* refers to a graph vertex, whose neighbors are scarcely connected with each other. *Near clique* refers to very densely connected subgraphs. *Heavy vicinity* refers to the case when most of the total weight of the vertex edges is distributed to only a few edges. Finally, *Dominant heavy link* refers to a single edge having extraordinarily large weight compared to other edges. A more extensive recent survey on anomaly detection that includes graph based algorithms can be found in [CBK09].

In fraud detection, the objective is to detect suspicious transactions in graph structures. It is reasonable to expect that fraudulent activities will be camouflaged by the perpetrators to look like legitimate transactions, to avoid raising suspicion. Three algorithms for the detection of suspicious presence or absence of graph elements, as well as suspicious changes in the labels of graph elements are presented in [EH07]. All three algorithms use breadth-first search and the Minimum Description Length (MDL), in order to extract a graph pattern that is considered normative, i.e., it can be used to best compress the graph in MDL terms. In the first algorithm, changes in the normative pattern are measured using the cost of the editing operation that causes a change in the description length of the entire graph, when it is compressed with the normative pattern. This algorithm detects anomalous vertex label changes. The second algorithm approximates the probability of additional vertices or edges being inserted into the normative pattern to produce a new pattern, according to whether they lead to patterns that match other possible changes, when compared with the total number of derivative patterns coming from all possible changes. Changes being below a threshold are considered anomalous additions. Another algorithm investigates sub-patterns of the normative pattern by removing elements, until the sub-pattern is no longer normative, at which point the cost of transformation and frequency of the sub-pattern is thresholded. Thus, anomalous absences are detected.

Graph similarity measures between two web graphs can also be used for anomaly detection [PDG10]. Three requirements must be satisfied by suitable measures: a) scalability, so that its time complexity is at most linear, b) sensitivity, so that it is easily affected by differences in the more important vertices and edges, and c) that it must be affected by additional data stored in graph elements, besides the graph connectivity information. Several similarity measures were proposed. Vertex and edge overlap is computed through edit distance. Vertex ranking is measured according to a formula inspired by Spearman's formula [WA79]. Vertex vector similarity is measured by comparing the quality values for vertices that are matched between the two graphs. Edge vector similarity is measured by comparing the quality of vertices adjacent to both edges. Sequence similarity is measured by how many short paths the two graphs share, taking advantage of the fact that the graphs are web graphs. Finally, a variation of SimHash [Cha02] is used to measure signature similarity.

Changes in compression ratio and graph entropy can also be used to detect graph anomalies [NC03], using an established graph compression technique [CH00], which iteratively finds a substructure that best compresses the graph in terms of MDL and replaces every instance of the substructure with a single vertex. An anomalous rating metric a is defined by:

$$a = 1 - \frac{1}{n} \sum_{i=1}^{n} \left((n - i + 1) \frac{DL_{i-1}(G) - DL_i(G)}{DL_0(G)} \right), \qquad (3.8.1)$$

where $DL_i(G)$ is the graph description length at iteration i. This metric ranges in $[0, 1]$

and the higher it is, the more anomalous the subgraph is considered to be. For any given subgraph, it starts at value 1 and, as the compression proceeds, drops, until no further compression is possible. The value is then thresholded to decide whether a subgraph is anomalous or not.

Another measure of graph irregularity can be derived from the conditional entropy $H(X|Y)$:

$$H(X|Y) = \sum_{y \in Y} \sum_{x \in X} P(y)P(x|y) \log P(x|y), \qquad (3.8.2)$$

measuring the remaining uncertainty of a variable X after variable Y is known, under the assumption that exactly one "event" from set X has occurred. The definition can be expanded to:

$$H(X|Y) = \sum_{y \in Y} \sum_{x \in X} ((P(y)P(x|y) \log P(x|y)) + ((1 - P(x|y)) \log(1 - P(x|y)))) \qquad (3.8.3)$$

that now suitably models multiple ways in which a given subgraph X can be extended to a bigger structure.

Using a related, edge-based definition of entropy, a similar approach is proposed in [SA05] for detecting important vertices in multi-graphs (graphs that allow for more than one edge between the same vertices). The vertex entropy is measured as the sum of the entropy values of its edges:

$$E(i) = H(v_i, P) = \sum_{e \in N(v_i)} p(e) \log(\frac{1}{p(e)}), \qquad (3.8.4)$$

where the probability of each edge e can be calculated from the multi-graph. Important vertices are assumed to influence their neighbors, paths of edges can be included in the computation of each vertex entropy. The suitability of combining two or more edges in a path can be measured using edge data similarity, e.g., the similar subjects in emails, or proximity in time (if such information is available). In order to rank the vertex importance, the so-called vertex effect is determined by:

$$\text{Effect}_i = \frac{EN(i)}{\log(\frac{EN(i)}{E(i)})}, \qquad (3.8.5)$$

where $E(i)$ is the entropy of vertex u_i and $EN(i)$ is the entropy of the rest of the graph, when u_i and all its edges are removed from the graph. The larger the effect value is, the more important the vertex is considered to be.

3.8.1 Spectral anomaly detection

Anomalous vertices in weighted graphs can be detected based on their *egonet* [AMF10], i.e., the induced subgraph of a vertex and its neighborhood. The anomalies that can be identified are near star, near clique, heavy vicinity and dominant heavy links. Four basic vertex-based features are computed, namely the neighbor number N_i, the number of edges in the egonet E_i, the total edge weight of the egonet W_i and the principal eigenvalue λ_i of the egonet weighted adjacency matrix. These features have been observed to follow power laws, as the graph increases in size and are, thus, considered desirable features. The appropriate pair of features to use for each anomaly type has been determined to be N_i and E_i for near-star and near-clique, W_i and E_i for heavy vicinity and λ_i and W_i for dominant edge.

Since the Probability Density Function (PDF) of every feature follows a power law, then, for each pair of features x and y, the following equation holds [AMF10]:

$$\exists x, \theta \in \mathbb{R} : y = cx^{\theta}. \tag{3.8.6}$$

The outlier score of a vertex u_i can be evaluated by employing the appropriate features x and y as follows:

$$o(u_i) = \frac{\max\left(y, cx^{\theta}\right)}{\min\left(y, cx^{\theta}\right)} \log |y - cx^{\theta}| + 1. \tag{3.8.7}$$

Finally, this measure can be complemented by a probability density based outlier detection measure for improved performance [BKNS00].

Anomalous behavior detection in a time series of weighted graphs, each graph representing the state of a graph at different times, is presented in [Ide04]. The basic feature vector used is the eigenvector corresponding to the highest eigenvalue of the weight matrix, referred to as the activity vector. An attempt to cluster the graph vertices is made using the activity vector. If graph clustering produces more than one cluster, each cluster is handled separately. In order to establish the typical, non-anomalous behavior, the method requires training data, in which no anomalies are present. The activity vectors from the various training samples are collected to form the dependency matrix $\mathbf{U}$. The reference activity vector for non-anomalous behavior is selected to be the principal eigenvector of matrix $\mathbf{U}^T \mathbf{U}$. Graph anomaly is measured by comparing the angle (inner product) that the activity vector of the test graph forms with the reference vector for non-anomalous behavior. If the inner product is close to 0, the vectors are perpendicular and, thus, the test graph is extremely anomalous. If the inner product is close to 1, the test graph coincides with the average typical, non-anomalous graph.

The inability of graph spectra to take into account vertex features renders them unable to detect anomalies based on those features. However, they remain strong tools for detecting structural graph anomalies.

3.9 Conclusions

In this chapter, we have presented an overview to the wide field of the combination of graph theory and algebraic or statistical techniques. It is our hope that readers interested in exploring these fields will find this survey useful to quickly get a good idea of problem definitions, notably the ones based on graph spectrum.

We have chosen to focus this survey on statistical and algebraic approaches, because of their potential and efficiency. In contrast to other combinatorial or algorithmic approaches that are more likely to involve ad hoc elements, statistical and algebraic approaches have a solid theoretical framework guiding their application to the various graph related tasks. Spectral techniques, in particular, have found use in several graph related tasks, because the graph spectrum provides a powerful graph representation. It reveals several graph properties and can be used as an efficient graph descriptor.

We started by providing standard graph theory and the spectral graph theory background that is required knowledge for anyone interested in working in the field. Then, we have provided a quick overview of the applications of techniques from all the aforementioned subfields. We also covered graph compression, which is essential in handling large graphs and the theoretical background for the random generation of graphs. Finally, we overviewed graph clustering, community detection, graph matching, random walks, and anomaly detection.

Techniques for graph related tasks have been under constant development for relatively small applications. The proliferation and growing popularity of social media and social networks, however, has further rekindled scientific interest in these issues in the past several years. With the advances in computer processors, computer memory, and compression techniques, it is now feasible to work on big graphs, such as on Facebook or YouTube data. The material and scientific interest for future research seems abundant.

Bibliography

[ABL10] Y.-Y. Ahn, J. P. Bagrow, and S. Lehmann. Link communities reveal multiscale complexity in networks. *Nature*, 466(7307):761–764, 2010.

[AMF10] L. Akoglu, M. McGlohon, and C. Faloutsos. OddBall: Spotting anomalies in weighted graphs. In M. J. Zaki, J. X. Yu, B. Ravindran, and V. Pudi, editors, *Proc. Advances in Knowledge Discovery and Data Mining*, volume 6119, pages 410–421, 2010.

[Ans03] K. M. Anstreicher. Recent advances in the solution of quadratic assignment problems. *Mathematical Programming*, 97(1-2):27–42, 2003.

[AO01] R. K. Ahuja and J. B. Orlin. Inverse optimization. *Operations Research*, 49:771–783, 2001.

[APF+06] B. Adamcsek, G. Palla, I. J. Farkas, I. Derényi, and T. Vicsek. Cfinder: locating cliques and overlapping modules in biological networks. *Bioinformatics*, 22(8):1021–1023, 2006.

[AR13] C. C. Aggarwal and C. K. Reddy. *Data clustering: algorithms and applications*. CRC Press, 2013.

[Bab79] L. Babai. Monte-Carlo algorithms in graph isomorphism testing. *Université de Montréal Technical Report, DMS*, pages 79–10, 1979.

[BB03] A.-L. Barabási and E. Bonabeau. Scale-free networks. *Scientific American*, 288(5):50–59, 2003.

[BBHW10] E. Bodine-Baron, B. Hassibi, and A. Wierman. Distance-dependent Kronecker graphs for modeling social networks. *IEEE Journal of Selected Topics in Signal Processing*, 4(4):718–731, 2010.

[Ber09] R. Beraldi. Biased random walks in uniform wireless networks. *IEEE Transactions on Mobile Computing*, 8(4):500–513, 2009.

[BGK+05] J. Baumes, M. K. Goldberg, M. S. Krishnamoorthy, M. M. Ismail, and N. Preston. Finding communities by clustering a graph into overlapping subgraphs. In *Proc. IADIS Applied Computing*, pages 97–104, 2005.

[BGMI05] J. Baumes, M. Goldberg, and M. Magdon-Ismail. Efficient identification of overlapping communities. In *Proc. Intelligence and Security Informatics*, pages 27–36. 2005.

[Big93]   N. Biggs. *Algebraic graph theory*. Cambridge University Press, 1993.

[BKNS00]   M. M. Breunig, H.-P. Kriegel, R. T. Ng, and J. Sander. LOF: Identifying density-based local outliers. In *Proc. of the 2000 ACM SIGMOD International Conference on Management of Data*, volume 29, pages 93–104, 2000.

[BL06]   K. Bryan and T. Leise. The $25,000,000,000 eigenvector: the linear algebra behind Google. *SIAM Review*, 48(3):569–581, 2006.

[BP98]   S. Brin and L. Page. The anatomy of a large-scale hypertextual web search engine. *Computer Networks and ISDN Systems*, 30(1-7):107–117, 1998.

[Bro87]   M. Browne. The young-householder algorithm and the least squares multidimensional scaling of squared distances. *Journal of Classification*, 4(2):175–190, 1987.

[BRSV11]   P. Boldi, M. Rosa, M. Santini, and S. Vigna. Layered label propagation: a multiresolution coordinate-free ordering for compressing social networks. In *Proc. of the 20th International Conference on World Wide Web*, pages 587–596, 2011.

[BV03]   P. Boldi and S. Vigna. The webgraph framework I: Compression techniques. In *Proc. of the 13th International World Wide Web Conference*, pages 595–601, 2003.

[CBK09]   V. Chandola, A. Banerjee, and V. Kumar. Anomaly detection: A survey. *ACM Computing Surveys*, 41(3):15:1–15:58, 2009.

[CCLS07]   T. S. Caetano, L. Cheng, Q. V. Le, and A. J. Smola. Learning graph matching. In *Proc. of the 11th IEEE International Conference on Computer Vision*, pages 1–8, 2007.

[CFSV04]   D. Conte, P. Foggia, C. Sansone, and M. Vento. Thirty years of graph matching in pattern recognition. *International Journal of Pattern Recognition and Artificial Intelligence*, 18(03):265–298, 2004.

[CH98]   A. D. J. Cross and E. R. Hancock. Graph matching with a dual-step EM algorithm. *IEEE Transactions on Pattern Analysis and Machine Intelligence*, 20:1236–1253, 1998.

[CH00]   D. J. Cook and L. B. Holder. Graph-based data mining. *IEEE Intelligent Systems*, 15:32–41, 2000.

[CH03]   M. Carcassoni and E. Hancock. Spectral correspondence for point pattern matching. *Pattern Recognition*, 36(1):193–204, 2003.

[Cha02]   M. S. Charikar. Similarity estimation techniques from rounding algorithms. In *Proc. of the 34th annual ACM Symposium on Theory of Computing*, pages 380–388, 2002.

[Chu97]   F. R. K. Chung. *Spectral Graph Theory*. American Mathematical Society, 1997.

[CJHJ11]   R. Chitta, R. Jin, T. C. Havens, and A. K. Jain. Approximate kernel k-means: Solution to large scale kernel clustering. In *Proc. of the 17th ACM SIGKDD International Conference on Knowledge Discovery and Data Mining*, pages 895–903, 2011.

[CKP95]   W. J. Christmas, J. Kittler, and M. Petrou. Structural matching in computer vision using probabilistic relaxation. *IEEE Transactions on Pattern Analysis Machine Intelligence*, 17:749–764, 1995.

[CL90]   L.-H. Chen and J.-R. Lieh. Handwritten character recognition using a 2-layer random graph model by relaxation matching. *Pattern Recognition*, 23:1189–1205, 1990.

[CMP+10]   D. Cordeiro, G. Mounié, S. Perarnau, D. Trystram, J.-M. Vincent, and F. Wagner. Random graph generation for scheduling simulations. In *Proc. of the 3rd International Conference on Simulation Tools and Techniques*, pages 1–10, 2010.

[CSN09]   A. Clauset, C. R. Shalizi, and M. E. Newman. Power-law distributions in empirical data. *SIAM Review*, 51(4):661–703, 2009.

[CSRL01]   T. H. Cormen, C. Stein, R. L. Rivest, and C. E. Leiserson. *Introduction to Algorithms*. McGraw-Hill, 2nd edition, 2001.

[DF88]   M. E. Dyer and A. M. Frieze. On the complexity of computing the volume of a polyhedron. *SIAM Journal on Computing*, 17:967–974, 1988.

[DGK04]   I. S. Dhillon, Y. Guan, and B. Kulis. Kernel k-means: Spectral clustering and normalized cuts. In *Proc. of the 10th ACM SIGKDD International Conference on Knowledge Discovery and Data Mining*, pages 551–556, 2004.

[DPZ01]   S. Dickinson, M. Pelillo, and R. Zabih. Introduction to the special section on graph algorithms in computer vision. *IEEE Transactions on Pattern Analysis Machine Intelligence*, 23:1049–1052, 2001.

[DRSS96]   A. A. Diwan, S. Rane, S. Seshadri, and S. Sudarshan. Clustering techniques for minimizing external path length. In *Proc. of the International Conference on Very Large Databases*, pages 342–353, 1996.

[EB98]   M. G. Everett and S. P. Borgatti. Analyzing clique overlap. *Connections*, 21(1):49–61, 1998.

[EF86]   M. A. Eshera and K.-S. Fu. An image understanding system using attributed symbolic representation and inexact graph-matching. *IEEE Transactions on Pattern Analysis and Machine Intelligence*, (5):604–618, 1986.

[EH07]   W. Eberle and L. Holder. Anomaly detection in data represented as graphs. *Intelligent Data Analysis*, 11(6):663–689, 2007.

[EL09]   T. S. Evans and R. Lambiotte. Line graphs, link partitions, and overlapping communities. *Physical Review E*, 80(1):016105, 2009.

[Ele86]   G. Elekes. A geometric inequality and the complexity of computing volume. *Discrete & Computational Geometry*, 1:289–292, 1986.

[ER60]   P. Erdos and A. Renyi. On the evolution of random graphs. *Publication of the Mathematical Institute of the Hungarian Academy of Sciences*, 5:17–61, 1960.

[FE73]   M. A. Fischler and R. A. Elschlager. The Representation and Matching of Pictorial Structures. *IEEE Transactions on Computers*, C-22(1):67–92, 1973.

[FGK95]   A. Filatov, A. Gitis, and I. Kil. Graph-based handwritten digit string recognition. In *Proc. of the 3rd International Conference on Document Analysis and Recognition*, volume 2, pages 845–848, 1995.

[Fie73]   M. Fiedler. Algebraic connectivity of graphs. *Czechoslovak Mathematical Journal*, 23(2):298–305, 1973.

[For10]   S. Fortunato. Community detection in graphs. *Physics Reports*, 486(3):75–174, 2010.

[Fra00]   J. Franklin. *Matrix Theory*. Dover Publications, 2000.

[Fre77]   L. C. Freeman. A Set of Measures of Centrality Based on Betweenness. *Sociometry*, 40(1):35–41, 1977.

[Gil59]   E. N. Gilbert. Random graphs. *Annals of Mathematical Statistics*, 30(4):1141–1144, 1959.

[GJ90]   M. R. Garey and D. S. Johnson. *Computers and Intractability; A Guide to the Theory of NP-Completeness*. W. H. Freeman & Co., 1990.

[GM98]   S. Guattery and G. L. Miller. On the quality of spectral separators. *SIAM Journal on Matrix Analysis and Applications*, 19:701–719, 1998.

[Gra06]   L. Grady. Random walks for image segmentation. *IEEE Transactions on Pattern Analysis Machine Intelligence*, 28:1768–1783, 2006.

[GvL96]   G. H. Golub and C. F. van Loan. *Matrix computations (3rd ed.)*. Johns Hopkins University Press, 1996.

[GY99]   J. Gross and J. Yellen. *Graph theory and its applications*. CRC Press, 1999.

[HH98]   B. Huet and E. R. Hancock. Fuzzy relational distance for large-scale object recognition. In *Proc. IEEE Computer Society Conference on Computer Vision and Pattern Recognition*, pages 138–143, 1998.

[HK92]   L. Hagen and A. B. Kahng. New spectral methods for ratio cut partitioning and clustering. *IEEE Transactions on Computer-Aided Design of Integrated Circuits and Systems*, 11(9):1074–1085, 1992.

[Ide04]   T. Ide. Eigenspace-based anomaly detection in computer systems. In *Proc. of the 10th ACM SIGKDD International Conference on Knowledge Discovery and Data Mining*, pages 440–449, 2004.

[JB03]   M. I. Jordan and F. R. Bach. Learning spectral clustering. In *Proc. Advances in Neural Information Processing Systems*, 2003.

[JCZ13]   C. Jiang, F. Coenen, and M. Zito. A survey of frequent subgraph mining algorithms. *The Knowledge Engineering Review*, 28:75–105, 2013.

[Kel09]   S. Kelley. *The existence and discovery of overlapping communities in large-scale networks*. Rensselaer Polytechnic Institute, 2009.

[KJ11]   Y. Kim and H. Jeong. Map equation for link communities. *Physical Review E*, 84(2):026110, 2011.

[KKKS08] J. M. Kumpula, M. Kivelä, K. Kaski, and J. Saramäki. Sequential algorithm for fast clique percolation. *Physical Review E*, 78(2):026109, 2008.

[KKR06]   E. Kranakis, D. Krizanc, and S. Rajsbaum. Mobile agent rendezvous: A survey. In *Structural Information and Communication Complexity*, pages 1–9. Springer, 2006.

[Kle00]   J. Kleinberg. The small-world phenomenon: An algorithmic perspective. In *Proc. of the 32nd ACM Symposium on Theory of Computing*, pages 163–170, 2000.

[KP07]   I. Kotsia and I. Pitas. Facial expression recognition in image sequences using geometric deformation features and support vector machines. *IEEE Transactions on Image Processing*, 16(1):172–187, 2007.

[KPBV09]   D. Krioukov, F. Papadopoulos, M. Boguna, and A. Vahdat. Greedy Forwarding in Scale-Free Networks Embedded in Hyperbolic Metric Spaces. In *Proc. of the 11th ACM SIGMETRICS Workshop on Mathematical Performance Modeling and Analysis*, 2009.

[KVV04]   R. Kannan, S. Vempala, and A. Vetta. On clusterings: Good, bad and spectral. *Journal of the ACM*, 51(3):497–515, 2004.

[LCKF05]   J. Leskovec, D. Chakrabarti, J. Kleinberg, and C. Faloutsos. Realistic, mathematically tractable graph generation and evolution, using Kronecker multiplication. In *Proc. Knowledge Discovery in Databases*, pages 133–145, 2005.

[LKF05]   J. Leskovec, J. Kleinberg, and C. Faloutsos. Graphs over time: densification laws, shrinking diameters and possible explanations. In *Proc. of the 11th ACM SIGKDD International Conference on Knowledge Discovery in Data Mining*, pages 177–187, 2005.

[LMV01]   J. Lladoós, E. Martí, and J. J. Villanueva. Symbol recognition by error-tolerant subgraph matching between region adjacency graphs. *IEEE Transactions on Pattern Analysis Machine Intelligence*, 23:1137–1143, 2001.

[Lov93]   L. Lovász. Random walks on graphs: A survey. *Combinatorics: Paul Erdos is Eighty*, 2:1–46, 1993.

[MGSZ02]   M. Mihail, C. Gkantsidis, A. Saberi, and E. Zegura. On the semantics of Internet topologies. In *Proc. IPAM Workshop on Large Scale Communication Networks.*, 2002.

[Mil67]   S. Milgram. The small-world problem. *Psychology Today*, 1(1):61–67, 1967.

[MM13]   M. H. M. Maier, U. von Luxburg. How the result of graph clustering methods depends on the construction of the graph. *ESAIM: Probability and Statistics*, (17):370–418, 2013.

[MRR$^+$53] N. Metropolis, A. W. Rosenbluth, M. N. Rosenbluth, A. H. Teller, and E. Teller. Equation of State Calculations by Fast Computing Machines. *The Journal of Chemical Physics*, 21(6):1087–1092, 1953.

[MX07]   M. Mahdian and Y. Xu. Stochastic Kronecker graphs. In *Proc. of the 5th International Conference on Algorithms and Models for the Web-graph*, pages 179–186, 2007.

[NC03]   C. C. Noble and D. J. Cook. Graph-based anomaly detection. In *Proc. of the 9th ACM SIGKDD International Conference on Knowledge Discovery and Data Mining*, pages 631–636, 2003.

[OTNP14] G. Orfanidis, A. Tefas, N. Nikolaidis, and I. Pitas. Facial image clustering in stereo videos using local binary patterns and double spectral analysis. In *Proc. IEEE Symposium on Computational Intelligence and Data Mining*, pages 217–221, 2014.

[PDFV05] G. Palla, I. Derényi, I. Farkas, and T. Vicsek. Uncovering the overlapping community structure of complex networks in nature and society. *Nature*, 435(7043):814–818, 2005.

[PDG10] P. Papadimitriou, A. Dasdan, and H. Garcia-Molina. Web graph similarity for anomaly detection. *Journal of Internet Services and Applications*, 1(1):19–30, 2010.

[PF97] E. G. M. Petrakis and C. Faloutsos. Similarity searching in medical image databases. *IEEE Transactions on Knowledge and Data Engineering*, 9(3):435–447, 1997.

[Pit00] I. Pitas. *Digital Image Processing Algorithms and Applications*. Wiley, 2000.

[PKV10] S. Papadopoulos, Y. Kompatsiaris, and A. Vakali. A graph-based clustering scheme for identifying related tags in folksonomies. In *Proc. of the 12th International Conference on Data Warehousing and Knowledge Discovery*, pages 65–76, 2010.

[QH04] H. Qiu and E. R. Hancock. Graph matching and clustering using spectral partitions. *Pattern Recognition*, 39:22–34, 2004.

[RB02] A. Réka and A.-L. Barabási. Statistical mechanics of complex networks. *Reviews of Modern Physics*, 74:47–97, 2002.

[RB08] M. Rosvall and C. T. Bergstrom. Maps of random walks on complex networks reveal community structure. *Proceedings of the National Academy of Sciences*, 105(4):1118–1123, 2008.

[SA05] J. Shetty and J. Adibi. Discovering important nodes through graph entropy the case of Enron email database. In *Proc. of the 3rd International Workshop on Link Discovery*, pages 74–81, 2005.

[SB92] L. S. Shapiro and J. M. Brady. Feature-based correspondence: an eigenvector approach. *Image and Vision Computing*, 10(5):283–288, 1992.

[Sch07] S. E. Schaeffer. Graph clustering. *Computer Science Review*, 1(1):27–64, 2007.

[SF83] A. Sanfeliu and K. Fu. A distance measure between attributed relational graphs for pattern recognition. *IEEE Transactions on Systems, Man, and Cybernetics*, 13:353–362, 1983.

[SH85] L. G. Shapiro and R. M. Haralick. A metric for comparing relational descriptions. *IEEE Transactions on Pattern Analysis Machine Intelligence*, 7:90–94, 1985.

[SLh90] G. L. Scott and H. C. Longuet-Higgins. Feature grouping by relocalisation of eigenvectors of proximity matrix. In *Proc. of British Machine Vision Conference*, pages 103–108, 1990.

[SM00] J. Shi and J. Malik. Normalized cuts and image segmentation. *IEEE Transactions on Pattern Analysis and Machine Intelligence*, 22(8):888–905, 2000.

[TJHA05]   I. Tsochantaridis, T. Joachims, T. Hofmann, and Y. Altun. Large margin methods for structured and interdependent output variables. *Journal of Machine Learning Research*, 6:1453–1484, 2005.

[TKP01]   A. Tefas, C. Kotropoulos, and I. Pitas. Using support vector machines to enhance the performance of elastic graph matching for frontal face authentication. *IEEE Transactions on Pattern Analysis Machine Intelligence*, 23(7):735–746, 2001.

[TPN12]   N. Tsapanos, I. Pitas, and N. Nikolaidis. Graph representations using adjacency matrix transforms for clustering. In *Proc. 16th IEEE Mediterranean Electrotechnical Conference*, pages 383–386, 2012.

[TPP14]   I. Tsingalis, I. Pipilis, and I. Pitas. A statistical and clustering study on YouTube 2D and 3D video recommendation graph. In *Proc. of the 6th IEEE International Symposium on Communications, Control and Signal Processing*, pages 294–297, 2014.

[TTNP15]   N. Tsapanos, A. Tefas, N. Nikolaidis, and I. Pitas. A distributed framework for trimmed kernel k-means clustering. *Pattern Recognition*, 48(8):2685 – 2698, 2015.

[vD00]   S. van Dongen. *Graph Clustering by Flow Simulation*. PhD thesis, University of Utrecht, 2000.

[vL07]   U. von Luxburg. A Tutorial on Spectral Clustering. *Statistics and Computing*, (17):395–416, 2007.

[VSP11]   N. Vretos, V. Solachidis, and I. Pitas. A mutual information based face clustering algorithm for movie content analysis. *Image Vision Computing*, 29:693–705, 2011.

[WA79]   J. B. Wilcox and L. M. Austin. A method for computing the average spearman rank correlation coefficient from ordinally structured confusion matrices. *Journal of Marketing Research*, 16(3):p426–428, 1979.

[WH97]   R. C. Wilson and E. R. Hancock. Structural matching by discrete relaxation. *IEEE Transactions on Pattern Analysis and Machine Intelligence*, 19:634–648, 1997.

[WS98]   D. J. Watts and S. H. Strogatz. Collective dynamics of 'small-world' networks. *Nature*, 393:440–442, 1998.

[WS05]   S. White and P. Smyth. A spectral clustering approach to finding communities in graph. In *Proc. SIAM International Conference on Data Mining*, pages 76–84, 2005.

[WW07]   D. White and R. C. Wilson. Spectral generative models for graphs. In *Proc. of the 14th International Conference on Image Analysis and Processing*, pages 35–42, 2007.

[XH06]   B. Xiao and E. Hancock. A spectral generative model for graph structure. In *Structural, Syntactic, and Statistical Pattern Recognition*, volume 4109, pages 173–181. Springer, 2006.

[XK01]   L. Xu and I. King. A PCA approach for fast retrieval of structural patterns in attributed graphs. *IEEE Transactions on Systems, Man and Cybernetics, Part B: Cybernetics*, 31(5):812–817, 2001.

[XKS13]   J. Xie, S. Kelley, and B. K. Szymanski. Overlapping community detection in networks: The state-of-the-art and comparative study. *ACM Computing Surveys*, 45(4):43, 2013.

[YS03]   S. X. Yu and J. Shi. Multiclass spectral clustering. In *Proc. of the 9th IEEE International Conference on Computer Vision*, pages 313–319, 2003.

[ZD13]   F. Zhou and F. De la Torre. Deformable graph matching. In *Proc. IEEE Conference on Computer Vision and Pattern Recognition*, 2013.

[ZWZ07]   S. Zhang, R.-S. Wang, and X.-S. Zhang. Identification of overlapping community structure in complex networks using fuzzy c-means clustering. *Physica A: Statistical Mechanics and its Applications*, 374(1):483–490, 2007.

Chapter 4

Web Search Based on Ranking

Andrea Tagarelli

University of Calabria, Italy

Santosh Kabbur and George Karypis

University of Minnesota, U.S.A.

4.1 Introduction

Web search is the process of identifying and ranking the web pages that are the most *relevant* to a user's query. Though this is similar to the task performed by traditional Information Retrieval (IR) systems, the nature of the underlying document collection (i.e., the Web) and the widely varying needs and characteristics of its users, have made web search a research field of its own.

In traditional IR systems, the document collection consists of a set of text-based documents, whereas web pages have significantly richer content such as layout markup tags, scripts, images, and videos. This additional information can be leveraged during web search in order to identify the important aspects of a web page. Related to that is the fact that the web pages are linked with each other via *hyperlinks*. This provides valuable information as to what a web page is all about; particularly, by inspecting the content of the web pages that are connected to it via *in-links* or *out-links*, one can go beyond the content that is provided by the web page itself. In addition, by analyzing the network formed by the web pages, web search systems can identify which are the important query-relevant web pages and thus produce better rankings. The fact that the web pages belong to different websites also leads to a document collection that has a hierarchical organization. This within-website hierarchy can be leveraged to transfer information across web pages from the same website (and as such, influence the ranking) and to organize the results in more meaningful ways. Finally, another important difference of web pages is that their content is not always curated and it can be of low quality, intentionally misleading, and span many topics that are sometimes unrelated to each other. This creates many challenges associated with how to determine the web pages that are both relevant to a user's query and at the same time are of high-quality and do not contain unnecessary information.

Unlike traditional IR systems, whose users are often quite familiar with the query language supported by them and the various operators and query reformulations that are required to retrieve the desired set of documents, web search systems need to support users with significantly varying levels of sophistication and expertise. To address this challenge, a significant amount of effort has been devoted to developing search interfaces that allow users to efficiently navigate through the results and easily reformulate their queries. In addition, users' information needs vary significantly and include (i) informational queries that cover a broad topic (e.g., cars, Minnesota), (ii) navigational queries that seek a specific website or a web page (e.g., Facebook, YouTube), and (iii) transactional queries that show an intent of performing an action (e.g., purchase a product, download software). In order to retrieve the most relevant results, web search systems must identify how the different web pages relate to these needs, and for each query it must automatically identify the users' (latent) needs. Though this is a difficult problem to solve for an isolated query and an isolated user, the problem becomes tractable by leveraging information across different users with the same (or similar) query and across the different queries from the same user.

In this chapter we discuss various methods that have been developed to address the above web structure, content, and user-related aspects of web search systems. Note that we will restrict our discussion to text-based queries and web pages, although many of the methods and models here discussed are applicable to media rich web pages consisting of images and videos. Note also that this chapter will not cover the important issue of the evaluation of ranking methods; the interested reader is referred to works that discuss assessment criteria for ranking problems, such as mean average precision (MAP) and precision@n, mean reciprocal rank, binary preference function (Bpref) [BV04], Kendall-tau rank correlation [Abd07], normalized discounted cumulative gain (nDCG) [JK02], and Fagin's intersection metric [FKS03].

We organize the contents of the rest of this chapter into seven sections. Section 4.2 provides background information on classic document representation and retrieval models. Section 4.3 discusses essential hyperlink-related aspects and implications in web search, with emphasis on query expansion techniques, such as latent semantic analysis (LSI), regularization frameworks (e.g., RLSI, RRMF), and (pseudo-)relevance feedback. Section 4.4 focuses on link analysis research, thus offering an overview of prominent centrality and prestige methods, including PageRank and HITS; note that while most of the centrality and prestige methods have been originally conceived for web search, they have also applied equally to related environments such as, e.g., collaboration networks, bibliographic networks, and social media networks. Section 4.5 discusses topic-sensitive web search and

TABLE 4.1.1: Road map of this chapter.

Aspect/Section	Section 4.2	Section 4.3	Section 4.4	Section 4.5	Section 4.6	Section 4.7
Web page Structure	–	Anchor Text, Query Expansion	–	–	–	–
Hyperlinking, Network Topology	–	Anchor Text	Centrality, Prestige, PageRank	–	Heterogeneous Networks	–
Low Quality Content	LSI	RRMF	Hubs & Authorities	TrustRank, Topic-sensitive Ranking	–	–
Organizational Structure	–	–	–	–	RankClus, NetClus	Clustering Results
Diverse Topics	–	–	–	Topic-sensitive Ranking	–	Clustering Results
User Needs	LSI	RRMF, Query Expansion	–	–	–	Clustering Results

ranking approaches, from both the natural perspective of "content as topic" (e.g., topic-biased PageRank, TwitterRank) and the alternative perspective of "trust as topic" (e.g., TrustRank). The subject of Section 4.6 is heterogeneous information networks (and classic related methods such as, e.g., ObjectRank, PopRank, RankClus, NetClus), which can enhance searching and ranking solutions in order to better capture the manifold structure and semantics underlying different types of networked entities and their relationships. Section 4.7 focuses on how web search results can be organized, e.g., through a clustering task. A mapping of the aforementioned sections and the various aspects that they apply to is shown in Table 4.1.1. Section 4.8 finally provides concluding remarks and a discussion on emerging research trends.

4.2 Information Retrieval Background

In this section, a brief introduction of the different representation of the documents and the queries is provided. The interested reader is referred to textbooks, such as, e.g., [BYBRN99, MRS08] for extensive overviews of document modeling and processing techniques.

4.2.1 Document representation

The most widely used approach to document representation is known as the *bag-of-words* model. According to this model, documents are represented as sets of terms that they are comprised of, therefore the appearance order of the terms is ignored. A *term-document matrix* (also known as a matrix *inverted index*) is used for representing a collection of documents, where rows of the matrix correspond to the terms in the collection and the columns correspond to the documents. Various schemes are used to determine the value of the each entry (called a *weight*) in the matrix. Such weights are then used to quantify the *importance* of the term on the corresponding page for ranking. One such weighting scheme is called *term-frequency*, $tf(t, d)$, where the frequency of terms in documents, $f(t, d)$, is used as the weight in the matrix. The frequencies of terms are used directly (i.e., *raw* frequencies), or the raw frequencies are scaled using a function (e.g., $\log(f(t,d))$), or augmented with a constant value to prevent bias toward the length of the document, or converted to binary frequencies (i.e., $tf(t,d) = 1$ if $f(t,d) > 0$, else 0). One downside of using only the term-frequency for weighting is that the weights of commonly occurring terms will be high, and it might not be useful to rank documents when the query contains one of these terms. In such a case, most of the documents in the corpus will be returned as relevant.

TF-IDF weighting. One way to handle this limitation of term-frequency is by down weighting the terms that occur very frequently in the documents and increasing the weight of terms that occur rarely. This is achieved by using a factor called *inverse-document-frequency* (IDF) [Jon72]. The IDF of a term is a measure of how much information the term provides. IDF is formally defined as:

$$idf(t, D) = \log \frac{|D|}{1 + |d \in D : t \in d|} \ , \tag{4.2.1}$$

where D is the set of documents in the corpus. The term $|d \in D : t \in d|$ gives the count of documents that contain the term t. 1 is added to count, to handle cases where the given term is not present in any of the documents.

The IDF is used along with the term-frequency to compute the weights in the term-document matrix. This measure is called *TF-IDF* (term-frequency inverse-document-frequency) [Jon72, SB88]. TF-IDF for a given pair of term and document is defined as the product of the term-frequency of the term in the document and the IDF score of the term in the corpus, i.e., $tfidf(t, d, D) = tf(t, d) \times idf(t, D)$. Thus, a high TF-IDF score for a given term is achieved by having a high frequency in the given document and a low frequency (i.e., occurs rarely) in the corpus of documents. In recent years, different variants of TF-IDF weighting schemes have been adopted. Notable ones include using a logarithm scale for computing TF and normalizing the TF relative to the maximum frequency term in the document.

Dimensionality reduction. The number of unique terms in a text (also termed *features*) corpus is normally very high. *Dimensionality reduction* techniques aim to alleviate this problem by decreasing noise in the term space. This can be accomplished by a *feature selection* approach, which aims to choose an optimal subset of features given some objective function, or a *feature extraction* approach, which seeks a lower-dimensional space mapping of the original feature space. The simplest selection technique prunes features with low or high document frequency: frequently occurring terms are deemed uninformative, while rare terms constitute noise; stop-words, which are lexicon-specific, frequent, non-content-bearing terms, are also removed.

Feature extraction technique algorithms project the data to some lower dimensional space. The earliest approach to linear projection, namely Principal Component Analysis (PCA) [Hot33], provides a low-rank representation of the covariance matrix of the features. PCA utilizes an orthogonal transformation based on Singular Value Decomposition (SVD) to convert a set of observations of possibly correlated variables into a set of linearly uncorrelated variables called *principal components*. In the text domain, a well-known technique is Latent Semantic Indexing (LSI) [DDL+90]. LSI was designed as an indexing and retrieval method that takes advantage of the *semantic structure* (implicit higher-order structure) in the association of terms with documents. LSI is in fact based on the principle that terms present in the same contexts tend to have similar meanings. Therefore, LSI aims to overcome problems associated with synonymies. Similar to PCA, LSI applies a rank reduced SVD on the constructed term-document matrix to identify the relationship between the terms and concepts present in the text. Given a corpus of documents containing m unique terms and n documents, a weighted term-document matrix $\mathbf{C}$ is constructed with each row representing a term and each column representing a document, i.e., $\mathbf{C} \in \mathbb{R}^{m \times n}$. SVD is then used to decompose $\mathbf{C}$ into three matrices:

$$\mathbf{C} = \mathbf{U}\mathbf{S}\mathbf{V}^{\mathrm{T}}, \tag{4.2.2}$$

such that $\mathbf{U}$ and $\mathbf{V}$ have orthonormal columns, i.e., $\mathbf{U}^{\mathrm{T}}\mathbf{U} = \mathbf{I}, \mathbf{V}^{\mathrm{T}}\mathbf{V} = \mathbf{I}$ and $\mathbf{S}$ is a diagonal matrix containing the singular values which are positive and are in decreasing order along the primary diagonal, i.e., $S(1,1) \geq S(2,2) \geq \ldots \geq S(r,r) > 0, S(i,j) = 0$ where $i \neq j$ and r is the rank of the matrix $\mathbf{C}$. The matrices obtained by decomposition are then truncated by retaining only the top k singular values in $\mathbf{S}$, where $k \ll m, n$. Correspondingly, only the first k columns of $\mathbf{U}$ and $\mathbf{V}$ are retained. Thus, the given matrix $\mathbf{C}$ is approximated as the product of these truncated matrices. That is,

$$\mathbf{C} \approx \hat{\mathbf{C}} = \mathbf{U}_k \mathbf{S}_k \mathbf{V}^{\mathrm{T}}{}_k. \tag{4.2.3}$$

$\hat{\mathbf{C}}$ is called the rank-k approximation of $\mathbf{C}$. It can be shown that $\hat{\mathbf{C}}$ is the rank-k model with the best possible least squares fit to $\mathbf{C}$. The geometric interpretation of the SVD model allows the rows the reduced matrices to be considered as the points corresponding to the documents and terms in a k-dimensional vector space. This representation helps to compute distances (or similarities) between the points in the space. The amount of dimension reduction, that is the value of k, is typically small compared to m and n, and the exact value is obtained based on the value which gives the best retrieval performance. Some empirical studies [Bra08a] on LSI have shown that a small value of k in the range of 300 to 400 for moderate (hundreds of thousands) to large (millions) sized document collections is sufficient to obtain the best retrieval performance.

4.2.2 Retrieval models

An information retrieval model typically encompasses three elements, namely, documents, queries, and ranking functions. The goal of the model is to predict which documents are relevant to the user's query and to rank the relevant documents in the order of the predicted likelihood of relevance to the user. Using the representation of documents presented earlier, this section explains various classical retrieval models. The inputs to these models are the set of documents (in the form of an inverted index) and a query (in the form of a list of words). The output is the ranked list of matched documents in the order of their relevance to the given query. All of these methods assume that the underlying representation for documents is bag-of-words.

Boolean retrieval model. The Boolean retrieval model is one of the earliest models proposed for matching documents and queries. In this model, the documents are represented as sets of terms and the queries are the boolean expressions of the terms. Rules from boolean algebra are then used to extract the *exact-match* documents with the query represented using boolean operators *AND*, *OR*, and *NOT*. A boolean result (i.e., either 0 or 1) is obtained by applying a boolean expression on a document. Thus, there is no notion of ranking for the retrieved documents. A document is either relevant or not relevant. There is no mechanism to quantify to what extent the retrieved documents are relevant. Also, there is no notion of importance for different words in the queries, i.e., all words are treated with equal weight.

Vector space model. The vector space model is an algebraic model where the documents and queries are represented as *vectors*. The number of dimensions of such a vector space is determined by the number of distinct terms (also called *index terms*) in the document corpus. Each distinct term of the corpus is represented as one dimension. The documents and queries are then represented as vectors in such a space based on the terms that they contain. A document d_j and a query q_l are represented as column vectors,

$$\mathbf{d}_j^{\mathrm{T}} = (w_{1,j}, w_{2,j}, \ldots, w_{m,j})$$
$$\mathbf{q}_l^{\mathrm{T}} = (w_{1,l}, w_{2,l}, \ldots, w_{m,l}),$$

where m is the number of unique terms in the corpus. If a term t is present in a document d_j or a query q_l, its corresponding value $w_{t,j}$ and $w_{t,l}$ is non-zero. Different weighting schemes (Section 4.2.1) are used to set the non-zero value corresponding to a term present in the document or the query.

Representation of documents as vectors helps in utilizing vector algebra techniques to compute similarities between the documents. Vector similarity measures can be utilized to compute the similarity between the documents corresponding to the vectors. One such popular vector similarity measure is the *cosine similarity*. Given two documents, $\mathbf{d}_i$ and $\mathbf{d}_j$, the cosine similarity between them is given by:

$$\cos(\mathbf{d}_i, \mathbf{d}_j) = \frac{\mathbf{d}_i \cdot \mathbf{d}_j^{\mathrm{T}}}{\|\mathbf{d}_i\| \|\mathbf{d}_j\|} = \frac{\sum_{k=1}^{m} d_{k,i} \times d_{k,j}}{\sqrt{\sum_{k=1}^{m} (d_{k,i})^2} \sqrt{\sum_{k=1}^{m} (d_{k,j})^2}}, \qquad (4.2.4)$$

where the numerator term is the vector *dot product* (or inner product) between the two vectors and the denominator is the product of the Euclidean lengths of the vectors. The denominator normalizes the effect of different document lengths. Apart from the cosine similarity, other vector similarity measures can be utilized to compute document similarities, such as the Pearson correlation, the Jaccard similarity, and the normalized Euclidean similarity. A detailed comparison of different vector similarity measures can be found in [TSK06].

Vector similarity can also be extended to compute similarities between documents and queries. For a given query, it can be represented as a bag-of-words document and the cosine similarity can be used to compute the similarity scores with each of the documents in the corpus. The obtained similarity scores can then be used to rank the documents. Unlike boolean retrieval models which retrieve only exact-matched documents, using a model which computes continuous ranking score like vector similarities, allows documents with partial match also to be retrieved.

4.3   Relevance Beyond the Web Page Text

A typical web document (or web page) is an HTML document consisting of nested HTML elements. There are many elements beyond the text which are associated with web documents. One such important aspect is *hyperlinks*. Hyperlinks provide a navigational structure to the collection of web documents. Hyperlinks are created by using the *anchor* tag ($\langle$a$\rangle$) coded in HTML. This linking structure of the web pages can be utilized to improve the retrieval and ranking of web pages. The rest of this section will provide a brief overview of various methods which are used to compute the relevance of a web page for a query by utilizing information beyond the textual content of the given web page.

4.3.1   Anchor text

The *anchor text* is the visible and clickable text in a hyperlink. Anchor text is meant to give the user a context about the hyperlink's target destination. Thus, the anchor text in most hyperlinks can be considered as a good description of the target web page.

The textual content present in a web page can be thought of as the self-description of that page. Many times, the self-description is not sufficient or complete. For a given web page, collecting all the anchor texts from the inlinking pages provides a good description of the target web page. It also has been observed that users tend to submit very short web search queries consisting of few search terms on average, and in many ways the anchor text associated with the inlinks for a web page also shares this characteristic. Thus, the anchor text can be deemed a short summarization of the contents of the target web page in the context of the corresponding inlinking documents [EM03]. There can even be cases where a given web page is the most relevant page for a query, even though none of the words in the query are present in the web page. The relevance in such cases is ascertained with the help of anchor text.

Thus, associating the anchor text with the target web pages has many advantages for web search. First, by augmenting the web page content [McB94] with the anchor text helps to better represent the web page. Second, the summarization of the web pages provided by anchor texts are often more accurate than the pages themselves [BP98]. Third, anchor text helps to summarize different content type on the web (e.g., images, videos, and programs) which cannot be directly indexed by the text-based web search methods [BP98].

However, the usage of anchor text from the web to rank web pages can potentially lead to gaming the search engine. Groups of people with a malicious intent can potentially spam by linking a specific site with a spam keyword. There have been many instances where orchestrated spam campaigns are run against specific sites. Web search methods have evolved to detect and act on such anchor text spamming attempts.

Extending the usage of anchor text to compute the relevance, and text, surrounding the anchor text is also utilized as part of the anchor text to augment the contents of the target web page [CDR+98]; specifically, in cases where the anchor text is not really informative, while the neighboring text can potentially have valuable information (e.g., "*to know more about Web Search,* ⟨**a**⟩ *click here*⟨**/a**⟩").

4.3.2 Query expansion

Query expansion is the process of reformulating the user query to improve the quality of the search results retrieved by the search engine. Given the diversity of the information sources on the Web, the vocabularies of the authors of the content vary greatly. While formulating the query, users might use a vocabulary different from the one used by the authors of the content. This raises a fundamental problem of term mismatch. This issue is also known as *lexical ambiguity*, and it involves synonymy and polysemy of terms. Query expansion aims to solve this problem and is primarily employed with the assumption that the user may not always formulate the search query using the appropriate terms. In this process, given a user search query, the terms constituting the query are examined and new words or phrases are added to generate an *expanded query*. This expanded query is then used to match and retrieve the documents from the index.

Many different techniques are employed to expand the query. These techniques can be broadly divided into two classes: global analysis and local analysis.

Global analysis. Global analysis consists of techniques which are independent of the query and the results returned from it. These techniques compute a global (corpus-wide) term similarity matrix using various methods. The most widely used methods among them use the co-occurrence of the terms in the corpus, synonyms from authoritative sources like thesauri or lexical ontologies (e.g., WordNet [Fel98]), or computing the related terms using the query logs. To expand a query, the similarity matrix is used to add terms which are most similar to the ones present in the query. One of the earliest techniques in this category is *term clustering* [Jon71], which clusters document terms based on their co-occurrence. The query is then expanded using the terms present in the same cluster. Techniques that involve correcting the misspelled query term are also used in practice.

Another popular global analysis approach is based on linear projection techniques, such as Latent Semantic Indexing (LSI), which has been previously discussed in Section 4.2. In effect, semantic connections among documents (and queries) may exist even if they do not share terms. For example, assuming that "car" and "auto" co-occurring in a document are related to each other, they not only are expected to occur in similar sets of documents but also make other co-occurring terms indirectly related to each other, e.g., if a document contains "car" and "engine" and another document contains "auto" and "motor," then "engine" and "motor" have some reciprocal relatedness. Generally speaking, some rows and/or columns of the term-document matrix of a corpus may be somewhat "redundant", thus the matrix may have a rank far lower than its original feature dimension. In LSI, the obtained truncated matrices $\mathbf{U}_k$ and $\mathbf{V}_k$ can then be used to compute similarity between two terms, two documents, or a term and a document. From the geometric interpretation of the rows of the truncated matrices as vectors in the k-dimensional space, the dot product between corresponding rows of $\mathbf{U}_k$ gives the similarity between the two terms. Similarly, $\mathbf{V}_k$ can be used to compute the similarity between two documents by taking the dot product of the rows corresponding to the two documents. The similarity between a term and a document, is given by the dot product between the row corresponding to the term in $\mathbf{U}_k$ and the row corresponding to the document in $\mathbf{V}_k$. Such similarities can then be used for various tasks such as query expansion, document clustering, and finding representation for pseudo-documents. Each of the k latent dimensions can be thought of as the latent (or abstract) topic through which the terms and the documents are related.

The above point makes the concept of dimensionality reduction similar to *topic modeling*. Topic modeling is a statistical modeling technique used to discover the abstract "topics" in a collection of documents. The basic assumption is that a document can be represented as a mixture of probability distributions over its constituent terms, where each component of the mixture refers to a main topic. The document representation is obtained by a *generative process*, i.e., a probabilistic process that expresses document features as being generated by a number of latent variables. Topic modeling methods, such as the popular *Latent Dirichlet Allocation* (LDA) [BNJ03], can be directly employed for performing global analysis based query expansion. Just as a brief mention about LDA, the generative process performed by LDA consists of three levels that involve the whole corpus, the documents, and the terms of each document. The algorithm first samples, for each document, a distribution over collection topics from a Dirichlet distribution, and for each topic, it samples a distribution over terms from a Dirichlet distribution. For each document and term position, it selects a single topic according to the topic distribution specific to the sampled document. Finally, each term is then sampled from a multinomial (i.e., categorical) distribution over terms specific to the sampled topic.

One of the main challenges associated with LSI is scalability and performance. LSI requires relatively high computational and memory performance. Given the size of the Web (i.e., billions of documents), scaling LSI is a huge challenge. In one of the recent methods called *Regularized Latent Semantic Indexing* (RLSI) [WXLC11], parallelization is employed to scale the topic modeling. RLSI formalizes the problem of topic modeling as a problem of minimizing a quadratic loss function regularized by ℓ_1 and ℓ_2 norms. This regularized formulation helps to decompose the problem into multiple sub-optimization problems, which can be solved in parallel.

Relation Regularized Matrix Factorization (RRMF) [LY09] is a technique which extends the LSI to utilize the web page characteristics. It uses hyperlinks (relations) between the web pages and simultaneously models both the links and the content information into a lower-dimensional latent space. The dimensions corresponding to this latent space are known as latent factors, and the terms and documents are modeled as vectors (known as latent vectors) in this low dimensional latent space. The intuition behind this method is to make the latent representation of two web pages which are linked, to be as close as possible. RRMF uses the links between the pages to regularize their latent factors. Given the term-document matrix $\mathbf{C} \in \mathbb{R}^{m \times n}$, adjacency matrix $\mathbf{A} \in \mathbb{R}^{n \times n}$ containing the links between the web pages, i.e., $A(i,j) = 1$ if a link exists between web page i and web page j, and $A(i,j) = 0$ otherwise, the latent factors for terms and web pages are learned by solving the following regularized optimization problem:

$$\min_{\mathbf{U},\mathbf{V}} \frac{1}{2}\|\mathbf{C} - \mathbf{U}\mathbf{V}^{\mathrm{T}}\|^2 + \frac{\alpha}{2}(\|\mathbf{U}\|^2 + \|\mathbf{V}\|^2) + \frac{\beta}{2}tr(\mathbf{V}\boldsymbol{\mathcal{L}}\mathbf{V}^{\mathrm{T}}), \tag{4.3.1}$$

where $\mathbf{U}$ and $\mathbf{V}$ are the latent factor matrices for terms and web pages respectively, i.e., $\mathbf{U} \in \mathbb{R}^{m \times k}$ represents the latent k-dimensional representations of all the terms in the form of latent vectors and $\mathbf{V} \in \mathbb{R}^{n \times k}$, the latent k-dimensional representations of all the documents, and $tr(\mathbf{V}\boldsymbol{\mathcal{L}}\mathbf{V}^{\mathrm{T}})$ is the link based regularization term and $\boldsymbol{\mathcal{L}}$ is the Laplacian matrix of the adjacency matrix, $\mathbf{A}$. This regularization term is obtained by minimizing the difference between the latent factors of the linked web pages. The link regularization term is computed as:

$$\begin{aligned}
f &= \frac{1}{2}\sum_{i=1}^{n}\sum_{j=1}^{n}a_{ij}\|\mathbf{v}_i - \mathbf{v}_j\|^2 \\
&= tr(\mathbf{V}^{\mathrm{T}}\boldsymbol{\mathcal{L}}\mathbf{V}),
\end{aligned} \tag{4.3.2}$$

where $\mathbf{v}_i$ and $\mathbf{v}_j$ are the vectors corresponding to the documents i and j, $\boldsymbol{\mathcal{L}} = \mathbf{D} - \mathbf{A}$ is the Laplacian of adjacency matrix $\mathbf{A}$ and $\mathbf{D}$ is the diagonal matrix whose diagonal elements $d_{ii} = \sum_j a_{ij}$. RRMF learns the parameters using a linear time learning algorithm, thus it is suitable for large datasets of web scale.

Local analysis. Local analysis methods are designed to adjust a query relative to the results returned to the initial query. Local analysis uses only some of the initially retrieved documents for expanding the subsequent queries by the same user. A well known class of local analysis techniques is called the *relevance feedback* method [Roc71, SB97]. In this class of methods, the expansion terms are extracted from relevant documents of the initial set of results. The relevance of the initial set of retrieved documents is obtained via user feedback, i.e., the user marks the initial set of retrieved documents as relevant or not relevant. In a real-world commercial search engine case, users rarely provide such relevance feedback. Thus, relevance feedback methods are rarely used in present day search engines.

To overcome the shortcoming of lack of relevance feedback from the users, a *pseudo-relevance feedback* (also known as *blind relevance feedback*) method is used [BYBRN99]. This method improves on manual relevance feedback, but automates the feedback loop. The top-ranked documents of the initial query are assumed to be relevant and the expansion terms for the subsequent queries are extracted from such top-ranked documents. This approach has the drawback that, if the initial set of retrieved top-ranked documents are actually irrelevant, then the words drawn from these documents and added to subsequent queries are likely to be unrelated to the topic. Thus, it can potentially worsen the quality of the results. However, in practice it is shown to work reasonably well and performs better than the global analysis methods [XC96]. In one of the pseudo-relevance feedback based methods [XJW09], Wikipedia pages have been used to improve the retrieval performance. Specifically, the pseudo relevant documents are constructed using the top ranked Wikipedia pages for the given query and the Wikipedia entity pages corresponding to the query. Then, the term distributions and the structure of Wikipedia pages is used to select the specific expansion terms. In a related method [AECC08], Wikipedia's anchor texts and hyperlinks are used to expand user's initial query. In another method [KZ04], anchor texts from the web page corpus are ranked using several criteria such as the number of occurrences of the anchor text, and whether the link originates from the same domain or a different domain. The highest-ranked anchor texts which have common terms with the given query are used for the query expansion.

More recently, indirect sources of evidence are used as a surrogate for users' relevance feedback. This is called *indirect* or *implicit relevance feedback*. Although indirect feedback in general is less reliable compared to explicit feedback, implicit feedback can be collected in large quantities at no extra cost to the user. Generally, implicit feedback is more useful than the pseudo-relevance feedback, as users' actions are used as feedback. Given the large scale of web search queries and user interaction with the results, collecting implicit feedback is easy. The feedback thus collected can be used for query expansion. In one of the methods, used by commercial search engines, user actions (e.g., browsing) and other interactions on the retrieved documents for the initial query are assumed to be relevant to the user query [KB01]. These actions are used to deduce the relevancy of the retrieved documents and the conclusions are then used to expand the subsequent queries.

User interactions based on web search logs are used to compute query expansion terms. There are two main classes of techniques which are based on web search logs [CR12]. The first class of methods treats the individual queries as documents and extracts features which are related to the original user query. Some methods make use of associated retrieved results (pseudo-relevance feedback), while some do not [HCO03, JRMG06, YSC09]. The second class of techniques exploits the relation of queries and retrieval results to provide additional or greater context in finding expansion features. Example approaches include finding queries associated with the same documents [BSWZ03] or user clicks [BB00], and extracting terms directly from clicked results [CWNM03, RVT+07]. In [CWNM03], correlation between query terms and document terms is extracted by analyzing user query logs. The computed correlations are then used to select high-quality expansion terms for new queries. In another method [XZC+04], the user click-through data is used to associate the queries with the clicked web pages. These query terms can be taken as user short summarization of the web page and similar to anchor text, these query terms are augmented to the web page content of the clicked pages while indexing.

These methods can be further extended to exploit the characteristics of the web pages for finding the query expansion terms. A personalized query expansion technique [CFN07] selects the query expansion terms based on the user's personal information repository which stores the user's personal collection of tracked behavior (e.g., previous search queries, search results clicked, web pages visited). Clickstream mining is also used to suggest queries (a form of query expansion) based on the user queries, click-through data and the search context [MYKL08, CJP+08]. Using the web search query logs, the user query reformulation strategies are studied [HE09] to better understand how web searchers refine queries and form a theoretical foundation for query reformulation.

4.4 Centrality and Prestige

As we previously discussed, hyperlinks provide a valuable source of information for web search. In fact, the analysis of the hyperlink structure of the Web, commonly called *link analysis*, has been successfully used for improving both the retrieving of web documents (i.e., which web pages to crawl) and the scoring of web documents according to some notion of "quality" (i.e., how to rank web pages). The latter is in general meant either as the relevance of the documents with respect to a user query or as some query-independent, intrinsic notion of *centrality*. In network theory and analysis, the identification of the "most central" nodes in the network (e.g., documents in the web network, actors in a social network, etc.) represents a core task. The term centrality commonly resembles that of importance or prominence of a vertex in a network, i.e., the status of being located in strategic locations within the network. However, there is no unique definition of centrality, as for instance one may postulate that a vertex is important if it is involved in many direct interactions, or if it connects two large components (i.e., if it acts as a bridge), or if it allows for quick transfer of the information also by accounting for indirect paths that involve intermediaries. Consequently, there are only very few desiderata for a centrality measure, which can be expressed as follows:

- A vertex centrality is a function that assigns a real-valued score to each vertex in a network. The higher the score, the more important or prominent the vertex is for the network.

- If two graphs G_1, G_2 are isomorphic and $m(v)$ denotes the mapping function from a node v in G_1 to some node v' in G_2, then the centrality of v in G_1 needs to be the same as the centrality of $m(v) = v'$ in G_2. In other words, the centrality of a vertex depends on the structure of the network.

The term centrality is originally designed for undirected networks. In the case of directional relations, which imply directed networks, the term centrality is still used and refers to the "choices made," or out-degrees of vertices, while the term *prestige* is introduced to examine the "choices received," or in-degrees of vertices [WF94]. Moreover, the vertex centrality scores can be aggregated over all vertices in order to obtain a single, network-level measure of centrality, or alternatively *centralization*, which aims to provide a clue to the variability of the individual vertex centrality scores with respect to a given centrality notion. In the following, we will provide an overview of the most prominent measures of centrality and prestige, and their definitions for undirected and directed networks. Particularly, we will focus on two well-known methods, namely PageRank and Hubs & Authorities, which have been widely applied to web search contexts.

Through the rest of this section and in the subsequent sections of this chapter, we will denote with $G = (\mathcal{V}, \mathcal{E})$ a *network graph*, which consists of two sets, $\mathcal{V}$ and $\mathcal{E}$, such that $\mathcal{V} \neq \emptyset$ and $\mathcal{E}$ is a set of pairs of elements of $\mathcal{V}$. If the pairs in $\mathcal{E}$ are ordered, the graph is said to be *directed*, otherwise it is *undirected*. The elements in $\mathcal{V}$ are the vertices (or nodes) of G, while the elements in $\mathcal{E}$ are the edges (or links) of G.

4.4.1 Basic measures

Vertex-level centrality. The most intuitive measure of centrality for any vertex $v \in \mathcal{V}$ is the *degree centrality*, which is defined as the number of edges incident with v, or degree of v:

$$c_D(v) = deg(v). \tag{4.4.1}$$

Being dependent only on adjacent neighbors of a vertex, this type of centrality focuses on the most "visible" vertices in the network, as those that act as the major point of relational information; by contrast, vertices with low degrees are peripheral in the network. Moreover, the degree centrality depends on the graph size: indeed, since the highest degree for a network (without loops) is $|\mathcal{V}| - 1$, the *relative degree centrality* is:

$$\widehat{c_D}(v) = \frac{c_D(v)}{|\mathcal{V}| - 1} = \frac{deg(v)}{|\mathcal{V}| - 1}. \tag{4.4.2}$$

The above measure is independent of the graph size, and hence it can be compared across networks of different sizes.

The definitions of both absolute and relative degree centrality and degree prestige of a vertex in a directed network are straightforward. In that case, the degree and the set of neighbors have two components: we denote with $B_i = \{v_j | (v_j, v_i) \in \mathcal{E}\}$ the set of *in-neighbors* (or "backward" vertices) of v_i, and with $R_i = \{v_j | (v_i, v_j) \in \mathcal{E}\}$ the set of *out-neighbors* (or "reference" vertices) of v_i. The sizes of sets B_i and R_i are the *in-degree* and the *out-degree* of v_i, denoted as $in(v_i)$ and $out(v_i)$, respectively.

Note also that the degree centrality is also the starting point for various other measures; for instance, the *span* of a vertex, which is defined as the fraction of links in the network that involves the vertex or its neighbors, and the *ego density*, which is the ratio of the degree of the vertex to the theoretical maximum number of links in the network.

Unlike degree centrality, *closeness centrality* also takes into account indirect links between vertices in the network, in order to score higher those vertices that can quickly interact with all others because of their shorter distance to the other vertices [Sab66]:

$$c_C(v) = \frac{1}{\sum_{u \in \mathcal{V}} d(v, u)}, \tag{4.4.3}$$

where $d(v, u)$ denotes the graph theoretic, or geodesic, distance (i.e., length of the shortest path) between vertices v, u. Since a vertex has the highest closeness if it has all the other vertices as neighbors, the *relative closeness centrality* is defined as:

$$\widehat{c_C}(v) = (|\mathcal{V}| - 1)c_C(v) = \frac{|\mathcal{V}| - 1}{\sum_{u \in \mathcal{V}} d(v, u)}. \tag{4.4.4}$$

In the case of directed networks, closeness centrality and prestige can be computed according to outgoing links (i.e., how many hops are needed to reach all other vertices from the selected one) or incoming links (i.e., how many hops are needed to reach the selected vertex from all other vertices), respectively. Note that the closeness centrality is only meaningful for a connected network—in fact, the geodesics to a vertex that is not reachable from any other vertex are infinitely long. One remedy to this issue is to define closeness by focusing on distances from the vertex v to only the vertices that are in the *influence range* of v (i.e., the set of vertices reachable from v) [WF94].

Besides (shortest) distance, another important property refers to the ability of a vertex to have control over the flow of information in the network. The idea behind *betweenness centrality* is to compute the centrality of a vertex v as the fraction of the shortest paths between all pairs of vertices that pass through v [Fre77]:

$$c_B(v) = \sum_{u, z \in \mathcal{V}, u \neq v, z \neq v} \frac{m_{u,z}(v)}{m_{u,z}(\mathcal{V})}, \tag{4.4.5}$$

where $m_{u,z}(v)$ is the number of shortest paths between u and z and passing through v, and $m_{u,z}(\mathcal{V})$ is the total number of shortest paths between u and z. This centrality is minimum (zero) when the vertex does not fall on any geodesic, and maximum when the vertex falls on all geodesics, which is equal to $(|\mathcal{V}| - 1)(|\mathcal{V}| - 2)/2$. Analogously to the other centrality measures, it's recommended to standardize the betweenness to obtain a *relative betweenness centrality*:

$$\widehat{c_B}(v) = \frac{2c_B(v)}{(|\mathcal{V}| - 1)(|\mathcal{V}| - 2)}, \tag{4.4.6}$$

which should be divided by 2 for directed networks. Note that, unlike closeness, betweenness can be computed even if the network is disconnected.

It should be noted that the computation of betweenness centrality is the most resource-intensive among the above discussed measures: while standard algorithms based on Dijkstra's or breadth-first search methods require $\mathcal{O}(|\mathcal{V}|^3)$ time and $\mathcal{O}(|\mathcal{V}|^2)$ space, algorithms designed for large, sparse networks require $\mathcal{O}(|\mathcal{V}| + |\mathcal{E}|)$ space and $\mathcal{O}(|\mathcal{V}||\mathcal{E}|)$ and $\mathcal{O}(|\mathcal{V}||\mathcal{E}| + |\mathcal{V}|^2 \log |\mathcal{V}|)$ time on unweighted and weighted networks, respectively [Bra01]. A number of variants of betweenness centrality has also been investigated; for instance, in [Bra08b], an extension of betweenness to edges is obtained by replacing the term $m_{u,z}(v)$ in equation (4.4.5) by a term $m_{u,z}(e)$ calculating the number of shortest (u, z)-paths containing the edge e. An application of this version of edge betweenness is the clustering approach by Newman and Girvan [NG04], where edges of maximum betweenness are removed iteratively to decompose a graph into relatively dense subgraphs.

Besides the computational complexity issue, a criticism of betweenness centrality is that it assumes that all geodesics are equally likely when calculating if a vertex falls on a particular geodesic. However, a vertex with a large in-degree is more likely to be found on a geodesic. Moreover, in many contexts, it may be equally likely that other paths than geodesics are chosen for the information propagation; therefore the paths between vertices should be weighted depending on their length. The index defined by Stephenson and Zelen [SZ89] builds upon the above generalization, by accounting for all paths, including geodesics, and assigning them with weights, which are computed as the inverse of the path lengths (geodesics are given unitary weights). The same researchers also developed an *information centrality* measure, which focuses on the information contained in all paths that originate and end at a specific vertex. The information of a vertex is a function of all the information for paths flowing out from the vertex, which in turn is inversely related to the variance in the transmission of a signal from one vertex to another. Formally, given an undirected network, possibly with weighted edges, a $|\mathcal{V}| \times |\mathcal{V}|$ matrix $\mathbf{X}$ is computed as follows: the i-th diagonal entry is equal to 1 plus the sum of weights for all incoming links to vertex v_i, and the (i, j)-th off-diagonal entry is equal to 1, if v_i and v_j are not adjacent, otherwise it is equal to 1 minus the weight of the edge between v_i and v_j. For any vertex v_i, the *information centrality* is defined as:

$$c_I(v_i) = \frac{1}{y_{ii} + \frac{1}{|\mathcal{V}|}\left(\sum_{v_j \in \mathcal{V}} y_{jj} - 2\sum_{v_j \in \mathcal{V}} y_{ij}\right)}, \tag{4.4.7}$$

where $\{y_{ij}\}$ are the entries of the matrix $\mathbf{Y} = \mathbf{X}^{-1}$. Since function c_I is only lower bounded (the minimum is zero), the *relative information centrality* for any vertex v_i is obtained by dividing $c_I(v_i)$ by the sum of the c_I values for all vertices.

Network-Level centrality. A basic network-level measure of degree centrality is simply derived by taking into account the (standardized) average of the degrees:

$$\frac{\sum_{v \in \mathcal{V}} c_D(v)}{|\mathcal{V}||\mathcal{V} - 1|} = \frac{\sum_{v \in \mathcal{V}} \widehat{c_D}(v)}{|\mathcal{V}|}, \tag{4.4.8}$$

which is exactly the *density* of the network.

Focusing on a global notion of closeness, a simplification of this type of centrality stems from the graph-theoretic *center* of a network. This is in turn based on the notion of *eccentricity* of a vertex v, i.e., the distance to a vertex farthest from v. Specifically, the *Jordan center* of a network is the subset of vertices that have the lowest maximum distance to all other vertices, i.e., the subset of vertices within the *radius* of a network.

A unifying view of network-level centrality is based on the notion of *network centraliza-*

tion, which expresses how the vertices in the network graph G differ in centrality [Fre79]:

$$\mathsf{C}(G) = \frac{\sum_{v \in \mathcal{V}} \mathsf{maxC} - \mathsf{C}(v)}{\max \sum_{v \in \mathcal{V}} \mathsf{maxC} - \mathsf{C}(v)}, \tag{4.4.9}$$

where $\mathsf{C}(\cdot)$ is a function that expresses a selected measure of relative centrality, and maxC is the maximum value of relative centrality over all vertices in the network graph. Therefore, centralization is lower when more vertices have similar centrality, and higher when one or a few vertices dominate the other vertices; as extreme cases, a star network and a regular (e.g., cycle) network have centralization equal to 1 and 0, respectively. According to the type of centrality considered, the network centralization assumes a different form. More specifically, considering the degree, closeness, and betweenness centralities, the denominator in equation (4.4.9) is equal to $(n-1)(n-2)$, $(n-1)(n-2)/(2n-3)$, and $(n-1)$, respectively.

4.4.2   Eigenvector centrality and prestige

None of the previously discussed measures reflects the importance of the vertices that interact with the target vertex when looking at (in)degree or distance aspects. Intuitively, if the influence range of a vertex involves many prestigious vertices, then the prestige of that vertex should also be high; conversely, the prestige should be low if the involved vertices are peripheral. Generally speaking, a vertex's prestige should depend on the prestige of the vertices that point to it, and their prestige should also depend on the vertices that point to them, and so on "ad infinitum" [See49]. It should be noted that the literature usually refers to the above property as *status*, or *rank*.

The idea behind status or rank prestige by Seeley, denoted by function $r(\cdot)$, can be formalized as follows:

$$r(v) = \sum_{u \in V} A(u, v) r(u), \tag{4.4.10}$$

where $A(u, v)$ is equal to 1 if u points to v (i.e., u is an in-neighbor of v), and 0 otherwise. Equation (4.4.10) corresponds to a set of $|\mathcal{V}|$ linear equations (with $|\mathcal{V}|$ unknowns) which can be rewritten as:

$$\mathbf{r} = \mathbf{A}^{\mathrm{T}} \mathbf{r}, \tag{4.4.11}$$

where $\mathbf{r}$ is a vector of size $|\mathcal{V}|$ storing all rank scores, and $\mathbf{A}$ is the adjacency matrix. Or, rearranging terms, we obtain $(\mathbf{I} - \mathbf{A}^{\mathrm{T}})\mathbf{r} = \mathbf{0}$, where $\mathbf{I}$ is the identity matrix of size $|\mathcal{V}|$.

Katz [Kat53] first recommended to manipulate the matrix $\mathbf{A}$ by constraining every row in $\mathbf{A}$ to have sum equal to 1, thus enabling equation (4.4.11) to have finite solution. In effect, equation (4.4.11) is a characteristic equation used to find the eigensystem of a matrix, in which $\mathbf{r}$ is an eigenvector of $\mathbf{A}^{\mathrm{T}}$ corresponding to an eigenvalue of 1. In general, equation (4.4.11) has no non-zero solution unless $\mathbf{A}^{\mathrm{T}}$ has an eigenvalue of 1.

A generalization of equation (4.4.11) was suggested by Bonacich [Bon72], where the assumption is that the status of each vertex is proportional (but not necessarily equal) to the weighted sum of the vertices to whom it is connected. The result, known as *eigenvector centrality*, is expressed as follows:

$$\lambda \mathbf{r} = \mathbf{A}^{\mathrm{T}} \mathbf{r}. \tag{4.4.12}$$

Note that the above equation has $|\mathcal{V}|$ solutions corresponding to $|\mathcal{V}|$ values of λ. Therefore, the general solution can be expressed as a matrix equation:

$$\lambda \mathbf{R} = \mathbf{A}^{\mathrm{T}} \mathbf{R}, \tag{4.4.13}$$

where $\mathbf{R}$ is a $|\mathcal{V}| \times |\mathcal{V}|$ matrix whose columns are the eigenvectors of $\mathbf{A}^{\mathrm{T}}$ and λ is a diagonal matrix of eigenvalues.

Katz [Kat53] also proposed to introduce in equation (4.4.11) an "attenuation parameter" $\alpha \in (0, 1)$ to adjust for the lower importance of longer paths between vertices. The result, known as *Katz centrality*, measures the prestige as a weighted sum of all the powers of the adjacency matrix:

$$\mathbf{r} = \sum_{i=1}^{\infty} \alpha^i (\mathbf{A}^T)^i \mathbf{r}. \tag{4.4.14}$$

When α is small, Katz centrality tends to probe only the local structure of the network; as α grows, more distant vertices contribute to the centrality of a given vertex. Note also that the infinite sum in the above equation converges to $\mathbf{r} = [(\mathbf{I} - \alpha\mathbf{A}^T)^{-1} - \mathbf{I}]\mathbf{1}$ as long as $|\alpha| < 1/\lambda_1$, where λ_1 is the first eigenvalue of $\mathbf{A}^T$.

All the above measures may fail in producing meaningful results for networks that contain vertices with null in-degree: in fact, according to the assumption that a vertex has no status if it does not receive choices from other vertices, vertices with null in-degree do not contribute to the status of any other vertex. A solution to this problem is to allow every vertex some status that is independent of its connections to other vertices. The Bonacich & Lloyd centrality [BL01], probably better known as *alpha-centrality*, is defined as:

$$\mathbf{r} = \alpha\mathbf{A}^T\mathbf{r} + \mathbf{e}, \tag{4.4.15}$$

where $\mathbf{e}$ is a $|\mathcal{V}|$-dimensional vector reflecting *exogenous* source of information or status, which is assumed to a vector of ones. Moreover, parameter α here reflects the relative importance of endogenous versus exogenous factors in determining the vertex prestiges. The solution of equation (4.4.15) is:

$$\mathbf{r} = (\mathbf{I} - \alpha\mathbf{A}^T)^{-1}\mathbf{e}. \tag{4.4.16}$$

It can easily be proved that equation (4.4.16) and equation (4.4.14) differ only by a constant (i.e., one) [BL01].

4.4.3 PageRank

In [BP98], Brin and Page presented *PageRank*, the Google's patented ranking algorithm. There are four key ideas behind PageRank. The first two are also shared with the previously discussed eigenvector centrality methods, that is: a page is prestigious if it is chosen (pointed to) by other pages, and the prestige of a page is determined by summing the prestige values of all pages that point to that page. The third idea is that the prestige of a page is propagated to its out-neighbors as distributed proportionally. Let $\mathbf{W}$ be a $|\mathcal{V}| \times |\mathcal{V}|$ matrix such that columns refer to those vertices whose status is determined by the connections received from the row vertices:

$$W(i, j) = \begin{cases} 1/out(v_i) & \text{if } (v_i, v_j) \in \mathcal{E} \\ 0 & \text{otherwise.} \end{cases} \tag{4.4.17}$$

Note that $\mathbf{W} = \mathbf{D}_{\text{out}}^{-1}\mathbf{A}$, where $\mathbf{A}$ is the adjacency matrix and $\mathbf{D}_{\text{out}}$ is a diagonal matrix storing the out-degrees of the vertices (i.e., $\mathbf{D}_{\text{out}} = diag(\mathbf{A1})$). Using matrix $\mathbf{W}$, the first three ideas underlying the PageRank can be expressed as $\mathbf{r} = \mathbf{W}^T\mathbf{r}$, or equivalently, for every $v_i \in \mathcal{V}$:

$$r(v_i) = \sum_{v_j \in B_i} \frac{r(v_j)}{out(v_j)}. \tag{4.4.18}$$

Therefore, vector $\mathbf{r}$ is the unique eigenvector of the matrix corresponding to eigenvalue 1. It should be noted that equation (4.4.18) is well-defined only if the graph is strongly connected (i.e., every vertex can be reached from any other vertex). Under this assumption, this equation has an interpretation based on random walks, called the *random surfer* model [BP98]. It can be shown that vector $\mathbf{r}$ is proportional to the stationary probability distribution of the random walk on the underlying graph. It should be remarked that, in contrast to PageRank, alpha-centrality does not have a natural interpretation in terms of probability distribution, i.e., the sum of the values in the alpha-centrality vector (cf. equation (4.4.15)) is not necessarily equal to 1.

However, the assumption of graph connectivity behind equation (4.4.18) needs to be relaxed for the practical application of PageRank, since the Web and, in general, real-world networks, are far from being strongly connected. It might be useful to recall here that the Web and many other directed networks have a structure which is characterized by five types of components (cf., e.g., [RU11]): (i) a large strongly connected component (SCC), (ii) an in-component, which contains vertices that can reach the SCC but are not reachable from the SCC, and an out-component, which contains vertices that are reachable from the SCC but cannot reach the SCC, (iii) in-tendrils and out-tendrils, which are vertices that are only connected to the out-component (via out-links) and vertices that are only connected to the in-component (via in-links), (iv) tubes, which are vertices reachable from the in-component and able to reach the out-component, but have neither in-links nor out-links with the SCC, and (v) isolated components, which contain vertices that are disconnected from each of the previous components. Most of these components violate the assumptions needed for the convergence of a Markov process. In particular, when a random surfer enters the out-component, she will eventually get stuck in it; as a result, vertices that are not in the out-component will receive a zero rank, i.e., one cannot distinguish the prestige of such vertices. More specifically, equation (4.4.18) needs to be modified to prevent anomalies that are caused by two types of structures: *rank sinks*, or "spider traps," and *rank leaks*, or "dead ends." The former are sets of vertices that have no links outwards, the latter are individual vertices with no out-links.

If leak vertices were directly represented in matrix $\mathbf{W}$, they would correspond to rows of zero, thus making $\mathbf{W}$ substochastic: as a result, by reiterating equation (4.4.18) for a certain number k of times (i.e., by computing $\mathbf{W}^{T^k}\mathbf{r}$), then some or all of the entries in $\mathbf{r}$ will go to 0. To solve this issue, two approaches can be suggested: (i) modification of the network structure, and (ii) modification of the random surfer behavior. In the first case, leak vertices could be removed from the network so that they will receive zero rank; alternatively, leak vertices could be "virtually" linked back to their in-neighbors, or even to all other vertices. The result will be a row-stochastic matrix, that is, a matrix that is identical to $\mathbf{W}$ except that it will have the columns corresponding to leak vertices that sum to 1. If we denote with $\mathbf{d}$ a vector indexing the leak vertices (i.e., $d(i) = 1$ if v_i has no outlinks, and $d(i) = 0$ otherwise), this row-stochastic matrix $\mathbf{S}$ is defined as:

$$\mathbf{S} = \mathbf{W} + \mathbf{d}\mathbf{1}^{\mathrm{T}}/|\mathcal{V}|. \tag{4.4.19}$$

However, equation (4.4.19) will not solve the problem of sinks. Therefore, Page and Brin [BP98] also proposed to modify the random surfer behavior by allowing for *teleportation*, i.e., the random surfer who gets stuck in a sink, or simply gets "bored" occasionally, can move by randomly jumping to any other vertex in the network. This is the fourth idea behind the PageRank measure, which is implemented by a damping factor $\alpha \in (0, 1)$ that enables to weigh the mixture of random walk and random teleportation:

$$\mathbf{r} = \alpha \mathbf{S}^{\mathrm{T}}\mathbf{r} + (1 - \alpha)\mathbf{p}. \tag{4.4.20}$$

Above, vector $\mathbf{p}$, usually called a *personalization vector*, is by default set to $\mathbf{1}/|\mathcal{V}|$, but it can be any probability vector. Equation (4.4.20) can be rewritten as:

$$\mathbf{G} = \alpha \mathbf{S} + (1 - \alpha)\mathbf{E}, \tag{4.4.21}$$

where $\mathbf{E} = \mathbf{1p}^{\mathrm{T}} = \mathbf{11}^{\mathrm{T}}/|\mathcal{V}|$. The convex combination of $\mathbf{S}$ and $\mathbf{E}$ makes the resulting "Google matrix" $\mathbf{G}$ to be both *stochastic and irreducible*.[1] This is important to ensure (i) the existence and uniqueness of the PageRank vector as stationary the probability distribution $\boldsymbol{\pi}$, and (ii) the convergence of the underlying Markov chain (at a certain iteration k, i.e., $\boldsymbol{\pi}^{(k+1)} = \mathbf{G}\boldsymbol{\pi}^{(k)}$) independently of the initialization of the rank vector.[2]

Computing PageRank. As previously indicated, the computation of PageRank requires solving equation (4.4.21), which is equivalent to finding the principal eigenvector of matrix $\mathbf{G}$. Therefore, similar to other eigenvector centrality methods, the power iteration algorithm is commonly used. Starting from any random vector $\mathbf{r}^{(0)}$, it iterates through equation (4.4.20) until some termination criterion is met; typically, the power method is assumed to terminate when the residual (as measured by the difference of successive iterations) is below some predetermined threshold. Actually, as first observed by Haveliwala (cf. [LM05]), the ranking of the PageRank scores are more important than the scores themselves, that is, the power method can be iterated until ranking stability is achieved, thus leading to a significant saving of iterations on some datasets.

The power iteration method lends itself to efficient implementation thanks to the sparsity of real-world network graphs. Indeed, computing and storing matrix $\mathbf{S}$ (cf. equation (4.4.19)), and hence $\mathbf{G}$, is not required, since the power method can be rewritten as [LM05]:

$$\boldsymbol{\pi}^{\mathrm{T}(k+1)} = \boldsymbol{\pi}^{\mathrm{T}(k)}\mathbf{G} = \alpha\boldsymbol{\pi}^{\mathrm{T}(k)}\mathbf{W} + (\alpha\boldsymbol{\pi}^{\mathrm{T}(k)}\mathbf{d})\mathbf{1}^{\mathrm{T}}/|\mathcal{V}| + (1-\alpha)\mathbf{p}^{\mathrm{T}}, \qquad (4.4.22)$$

which indicates that only sparse vector/matrix multiplications are required. When implemented in this way, each step of the power iteration method requires *nonzero*($\mathbf{W}$) operations, where *nonzero*($\mathbf{W}$) is the number of nonzero entries in $\mathbf{W}$, which approximates to $\mathcal{O}(|\mathcal{V}|)$.

Choosing the damping factor. The damping factor α is by default set to 0.85. This choice actually has several explanations. One is intuitively based on the empirical observation that a web surfer is likely to navigate by following 6 hyperlinks (before discontinuing this navigation chain and randomly jumping on another page), which corresponds to a probability $\alpha = 1 - (1/6) \approx 0.85$. In addition, there are also computational reasons. With the default value of 0.85, the power method is expected to converge in about 114 iterations for a termination tolerance threshold of 1.0E-8 [LM05]. Moreover, since the second largest eigenvalue of $\mathbf{G}$ is α [Mey00], it can be shown that the asymptotic rate of convergence of the power method is $-\log_{10} 0.85 \approx 0.07$, which means that about 14 iterations are needed for each step of accuracy improvement (in terms of digits).

In general, higher values of α imply that the hyperlink structure is more accurately taken into account, however, along with slower convergence and higher sensitivity issues. In fact, experiments with various settings of α have shown that there can be significant variation in rankings produced by different values of α, especially when α approaches 1; more precisely, significant variations are usually observed for the mid-low ranks, while the top of the ranking is usually only slightly affected [Pre02, LM05].

Choosing the personalization vector. As previously discussed, the personalization vector $\mathbf{p}$ can be replaced with any vector whose non-negative components sum up to 1. This

---

[1]A matrix is said *irreducible* if every vertex in its graph is reachable from every other vertex.

[2]Recall that the property of irreducibility of a matrix is related to those of primitivity and aperiodicity. A nonnegative, irreducible matrix is said to be *primitive* if it has only one eigenvalue on its spectral circle; a simple test by Frobenius states that a matrix $\mathbf{X}$ is primitive if and only if $\mathbf{X}^k > 0$ for some $k > 0$, which is useful to determine whether the power method applied to $\mathbf{X}$ will converge [Mey00]. An irreducible Markov chain with a primitive transition matrix is called an *aperiodic chain*.

hence includes the possibility that the vertices in $\mathcal{V}$ might be differently considered when the random surfer restarts her chain by selecting a vertex v with probability $p(v)$, which is not necessarily uniform over all the vertices.

The teleportation probability $p(v)$ can be determined to be proportional to the score the vertex v obtains with respect to an external criterion of importance, or to the contribution that the vertex gives to a certain topological characteristic of the network. For instance, one may want to assign any vertex with a teleportation probability that is proportional to the in-degree of the vertex, i.e.,

$$p(v) = \frac{in(v)}{\sum_{u \in \mathcal{V}} in(u)}.$$

The personalization vector can also be used to boost the PageRank score for a specific *subset of vertices that are relevant to a certain topic*, thus making the PageRank *topic-sensitive*. We shall elaborate on this point later in Section 4.5.

4.4.4   Hubs and authorities

A different approach to the computation of vertex prestige is based on the notions of *hubs* and *authorities*. In a web search context, given a user query, authority pages are ones most likely to be relevant to the query, while hub pages act as indices of authority pages without necessarily being authorities themselves. These two types of web pages are related to each other by a mutual reinforcement mechanism: in fact, if a page is relevant to a query, one would expect that it will be pointed to by many other pages; moreover, pages pointing to a relevant page are likely to point as well to other relevant pages, thus inducing a kind of bipartite graph where pages that are relevant by content (authorities) are endorsed by special pages that are relevant because they contain hyperlinks to locate relevant contents (hubs)—although, it may be the case that a page is both an authority and a hub.

The above intuition is implemented by the Kleinberg's *HITS* (Hyperlink Induced Topic Search) algorithm [Kle98, Kle99]. Like PageRank and other eigenvector centrality methods, HITS still handles an iterative computation of a fixpoint involving eigenvector equations; however, it originally views the prestige of a page as a two-dimensional notion, thus resulting in two ranking scores for every vertex in the network. Also in contrast to PageRank, HITS, produces ranking scores that are query-dependent. In fact, HITS assumes that hubs and authorities are identified and ranked for vertices that belong to a *query-focused subnetwork*. This is usually formed by an initial set of randomly selected pages containing the query terms, which is expanded by also including the neighborhoods of those pages.

Let **a** and **h** be two vectors storing the authority and hub scores, respectively. The hub score of a vertex can be expressed as proportional to the sum of the authority scores of its out-neighbors; analogously, the authority score of a vertex can be expressed as proportional to the sum of the hub scores of its in-neighbors. Formally, *HITS equations* are defined as:

$$\mathbf{a} = \mu \mathbf{A}^{\mathsf{T}} \mathbf{h} \tag{4.4.23}$$

$$\mathbf{h} = \lambda \mathbf{A} \mathbf{a}, \tag{4.4.24}$$

where μ, λ are two (unknown) scaling constants that are needed to avoid the possibility that the authority and hub scores will grow beyond bounds; in practice, **a** and **h** are normalized, so that the largest value in each of the vectors equals 1 (or, alternatively, all values in each of the vectors sum up to 1). Therefore, HITS works as follows:

1. For every vertex in the expanded query-focused subnetwork, initialize hub and authority score (e.g., to 1).

2. Compute the following steps until convergence (i.e., a termination tolerance threshold is reached):

    (a) authority vector **a** using equation (4.4.23);

    (b) hub vector **h** using equation (4.4.24);

    (c) normalize **a** and **h**.

Note that, at the first iteration, **a** and **h** are none other than the vertex in-degrees and the out-degrees, respectively.

By substituting equation (4.4.23) and equation (4.4.24) in each other, hub and authority can in principle be computed independently of each other, through the computation of $\mathbf{AA}^{\mathrm{T}}$ (for the hub vector) and $\mathbf{A}^{\mathrm{T}}\mathbf{A}$ (for the authority vector). Note that, the (i,j)-th entry in matrix $\mathbf{AA}^{\mathrm{T}}$ corresponds to the number of pages jointly referred by pages i and j; analogously, the (i,j)-th entry in matrix $\mathbf{A}^{\mathrm{T}}\mathbf{A}$ corresponds to the number of pages that jointly point to pages i and j. However, both matrix products lead to matrices that are not as sparse, hence the only convenient way to compute **a** and **h** is iteratively in a mutual fashion as described above. In this regard, just as in the case of PageRank, the rate of convergence of HITS depends on the eigenvalue gap, and the ordering of hubs and authorities becomes stable with much less iterations than the actual scores.

It should be noted that the assumption of identifying authorities by means of hubs might not hold in information networks other than the Web; for instance, in citation networks, important authors typically acknowledge other important authors. This has somehow impacted on the probably less popularity of HITS with respect to PageRank—which, conversely, has been successfully applied in many other contexts, including citation and collaboration networks, lexical/semantic networks inferred from natural language texts, recommender systems, and social networks.

The TKC effect. Beyond limited applicability, HITS seems to suffer from two issues that are related to both the precision and coverage of the query search results. More precisely, while the coverage of search results directly affects the size of the subnetwork, the precision can significantly impact on the *tightly knit communities* (TKC) effect, which occurs when relatively many pages are identified as authoritative via link analysis although they actually pertain to only one aspect of the target topic; for instance, this is the case when hubs point both to actual relevant pages and to pages that are instead relevant to "related topics" [LM00]. The latter phenomenon is also called *topic drift*.

While the TKC effect can be attenuated by accounting for the analysis of contents and/or the anchor texts of the web pages (e.g., [BH98, CDR+98]), other link analysis approaches have been developed to avoid overly favoring the authorities of tightly knit communities. Lempel and Morgan [LM00] propose the Stochastic Approach for Link Structure Analysis, dubbed *SALSA*. This is a variation of Kleinberg's algorithm: it constructs an expanded query-focused subnetwork in the same way as HITS, and likewise it computes an authority and a hub score for each vertex in the neighborhood graph (and these scores can be viewed as the principal eigenvectors of two matrices). However, instead of using the straight adjacency matrix, SALSA weighs the entries according to their in- and out-degrees. More precisely, the authority scores are determined by the stationary distribution of a two-step Markov chain through random walking over in-neighbors of a page and then random walking over out-neighbors of a page, while the hub scores are determined similarly with inverted order of the two steps in the Markov chain. Formally, the Markov chain for authority scores has transition probabilities:

$$p_a(i,j) = \sum_{v_q \in B_i \cap B_j} \frac{1}{in(v_i)} \frac{1}{out(v_k)} \tag{4.4.25}$$

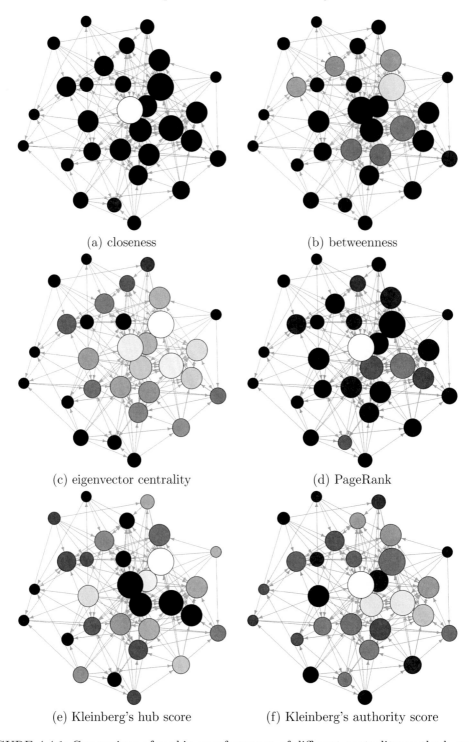

(a) closeness

(b) betweenness

(c) eigenvector centrality

(d) PageRank

(e) Kleinberg's hub score

(f) Kleinberg's authority score

FIGURE 4.4.1: Comparison of ranking performance of different centrality methods on the same example graph. Node size is proportional to the node degree. Lighter gray-levels correspond to higher rank scores.

and the Markov chain for hub scores has transition probabilities:

$$p_h(i,j) = \sum_{v_q \in R_i \cap R_j} \frac{1}{out(v_i)} \frac{1}{in(v_k)}. \tag{4.4.26}$$

Lempel and Morgan proved that the authority stationary distribution $\mathbf{a}$ is such that $a(v_i) = in(v_i)/\bigcup_{v \in \mathcal{V}} in(v)$, and that the hub stationary distribution $\mathbf{h}$ is such that $h(v_i) = out(v_i)/\bigcup_{v \in \mathcal{V}} out(v)$. Therefore, SALSA does not follow the mutual reinforcement principle used in HITS, since hub and authority scores of a vertex depend only on the local links of the vertex. Also, in the special case of a single-component network, SALSA can be seen as a one-step truncated version of HITS [BRRT01]. Nevertheless, the TKC effect is overcome in SALSA through random walks on the hub-authority bipartite network, which imply that authorities can be identified by looking at different communities.

Figure 4.4.1 shows an illustrative comparison of various centrality methods discussed in this section, on the same example network graph. The nodes in the graph are colored using a gray palette, such that lighter gray-levels correspond to higher ranking scores that a particular centrality method has produced over that graph.

4.4.5 SimRank

SimRank [JW02] is a general, iteratively mutual reinforced similarity measure on a *link graph*, which is applicable in any domain with object-to-object relationships. The main intuition behind SimRank is that "two objects are similar if they are related to similar objects." For instance, on a hyperlinked document domain like the Web, two web pages can be regarded as similar if there exist hyperlinks between them, or in a recommender system, we might say that two users are similar if they rate similar items (and, in a mutual reinforcement fashion, two items are similar if they are rated by similar users). The underlying model of SimRank is the "random surfer-pairs model", i.e., SimRank yields a ranking of vertex pairs. The basic SimRank equation formalizes the intuition that two objects are similar if they are referenced by similar objects. Given any two vertices u and v, their similarity, denoted as $S(u,v)$, is defined as 1 if $u = v$, otherwise an iterative process is performed, in which the similarity between u and v is recursively calculated in terms of in-neighbors of u and v, respectively. The generic step of random walk of this process is defined as:

$$S(u,v) = \alpha \frac{1}{|B_u||B_v|} \sum_{i \in B_u} \sum_{j \in B_v} S(i,j), \tag{4.4.27}$$

where α is a constant between 0 and 1. As a particular case, if either u or v has no in-neighbors, then $S(u,v) = 0$. It should be noted that equation (4.4.27) expresses the average similarity between in-neighbors of u and in-neighbors of v. Moreover, it is easy to see that SimRank scores are symmetric.

The basic SimRank equation lends itself to several variations, which account for different contingencies in a network graph. One of these variations allows for resembling the HITS algorithm (cf. Section 4.4.4), since it considers that vertices in a graph may take on different roles, like hub and authority for importance. Within this view, equation (4.4.27) can be replaced by two mutually reinforcing functions that express the similarity of any two vertices in terms of either their in-neighbors or out-neighbors:

$$S_1(u,v) = \alpha_1 \frac{1}{|R_u||R_v|} \sum_{i \in R_u} \sum_{j \in R_v} S_2(i,j) \tag{4.4.28}$$

and

$$S_2(u, v) = \alpha_2 \frac{1}{|B_u||B_v|} \sum_{i \in B_u} \sum_{j \in B_v} S_1(i, j), \qquad (4.4.29)$$

where constants α_1, α_2 have the same semantics as α in equation (4.4.27). Another variation of SimRank is the *min-max* variation, which captures the commonality underlying two similarity notions that express the endorsement of one vertex towards the choices of another vertex, and vice versa. Given vertices u, v, two intermediate terms are defined as:

$$S_u(u, v) = \alpha \frac{1}{|R_u|} \sum_{i \in R_u} \max_{j \in R_v} S(i, j) \qquad (4.4.30)$$

and

$$S_v(u, v) = \alpha \frac{1}{|R_v|} \sum_{i \in R_v} \max_{j \in R_u} S(i, j), \qquad (4.4.31)$$

with the final similarity score computed as $S(u, v) = \min\{S_u(u, v), S_v(u, v)\}$, which ensures that each vertex chooses the other's choices.

SimRank is a computationally expensive method. The space required for each iteration is simply $O(|\mathcal{V}|^2)$, whereas the time required is $O(I|\mathcal{V}|^2 d)$, where d denotes the average of $|B_u||B_v|$ over all vertex-pairs u, v, and I is the number of iterations. One way to reduce the computational burden is to prune the link graph, which avoids computing the similarity for every vertex-pair by considering only vertex-pairs within a certain radius from each other [JW02]. Many other methods to speed up the SimRank computation have been developed in the literature. For instance, Fogaras and Racz [FR05] proposed a probabilistic approximation based on the Monte Carlo method. Lizorkin et al. [LVGT08] proposed different optimization techniques, including partial sums memorization that can reduce repeated calculations of the similarity among different pairs by caching part of similarity summations for later reuse. Antonellis et al. [AGC08] extended SimRank using evidence factor for incident nodes and link weights. More recently, Yu et al. [YLZ13] proposed a fine-grained memoization method to share the common parts among different partial sums; the same authors also studied efficient incremental SimRank computation over evolving graphs. At the time of the writing of this chapter, the most recent study is that by Du et al. [DLC+15], which has focused on SimRank problems in uncertain graphs.

4.5 Topic-Sensitive Ranking

In this section we discuss main approaches to make the process of ranking web documents, or similarly their corresponding users, *topic-sensitive*. The general goal is to drive the ranking mechanism in such a way that the obtained ordering and scoring reflects a target scenario in which the vertices in the network are to be evaluated based on their relevance to a topic of interest. The term topic is here intentionally used with two different meanings, which correspond to different perspectives of quality of web resources: the one normally refers to the *content* of web documents, whereas the other one refers to the relation of web documents with web spammers, and more specifically to their trustworthiness, or likelihood of not being a spammer's target.

4.5.1 Content as topic

As previously mentioned in Section 4.4, the PageRank personalization vector $\mathbf{p}$ can be replaced with any probability vector defined to boost the PageRank score for a specific subset of vertices that are relevant to a particular topic of interest.

A natural way to implement the above idea is to make the teleportation query-dependent, in such a way that a vertex (page) is more likely to be chosen if it covers the query terms. More precisely, if we denote with $\mathcal{B} \subseteq \mathcal{V}$ a subset of vertices of interest, then $\mathbf{p} = \mathbf{1}/|\mathcal{V}|$ is replaced with another vector biased by $\mathcal{B}$, $\mathbf{p}_B$ whose entries are set to $\mathbf{1}/|\mathcal{B}|$ only for those vertices that belong to $\mathcal{B}$, and zero otherwise. Because of the concentration of random walk restarts at vertices from $\mathcal{B}$, these vertices will obtain a higher PageRank score than they obtained using a conventional (non-topic-biased) PageRank.

Intuitively, this way of altering the behavior of random surfing reflects the different preferences and interests that the random surfer may have and specify as terms in a query. Moreover, for efficient indexing and computation, the subset $\mathcal{B}$ usually corresponds to a relatively small selection of vertices that cover some small number of topics. For instance, one might want to constrain the restart of the random walk to select only pages that are classified under one or more categories (e.g., "politics," "sports," "technology," and so on) of the Wikipedia topic classification system[3] or any other publicly available web directory. The consequence of this topic-biased selection is that not only will the random surfer be at pages that are identified as relevant to the selected topics, but also any neighboring page, and any page reachable along a short path (from one of the known relevant pages) will be likely relevant as well.

Topic affinity and user ranking. Determining topic affinity in web sources is central to identifying web pages that are related to a set of target pages. Topic affinity can be measured by using one or a combination of the following three main approaches.

- *Text-Based methods.* Besides cosine similarity in vector space models (e.g., [WVS96]), text-based approaches can also involve *resemblance* measures (e.g., Jaccard or Dice coefficients) which are defined in terms of the overlap between two documents modeled as sets of text chunks [BGMZ97].

- *Link-Based methods.* The link-based topic affinity approach has traditionally borrowed from citation analysis, since the hyperlinking system of endorsement is in analogy with the citation mechanism in research collaboration networks. Particularly, *co-citation analysis* is effective in detecting cores of articles or authors given a particular subject matter. Early applications of co-citation analysis to topical affinity detection and web document clustering include [ER90, Lar96, PP97]. Essentially, a citation-based measure of topic affinity can be formalized as a co-citation strength or, alternately, as a bibliographic-coupling (also called co-reference) strength; i.e., two documents are related proportionally to the frequency with which they are cited together (resp., the frequency with which they have references in common). Furthermore, similarity between two documents can also be evaluated in terms of the number of direct paths between the two documents.

- *Usage-Based methods.* Finally, the usage-based topic affinity approach is based on the assumption that the interaction of users with web resources (stored via user access and activity logs) can aid in improving the quality of content, thus increasing the performance of web search systems. This approach is strictly related to techniques of web personalization and adaptivity based on customization or optimization of the

---

[3]http://en.wikipedia.org/wiki/Category:Main_topic_classifications.

users' navigational experiences, as originally studied by Perkowitz and Etzioni [PE97, PE99].

Topic affinity detection is, however, not only essential to characterize similarity and relatedness of web pages by content, but is also helpful in driving the topic-sensitive ranking of web sources and their users.

TwitterRank. Topic-sensitive ranking in combination with topic affinity measures is in fact widely used in online user communities, such as social media networks and collaboration networks. An exemplary method is represented by the *TwitterRank* algorithm [WLJH10], which was originally designed to compute the prestige or influence of users in the Twitter environment; the algorithm can, however, be applied to other platforms similar to Twitter, or in general to any social network providing microblogging services. A directed graph $G = (\mathcal{V}, \mathcal{E})$ is used to model *followship* relations among users of Twitter, i.e., there is an edge from vertex v_i to vertex v_j if the i-th user follows the j-th user. Therefore, according to PageRank, the higher the number and influence of the followers, the higher the influence of a user of Twitter. Besides the PageRank principle, a key assumption in TwitterRank is that, since followship presumes content consumption (i.e., reading, or replying to tweets), the influence a user has on each follower is determined by the relative amount of content the follower received from her. This means that in TwitterRank a random surfer performs a topic-specific random walk. Moreover, since users generally have different interests in various topics, the influence of Twitter users also varies with respect to different topics. Formally, the (stochastic) transition probability matrix $\mathbf{P}_t$ used in TwitterRank is specified contextually to a topic t, and defined in such a way that, for any users v_i, v_j:

$$P_t(i, j) = \frac{|T_j|}{\sum_{(v_i, v_k) \in \mathcal{E}} |T_k|} sim_t(i, j), \tag{4.5.1}$$

where T_j is the set of tweets published by user v_j, and $sim_t(i, j)$ is the similarity between v_i and v_j with respect to topic t.

To compute the similarity between two users conditionally to a given topic, TwitterRank evaluates the difference between the probability that the two users are interested in the same topic t:

$$sim_t(i, j) = 1 - |\widehat{DT}_{it} - \widehat{DT}_{jt}|, \tag{4.5.2}$$

where $\widehat{\mathbf{DT}}$ is the row-normalized version of the *document-topic matrix* $\mathbf{DT}$, whose (i, t)-th entry stores the number of times a word in the tweets by user v_i has been assigned to topic t. The document-topic matrix represents a low-dimensional representation of a collection of documents, where a document corresponds to the set of tweets of a user. This document representation can in principle be obtained by using some linear projection technique, such as LSI (cf. Section 4.3), which is able to provide a low dimensional mapping from a high dimensional vector space, using an orthogonal transformation based on singular value decomposition to convert a set of observations of possibly correlated variables into a set of linearly uncorrelated variables (or components). A document-topic representation can also be obtained via *statistical topic modeling*, which assumes that a document can be represented as a mixture of probability distributions over its constituent terms, where each component of the mixture refers to a main topic. While still using a "bag of words" assumption (a document is treated as a vector of word counts), the document representation is obtained by a generative process, i.e., a probabilistic process that expresses document features as being generated by a number of latent variables. Compared to conventional vector-space modeling, statistical topic models are generally able to involve (latent) semantic aspects underlying correlations between words to leverage the structure of topics within a document. TwitterRank utilizes the well-known Latent Dirichlet Allocation (LDA) method [BNJ03].

As a variant of topic-biased PageRank, the *TwitterRank* equation, for a given topic t, is defined as:

$$\mathbf{r}_t = \alpha \mathbf{P}_t \mathbf{r}_t + (1 - \alpha)\mathbf{e}_t, \tag{4.5.3}$$

where $\mathbf{e}_t$ is the t-th (normalized) column vector of the $\mathbf{DT}$ matrix. By aggregating over all topics, the global TwitterRank vector is given as: $\mathbf{r} = \sum_t \omega_t \mathbf{r}_t$, where ω_t is the weight associated to topic t. The authors of TwitterRank suggest a number of ways to compute these topic weights. One of the ways is to set ω_t as the prior probabilities of the various topics, estimated proportionally to the number of times unique words have been assigned to corresponding topics. Alternatively, r_t can be set as the probabilities that a particular user v_i is interested in different topics, which are calculated according to the number of times words in v_i's tweets have been assigned to corresponding topics as captured in $\mathbf{DT}$.

4.5.2 Trust as topic

PageRank is vulnerable to adversarial information retrieval. In fact, link spamming techniques can enable web pages to achieve higher scores than what they actually deserve. To do this, spammers normally create the so-called spam farms, i.e., collections of pages whose role is to support the artificial increase of the PageRank score of target pages. In a typical scenario, a spam farm owns a certain number n of supporting pages, each of which has a bidirectional connection only with the target page. Moreover, the target page also has incoming links from outside the spam farm; this is made possible by applying one or more link spamming strategies, such as inviting others to post comments on the spammer site (target page). It can easily be demonstrated that the PageRank score $r^{(s)}$ of the spammer's target page can be computed as:

$$r^{(s)} = \frac{r^{(ns)}}{1 - \alpha^2} + \frac{\alpha}{1 + \alpha}\frac{n}{N}, \tag{4.5.4}$$

assuming that a certain amount $r^{(ns)}$ of PageRank comes from outside the spam farm (i.e., from the pages not owned by the spammer but linked to the target page), and that there are N pages on the Web. Therefore, the size and structure of the spam farm can be manipulated to amplify the PageRank score of the spammer's target page.

Combating link spam has been a necessary task for developers of web search systems in the last few years. One approach is to locate the spam farms, knowing that, as we previously discussed, they may have a typical structure where one page links to a very large number of pages, each of which links back to it. However, this approach is not scalable since spammers can always develop farm structures that differ from the known ones.

TrustRank. A different approach is instead to make the PageRank aware of spam or, in general, untrustworthy pages, by inducing topic-sensitivity in order to lower the score of those pages. Within this view, a well-known method that was introduced to combat web spam and finally detect trustworthy pages is *TrustRank* [GGMP04]. Basically, the algorithm first selects a small seed set of pages whose "spam status" needs to be determined. A human expert then examines the seed pages, and tells the algorithm if they are spam (bad pages) or not (good pages). Finally, the algorithm identifies other pages that are likely to be good based on their connectivity with the good seed pages. The pages in the seed set are classified using a function called *oracle*, which is as:

$$O(p) = \begin{cases} 0 & \text{if } p \text{ is bad} \\ 1 & \text{if } p \text{ is good.} \end{cases} \tag{4.5.5}$$

However, at a large scale, oracle invocations are expensive, and in fact they are used only over the seed set. Therefore, to evaluate pages without relying on the oracle function, the likelihood that a given page p is good will be estimated. A key assumption used in TrustRank to identify good pages is the so-called *approximate isolation* principle, that is, "high-quality pages are unlikely to point to spam or low-quality pages." Upon this principle, a *trust* function T is defined that yields a range of values between 0 (bad) and 1 (good). Ideally, for any page p, $T(p)$ gives the probability that p is good: $T(p) = \Pr[O(p) = 1]$. Desirable properties for the trust function are:

- Ordered Trust Property:

$$T(p) < T(q) \Leftrightarrow Pr[O(p) = 1] < Pr[O(q) = 1]$$

$$T(p) = T(q) \Leftrightarrow Pr[O(p) = 1] = Pr[O(q) = 1]$$

- Threshold Trust Property:
$$T(p) > \delta \Leftrightarrow O(p) = 1.$$

Given the network graph $G = (\mathcal{V}, \mathcal{E})$, TrustRank first computes the seed set, characterized by a vector $\mathbf{s}$, via the following iterative equation:

$$\mathbf{s} = \beta \mathbf{U}\mathbf{s} + (1 - \beta)\frac{1}{|\mathcal{V}|}\mathbf{1}, \qquad (4.5.6)$$

where β is a decaying factor ranging between 0 and 1, and $\mathbf{U}$ is the "inverse" connectivity matrix of the graph:

$$U(i, j) = \begin{cases} 1/in(v_j) & \text{if } (v_i, v_j) \in \mathcal{E} \\ 0 & \text{otherwise.} \end{cases} \qquad (4.5.7)$$

Note that pages in the seed set should be well-connected to other pages in order to propagate trust to many pages quickly. Therefore, they are chosen among those that have a large out-degree. For this purpose, PageRank is computed by reversing the in-links with the out-links in the graph; here a high PageRank score will indicate that trust can flow with a small number of hops along the out-links.

Once the $\mathbf{s}$ vector is computed, it is sorted in decreasing order according to the probability that every vertex belongs to the seed set. Only the top-L vertices are retained in the seed set. The next step consists of the computation of the personalization vector $\mathbf{p}$, such that $p(v_i) = 1$ if the vertex v_i belongs to the seed set and is a good page, and 0 otherwise. The *TrustRank* vector is finally computed using the basic PageRank equation (cf. Section 4.4, equation (4.4.20)), with personalization vector set to the normalized $\mathbf{p}$.

4.6 Ranking in Heterogeneous Networks

So far we have discussed information networks under the common assumption of representation as *homogeneous* networks, i.e., nodes are objects of the same entity type (e.g., web pages, users) and links are relationships of the same type (e.g., hypertext linkage, friendship). However, nodes and node relations can be of different types. For instance, in a research publication network context, nodes can represent authors, publications, and venues, while relations can be of type "written by" (between publication nodes and author nodes), "cited by" (between publication nodes), co-authorship (between author nodes), and so on. As another example, an online social network consists not only of persons, but also of different objects like photos, tags, texts, and so on; moreover, different kinds of relations may occur among different objects (e.g., a photo may be labeled with a certain tag, a person can upload a photo, write a text, or request friendship to another person). Similar scenarios can be found in a variety of application domains, including online e-commerce systems, medical systems, and many others. Consequently, such real-world networks might be conveniently modeled as *heterogeneous* or *typed* networks, in order to better capture the (possibly subtly) different semantics underlying the different types of entities and relationships.

Following [SH12], given a set of vertex types $\mathcal{T}$ and a set of edge types $\mathcal{R}$, a *heterogeneous information network* (HIN) is defined as a directed graph $G = (\mathcal{V}, \mathcal{E})$ with a vertex type mapping function $\tau : \mathcal{V} \to \mathcal{T}$ and an edge type mapping function $\phi : \mathcal{E} \to \mathcal{R}$, where each vertex $v \in \mathcal{V}$ belongs to one particular vertex type $\tau(v) \in \mathcal{T}$, each edge $e \in \mathcal{E}$ belongs to a particular relation $\phi(e) \in \mathcal{R}$. If two edges belong to the same relation type, they share the same starting vertex type as well as the ending vertex type. Moreover, it holds that either $|\mathcal{T}| > 1$ or $|\mathcal{R}| > 1$; otherwise, as a particular case, the information network is *homogeneous*.

The *network schema*, denoted as $S_G = (\mathcal{T}, \mathcal{R})$, is a meta template for a heterogeneous network $G = (\mathcal{V}, \mathcal{E})$ with set of vertex types $\mathcal{T}$ and set of edge types $\mathcal{R}$.

In [SH12], Sun and Han provide a set of suggestions to guide systematic analysis of HINs, which are reported as follows.

1. *Information propagation.* A first challenge is how to propagate information across heterogeneous types of nodes and links; in particular, how to compute ranking scores, similarity scores, and clusters, and how to make good use of class labels, across heterogeneous nodes and links. Objects in HINs are interdependent and knowledge can only be mined using the holistic information in a network.

2. *Exploring network meta structures.* The network schema provides a meta structure of the information network. It provides guidance on the search and mining of the network and helps analyze and understand the semantic meaning of the objects and relations in the network. Meta-path-based similarity searches and mining methods can be useful to explore network meta structures.

3. *User-Guided exploration of information networks.* A certain weighted combination of relations or meta-paths may best fit a specific application for a particular user. Therefore, it is often desirable to automatically select the relation (or meta-path) combinations with appropriate weights for a particular search or mining task based on a user's guidance or feedback. User-Guided or feedback-based network exploration can be a useful strategy.

4.6.1 Ranking in heterogeneous information networks

Ranking models are central to address the new challenges in managing and mining large-scale heterogeneous information networks. In fact, many proposals have been developed for a variety of tasks such as keyword search in databases (e.g., [BHP04]), Web object ranking (e.g., [NZWM05]), expert search in digital libraries (e.g., [GMG11, ZFT+11, DHLK12]), link prediction (e.g., [LK07, DLC11]), recommender systems and Web personalization (e.g., [HSG10, LSK+11, VNL+11]), and sense ranking in tree-structured data [IT14]. Some work has also been developed using path-level features in the ranking models, such as the path-constrained random walk [LC10] and PathSim [SHY+11] for top-k similarity searches based on meta-paths. Moreover, there has been an increasing interest in integrating ranking with mining tasks, like the case of ranking-based clustering addressed by the

RankClus [SHZ$^+$09] and NetClus [SYH09] methods. In the following, we focus on ranking in heterogeneous information networks and provide a brief overview of the main methods.

ObjectRank. One of the first attempts to use a random-walk model over a heterogeneous network is represented by ObjectRank [BHP04]. The algorithm is an adaptation of topic-sensitive PageRank to a keyword search task in databases modeled as labeled graphs.

The HIN framework in ObjectRank consists of a *data graph*, a *schema graph* and an *authority transfer graph*. The data graph $G_D(\mathcal{V}_D, \mathcal{E}_D)$ is a labeled directed graph where every node v has a label $\lambda(v)$ and a set of keywords. Nodes in $\mathcal{V}_D$ represent database objects which may have a sub-structure (i.e., each node has a tuple of attribute name/attribute value pairs). Moreover, each edge $e \in \mathcal{E}_D$ is labeled with a "role" $\lambda(e)$ which describes the relation between the connected nodes.

The *schema graph* $G_S(\mathcal{V}_S, \mathcal{E}_S)$, is a directed graph which describes the structure of G_D, i.e., it defines the set of node and edge labels. A data graph $G_D(\mathcal{V}_D, \mathcal{E}_D)$ conforms to a schema graph $G_S(\mathcal{V}_S, \mathcal{E}_S)$ if there is a unique assignment μ such that:

1. for every node $v \in \mathcal{V}_D$ there is a node $\mu(v) \in \mathcal{V}_S$ such that $\lambda(v) = \lambda(\mu(v))$;

2. for every edge $e \in \mathcal{E}_D$ from node u to node v there is an edge $\mu(e) \in \mathcal{E}_S$ from $\mu(u)$ to $\mu(v)$ and $\lambda(e) = \lambda(\mu(e))$.

The *authority transfer graph* can refer to both a schema graph or a data graph. The *authority transfer schema graph* $G_A(\mathcal{V}_S, \mathcal{E}_A)$ reflects the authority flow through the edges of the graph. In particular, for each edge in $\mathcal{E}_S$ two *authority transfer edges* are created, which carry the label of the schema graph edge forward and backward and are annotated with a (potentially different) *authority transfer rate*. The authority transfer schema graph can be based on a trial and error process or on a domain expert task.

A data graph conforms to an authority transfer schema graph if it conforms to the corresponding schema graph. From a data graph $G_D(\mathcal{V}_D, \mathcal{E}_D)$ and a conforming authority transfer schema graph $G_A(\mathcal{V}_S, \mathcal{E}_A)$ a *authority transfer data graph* $G_{AD}(\mathcal{V}_D, \mathcal{E}_{AD})$ can be derived. Edges of the authority transfer data graph are annotated with authority transfer rates as well, controlled by a formula which propagates the authority from a node based on the number of its outgoing edges.

ObjectRank can be used to obtain a keyword-specific ranking as well as a global ranking. Given a keyword w, the *keyword-specific ObjectRank* is a biased PageRank in which the base set is built upon the set of nodes containing the keyword w:

$$\mathbf{r}^w = \alpha \mathbf{A} \mathbf{r}^w + \frac{1-\alpha}{|S(w)|} \mathbf{s}, \tag{4.6.1}$$

where $S(w)$ denotes the base set specific to w, and $s_i = 1$ if $v_i \in S(w)$ and $s_i = 0$ otherwise. The *global ObjectRank* is basically a standard PageRank. The final score of a node given a keyword w is then obtained by combining the keyword-specific rank and the global rank.

In [BHP04], Balmin et al. also discussed an optimization of the ranking task in the case of directed acyclic graphs (DAGs). More specifically, the authors showed how to serialize the ObjectRank evaluation over single-pass ObjectRank calculations for disjoint, non-empty subsets $L_1, \ldots, L_q$ obtained by partitioning the original set of vertices in a DAG. Upon a topological ordering of L_h ($h = 1..q$) that imposes no backlink from every vertex in L_j to any vertex in L_i, with $i < j$, the ranking of nodes is first computed on L_1 ignoring the rest of the graph, then only the ranking scores of vertices in L_1 connected to vertices in L_2 are reused to calculate ObjectRank for L_2, and so on.

PopRank. In [NZWM05], PopRank is proposed to rank heterogeneous web objects of a specific domain by using both web links and object relationship links. The rank of an object is calculated based on the ranks of objects of different types connected to it, and a parameter called the *popularity propagation factor* is associated with every type of relation between objects of different types.

The PopRank score vector $\mathbf{r}_\tau$ for objects of type τ_0 is defined as a combination of the individual popularity $\mathbf{r}$, and the influence from objects of other types:

$$\mathbf{r}_\tau = \alpha \mathbf{r} + (1 - \alpha) \sum_{\tau_t} \gamma_{\tau_t \tau_0} \mathbf{M}_{\tau_t \tau_0}^{\mathrm{T}} \mathbf{r}_{\tau_t}, \tag{4.6.2}$$

where $\gamma_{\tau_t \tau_0}$ is the *popularity propagation factor* of the relationship link from an object of type τ_t to an object of type τ_0 and $\sum_{\tau_t} \gamma_{\tau_t \tau_0} = 1$, $\mathbf{M}_{\tau_t \tau_0}$ is the row-normalized adjacency matrix between type τ_t and type τ_0, and $\mathbf{r}_{\tau_t}$ is the PopRank score vector for type τ_t. In order to learn the popularity propagation factor $\gamma_{\tau_t \tau_0}$, a simulated annealing-based algorithm is proposed, according to partial ranking lists given by domain experts.

Bipartite SimRank. The *SimRank* algorithm discussed in Section 4.4.5 can naturally be extended to bipartite networks such as, e.g., user-item rating networks. Intuitively, the similarity of users and the similarity of items are mutually reinforcing notions that can be formalized by two equations analytically similar to equation (4.4.28) and equation (4.4.29). More precisely, assuming that edges are directed from vertices of type 1 (e.g., users) to vertices of type 2 (e.g., items), and using superscripts (1) and (2) to denote the two types of vertices, respectively, we have the following equations:

$$S(u^{(1)}, v^{(1)}) = \alpha_1 \frac{1}{|R_{u^{(1)}}||R_{v^{(1)}}|} \sum_{i \in R_{u^{(1)}}} \sum_{j \in R_{v^{(1)}}} S(i, j), \tag{4.6.3}$$

and

$$S(u^{(2)}, v^{(2)}) = \alpha_2 \frac{1}{|B_{u^{(2)}}||B_{v^{(2)}}|} \sum_{i \in B_{u^{(2)}}} \sum_{j \in B_{v^{(2)}}} S(i, j), \tag{4.6.4}$$

where constants α_1, α_2 have the same semantics as α in equation (4.4.27). Equation (4.6.3) corresponds to the similarity between vertices of type-1 is the average similarity between the vertices (of type-2) that they refer to (e.g., items that the two users rated), whereas equation (4.6.4) corresponds to the similarity between vertices of type-2 is the average similarity between the vertices (of type-1) that they are referred to (e.g., the users who rated the two items).

4.6.2 Ranking-Based clustering

Given the diversity of node and link types that characterizes HINs, it is also important to understand how the various nodes of different types can be grouped together. An effective solution is to "integrate" *ranking* and *clustering* tasks. This is the basic idea behind methods such as RankClus and NetClus, which will be discussed next.

RankClus. In [SHZ$^+$09], the RankClus algorithm is introduced, which integrates clustering and ranking on a bi-typed information network $G = (\mathcal{V}, \mathcal{E})$, such that $\mathcal{V} = \mathcal{V}_0 \cup \mathcal{V}_1$, with $\mathcal{V}_0 \cap \mathcal{V}_1 = \emptyset$. Hence, the nodes in the network belong to one of two predetermined types, hereinafter denoted as τ_0, τ_1. The authors use a bibliographic network as a running example, which contains venues and authors as nodes. Two types of links are considered: author-venue publication links, with edge weights indicating the number of papers an author has published in a venue, and co-authorship links, with edge weights indicating the number of times two authors have collaborated. A formal definition of a bi-typed information network is reported as follows.

A key issue in clustering tasks over network objects is that, unlike in traditional attribute based datasets, object features are not explicit. RankClus explores rank distribution for each cluster to generate new measures for target objects, which are low-dimensional. The clusters are improved under the new measure space. More importantly, this measure can be further enhanced during the iterations of the algorithm, so that the quality of clustering and ranking can be mutually enhanced in RankClus.

Two ranking functions over bi-typed bibliographic network are defined in [SHZ+09]: *Simple Ranking* and *Authority Ranking*. *Simple Ranking* is based on the number of publications, which is proportional to the number of papers accepted by a venue or published by an author. Using this measure, authors publishing more papers will have a higher rank score, even if these papers are all in junk venues. *Authority Ranking* is defined to give an object a higher rank score if it has more authority. Iterative rank score formulas for authors and venues are defined based on two principles: (i) highly ranked authors publish many papers in highly ranked venues, and (ii) highly ranked venues attract many papers from highly ranked authors. When considering the co-author information, the rank of an author is enhanced if s/he co-authors with many highly ranked authors.

Differently from *Simple Ranking* (which takes into account only the neighborhood of a node), the score of an object with *Authority Ranking* is based on the score propagation over the whole network. Assuming initial (e.g., random) partition of K clusters $\{C_k\}_{k=1}^K$ of nodes of target type τ_0 of a bi-typed information network, the conditional rank of τ_1-type nodes should be very different for each of the K clusters of τ_0-type nodes (e.g., in the bibliographic network case, the rank of authors should be different for each venue-cluster). The idea is that, for each cluster C_k, conditional rank of $\mathcal{V}_1$, $\mathbf{r}_{\mathcal{V}_1|C_k}$, can be viewed as a rank distribution of $\mathcal{V}_1$, which in fact is a measure for cluster C_k. Then, for each node $v \in \mathcal{V}_0$, the distribution of object $u \in V_1$ can be viewed as a mixed model over K conditional ranks of $\mathcal{V}_1$, and thus can be represented as a K dimensional vector in the new space [SHZ+09]. The authors use an expectation-maximization algorithm to estimate parameters of the mixed model for each target object, and then define a cosine similarity based distance measure between an object and a cluster.

Given a bi-typed information network $G = (\mathcal{V}_0 \cup \mathcal{V}_1, \mathcal{E})$, the ranking functions for $\mathcal{V}_0$ and $\mathcal{V}_1$, and a number K of clusters, RankClus produces K clusters over $\mathcal{V}_0$ with conditional rank scores for each $v \in \mathcal{V}_0$, and conditional rank scores for each $u \in \mathcal{V}_1$. The main steps of the RankClus algorithm are summarized as follows [SH12].

- Step 0 - Initialization: Assign each target node a cluster label from 1 to K randomly.

- Step 1 - Ranking for each cluster: Calculate conditional ranks for nodes of type $\mathcal{V}_1$ and $\mathcal{V}_0$ and within-cluster ranks for nodes of type $\mathcal{V}_0$. If any cluster is empty, the algorithm needs to restart in order to produce K clusters.

- Step 2 - Estimation of the cluster membership vectors for the target objects: Estimate the parameter of the mixted model, obtain new representations for each target object and centers for each target cluster: $\mathbf{s}_v$ and $\mathbf{s}_{C_k}$.

- Step 3 - Cluster adjustment: Calculate the distance from each object to each cluster center and assign it to the closest cluster.

- Repeat Steps 1, 2, and 3 until clusters are stable or change by a very small ratio ϵ, or until a predefined maximum number of iterations is reached.

NetClus.   NetClus [SYH09] extends RankClus from bi-type information networks to multi-typed heterogeneous networks with a *star network schema*, where the objects of different types are connected via a unique "center" type. An information network, $G = (\mathcal{V}, \mathcal{E}, W)$, with $T + 1$ types of objects such that $\mathcal{V} = \{\mathcal{V}_t\}_{t=0}^{T}$, has a star network schema if $\forall\, e = (v_i, v_j) \in \mathcal{E}, v_i \in \mathcal{V}_0 \land v_j \in \mathcal{V}_t (t \neq 0)$ or vice versa. Type τ_0 is called the center or target type, whereas $\tau_t (t \neq 0)$ are attribute types.

Examples of star networks are tagging networks, usually centered on a tagging event, and bibliographic networks, which are centered on papers. In general, a star network schema can be used to map any n-nary relation set (e.g., records in a relational database, with each tuple in the relation as the center object and all attribute entities linking to the center object).

NetClus aims to discover a set of sub-network clusters, and within each cluster a generative model for target objects is built given the ranking distributions of attribute objects in the network. This ranking distribution is calculated using an authority ranking process based on a power iteration method that combines the weight matrices defined between the various types and the center type. The clusters generated are not groups of single typed objects but a set of sub-networks with the same topology as the input network, called *net-clusters*. Each net-cluster is a sub-layer representing a concept of community of the network, which is an induced network from the clustered target objects, and attached with statistical information for each object in the network.

NetClus maps each target object, i.e., that from the center type, into a K-dimensional vector measure, where K is the number of clusters specified by the user. The probabilistic generative model for the target objects in each net-cluster is ranking-based, which factorizes a net-cluster into T independent components, where T is the number of attribute types. NetClus uses the same ranking functions defined for RankClus (*Simple Ranking* and *Authority Ranking*) adapted to the star network case. The core steps of the NetClus algorithm, given the desired number of clusters K, are summarized as follows [SH12].

- Step 0: Generate initial partitions for target objects and induce initial net-clusters from the original network according to the partitions, i.e., $\{C_k^0\}_{k=1}^{K}$.

- Step 1: For each net-cluster, build ranking-based probabilistic generative model, i.e., $\{P(v|C_k^t)\}_{k=1}^{K}$.

- Step 2: For each target object, calculate the posterior probabilities $(P(C_k^t|v))$ and update their cluster assignment according to the new measure defined by the posterior probabilities to each cluster.

- Step 3: Repeat Step 1 and 2 until the clusters do not change significantly, i.e., $\{C_k^*\}_{k=1}^{K} = \{C_k^t\}_{k=1}^{K} = \{C_k^{t-1}\}_{k=1}^{K}$.

- Step 4: Calculate the posterior probabilities for each attribute object $(P(C_k^*|v))$ in each net-cluster.

4.7   Organizing Search Results

The utility of web search methods depends on multiple factors. The primary ones are the underlying retrieval method and the ranking function. The organization of search results and the way the search results are presented also form an important aspect of the utility of a web search method. If the retrieved results for a user query are homogeneous, then simply presenting the ranked list as it is, is a fairly good strategy. In case of queries which are ambiguous the result set can be diverse (e.g., queries which span multiple topics), a simple ranked list presentation of the results will not be effective. A better organization of the results is needed to assist the users to quickly find the information they are looking for. This section deals with the different methods used to organize the search results.

Clustering Results

One way to reduce the ambiguity of the search results, which span multiple topics, is by clustering the top ranked documents into multiple natural clusters. These clusters are expected to correspond to different subtopics associated with the general query topic. A label is generated for each of the identified clusters, and results grouped according to their cluster are presented along with the identified cluster label. The clustering algorithms that can be employed for clustering results on-the-fly need to have a few specific properties, the ability to cluster results efficiently based on the snippets associated with them instead of the whole document, so that the user experience is not affected; fast enough so that it can be used in an online setting, so that the model can be built incrementally utilizing the learning done so far; and the ability to generate a label for each of the clusters, so that the user can relate to the clusters produced.

Text-based clustering. One of the earliest methods is called *Suffix Tree Clustering* (STC) [ZE98], which creates clusters based on phrases shared between the documents. STC treats documents as strings instead of bags-of-words. This allows the algorithm to make use of proximity information between words. A *suffix tree* is then created to efficiently identify sets of documents that share common phrases and this information is used to create clusters. It was shown that STC was both faster and more efficient than the standard clustering algorithms, and using snippets was as effective as using whole documents. Computing meaningful names for the clusters is a challenge. A supervised learning was proposed by Zeng et. al. [ZHMM04]. In this method, the clustering problem was re-formalized as a phrase ranking problem. Given a query and the ranked list of documents from the web search engine results, the method builds a regression model, and then extracts and ranks salient phrases as candidate cluster names. The documents are assigned to relevant salient phrases to form candidate clusters, and the final clusters are then generated by merging these candidate clusters.

Utilizing hyperlinking information. A web hyperlink can be thought of as a statement to indicate that the linked document is related to the document linking to it. Thus, clustering of search results can also be performed by utilizing the hyperlinking structure between the documents. The classical approach of clustering documents is by computing similarities using the content based features (i.e., the features derived from the textual content of the document). The linking structure can be used alone or along with content based features to compute the similarities. Many methods have been developed which utilize the co-citation, bibliographic coupling, and direct paths based similarity between the web pages to compute the clusters, as mentioned in Section 4.5. Such link based features can be used to compute correlation coefficients, which can then be used to do cluster analysis. Clustering based on linking features' information has been shown to produce high quality clusters [Lar96].

Many hybrid approaches have also been developed which utilize both text based features and linking features to cluster the web pages. In one of the simple extensions termed a *content-link clustering* [WVS96] algorithm, the similarity between the documents is computed as the maximum between the text similarity and the link based similarity. The computed similarity is then used to do the cluster analysis. In another approach [PPR96], the combined features are represented as feature vectors for the documents. An activation spreading technique is used to cluster the collection of documents. Activation spreading techniques start by *activating* a node and then *spreading* the value of the node to all its connected nodes. This process is continued and eventually the nodes which have the highest scores are considered to be the nodes relevant to the initial node.

Other hybrid approaches can utilize hyperlink information in combination with usage-based features (cf. Section 4.5). A typical clustering framework in this category, known as *PageGather* [PE97, PE99], requires the calculation of co-occurrence frequencies between web pages. This information is then used to identify the maximal cliques or connected components in the graph built over the page similarity graph. A ranking of clusters is finally obtained based on the average co-occurrence frequency between all pairs of web pages in a cluster.

4.8   Conclusion

The web search methods presented in this chapter focused on providing an overview of existing information retrieval methods, followed by methods which relate to utilizing the hyperlink structure and network topology of the web pages, identifying the high quality and diversity of the web content, and addressing the different user needs. Some methods designed for web-related contexts, like ranking in heterogeneous information networks, were also presented in this chapter; the ever increasing closeness of the web with such different information network types makes the application and extensions of web search methods relevant in the context of various information networks. This chapter concludes by providing a brief discussion on some of the emerging trends in the area of web search.

Emerging Trends

The ever increasing utility of web search has enabled many emerging trends in this area. One such area is *Learning to Rank* (LTR), which specializes in constructing ranking models based on different machine learning methods. In these methods, the query-document pairs are usually represented by a bag-of-words model using numerical vectors (feature vectors). LTR methods utilize various features associated with web pages and queries, such as structural features (based on HTML tags), language modeling features, PageRank, and query popularity. These features are divided into three groups, namely: (i) query-independent or static features, i.e., features that depend only on the document and not on the query (e.g., PageRank), (ii) query-dependent or dynamic features, i.e., features that depend both on the document and the query (e.g., TF-IDF score), and (iii) query features, i.e., features that depend only on the query (e.g., query popularity). Based on the input training data and the loss function used, LTR approaches are broadly categorized into three types, namely: (i) the *pointwise approach*, which approximates the LTR problem as a regression problem by representing each query-document pair with a real valued or ordinal score; (ii) the *pairwise approach*, which approximates the LTR problem as a classification problem and a binary classifier is used to classify the documents in pairs as relevant or not relevant; and (iii) the *listwise approach*, which optimizes either pointwise or pairwise evaluation criteria on the complete list of queries in the training data. In recent years, several approaches have been proposed corresponding to optimizing different metrics. For further reading, Liu [Liu09] has provided a good survey of the different LTR approaches.

Another area of focus in web search is on user personalization, i.e., different users searching the same query may observe different results. Given the diverse set of user needs, modeling the users based on their historical data helps to better customize the results for users. The web search systems typically collect various user data like queries, clicks, browsing behavior, etc. Therefore, the challenge lies in fusing these different information sources to provide better personalized results. To introduce personalization into web search, many new methods and extensions to existing methods have been proposed in the recent years. Dou et al. have provided in [DSW07] an extensive evaluation of various web search personalization methods.

The increasing popularity of recommender systems and its similarity to personalized web search has attracted significant attention in recent years. *Collaborative Filtering* (CF) based methods, which utilize the preferences from different users to compute recommendations, are currently the most effective method for recommender systems. Matrix factorization based methods (LSI, RRMF), which were discussed earlier in this chapter, are one of the most popular tools for CF based tasks. Fusing the CF and the retrieval methods is one of the active research areas. A major goal of these methods is to personalize the web search by utilizing preferences from other users [WWWB12].

Bibliography

[Abd07] H. Abdi. The Kendall Rank Correlation Coefficient. In *Encyclopedia of Measurement and Statistics*. 2007.

[AECC08] J. Arguello, J. L. Elsas, J. Callan, and J. G. Carbonell. Document representation and query expansion models for blog recommendation. *Proc. International Conference Weblogs and Social Media*, 2008(0):1, 2008.

[AGC08] I. Antonellis, H. Garcia-Molina, and C. Chang. SimRank++: query rewriting through link analysis of the click graph. *Proc. of the Very Large Data Bases Endowment*, 1(1):408–421, 2008.

[BB00] D. Beeferman and A. Berger. Agglomerative clustering of a search engine query log. In *Proc. ACM International Conference on Knowledge Discovery and Data Mining*, pages 407–416, 2000.

[BGMZ97] A. Broder, S. Glassman, M. Manasse, and G. Zweig. Syntactic clustering of the web. In *Proc. ACM Conference on World Wide Web*, 1997.

[BH98] K. Bharat and M. R. Henzinger. Improved algorithms for topic distillation in hyperlinked environments. In *Proc. of the ACM Conference on Research and Development in Information Retrieval*, pages 104–111, 1998.

[BHP04] A. Balmin, V. Hristidis, and Y. Papakonstantinou. ObjectRank: Authority-Based Keyword Search in Databases. In *Proc. International Conference on Very Large Data Bases*, pages 564–575, 2004.

[BL01] P. Bonacich and P. Lloyd. Eigenvector-like measures of centrality for asymmetric relations. *Social Networks*, 23:191–201, 2001.

[BNJ03] D. M. Blei, A. Y. Ng, and M. I. Jordan. Latent Dirichlet Allocation. *Journal of Machine Learning Research*, 3(4–5):993–1022, 2003.

[Bon72] P. Bonacich. Factoring and weighing approaches to status scores and clique identification. *Journal of Mathematical Sociology*, 2:113–120, 1972.

[BP98] S. Brin and L. Page. The anatomy of a large-scale hypertextual Web search engine. *Computer Networks*, 30(1-7):107–117, 1998.

[Bra01] U. Brandes. A faster algorithm for betweenness centrality. *Journal of Mathematical Sociology*, 25(2):163–177, 2001.

[Bra08a] R. B. Bradford. An empirical study of required dimensionality for large-scale latent semantic indexing applications. In *Proc. ACM Conference on Information and Knowledge Management*, pages 153–162, 2008.

[Bra08b] U. Brandes. On variants of shortest-path betweenness centrality and their generic computation. *Social Networks*, 30(2):136–145, 2008.

[BRRT01] A. Borodin, G. O. Roberts, J. S. Rosenthal, and P. Tsaparas. Finding authorities and hubs from link structures on the World Wide Web. In *Proc. of the ACM Conference on World Wide Web*, pages 415–429, 2001.

[BSWZ03] B. Billerbeck, F. Scholer, H. E. Williams, and J. Zobel. Query expansion using associated queries. In *Proc. ACM Conference on Information and Knowledge Management*, pages 2–9, 2003.

[BV04] C. Buckley and E. M. Voorhees. Retrieval evaluation with incomplete information. In *Proc. ACM Conference on Research and Development in Information Retrieval*, pages 25–32, 2004.

[BYBRN99] R. Baeza-Yates and B. B. Ribeiro-Neto. *Modern information retrieval*, volume 463. ACM press New York, 1999.

[CDR+98] S. Chakrabarti, B. Dom, P. Raghavan, S. Rajagopalan, D. Gibson, and J. M. Kleinberg. Automatic resource compilation by analyzing hyperlink structure and associated text. In *Proc. of the ACM Conference on World Wide Web*, pages 65–74, 1998.

[CFN07] P.-A. Chirita, C. S. Firan, and W. Nejdl. Personalized query expansion for the web. In *Proc. ACM Conference on Research and Development in Information Retrieval*, pages 7–14, 2007.

[CJP+08] H. Cao, D. Jiang, J. Pei, Q. He, Z. Liao, E. Chen, and H. Li. Context-aware query suggestion by mining click-through and session data. In *Proc. ACM International Conference on Knowledge Discovery and Data Mining*, pages 875–883. ACM, 2008.

[CR12] C. Carpineto and G. Romano. A survey of automatic query expansion in information retrieval. *ACM Computing Surveys*, 44(1):1, 2012.

[CWNM03] H. Cui, J.-R. Wen, J.-Y. Nie, and W.-Y. Ma. Query expansion by mining user logs. *IEEE Knowledge and Data Engineering*, 15(4):829–839, 2003.

[DDL+90] S. C. Deerwester, S. T. Dumais, T. K. Landauer, G. W. Furnas, and R. A. Harshman. Indexing by latent semantic analysis. *Journal of the American Society for Information Science*, 41(6):391–407, 1990.

[DHLK12] H. Deng, J. Han, M. R. Lyu, and I. King. Modeling and exploiting hetero-geneous bibliographic networks for expertise ranking. In *Proc. International Joint Conference on Digital Libraries*, pages 71–80, 2012.

[DLC11] D. A. Davis, R. Lichtenwalter, and N. V. Chawla. Multi-relational link predic-tion in heterogeneous information networks. In *Proc. International Conference on Advances in Social Networks Analysis and Mining*, pages 281–288, 2011.

[DLC+15] L. Du, C. Li, H. Chen, L. Tan, and Y. Zhang. Probabilistic SimRank compu-tation over uncertain graphs. *Information Sciences*, 295:521–535, 2015.

[DSW07] Z. Dou, R. Song, and J.-R. Wen. A large-scale evaluation and analysis of personalized search strategies. In *Proc. ACM Conference on World Wide Web*, pages 581–590, 2007.

[EM03] N. Eiron and K. S. McCurley. Analysis of anchor text for web search. In *Proc. ACM Conference on Research and Development in Information Re-trieval*, pages 459–460, 2003.

[ER90] L. Egghe and R. Rousseau. *Introduction to Informetrics*. Elsevier Science Publishers. Amsterdam, The Netherlands, 1990.

[Fel98] C. Fellbaum, editor. *WordNet: An Electronic Lexical Database*. Cambridge, MA: MIT Press, 1998.

[FKS03] R. Fagin, R. Kumar, and D. Sivakumar. Comparing Top k Lists. *SIAM Journal on Discrete Mathematics*, 17(1):134–160, 2003.

[FR05] D. Fogaras and B. Racz. Scaling link-based similarity search. In *Proc. ACM Conference on World Wide Web*, pages 641–650, 2005.

[Fre77] L. C. Freeman. A set of measures of centrality based on betweenness. *Sociom-etry*, 40:35–41, 1977.

[Fre79] L. C. Freeman. Centrality in social networks conceptual clarification. *Social Networks*, 1(3):215–239, 1979.

[GGMP04] Z. Gyöngyi, H. Garcia-Molina, and J. O. Pedersen. Combating Web Spam with TrustRank. In *Proc. International Conference on Very Large Data Bases*, pages 576–587, 2004.

[GMG11] S. D. Gollapalli, P. Mitra, and C. L. Giles. Ranking authors in digital libraries. In *Proc. International Joint Conference on Digital Libraries*, pages 251–254, 2011.

[HCO03] C.-K. Huang, L.-F. Chien, and Y.-J. Oyang. Relevant term suggestion in in-teractive web search based on contextual information in query session logs. *Journal of the American Society for Information Science and Technology*, 54(7):638–649, 2003.

[HE09] J. Huang and E. N. Efthimiadis. Analyzing and evaluating query reformulation strategies in web search logs. In *Proc. ACM Conference on Information and Knowledge Management*, pages 77–86, 2009.

[Hot33] H. Hotelling. Analysis of a complex of statistical variables into principal com-ponents. *Journal of Educational Psychology*, 24:417–441, 1933.

[HSG10]   S. E. Helou, C. Salzmann, and D. Gillet. The 3A Personalized, Contextual and Relation-Based Recommender System. *J. UCS*, 16(16):2179–2195, 2010.

[IT14]   R. Interdonato and A. Tagarelli. Multi-Relational PageRank for Tree Structure Sense Ranking. *World Wide Web Journal*, 2014.

[JK02]   K. Järvelin and J. Kekäläinen. Cumulated gain-based evaluation of IR techniques. *ACM Tsansactions on Information Systems*, 20(4):422–446, 2002.

[Jon71]   K. S. Jones. *Automatic keyword classification for information retrieval*. Archon Books, 1971.

[Jon72]   K. S. Jones. A statistical interpretation of term specificity and its application in retrieval. *Journal of Documentation*, 28:11–21, 1972.

[JRMG06]   R. Jones, B. Rey, O. Madani, and W. Greiner. Generating query substitutions. In *Proc. ACM Conference on World Wide Web*, pages 387–396, 2006.

[JW02]   G. Jeh and J. Widom. SimRank: a measure of structural-context similarity. In *Proc. ACM International Conference on Knowledge Discovery and Data Mining*, pages 538–543, 2002.

[Kat53]   L. Katz. A new status index derived from sociometric analysis. *Psychometrika*, 18(1):39–43, 1953.

[KB01]   D. Kelly and N. J. Belkin. Reading time, scrolling and interaction: exploring implicit sources of user preferences for relevance feedback. In *Proc. ACM Conference on Research and Development in Information Retrieval*, pages 408–409, 2001.

[Kle98]   J. M. Kleinberg. Authoritative sources in a hyperlinked environment. In *Proc. of the ACM-SIAM Symposium on Discrete Algorithms*, pages 668–677, 1998.

[Kle99]   J. M. Kleinberg. Authoritative sources in a hyperlinked environment. *Journal of ACM*, 46(5):604–632, 1999.

[KZ04]   R. Kraft and J. Zien. Mining anchor text for query refinement. In *Proc. ACM Conference on World Wide Web*, pages 666–674, 2004.

[Lar96]   R. R. Larson. Bibliometrics of the world wide web: An exploratory analysis of the intellectual structure of cyberspace. In *Proc. Annual Meeting American Society for Information Science*, volume 33, pages 71–78, 1996.

[LC10]   N. Lao and W. W. Cohen. Relational retrieval using a combination of path-constrained random walks. *Machine Learning*, 81(1):53–67, 2010.

[Liu09]   T.-Y. Liu. Learning to rank for information retrieval. *Foundations and Trends in Information Retrieval*, 3(3):225–331, 2009.

[LK07]   D. Liben-Nowell and J. M. Kleinberg. The link-prediction problem for social networks. *Journal of the American Society for Information Science and Technology*, 58(7):1019–1031, 2007.

[LM00]   R. Lempel and S. Moran. The stochastic approach for link-structure analysis (SALSA) and the TKC effect. *Computer Networks*, 33(1-6):387–401, 2000.

[LM05]   A. N. Langville and C. D. Meyer. Deeper inside PageRank. *Internet Mathematics*, 1(3):335–400, 2005.

[LSK$^+$11] S. Lee, S. Song, M. Kahng, D. Lee, and S. Lee. Random walk based entity ranking on graph for multidimensional recommendation. In *Proc. ACM Conference on Recommender Systems*, pages 93–100, 2011.

[LVGT08]   D. Lizorkin, P. Velikhov, M. N. Grinev, and D. Turdakov. Accuracy estimate and optimization techniques for SimRank computation. *Proc. of the Very Large Data Bases Endowment*, 1(1):422–433, 2008.

[LY09]   W.-J. Li and D.-Y. Yeung. Relation regularized matrix factorization. In *Proc. International Joint Conference on Artificial Intelligence*, pages 1126–1131, 2009.

[McB94]   O. A. McBryan. GENVL and WWWW: Tools for taming the web. In *Proc. ACM Conference on World Wide Web*, volume 341. Geneva, 1994.

[Mey00]   C. D. Meyer. *Matrix Analysis and Applied Linear Algebra*. SIAM, 2000.

[MRS08]   C. D. Manning, P. Raghavan, and H. Schütze. *Introduction to information retrieval*. Cambridge University Press, 2008.

[MYKL08]   H. Ma, H. Yang, I. King, and M. R. Lyu. Learning latent semantic relations from clickthrough data for query suggestion. In *Proc. ACM Conference on Information and Knowledge Management*, pages 709–718, 2008.

[NG04]   M. E. J. Newman and M. Girvan. Finding and evaluating community structure in networks. *Physical Review E*, 69, 026113, 2004.

[NZWM05]   Z. Nie, Y. Zhang, J.-R. Wen, and W.-Y. Ma. Object-level ranking: bringing order to Web objects. In *Proc. ACM Conference on World Wide Web*, pages 567–574, 2005.

[PE97]   M. Perkowitz and O. Etzioni. Adaptive web sites: an AI challenge. In *Proc. International Joint Conference on Artificial Intelligence*, 1997.

[PE99]   M. Perkowitz and O. Etzioni. Towards adaptive web sites: conceptual framework and case study. In *Proc. ACM Conference on World Wide Web*, 1999.

[PP97]   J. Pitkow and P. Pirolli. Life, death and lawfulness on the electronic frontier. In *Proc. ACM SIGCHI Conference on Human Factors in Computing Systems*, 1997.

[PPR96]   P. Pirolli, J. Pitkow, and R. Rao. Silk from a sow's ear: Extracting usable structures from the web. In *Proc. ACM SIGCHI Conference on Human Factors in Computing Systems*, pages 118–125, 1996.

[Pre02]   L. Pretto. A theoretical analysis of PageRank. In *Proc. of the International Symposium on String Processing and Information Retrieval*, pages 131–144, 2002.

[Roc71]   J. J. Rocchio. *Relevance feedback in information retrieval*. Prentice-Hall, Englewood Cliffs NJ, 1971.

[RU11]   A. Rajaraman and J. D. Ullman. *Mining of Massive Datasets*. Cambridge University Press, 2011.

[RVT+07]   S. Riezler, A. Vasserman, I. Tsochantaridis, V. Mittal, and Y. Liu. Statistical machine translation for query expansion in answer retrieval. In *Proc. Annual Meeting of the Association for Computational Linguistics*, volume 45, page 464, 2007.

[Sab66]   G. Sabidussi. The centrality index of a graph. *Psychometrika*, 31:581–603, 1966.

[SB88]   G. Salton and C. Buckley. Term-weighting approaches in automatic text retrieval. *Information Processing and Management*, 24(5):513–523, 1988.

[SB97]   G. Salton and C. Buckley. Improving retrieval performance by relevance feedback. *Readings in information retrieval*, 24(5), 1997.

[See49]   J. R. Seeley. The net of reciprocal influence: a problem in treating sociometric data. *Canadian Journal of Psychology*, 3:234–240, 1949.

[SH12]   Y. Sun and J. Han. *Mining Heterogeneous Information Networks: Principles and Methodologies*. Synthesis Lectures on Data Mining and Knowledge Discovery. Morgan & Claypool Publishers, 2012.

[SHY+11]   Y. Sun, J. Han, X. Yan, P. S. Yu, and T. Wu. PathSim: Meta Path-Based Top-K Similarity Search in Heterogeneous Information Networks. *Proceedings of the VLDB Endowment*, 4(11):992–1003, 2011.

[SHZ+09]   Y. Sun, J. Han, P. Zhao, Z. Yin, H. Cheng, and T. Wu. RankClus: integrating clustering with ranking for heterogeneous information network analysis. In *Proc. International Conference on Extending Database Technology*, pages 565–576, 2009.

[SYH09]   Y. Sun, Y. Yu, and J. Han. Ranking-based clustering of heterogeneous information networks with star network schema. In *Proc. ACM International Conference on Knowledge Discovery and Data Mining*, pages 797–806, 2009.

[SZ89]   K. Stephenson and M. Zelen. Rethinking centrality: Methods and applications. *Social Networks*, 11:1–37, 1989.

[TSK06]   P. N. Tan, M. Steinbach, and V. Kumar. *Introduction to Data Mining*. Pearson Addison-Wesley Boston, 2006.

[VNL+11]   V. Vasuki, N. Natarajan, Z. Lu, B. Savas, and I. S. Dhillon. Scalable Affiliation Recommendation using Auxiliary Networks. *ACM Tsansactions on Intelligent Systems and Technology*, 3(1):3, 2011.

[WF94]   S. Wasserman and K. Faust. *Social Network Analysis: Methods and Applications*. Cambridge University Press, 1994.

[WLJH10]   J. Weng, E. P. Lim, J. Jiang, and Q. He. TwitterRank: Finding Topic-Sensitive Influential Twitterers. In *Proc. ACM Conference on Web Search and Web Data Mining*, pages 261–270, 2010.

[WVS96]   R. Weiss, B. Vélez, and M. A. Sheldon. Hypursuit: A hierarchical network search engine that exploits content-link hypertext clustering. In *Proc. ACM Conference on Hypertext*, pages 180–193, 1996.

[WWWB12] J. Weston, C. Wang, R. Weiss, and A. Berenzweig. Latent collaborative retrieval. *arXiv preprint arXiv:1206.4603*, 2012.

[WXLC11] Q. Wang, J. Xu, H. Li, and N. Craswell. Regularized latent semantic index-ing. In *Proc. ACM Conference on Research and Development in Information Retrieval*, pages 685–694, 2011.

[XC96] J. Xu and W. B. Croft. Query expansion using local and global document anal-ysis. In *Proc. ACM Conference on Research and Development in Information Retrieval*, pages 4–11, 1996.

[XJW09] Y. Xu, G. J. F. Jones, and B. Wang. Query dependent pseudo-relevance feedback based on Wikipedia. In *Proc. ACM Conference on Research and Development in Information Retrieval*, pages 59–66, 2009.

[XZC+04] G.-R. Xue, H.-J. Zeng, Z. Chen, Y. Yu, W.-Y. Ma, W. S. Xi, and W. G. Fan. Optimizing web search using web click-through data. In *Proc. ACM Conference on Information and Knowledge Management*, pages 118–126, 2004.

[YLZ13] W. Yu, X. Lin, and W. Zhang. Towards efficient SimRank computation on large networks. In *Proc. International Conference on Data Engineering*, pages 601–612, 2013.

[YSC09] Z. Yin, M. Shokouhi, and N. Craswell. Query expansion using external evi-dence. In *Advances in Information Retrieval*, pages 362–374. Springer, 2009.

[ZE98] O. Zamir and O. Etzioni. Web document clustering: A feasibility demonstra-tion. In *Proc. ACM Conference on Research and Development in Information Retrieval*, pages 46–54, 1998.

[ZFT+11] M. Zhang, S. Feng, J. Tang, B. A. Ojokoh, and G. Liu. Co-Ranking Multiple Entities in a Heterogeneous Network: Integrating Temporal Factor and Users' Bookmarks. In *Proc. International Conference on Asian Digital Libraries*, pages 202–211, 2011.

[ZHMM04] H.-J. Zeng, Q.-C. He, Z. C. W.-Y. Ma, and J. Ma. Learning to cluster web search results. In *Proc. ACM Conference on Research and Development in Information Retrieval*, pages 210–217, 2004.

Chapter 5

Label Propagation and Information Diffusion in Graphs

Eftychia Fotiadou, Olga Zoidi and Ioannis Pitas

Aristotle University of Thessaloniki, Greece

5.1   Introduction

The present day is characterized by the ability to directly access a huge volume of any kind of information, especially through the Internet. Furthermore, an expansion of the on-line multimedia sharing communities and social network services, which enable users to interact with each other, as well as to upload, create and share multimedia content, such as images, videos and audio, has been observed in the last few years. This has led to an enormous increase in the volume of the on-line available multimedia data. The effective handling of information at such scales requires the development of special algorithms.

Multimedia objects, e.g. videos, images, and music, can be characterized (labeled) by labels (tags) that are assigned to them and describe their semantic content. This process, referred to as *annotation* (or *labeling*), is necessary for semantic search in multimedia or social media databases. In large data collections, such as the on-line multimedia sharing websites and social networks, the tags are usually user-contributed. However, manually annotating multimedia objects becomes infeasible as the volume of the data grows. Additionally, in many cases, users may provide incorrect or incomplete tags. Furthermore, expert domain knowledge may be required for assigning proper tags. For the aforementioned reasons, semi-automatic labeling of a large number of multimedia data items is of great importance.

To this end, one solution can be *label propagation*, whose objective is to disseminate the labels of a small set of annotated data to a larger set of unlabeled data. Label propagation algorithms make two common assumptions: first, the labels of the initially labeled data should remain unchanged. Second, data samples that are similar (or "close") to each other or lie in the same structure (e.g., cluster or manifold) should be assigned the same label. In order to describe the label propagation process, a graph-based approach is usually adopted. Graph nodes correspond to data samples, while graph edges reflect their relationships. For example, edge weights may represent pairwise similarities (or distances) between the data samples. Label inference can be subsequently performed through graph paths that connect labeled to unlabeled nodes. The graph construction method, namely the adopted pairwise similarity/distance function, is crucial to the label propagation performance, since it regulates the way the labels are spread through the graph paths. Figure 5.1.1a depicts the graph of a recommendation network, which models how the recommendation of an idea or a product is propagated through a network of individuals, represented by graph nodes. The graph of Figure 5.1.1b represents a network of "related" videos from the YouTube video-sharing website, where each video is related to other relevant videos. In this case, graph nodes correspond to videos, while graph edges connect a video with its related videos. Label propagation algorithms belong to the more general class of *semi-supervised classifiers*, since they exploit information from unlabeled data during the learning process, instead of relying merely on labeled data for training, as is the case in *supervised learning*. As a result, semi-supervised classifiers can provide a better understanding of the data structure and lead to improved classification performance in comparison to supervised classifiers. *Transductive* semi-supervised classifiers learn a local representation of the data feature space and can be applied in a specific dataset consisting of labeled and unlabeled data, while *inductive classifiers* learn a global representation of the data feature space and can be applied to "unseen" data that were not included in the original dataset. The majority of label propagation algorithms are transductive classifiers.

Label propagation can be thought of as an information (label) diffusion process over the data graph. In general, information diffusion studies how ideas are spread through a network, usually a social system. While label propagation is a classification problem, information diffusion in social networks focuses on different issues, such as the mechanisms

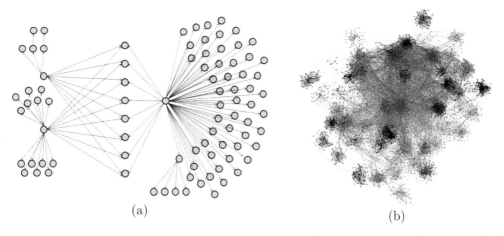

FIGURE 5.1.1: a) Recommendation network graph, b) YouTube video relation graph. (**See color insert.**)

governing the diffusion process, the influence of the network structure on the diffusion, or the maximization of the information spread.

In order to study the information diffusion within a social network, various types of models can be adopted, such as *epidemic, game theoretical, threshold* and *cascade* models. Information diffusion methods are commonly used in *viral marketing* applications or, in general, when maximizing the information spread is desirable. They are also employed in collaborative filtering systems, in community detection algorithms, as well as in the study of citation networks. Finally, information diffusion methods find application in emergency management, where providing the public with useful and valid information is crucial. Apart from studying information flow in social networks, label propagation methods find numerous applications in multimedia content annotation [TYH+09], medical imaging [HHA+06], biology [LK03, WLI+05, HK10] and language analysis [NJT05, YJZ+06, ZK09, RR09, SSUB11].

The current chapter focuses on the study of graph-based label propagation algorithms, also referred to as *graph-based semi-supervised learning* (SSL). Furthermore, diffusion concepts, with a focus on social networks, are discussed. The label propagation procedure consists of two steps: a) the construction of a graph representing the data and the relationships between them and b) label inference, based on the constructed graph. Graph construction methods suitable for label propagation, are discussed in Section 5.2, while label inference algorithms are presented in Section 5.3. In Section 5.4, the notions of diffusion in physics as well as in social networks are briefly reviewed. Moreover, in Section 5.5, models for diffusion in social networks, as well as the problem of influence maximization are discussed.

5.2 Graph construction approaches

In graph-based label propagation, the labels are diffused through a graph, where nodes correspond to entities, such as multimedia objects, while the edge weights represent their similarity. The *similarity graph* should reflect the relationships between the entities being labeled, with respect to the specific label propagation task. For this reason, the construction of the similarity graph is critical to label propagation performance. In the following

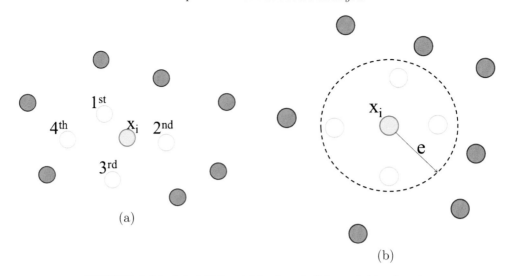

FIGURE 5.2.1: (a) 4-NN neighborhood, (b) e-neighborhood.

subsections, various graph construction methods are discussed. These methods are divided into three categories: neighborhood, local reconstruction and metric learning ones.

5.2.1   Neighborhood approaches

According to *neighborhood methods*, the graph is constructed by connecting each node to its closest ones, where "closeness" between nodes is determined by a *distance* or *similarity function*. Neighborhood methods construct sparse graphs. Two often utilized neighborhood graphs are the k-Nearest Neighbor (k-NN) and the e-neighborhood graphs [Tal09], which are illustrated in Figure 5.2.1.

In k-*NN graphs*, each node is connected to its k nearest neighbors. k-NN graphs are asymmetric, since the participation of a node j to the set of i-th node closest neighbors does not guarantee the participation of node i to the set of closest neighbors of node j. Furthermore, graph construction with the k-NN method produces irregular graphs, since certain nodes end up with a higher degree than others. Irregular graphs may affect label propagation negatively, leading to degenerate solutions, as most nodes may be assigned the same label to those of the high degree nodes. In several label propagation methods, [BSS$^+$08, TC09], this problem is dealt with by discounting the importance of high degree nodes.

In e-*neighborhood graphs*, a node is connected to all the nodes within a predefined distance $e > 0$. Such a graph construction is quite sensitive to parameter e selection. Additionally, it often leads to graphs having disconnected components. For the above reasons, the k-NN method shows advantages over the e-neighborhood one and is usually preferred in practice.

As already mentioned, with respect to the label propagation task, regular graphs are preferred. As opposed to k-NN and e-neighborhood graphs, the b-matching based method proposed in [JWC09] guarantees resulting graph regularity. The b-matching method involves two steps:

- *Graph sparsification*: during this step, starting with a complete graph, the edges are selected that will be present in the final, sparse graph.

- *Edge re-weighting*: weights are learned for the edges that were selected in the previous step.

The sparsification step generates a binary matrix $\mathbf{P} \in \{0,1\}^{N \times N}$, whose entries P_{ij} determine whether an edge will be present between data samples $\mathbf{x}_i$ and $\mathbf{x}_j$ in the final matrix ($P_{ij} = 1$) or not ($P_{ij} = 0$). The calculation of matrix $\mathbf{P}$ is performed by minimizing the following objective function:

$$\min_{\mathbf{P}} \sum_{ij} P_{ij} D_{ij}, \tag{5.2.1}$$

subject to (s.t. in short):

$$\sum_{j} P_{ij} = b, \quad P_{ii} = 0, \quad P_{ij} = P_{ji}, \quad \forall i, j = 1, \dots, N. \tag{5.2.2}$$

The distance matrix $\mathbf{D}$ is calculated from $D_{ij} = \sqrt{W_{ii} + W_{jj} - 2W_{ij}}$, where $\mathbf{W}$ denotes the weight matrix. During edge re-weighting, the weights for the selected edges are learned, using three different weighting schemes: a) binary weights ($\mathbf{W} = \mathbf{P}$), b) *Gaussian kernel* weights:

$$W_{ij} = P_{ij} \exp\left(-\frac{d(\mathbf{x}_i, \mathbf{x}_j)}{2\sigma^2}\right), \tag{5.2.3}$$

where $d(\mathbf{x}_i, \mathbf{x}_j)$ is some distance function between the node feature vectors $\mathbf{x}_i$, $\mathbf{x}_j$ and σ is the kernel bandwidth or, c) Locally Linear Reconstruction (LLR), motivated by the *Locally Linear Embedding* (LLE) algorithm [RS00], which seeks the coefficients that minimize the reconstruction error:

$$\min_{\mathbf{W}} \sum_{i} \|\mathbf{x}_i - \sum_{j} P_{ij} W_{ij} \mathbf{x}_j\|^2, \tag{5.2.4}$$

$$\text{s.t.} \sum_{j} W_{ij} = 1, W_{ij} \geq 0. \tag{5.2.5}$$

5.2.2 Local reconstruction approaches

In contrast to the aforementioned methods, that utilize pairwise distances between nodes to construct the similarity graph, local reconstruction methods take neighborhood information into consideration, by expressing each node as a combination of its neighboring nodes. In the *Linear Neighborhood Propagation* (LNP) method introduced in [WZ06], the neighborhood of each node is regarded as a linear patch and each node is reconstructed by a linear combination of its k nearest neighbors. The whole graph is, therefore, approximated by a number of overlapping linear patches. The edge weights in each patch are determined using a quadratic programming solver. The objective of the LNP method is to minimize the reconstruction error of each node by its k nearest neighbors:

$$\min_{\mathbf{W}} \sum_{i} \|\mathbf{x}_i - \sum_{j:\mathbf{x}_j \in \mathcal{N}(\mathbf{x}_i)} W_{ij} \mathbf{x}_j\|^2, \tag{5.2.6}$$

$$\text{s.t.} \sum_{j} W_{ij} = 1, \quad W_{ij} \geq 0, \tag{5.2.7}$$

where $\mathcal{N}(\mathbf{x}_i)$ denotes the neighborhood of node i, while W_{ij} expresses the contribution of $\mathbf{x}_j$ to the reconstruction of $\mathbf{x}_i$. The values of the reconstruction weights W_{ij} can be regarded as a similarity measure between nodes i and j. The reconstruction error of a node i can take the form:

$$
\begin{aligned}
\epsilon_i &= \|\mathbf{x}_i - \sum_{j:\mathbf{x}_j \in \mathcal{N}(\mathbf{x}_i)} W_{ij}\mathbf{x}_j\|^2 \\
&= \|\sum_{j:\mathbf{x}_j \in \mathcal{N}(\mathbf{x}_i)} W_{ij}(\mathbf{x}_i - \mathbf{x}_j)\|^2 \\
&= \sum_{j,k:\mathbf{x}_j,\mathbf{x}_k \in \mathcal{N}(\mathbf{x}_i)} W_{ij}W_{ik}(\mathbf{x}_i - \mathbf{x}_j)^T(\mathbf{x}_i - \mathbf{x}_k) \\
&= \sum_{j,k:\mathbf{x}_j,\mathbf{x}_k \in \mathcal{N}(\mathbf{x}_i)} W_{ij}G^i_{jk}W_{ik},
\end{aligned}
\tag{5.2.8}
$$

where $G^i_{jk} = (\mathbf{x}_i - \mathbf{x}_j)^T(\mathbf{x}_i - \mathbf{x}_k)$ denotes the (j,k)-th entry of the local *Gram matrix* $\mathbf{G}^i$ of point $\mathbf{x}_i$. The Gram matrix contains all the possible inner products between data samples. Therefore, the reconstruction weights for each node can be calculated by solving a quadratic programming problem (one for each node) of the form:

$$
\min_{W_{ij}} \sum_{j,k:\mathbf{x}_j,\mathbf{x}_k \in \mathcal{N}(\mathbf{x}_i)} W_{ij}G^i_{jk}W_{ik},
\tag{5.2.9}
$$

$$
\text{s.t.} \sum_j W_{ij} = 1, \quad W_{ij} \geq 0.
\tag{5.2.10}
$$

Once the reconstruction weights are computed, a sparse weight matrix $\mathbf{W}$ is constructed, corresponding to a sparse graph, where each node is connected only to its k nearest neighbors. In [THQ$^+$08], a *Kernel Linear Neighborhood Propagation* (KLNP) method is proposed, which is a non-linear extension of the LNP method. Using a kernel mapping of the form $\phi : \mathbf{X} \to \mathbf{\Phi}$, the data are mapped into a high dimensional space and the objective function in this case is expressed as:

$$
\min_{W_{ij}} \|\phi(\mathbf{x}_i) - \sum_{j:\phi(\mathbf{x}_j) \in \mathcal{N}(\phi(\mathbf{x}_i))} W_{ij}\phi(\mathbf{x}_j)\|^2,
\tag{5.2.11}
$$

$$
\tag{5.2.12}
$$

$$
\text{s.t.} \sum_j W_{ij} = 1, \quad W_{ij} \geq 0.
\tag{5.2.13}
$$

Another graph construction method based on LNP is the *Correlative Linear Neighborhood Propagation* (CLNP) method [THW$^+$09], which addresses the problem of video annotation. The CLNP method incorporates the semantic correlations among the data labels during graph construction.

In [DKS09], a hard and an *a-soft* method for graph construction are proposed. *Hard graphs* are defined as those where each node has a weighted degree d_i of at least 1, with $d_i = \sum_j W_{ij}$. The hard graph construction method introduced in [DKS09] minimizes the objective function:

$$
\min_{\mathbf{W}} \sum_i \|d_i\mathbf{x}_i - \sum_j W_{ij}\mathbf{x}_j\|^2.
\tag{5.2.14}
$$

The *a-soft graph* construction allows the weighted degrees of some outlier nodes to take a lower value. This is achieved by relaxing the constraint on the weighted degree d, which is now expressed by:

$$
\sum_i (\max(0, 1 - d_i))^2 \leq aN,
\tag{5.2.15}
$$

where a is a hyper-parameter.

In [TYH+09], a graph construction method is introduced as part of an image annotation framework. The proposed method, which is motivated by the study [WYG+09] is robust to noise. More importantly, it constrains the images used to reconstruct a sample image to be semantically similar to the sample image. Specifically, an one-vs-all sparse reconstruction of each sample, based on the l_1-norm minimization is proposed. Each sample $\mathbf{x}_i$ is reconstructed by solving the following minimization problem:

$$\min_{\mathbf{w}_i} \|\mathbf{w}_i\|_1, \tag{5.2.16}$$

$$\text{s.t.} \quad \mathbf{x}_i = \mathbf{B}_i \mathbf{w}_i, \tag{5.2.17}$$

where $\mathbf{B}_i$ is the matrix formed by all samples except $\mathbf{x}_i$ and $\mathbf{w}_i$ is the vector of the reconstruction coefficients. Subsequently, the edge weight from sample j to sample i is determined by:

$$w_{ij} = \begin{cases} \mathbf{w}_i(j), & \text{if } j < i \\ \mathbf{w}_i(j-1), & \text{if } j > i \\ 0, & \text{if } j = i, \end{cases} \tag{5.2.18}$$

where $\mathbf{w}_i(j)$ is the j-th element of vector $\mathbf{w}_i$. The aforementioned process removes a significant number of edges between nodes that are semantically unrelated, rendering the resulting graph more efficient for label propagation. In a later work [THY+11], a one-vs-kNN sparse graph construction approach is proposed, where each sample is reconstructed from its k nearest neighbors, instead of using all of the remaining dataset samples. The reconstruction is calculated as in equation (5.2.16), with the difference that, only the k nearest neighbors of sample i, denoted by i_p, $p \in \{1, 2, ..., k\}$, are used in the matrix $\mathbf{B}_i$. The weights are selected using the following rule:

$$w_{ij} = \begin{cases} \mathbf{w}_i(p), & \text{if } \mathbf{x}_j \in \mathcal{N}(\mathbf{x}_i) \text{ and } j = i_p \\ 0, & \text{if } \mathbf{x}_j \notin \mathcal{N}(\mathbf{x}_i), \end{cases} \tag{5.2.19}$$

where $\mathbf{w}_i(p)$ is the p-th element of vector $\mathbf{w}_i$.

The method introduced in [WWZ+09] extends the NLP algorithm to hypergraphs, where hyperedges connect two or more nodes, thus simultaneously representing multiple relationships between nodes. Given the graph $G = (\mathcal{V}, \mathcal{E})$ with nodes $\mathcal{V}$ and edges $\mathcal{E}$, the hypergraph $\mathcal{G}' = (\mathcal{V}, \mathcal{E}')$ is constructed, where $\mathcal{V}$ is the set of nodes of graph G and $\mathcal{E}'$ is the set of hyperedges. If graph G is cast into a first-order Intrinsic Gaussian Markov Random Field (IGMRF) framework [RH05], then the hypergraph $\mathcal{G}'$ can be cast into a second-order IGMRF framework, where the increment for hyperedge e'_i is calculated by:

$$d_i = y_i - \sum_{j \in \mathcal{N}_i} W_{ij} y_j, \quad \sum_{j \in \mathcal{N}_i} W_{ij} = 1, \tag{5.2.20}$$

where y_i is the label of node i and $\mathcal{N}_i$ is the set of neighboring nodes to node i. The hyperedges' weights W_{ij} are calculated by solving the quadratic problem expressed by (5.2.6).

5.2.3 Metric learning approaches

Graph construction methods involve the selection of a distance metric, which is used to estimate the similarity between samples (nodes). The choice of the distance metric strongly affects the resulting graph and, consequently, the performance of the label propagation process. When no further information about the samples is available, graph construction algorithms usually employ the Euclidean distance to calculate the distances between nodes.

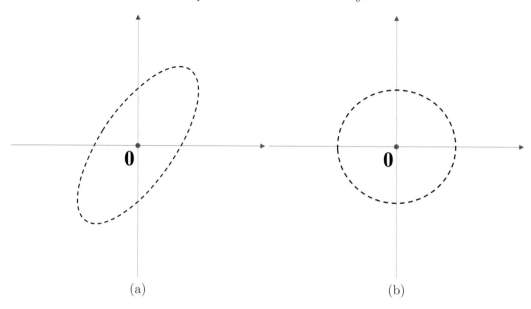

(a)   (b)

FIGURE 5.2.2: Equidistant points from the origin $\mathbf{0}$ for (a) Mahalanobis distance ($\mathbf{A} \neq \mathbf{I}$) and (b) Euclidean distance ($\mathbf{A} = \mathbf{I}$).

In this case, the metric is not able to capture the underlying structure (e.g., data clusters or manifolds) that may exist in the data. On the other hand, if prior knowledge regarding the similarity between data samples is available (labeled data or clustering information for example), it can be incorporated in the distance metric. In general, metric learning algorithms calculate a *Mahalanobis distance* between the samples, expressed by:

$$d_A(\mathbf{x}_i, \mathbf{x}_j) = \sqrt{(\mathbf{x}_i - \mathbf{x}_j)^T \mathbf{A}(\mathbf{x}_i - \mathbf{x}_j)}, \tag{5.2.21}$$

where $\mathbf{A} \in \mathbb{R}^{N \times N}$ is a positive semi-definite matrix, that incorporates the constraints derived from the prior knowledge on the data. By setting the matrix $\mathbf{A}$ equal to the identity matrix $\mathbf{I}_N \in \mathbb{R}^{N \times N}$, the distance expressed in equation (5.2.21) coincides with the *Euclidean distance*, as illustrated in Figure 5.2.2. The constraints imposed by matrix $\mathbf{A}$ may refer to pairs of similar samples (e.g., when the distance between two similar samples must be smaller than a threshold) or to dissimilar ones (e.g., when the distance between dissimilar samples must be greater than a threshold). The graph weights W_{ij} are subsequently calculated according to a function of the form:

$$W_{ij} \propto \exp\{-d_A(\mathbf{x}_i, \mathbf{x}_j)\}. \tag{5.2.22}$$

The objective of metric learning algorithms is to minimize a cost function $f(\mathbf{A})$ subject to a set of constraints.

The method introduced in [XNJR02] minimizes the squared sum of distances between similar samples, under the constraint that the sum of distances between dissimilar data samples is greater than a threshold. In [SSSN04], an on-line distance metric learning algorithm is proposed, where each sample is constrained to be closer to all samples with the same label than to any sample with a different label.

Information-Theoretic Metric Learning (ITML) [DKJ+07] assumes that prior knowledge of the distances between the data samples is available, denoted by the Mahalanobis distance parameter matrix $\mathbf{A}_0$. It considers similarity and dissimilarity constraints between pairs of data samples, such that, two data samples $\mathbf{x}_i, \mathbf{x}_j$ are similar, if their Mahalanobis distance is below an upper limit $d_A(\mathbf{x}_i, \mathbf{x}_j) \leq u$ and dissimilar, if it exceeds a lower limit $d_A(\mathbf{x}_i, \mathbf{x}_j) \geq l$. The objective is to find a Mahalanobis distance function parametrized by the matrix $\mathbf{A}$, which has to be as close as possible to the prior Mahalanobis distance function, parameterized by $\mathbf{A}_0$.

To measure the distance between two Mahalanobis functions, an information-theoretic approach can be followed. There exists a one-to-one correspondence between the two Mahalanobis distance functions and the set of equal-mean multivariate Gaussian distributions. Taking this bijection into account, the distance between two Mahalanobis distance functions parametrized by $\mathbf{A}$ and $\mathbf{A}_0$ can be expressed as the *Kullback-Leibler* (KL) *divergence* between the corresponding Gaussians $p(\mathbf{x}; \mathbf{A})$ and $p(\mathbf{x}; \mathbf{A}_0)$:

$$KL(p(\mathbf{x}; \mathbf{A}_0) || p(\mathbf{x}; \mathbf{A})) = \int p(\mathbf{x}; \mathbf{A}_0) \log \frac{p(\mathbf{x}; \mathbf{A}_0)}{p(\mathbf{x}; \mathbf{A})} d\mathbf{x}. \qquad (5.2.23)$$

The aforementioned relative entropy is shown to be equivalent to the *Log-Determinant* (Log-Det) *divergence* between the matrices $\mathbf{A}$ and $\mathbf{A}_0$, which is given by:

$$D_{ld}(\mathbf{A}, \mathbf{A}_0) = \mathrm{tr}(\mathbf{A}\mathbf{A}_0^{-1}) - \log \det(\mathbf{A}\mathbf{A}_0^{-1}) - N. \qquad (5.2.24)$$

The metric learning problem takes the form of the following minimization problem:

$$\begin{aligned}
\min_{\mathbf{A}} KL(p(\mathbf{x}; \mathbf{A}_0) || p(\mathbf{x}; \mathbf{A}) &= \min_{\mathbf{A}} \frac{1}{2} D_{ld}(\mathbf{A}, \mathbf{A}_0), \\
\text{s.t.} \quad d_A(\mathbf{x}_i, \mathbf{x}_j) &\leq u \quad (i, j) \in \mathcal{S} \\
d_A(\mathbf{x}_i, \mathbf{x}_j) &\geq l \quad (i, j) \in \mathcal{D},
\end{aligned} \qquad (5.2.25)$$

where $\mathcal{S}$ and $\mathcal{D}$ denote the sets of similar and dissimilar data sample pairs, respectively. A later study [DD08] deals with the problem of tractability of Mahalanobis distance learning, as data dimensionality grows. Two new algorithms are proposed, based on the Log-Det divergence, which are suitable for learning Mahalanobis distance functions in high dimensions.

In [GRHS04], a learning method called *Neighborhood Component Analysis* (NCA) is developed, under a k-Nearest Neighbor (k-NN) classification perspective. It seeks a linear transformation of the feature space, such that, in the transformed space, the k-NN objective is satisfied, i.e. the k-nearest neighbors belong to the same class. Matrix $\mathbf{A}$ takes the form $\mathbf{A} = \boldsymbol{\Phi}^T \boldsymbol{\Phi}$, where $\boldsymbol{\Phi}$ denotes the transformation matrix. The objective of the NCA method is to maximize the expected number of correctly classified nodes:

$$f(\boldsymbol{\Phi}) = \sum_i \sum_{j \in \mathcal{C}_i} \frac{\exp\left(-\|\boldsymbol{\Phi}\mathbf{x}_i - \boldsymbol{\Phi}\mathbf{x}_j\|_2^2\right)}{\sum_{k \neq i} \exp\left(-\|\boldsymbol{\Phi}\mathbf{x}_i - \boldsymbol{\Phi}\mathbf{x}_k\|_2^2\right)}, \qquad (5.2.26)$$

where $\mathcal{C}_i$ denotes the set of nodes with the same label as node $\mathbf{x}_i$. In [QTZ$^+$09], the problem of high data dimensionality in combination with a small number of available data samples is addressed, referred to often as the "curse of dimensionality." Departing from the observation that the *concentration matrix* $\boldsymbol{\Sigma}^{-1}$ (i.e., the inverse of the covariance matrix $\boldsymbol{\Sigma}$) is often sparse in the high dimensional space, since the correlation among the different dimensions is weak, *Sparse Distance Metric Learning* (SDML) is proposed, which imposes a sparse prior on the off-diagonal elements of the Mahalanobis matrix. The SDML problem is formulated as:

$$\min_{\mathbf{A}} \left\{ \mathrm{tr}(\mathbf{A}_0^{-1}\mathbf{A} + \gamma \mathbf{X}\mathbf{L}\mathbf{X}^T\mathbf{A}) - \log \det(\mathbf{A}) + \lambda \|\mathbf{A}\|_{1,off} \right\}, \qquad (5.2.27)$$

where λ is a trade-off parameter between the sparsity prior and the matrix $\mathbf{A}_0$, γ is a

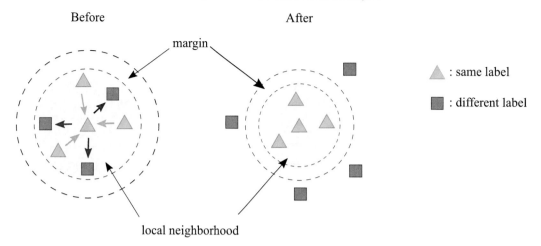

FIGURE 5.2.3: Schematic illustration of LMNN (adapted from [WS09]).

positive trade-off parameter, $\mathbf{L}$ is the graph Laplacian and $||\mathbf{A}||_{1,off} = \sum_{i \neq j} |A_{ij}|$ is the off-diagonal l_1-norm.

Another approach to learning a Mahalanobis distance metric using the k-NN rule is introduced in the *Large Margin Nearest Neighbor* (LMNN) algorithm [WS09]. The objective of the method is to learn a distance metric, such that the k nearest neighbors of a node $\mathbf{x}_i$ are forced to share the same label y_i with $\mathbf{x}_i$, while differently labeled nodes are separated by a large margin. The algorithm, which exhibits parallels to Support Vector Machines (SVMs), seeks for a linear transformation $\mathbf{\Phi}$ of the feature space, which fulfills the aforementioned objective. In more detail, the k-nearest neighbors of a node $\mathbf{x}_i$ with the same label y_i are defined as the *target* nodes. Ideally, these nodes should be closer to $\mathbf{x}_i$ than to any other differently labeled node, thus forming a perimeter around node $\mathbf{x}_i$, which should not be violated by differently labeled nodes. Nodes with a label other than y_i trying to enter this perimeter are called *impostors*. The goal of the LMNN algorithm is to pull the target nodes close to each other, while pushing the differently labeled nodes away from each other. This is achieved by minimizing the following cost function:

$$\epsilon(\mathbf{\Phi}) = (1 - \mu) \sum_{j \rightsquigarrow i} ||\mathbf{\Phi}(\mathbf{x}_i - \mathbf{x}_j)||^2 +$$
$$\mu \sum_{i,j \rightsquigarrow i} \sum_l (1 - y_{il}) \left[1 + ||\mathbf{\Phi}(\mathbf{x}_i - \mathbf{x}_j)||^2 - ||\mathbf{\Phi}(\mathbf{x}_i - \mathbf{x}_l)||^2 \right]_+ , \tag{5.2.28}$$

where the second term $[z]_+ = \max(z, 0)$ is the *hinge loss*. The indicator variable y_{il} is equal to 1, if i and l have the same label and equal to 0, otherwise, while $j \rightsquigarrow i$ indicates that j is a target node of i. The first term of the cost function penalizes large distances between each node and its target nodes, while the second one penalizes small distances between each node and all nodes having different labels. Parameter μ takes values in the interval $[0, 1]$ and controls the trade-off between the two terms in equation (5.2.28). It should be noted that the first term of the cost function penalizes large distances only between a data sample $\mathbf{x}_i$ and its target nodes, and not among all nodes having the same label as $\mathbf{x}_i$. A schematic illustration of LMNN algorithm is depicted in Figure 5.2.3.

Similarly to [GRHS04] and [WS09], a Support Vector Machine approach to metric learning, called *Metric Learning SVM* (MLSVM), was proposed in [NG08].

The distance metric learning method introduced in [BBM04] learns a different metric in each cluster, allowing, thus, different shapes for the clusters. The proposed algorithm, called

Metric Pairwise Constrained K-Means (MPCK-Means), assumes sets of pairwise *must-link* constraints, i.e. if $(\mathbf{x}_i, \mathbf{x}_j) \in \mathcal{S}$, then, data samples $\mathbf{x}_i$ and $\mathbf{x}_j$ must be in the same cluster, and *cannot-link* constraints, i.e. if $(\mathbf{x}_i, \mathbf{x}_j) \in \mathcal{D}$, then, $\mathbf{x}_i$ and $\mathbf{x}_j$ must be in different clusters. An individual weight matrix $\mathbf{A}_c$ is sought in each cluster c, which minimizes the sum of squared Euclidean distances between the samples of the cluster and the cluster centroids μ_c and maximizes the complete data log-likelihood:

$$\arg\min_{\mathbf{A}_c} \left\{ \sum_{\mathbf{x}_i \in \mathcal{X}} \left(d^2_{\mathbf{A}_c}(\mathbf{x}_i, \mu_c) - \log(\det(\mathbf{A}_c)) \right) + \sum_{(\mathbf{x}_i, \mathbf{x}_j) \in \mathcal{S}} w_{ij} \mathbf{1}[l_i \neq l_j] + \right.$$
$$\left. \sum_{(\mathbf{x}_i, \mathbf{x}_j) \in \mathcal{D}} \bar{w}_{ij} \mathbf{1}[l_i = l_j] \right\}, \tag{5.2.29}$$

where $\mathcal{X}$ denotes the whole dataset, $d^2_{\mathbf{A}_c}(\mathbf{x}_i, \mu_c)$ is given by equation (5.2.21), l_i is the cluster assignment (label) for sample i, and $\mathbf{1}$ is the indicator function $\mathbf{1}[true] = 1$ and $\mathbf{1}[false] = 0$. The constraints are incorporated in the aforementioned objective function through weights w_{ij} and $\bar{w}_{ij}$, which express the cost of violating the must-link and cannot-link constraints, respectively.

In [CTLC12], a *Two-Dimensional Smooth Metric Learning* (2DSML) framework suitable for visual analysis is proposed. Departing from the observation that the existing metric learning algorithms consider image pixels as independent and represent images of size $N_1 \times N_2$ pixels with vectors in the $\mathbb{R}^{N_1 \times N_2}$ space, the correlations existing between neighboring pixels in an image are taken into account and an image is treated as a function defined on an $N_1 \times N_2$ grid. The proposed regularization framework incorporates a discretized Laplacian penalty term, which ensures smoothness in two dimensions and incorporates the prior information regarding the spatial correlations of neighboring pixels.

A method for *Multi-Instance Multi-Label metric learning* is proposed in [JSWZHZ09]. Multi-Instance Multi-Label learning refers to problems where each example consists of a collection (or bag) of instances, e.g. key points, and can be assigned multiple labels. For example, in an image annotation application, each image may contain various objects belonging to various classes. The proposed algorithm involves two steps, which are applied iteratively: first, the relationships between the instances in the bags and the class labels assigned to the bags are estimated. Subsequently, a distance measure is learned based on the estimated relationships.

The metric learning algorithms discussed so far are supervised, since they utilize prior knowledge derived from the labeled data. In [DTC10], a new semi-supervised metric learning framework, called *Inference Driven Metric Learning* (IDML), is proposed, which exploits information from both labeled and unlabeled data. IDML combines supervised metric learning algorithms, such as the previously discussed ITML and LMNN, with transductive graph-based label propagation methods. Specifically, graph-based SSL is used to infer labels for the unlabeled samples and, subsequently, samples that exhibited high assigned label confidence during the label inference process are appended to the labeled set in the next iteration of metric learning. The aforementioned steps are executed iteratively, until no new samples can be added to the labeled data set. Another semi-supervised metric learning method is introduced in [HLC08]. The proposed algorithm, called *Laplacian Regularized Metric Learning* (LRML), incorporates information from unlabeled data through the graph Laplacian. Finally, a semi-supervised metric learning algorithm, called *Semi-Supervised Sparse Metric Learning* (S3ML), is introduced in [LMT$^+$10], based on *Sparse Distance Metric Learning* (SDML) [QTZ$^+$09].

In the case that label information is not available for the data, unsupervised metric learning is equivalent to *manifold learning* [LV07] (or dimensionality reduction methods [Fod02]). The objective of these techniques is to learn a low-dimensional manifold from the data, such that the geometric relationships between data samples are preserved [YR06]. Commonly used unsupervised dimensionality reduction methods include Principal Component Analysis (PCA) [Jol02], Multidimensional Scaling (MDS) [BG05], Locally Linear Embedding (LLE) [SR03], the Laplacian Eigenmap [BN03] and ISOMAP [TDSL00]. In [YR06], an extensive study on distance metric learning, as well as its relationship to dimensionality reduction is provided.

5.2.4 Scalable graph construction methods

A drawback common in the graph construction methods reviewed so far is that they are not scalable, i.e., they are not capable of handling large graphs. As the size of the dataset grows, their computational time becomes prohibitive. However, in many real-world applications, and especially in those concerning multimedia data, a large number of samples, e.g., images or videos, is usually involved. In order to address this issue, several methods have been recently proposed, which combine good graph approximations with computational efficiency.

In [ZHGL13], a method for approximate k-NN graph construction is proposed, which is based on a *locality sensitive hashing* (LSH) algorithm [GIM99]. The dataset is divided in smaller subsets of equal size and the k nearest neighbors of a sample are sought for within the subset it belongs to. Ideally, similar samples should be in the same subset, in order for the k nearest neighbors of a sample within the subset to be identical to its real k nearest neighbors. Additionally, the subset should have a small size, to ensure a small number of comparisons during the k-NN search. In order to divide the dataset, the LSH algorithm is adopted. LSH applies a hashing function on the data samples and maps them into buckets (subsets), according to their hash codes, such that similar samples are mapped to the same bucket. In order to improve the approximation performance, the process is repeated multiple times, by applying different hashing functions and the resulting approximate graphs are combined into a final graph. Typical LSH algorithms do not produce equal-sized subsets, which is a drawback in the case of k-NN graph construction. In order to alleviate this problem, a modification of the LSH is introduced, which is based on a linear projection of the data hash codes. The proposed method is fast, accurate and easy to paralellize, while it can be utilized along with any similarity measure.

In [HWLH12], a two-stage graph construction method for large image datasets is introduced. In the first stage, the samples are divided into overlapping subsets, called image pools, using an LSH method. Specifically, the MinHash [Bro97] algorithm is adopted to separate the samples into non-overlapping subsets. In order to enhance the hashing result, MinHash is applied repeatedly on the data, producing different divisions (hashing tables). While the subsets in the same hashing table do not overlap, overlapping may occur between subsets from different hashing tables. Image pools are subsequently formed by joining the overlapping subsets from different hashing tables. In the second stage of the method, the pair-wise similarities between samples belonging to the same pool are calculated and the graph is constructed. In order to handle the computational cost when dealing with large datasets, both steps are performed using the MapReduce parallel programming model [DG08]. Another LSH-based approach for large scale graph construction can be found in [CMYC10], where LSH is combined with l_1-graph construction in each individual bucket.

In [CFS09], a divide and conquer [Ben80] approach is proposed for k-NN graph computation in large scale and high-dimensional data. Each divide step separates the data samples in possibly overlapping subsets and computes k-NN graphs for each subset, while each conquer step combines the resulting graphs in a final kNN graph. In more detail, in the divide step, a Lanczos [Lan50] procedure is adopted, which performs spectral bisection on the data graph recursively. This procedure is combined with two different methods for dividing a set. According to the first one, referred to as the *overlap method*, the current set is divided into two overlapping subsets. The second method, referred to as the *glue method*, divides the current set into two disjoint subsets, while a third set consisting of data samples common to the two subsets is used in order to merge the two k-NN graphs in the conquer step. During the conquer step, the k-NN graphs computed by the divide step are combined into a final k-NN graph.

The algorithm proposed in [WWZ$^+$12] follows a divide and conquer process in order to partition the whole dataset into smaller subsets, within which neighborhood graphs are calculated. In more detail, the dataset is recursively divided into non-overlapping subsets, until the size of each subset reaches a pre-defined value. Since a single division is not capable of connecting a node with neighboring nodes belonging to different subsets, multiple random divisions are performed. As a result, a node is associated with multiple neighborhood sets, i.e. sets of neighboring nodes, each resulting from a different division. The neighbors of a node are defined as the k nearest nodes in the union of the neighborhood sets. Furthermore, a neighborhood propagation scheme is incorporated in the algorithm, in order to enhance the quality of the approximation. According to this scheme, true neighboring nodes of a node i are discovered within the neighborhoods of nodes that have already been identified as neighbors of i.

In [KTT$^+$12], a graph construction method is introduced, which calculates a *match graph* for large image databases, i.e. a graph where nodes correspond to images and edges express matching relationships between images. Two images are linked with an edge, if a correspondence, e.g., a similarity measure between them has been calculated. This process is referred to as *linking* or *verification*. The proposed algorithm departs from a sparse graph which gradually becomes dense, through an iterative process which predicts the existence of edges between candidate matching images (nodes). The initial sparse graph G_0 is formed by verifying a small number of images. Subsequently, the iterative process is applied on this graph, which alternates between two steps:

- **Prediction**: at time step t, for each unverified edge, a confidence measure is estimated, which expresses the probability that a candidate edge is a real edge.

- **Verification**: the candidate edges that exhibited the m highest confidence values are verified and graph G_t is updated accordingly.

Other approaches for large-scale match graph construction can be found in [HGO$^+$10, PSZ11].

In [YWZ12], the computational cost of the NCA method [GRHS04] is addressed, when dealing with large-scale or high dimensionality data. The proposed *Fast Neighborhood Component Analysis* (FNCA) exhibits a significantly lower computational complexity in comparison to NCA. Furthermore, it is modified, so that it can be applied to a non-linear metric learning scenario, by using the kernel trick.

The *k-NN fused Lasso graph* introduced in [ZLP13] is suitable for large scale high-dimensional datasets. The proposed algorithm extends the l_1 graph construction by incorporating structured sparsity. The local structure of the data is taken into consideration by assuming that the reconstruction coefficients of a data sample have values similar to those of its nearest neighbors. Furthermore, the k-NN method in combination with a kernel is adopted by the graph construction process, in order to render the algorithm scalable. The *NN-Descent*, proposed in [DML11], is an efficient algorithm for constructing an approximate k-NN graph. The main idea behind this algorithm is to improve the k-NN approximation for a sample by examining its neighbors' neighbors. In more detail, departing from a random approximate k-NN graph for each data sample, the approximation is improved through an iterative process, which performs comparisons between the sample and samples belonging to the k-neighborhood of its neighbors.

Another method for scalable graph construction, which is based on anchor points, is proposed in [LHC10]. The main idea behind this method is to use a small number of samples (anchors), which approximate the local neighborhood structure. These anchor points are also utilized during label inference. Finally, the *greedy filtering* method proposed in [PPLJ14] constructs an approximate *k*-NN graph by a *prefix filtering* method [LPSL10] on the data samples' feature vectors.

5.3 Label inference methods

Once the graph representing the data sample similarities has been constructed, label propagation is performed on the graph nodes, by means of a *label inference* method that determines how the labels are spread from the set of labeled nodes to the unlabeled nodes. Therefore, label inference is essentially an information diffusion process over the similarity graph.

Let us assume a set of labeled data, denoted with $\mathcal{X}_L = \{\mathbf{x}_i, i = 1, ..., n_l\}$, which are assigned labels from the set $\mathcal{L} = \{l_j\}_{j=1}^L$ and a set of unlabeled data $\mathcal{X}_U = \{\mathbf{x}_i, i = 1, ..., n_u\}$. The entire dataset, consisting of the labeled and the unlabeled data, can be denoted by $\mathcal{X} = \{\mathbf{x}_1, ..., \mathbf{x}_{n_l}, \mathbf{x}_{n_l+1}, ..., \mathbf{x}_N\}$, $N = n_l + n_u$. The label vector $\mathbf{y} = [y_1, ..., y_{n_l}, 0, ..., 0]^T = [\mathbf{y}_L, \mathbf{y}_U]^T \in \mathcal{L}^N$ contains the labels of the labeled data in the first n_l positions and takes the value 0 in the last n_u positions. In matrix notation, the label matrix $\mathbf{Y} \in \mathbb{R}^{N \times L}$ is defined as the matrix having entries $Y_{il} = 1$, if the i-th sample has the l-th label and $Y_{il} = 0$, otherwise. In correspondence to the label vector and the label matrix, we denote the estimated label vector and label matrix by $\hat{\mathbf{y}}$ and $\hat{\mathbf{Y}}$ respectively. With respect to the rules that govern label dissemination, as well as the type and the number of the graphs they apply to, label inference methods can be divided in various categories that are described subsequently.

5.3.1 Iterative algorithms

Iterative label propagation algorithms disseminate labels from labeled to unlabeled data gradually, in an iterative way. In each iteration, the labels are updated according to some rule, until the algorithm converges to a stationary state, as $t \to \infty$. Since the stationary state of the iterative algorithms can be estimated in advance, in practice, these methods can be applied in a single step.

An early iterative approach is proposed in [ZG02]. Initially, a fully connected graph is constructed from both labeled and unlabeled nodes, where the edge weights are calculated by:

$$W_{ij} = \exp(-\frac{\|\mathbf{x}_i - \mathbf{x}_j\|^2}{\sigma^2}), \qquad (5.3.1)$$

where parameter $\sigma > 0$ controls the weight values. In each iteration, the labels are propagated through the edges, according to a transition matrix $\mathbf{T}$ of size $N \times N$, whose entries are given by:

$$\mathbf{T}_{ij} \triangleq P(j \to i) = \frac{W_{ij}}{\sum_{k=1}^N W_{kj}}. \qquad (5.3.2)$$

Each entry T_{ij} of the transition matrix expresses the probability of propagating the label of the node j to node i. The node labels are estimated in each iteration according to the following update rule:

$$\hat{\mathbf{Y}}^{(t+1)} = \mathbf{T}\hat{\mathbf{Y}}^{(t)}, \tag{5.3.3}$$

where $\hat{\mathbf{Y}}^{(t)}$ and $\hat{\mathbf{Y}}^{(t+1)}$ denote the estimated labels at times t and $t+1$ respectively. In this algorithm, the estimated labels of the originally labeled data are constrained to be equal to the initial labels. In order to achieve improved accuracy and robustness, prior knowledge of class labels can be incorporated during label inference. The prior probabilities for each class j are estimated from the labeled examples as:

$$p_j = \frac{1}{n_l} \sum_{i=1}^{n_l} Y_{ij}. \tag{5.3.4}$$

Two different methods are proposed that utilize class prior knowledge, namely *class mass normalization* and *label bidding*. Assuming that each unlabeled example is assigned a vector $\hat{\mathbf{y}}_i = \{\hat{y}_{i1}, ..., \hat{y}_{iL}\}$, corresponding to a row in matrix $\hat{\mathbf{Y}}$, whose entries express the probability that the example belongs to each of the L classes, the mass of class j is defined as:

$$m_j = \frac{1}{n_u} \sum_{i=n_l+1}^{N} \hat{Y}_{ij}. \tag{5.3.5}$$

In class mass normalization, the j-th entry of $\hat{\mathbf{y}}_i$, $i = n_l + 1, ..., N$ ($\mathbf{x}_i \in \mathcal{X}_U$) is scaled by the factor $w_j = \frac{p_j}{m_j}$ and the label assignment for the unlabeled sample $\mathbf{x}_i$ is performed according to $\operatorname{argmax}_j \{w_j \hat{Y}_{ij}\}$. Class mass normalization does not require the knowledge of the exact label (class) proportions. However, if the exact class proportions are known, a label bidding heuristic can be employed. Let us denote by c_j the number of unlabeled examples that are assigned the label l_j, with $\sum_j c_j = n_u$. In each iteration, the unlabeled example $\mathbf{x}_i$ with the highest class probability $\max_j \{\hat{Y}_{ij}\}$ is found and assigned the label $l_{j_{max}}$, where $j_{max} = \operatorname{argmax}_j \{\hat{Y}_{ij}\}$, if the number of the already assigned labels $l_{j_{max}}$ does not exceed $c_{j_{max}}$. Otherwise it is ignored and the next highest class probability is searched.

In [BDR06], the proposed label propagation algorithm is based on the Jacobi iterative method for linear equations [BBC$^+$94]. The update rules for the labeled and unlabeled nodes take the form:

$$\begin{aligned}
\hat{y}_i^{(t+1)} &= \frac{\sum_j W_{ij}\hat{y}_j^{(t)} + \frac{1}{\mu}y_i^{(t)}}{\sum_j W_{ij} + \frac{1}{\mu} + \epsilon}, \quad i = 1, ..., n_l \\
\hat{y}_i^{(t+1)} &= \frac{\sum_j W_{ij}\hat{y}_j^{(t)}}{\sum_j W_{ij} + \epsilon}, \qquad\quad i = n_l + 1, ..., N,
\end{aligned} \tag{5.3.6}$$

respectively, where μ is a parameter that regulates the weight of the labeled node and $\epsilon > 0$ a regularization parameter which prevents numerical problems in case the denominator becomes too small. In this algorithm, the estimated labels for the originally labeled nodes are allowed to differ from the initial labels.

In [ZBL$^+$04], an iterative method that utilizes the graph Laplacian is introduced. It first constructs a weight matrix similar to the one used in [ZG02]. Subsequently, the normalized graph Laplacian is calculated by:

$$\mathcal{L} = \mathbf{D}^{-1/2}\mathbf{L}\mathbf{D}^{-1/2} = \mathbf{I} - \mathbf{D}^{-1/2}\mathbf{W}\mathbf{D}^{-1/2}. \tag{5.3.7}$$

In each iteration, the labels of both labeled and unlabeled nodes are updated, according to a classification matrix $\mathbf{F} \in \mathbb{R}^{N \times L}$ given by:

$$\mathbf{F}^{(t+1)} = \mu \mathbf{S} \mathbf{F}^{(t)} + (1-\mu)\mathbf{Y}, \text{ with } \mathbf{F}^0 = \mathbf{Y}, \tag{5.3.8}$$

where $\mathbf{S} = \mathbf{D}^{-1/2}\mathbf{W}\mathbf{D}^{-1/2}$ and μ is a parameter taking values in $(0, 1)$. If $\mathbf{F}^*$ is the limit of $\mathbf{F}$ upon convergence, a node i is finally labeled according to $\arg\max_j \{F_{ij}^*\}$. The aforementioned method was incorporated in the graph-based active learning method proposed in [LYZZ08]. An algorithm similar to [ZBL+04] is introduced by Wang et al. [WZ06], where a function $\mathbf{f} \in \mathbb{R}^N$ estimates the node labels, utilizing the following update rule:

$$\mathbf{f}^{(t+1)} = \alpha \mathbf{W}\mathbf{f}^{(t)} + (1-\alpha)\mathbf{y}, \text{ with } \mathbf{f}^0 = \mathbf{y}. \tag{5.3.9}$$

Parameter α takes values in the range $(0, 1)$ and expresses the fraction of label information that a node receives from its neighbors and $\mathbf{f}^{(t)}$ are the label predictions.

5.3.2 Random walks

Label inference can be performed through random walks on the similarity graph, starting from labeled nodes of different classes. In [SJ02], Markov random walks on the graph are considered, where the transition probability from node i to a neighbor k is calculated by:

$$p_{ik} = \frac{W_{ik}}{\sum_{j=1}^{N} W_{ij}}. \tag{5.3.10}$$

The probability that a walk, which started at some node with label y_{start}, reaches node k after t steps is given by:

$$P^t(y_{start}|k) = \sum_{i=1}^{N} P(y = y_{start}|i)P_{0|t}(i|k), \tag{5.3.11}$$

where $P(y|i)$ denotes the probability that node i has the label y and $P_{0|t}(i|k)$ is the probability of reaching node k starting from node i in t steps. The node k is assigned the label y_{start}, if the aforementioned probability is greater than 0.5. The probabilities $P(y|i)$ are estimated using two proposed alternatives: an iterative *Expectation Maximization* EM algorithm, or by maximizing a margin-based criterion that leads to a closed-form solution. The method proposed in [ZGL03] can be interpreted as a random walk in the graph, where an alternative formulation of the random walks to [SJ02] is considered. Specifically, a walk starts from an unlabeled node i and traverses the graph until it arrives at a labeled node k with a probability given by a function $f(i)$. Node i is, thus, assigned the same label as node k with a probability $f(i)$. Unlike [SJ02], the solution does not depend on the length t of the walk.

In [ZS04], a two-class classification scenario is examined, where the label assignment decision is taken according to the commute time G_{ij}, i.e., the number of steps required for a random walk initiated at node i to reach node j and then get back to i. The transition probability matrix is calculated as:

$$\mathbf{P} = (1-a)\mathbf{I} + a\mathbf{D}^{-1}\mathbf{W}, \tag{5.3.12}$$

where parameter a takes values in the range $(0, 1)$. The classification decision is taken by comparing the commute times $\bar{G}_{ij}$ from node i to the labeled nodes of different classes with labels $\mathcal{L} = \{+1, -1\}$:

$$p_+(\mathbf{x}_i) = \sum_{j|y_j=1} \bar{G}_{ij}, \ p_-(\mathbf{x}_i) = \sum_{j|y_j=-1} \bar{G}_{ij}, \tag{5.3.13}$$

where $\bar{G} = (\mathbf{I} - a\mathbf{D}^{-1/2}\mathbf{W}\mathbf{D}^{-1/2})^{-1}$.

The random walks introduced in [Azr07] start from unlabeled nodes, while labeled nodes serve as absorbing states, i.e., the random walk stops upon visiting a labeled node. The probability that an unlabeled node i is assigned the label l is calculated by the total probability that a random walk initiated at node i stops/is absorbed at a node labeled with l.

The *Adsorption* algorithm introduced in [BSS$^+$08] can be interpreted as a random walk on a graph. In order to estimate the label of a labeled or unlabeled node i , a random walk is initiated at node i and continues according to three alternative choices [TC09]:

- Injection: with a probability p_i^{inj}, the walk stops and the predefined label vector $\mathbf{y}_i$ is returned. For the unlabeled nodes, a probability $p_i^{inj} = 0$ is assumed.

- Continue: with a probability p_i^{cont}, the walk continues to another neighboring node j, according to the value of the transition probability $Pr[j|i]$ calculated by:

$$Pr[j|i] = \begin{cases} \dfrac{W_{ji}}{\sum\limits_{k:(k,i)\in\mathcal{E}} W_{ki}}, & \text{if } (j,i) \in \mathcal{E} \\ 0, & \text{otherwise.} \end{cases} \qquad (5.3.14)$$

- Abandon: with a probability p_i^{abnd}, the walk is abandoned and a dummy label is assigned, expressing uncertainty about the node label. The dummy label is introduced in order to reduce the effect of nodes with a large number of connections. High degree nodes are considered unreliable, since they may have connections to nodes which are not similar to each other. Therefore, the random walk should be abandoned upon visiting a high degree node. The higher the degree of a node, the higher the value of p^{abnd} is.

For the aforementioned probabilities, it holds that: $p_i^{inj}, p_i^{cont}, p_i^{abnd} \geq 0$ and $p_i^{inj} + p_i^{cont} + p_i^{abnd} = 1$.

5.3.3 Graph regularization

Graph regularization-based label propagation methods utilize a classification vector $\mathbf{f} \in \mathbb{R}^N$ (or a matrix $\mathbf{F} \in \mathbb{R}^{N \times L}$ alternatively), defined on labeled and unlabeled data (nodes), which disseminates the labels from the set of labeled nodes to the set of unlabeled nodes. This vector should ensure that the labels of the originally labeled nodes remain unchanged, while nodes that are similar to each other or belong to the same structure (e.g., cluster, manifold) are assigned the same label. According to the second assumption, vector $\mathbf{f}$ should be smooth over the graph. The aforementioned assumptions can be expressed by a regularization framework of the following form:

$$\min_{\mathbf{f}}\{\alpha \mathcal{C}(\mathbf{f}_L) + \beta \mathcal{S}(\mathbf{f})\}, \qquad (5.3.15)$$

where $\mathcal{C}(\mathbf{f}_L)$ denotes a cost function defined on the labeled nodes, which penalizes the difference between the estimated labels and the initial labels, while $\mathcal{S}(\mathbf{f})$ applies a smoothness constraint to the entire graph. The parameters α and β express the trade-off between the two terms. The smoothness constraint is usually expressed as:

$$\mathcal{S}(\mathbf{f}) = \mathbf{f}^T \mathbf{S} \mathbf{f}, \qquad (5.3.16)$$

where $\mathbf{S}$ is a smoothing matrix.

In [ZBL$^+$04], a simple method for label propagation is proposed, that minimizes the following cost function:

$$\mathcal{Q}(\mathbf{F}) = \frac{1}{2} \sum_{i,j=1}^N W_{ij} \left\| \frac{\mathbf{f}_i}{\sqrt{D_{ii}}} - \frac{\mathbf{f}_j}{\sqrt{D_{jj}}} \right\|^2 + \mu \sum_{i=1}^N \|\mathbf{f}_i - \mathbf{y}_i\|^2, \qquad (5.3.17)$$

where $\mu > 0$ is a regularization parameter and $\mathbf{f}_i$ the i-th row of $\mathbf{F}$. The closed form solution is calculated by:

$$\tilde{\mathbf{F}} = \beta(\mathbf{I} - \alpha\mathbf{S})^{-1}\mathbf{Y}. \tag{5.3.18}$$

This regularization framework is proven to be equivalent to the iterative process described by equation (5.3.8). In each iteration, the graph nodes receive information from their neighbors, while maintaining the information of their initial state (label). In the algorithms proposed in [WZ06] and [THY$^+$11], the cost function has the form of equation (5.3.15) and the smoothness matrix is $\mathbf{S} = \mathbf{I} - \mathbf{W}$, where $\mathbf{W}$ denotes the weight matrix.

The label propagation algorithm proposed in [ZGL03] is based on a Gaussian Random Field model, defined on the graph. The regularization problem consists of the minimization of the quadratic energy function, expressed as $\mathbf{f}^T\mathbf{L}\mathbf{f}$, while preserving the labels of the initially labeled nodes. This is achieved by setting parameter a in equation (5.3.15) to ∞. The minimum energy function satisfies the harmonic property, i.e., it is equivalent to the average energy of the neighboring nodes. The relationship between the Gaussian random fields and the Gaussian processes were studied in [ZLG03], using a spectral transformation on the graph Laplacian matrix.

Two different algorithms based on regularization frameworks were proposed in [BMN04]. The first one, referred to as *Tikhonov regularization*, assumes that labeled nodes may exhibit multiplicities, i.e. they may appear multiple times with the same or with a different label y. As a result, the number of the available labeled samples k may differ from the number of labeled nodes n_l. Furthermore, it is assumed that labels may be noisy. The proposed regularization framework minimizes the following objective function:

$$\min_{\mathbf{f}=\{f_1,f_2,...,f_N\}} \left\{ \frac{1}{k} \sum_i (f_i - \tilde{y}_i)^2 + \gamma\mathbf{f}^T\mathbf{S}\mathbf{f} \right\}, \tag{5.3.19}$$

$$\text{s.t. } \sum \mathbf{f}_i = 0,$$

where $\gamma \in \mathbb{R}$ and $\tilde{y}_i$ are elements of $\tilde{\mathbf{y}} = \{y_1 - \bar{y}, y_2 - \bar{y}, ..., y_k - \bar{y}\}$, with $\bar{y} = \frac{1}{k}\sum_i y_i$. Matrix $\mathbf{S}$ is a smoothness matrix, e.g., $\mathbf{S} = \mathbf{L}$ or $\mathbf{S} = \mathbf{L}^p$, $p \in \mathbf{N}$, where $\mathbf{L}$ is the graph Laplacian. The solution to the above minimization problem is obtained by:

$$\tilde{\mathbf{f}} = (k\gamma\mathbf{S} + \mathbf{I}_k)^{-1}(\mathbf{y}_\Sigma + \mu\mathbf{1}), \tag{5.3.20}$$

where $\mathbf{y}_\Sigma = \{\sum_i y_{1i}, \sum_i y_{2i}, ..., \sum_i y_{n_li}, 0, ..., 0\}$ is a vector of length N containing the sums of the labels assigned to the same node of the graph and $\mathbf{1}$ a vector of entries equal to one. $\mathbf{I}_k$ is a diagonal matrix:

$$\mathbf{I}_k \triangleq diag(m_1, m_2, ..., m_{n_l}, 0, ..., 0), \tag{5.3.21}$$

where m_i denotes the number of occurrences of node i in the labeled samples set. In the *interpolated regularization* algorithm proposed in [BMN04], the objective function takes the form:

$$\min_{\mathbf{f}=\{\tilde{y}_1,\tilde{y}_2,...,\tilde{y}_k,f_{k+1},...,f_n\}} \left\{ \mathbf{f}^T\mathbf{S}\mathbf{f} \right\}, \tag{5.3.22}$$

where $\mathbf{S}$ is a smoothness matrix as in (5.3.19). Unlike the Tikhonov regularization, in the interpolated regularization it is assumed that labels y do not contain noise. In addition, nodes are not allowed to appear multiple times in the sample ($k = n_l$). Matrix $\mathbf{S}$ can be written in the form:

$$\mathbf{S} = \begin{bmatrix} \mathbf{S}_1 & \mathbf{S}_2 \\ \mathbf{S}_2^T & \mathbf{S}_3 \end{bmatrix}, \tag{5.3.23}$$

where $\mathbf{S}_1$ is a $k \times k$ matrix, $\mathbf{S}_2$ a $k \times (N-k)$ matrix and $\mathbf{S}_3$ a $(N-k) \times (N-k)$ matrix. The solution to (5.3.22) is calculated by:

$$\tilde{\mathbf{f}} = \mathbf{S}_3^{-1}\mathbf{S}_2^T(\{\tilde{y}_1, ..., \tilde{y}_k\}^T + \mu\mathbf{1}). \tag{5.3.24}$$

In [WJC08], the *Graph Transduction via Alternating Minimization* (GTAM) algorithm is presented, that is capable of handling degenerate cases. They may occur, for example, when there are imbalances in the distribution of class labels or in the portion of the labeled data, or when there is noise or outliers in the dataset. This is achieved by minimizing an objective function, which is defined both on the classification function $\mathbf{F}$ and the label matrix $\mathbf{Y}$:

$$\min_{\mathbf{F},\mathbf{Y}} \left\{ \text{tr} \left\{ \mathbf{F}^T\mathcal{L}\mathbf{F} + \mu(\mathbf{F} - \mathbf{V}\mathbf{Y})^T(\mathbf{F} - \mathbf{V}\mathbf{Y}) \right\} \right\}, \tag{5.3.25}$$

where $\mu > 0$ is a trade-off parameter between the two terms, $\mathcal{L}$ is the normalized graph Laplacian and matrix $\mathbf{V}$ serves as a node regularizer that balances the influence of labels from different classes. The aforementioned objective function involves two variables, $\mathbf{F}$ and $\mathbf{Y}$, which are minimized through an alternating minimization scheme. Since $\mathbf{F}$ is a continuous classification matrix, optimization over $\mathbf{F}$ is obtained by:

$$\tilde{\mathbf{F}} = (\mathbf{L}/\mu + \mathbf{I})^{-1}\mathbf{V}\mathbf{Y}. \tag{5.3.26}$$

Minimizing with respect to $\mathbf{Y}$ is not straightforward, since it is a binary matrix. For this reason, optimization is obtained by replacing $\mathbf{F}$ in equation (5.3.25) with $\tilde{\mathbf{F}}$ and subsequently applying a greedy approach.

In the measure propagation algorithm proposed in [SB11], the objective function is based on the minimization of the Kullback-Leibler divergence between probability measures, which encode label membership probabilities. More precisely, two probability measures for the nodes of the graph are defined: $p_i(l), i = 1, .., N$ is defined for each node i of the graph and corresponds to the probability that node i is assigned the label l, $l \in \mathcal{L}$, while $r_j(l), j = 1, ..., n_l$ is defined only on the labeled nodes of the graph, and expresses the probability distribution of the labeled nodes. The objective function is expressed as:

$$\min_{\mathbf{P}} \sum_{i=1}^{n_l} D_{KL}(r_i||p_i) + \mu \sum_{i=1}^{N} \sum_{j \in \mathcal{N}(i)} w_{ij}D_{KL}(p_i||p_j) - \nu \sum_{i=1}^{N} H(p_i), \tag{5.3.27}$$

where $\mu, \nu > 0$ are hyper-parameters and D_{KL} is the Kullback-Leibler divergence between p_i and q_j, given by $D_{KL}(p||q) = \sum_y p(y) \log \frac{p(y)}{q(y)}$. H denotes the Shannon entropy of p:

$$H(p) = -\sum_y p(y) \log p(y). \tag{5.3.28}$$

The first two terms in the objective function represent the constraints expressed in equation (5.3.15). The additional constraint introduced by the third term, enforces the probability distributions p_i to be close to the uniform distribution. The aforementioned optimization problem does not admit a closed form solution. To this end, a *method of multipliers* (MOM) [Ber99] is employed instead.

A modification of the Adsorption algorithm [BSS$^+$08] is introduced in [TC09], where learning is stated as a convex optimization problem. The proposed method assumes matrices containing the predefined and the estimated labels, $\mathbf{Y}$ and $\mathbf{F} \in \mathbb{R}^{N \times (L+1)}$ (L denotes the number of labels) respectively. An alternative definition of the weight matrix is also introduced, where the weights between two nodes i and j are expressed as $W'_{ij} = p_i^{cont}W_{ij}$. Furthermore, a matrix $\mathbf{R} \in \mathbb{R}^{N \times (L+1)}$ is defined, where the first L columns are equal to zero and the last column contains the abandon probabilities p_i^{abnd}. The optimization problem minimizes the following cost function:

$$C(\mathbf{F}) = \sum_l \mu_1 \left(\mathbf{y}_l - \mathbf{f}_l\right)^T \mathbf{S} \left(\mathbf{y}_l - \mathbf{f}_l\right) + \mu_2 \mathbf{f}_l^T \mathbf{L} \mathbf{f}_l + \mu_3 \left\|\mathbf{f}_l - \mathbf{r}_l\right\|^2, \tag{5.3.29}$$

where $\mathbf{y}_l, \mathbf{f}_l$ and $\mathbf{r}_l$ denote the l-th column of the matrices $\mathbf{Y}$, $\mathbf{F}$ and $\mathbf{R}$ respectively, while $\mathbf{S} \in \mathbb{R}^{N \times N}$ is a diagonal matrix with $\mathbf{S}_{ii} = p_i^{inj}$. Matrix $\mathbf{L}$ is calculated as:

$$\mathbf{L} = \mathbf{D} + \bar{\mathbf{D}} + \mathbf{W}' + \mathbf{W}'^T, \tag{5.3.30}$$

where $\mathbf{D}$ and $\bar{\mathbf{D}}$ are $N \times N$ diagonal matrices with

$$\mathbf{D}_{ii} = \sum_j W'_{ji}, \quad \bar{\mathbf{D}}_{ii} = \sum_j W'_{ij}. \tag{5.3.31}$$

In the above optimization framework, the first two terms match the requirements of (5.3.15), while the third term discounts high degree nodes. The importance of each term is controlled by the weight parameters $\mu_1, \mu_2, \mu_3 > 0$. The solution is obtained by:

$$\mathbf{f}_l = (\mu_1 \mathbf{S} + \mu_2 \mathbf{L} + \mu_3 \mathbf{I})^{-1}(\mu_1 \mathbf{S} \mathbf{y}_l + \mu_3 \mathbf{R}_l). \tag{5.3.32}$$

Both the Adsorption algorithm [BSS⁺08], as well as its modified version [TC09], try to reduce the influence of high degree nodes, since they may connect dissimilar nodes. In contrast to these methods, the *Transduction Algorithm with Confidence* (TACO) algorithm [OC12] takes into account the level of agreement among the labels of node neighbors and utilizes a measure of confidence regarding the label assignment of each node. The TACO algorithm does not reduce the influence of all high degree nodes, but only of those that exhibit a low confidence in their predicted labels. Two parameter sets are defined on graph nodes: the first one consists of the score vectors $\mathbf{f}_i, \in \mathbb{R}^L$, expressing the extent to which a node i belongs to each of the L classes (labels). Furthermore, for each node i, a diagonal matrix $\boldsymbol{\Sigma}_i \in \mathbb{R}^{L \times L}$ is calculated, whose entries represent the uncertainties of the corresponding scores in $\mathbf{f}_i$. The objective function takes the form:

$$\min_{\mathbf{f}, \boldsymbol{\Sigma}} \left\{ \frac{1}{4} \sum_{i,j=1}^{N} w_{ij} \left[(\mathbf{f}_i - \mathbf{f}_j)^T (\boldsymbol{\Sigma}_i^{-1} + \boldsymbol{\Sigma}_j^{-1})(\mathbf{f}_i - \mathbf{f}_j) \right] \right.$$
$$+ \frac{1}{2} \sum_{i=1}^{n_l} \left[(\mathbf{f}_i - \mathbf{y}_i)^T (\boldsymbol{\Sigma}_i^{-1} + \frac{1}{\gamma}\mathbf{I})(\mathbf{f}_i - \mathbf{y}_i) \right] \tag{5.3.33}$$
$$\left. + \alpha \sum_{i=1}^{N} tr \boldsymbol{\Sigma}_i - \beta \sum_{i=1}^{N} \log \det \boldsymbol{\Sigma}_i \right\},$$

where α, β, γ are positive parameters. The solution to the aforementioned optimization is obtained through an iterative algorithm.

In [BC01], a two-class label propagation problem is regarded as a clustering problem and a *graph mincuts* algorithm is employed [CGK⁺97]. Specifically, this algorithm discovers the minimum total weight of a set of edges, whose removal partitions the graph into two sets of nodes: one consisting of nodes with label 1 and one consisting of nodes with label -1. The graph mincut is expressed by the following objective function:

$$\min_{\mathbf{f}} \{a(\mathbf{f}_L - \mathbf{Y}_L)^T(\mathbf{f}_L - \mathbf{Y}_L) + \frac{1}{2}\mathbf{f}^T \mathbf{L} \mathbf{f}\}, \tag{5.3.34}$$

where $a \to \infty$ and $\mathbf{L}$ is the graph Laplacian, under the constraint $\mathbf{f}_i \in \{0, 1\}$. In a later study, the aforementioned graph mincuts approach is extended, by adding artificial random noise on the edge weights [BLRR04]. The proposed algorithm exhibits improved performance and also provides a confidence score for the assigned labels.

In [Joa03], spectral graph partitioning is performed through the constrained ratiocut algorithm that adds a quadratic penalty to the objective function of the standard ratio cut [HK92]:

$$\min_{\mathbf{f}} \{ \mathbf{f}^T \mathbf{L} \mathbf{f} + c(\mathbf{f} - \mathbf{g})^T \mathbf{C} (\mathbf{f} - \mathbf{g}) \}, \tag{5.3.35}$$

$$\text{s.t. } \mathbf{f}^T \mathbf{1} = 0 \text{ and } \mathbf{f}^T \mathbf{f} = N, \tag{5.3.36}$$

where c is a regularization parameter, $\mathbf{L}$ is the graph Laplacian and $\mathbf{C}$ is a diagonal matrix, whose i-th diagonal element contains a misclassification cost for node i.

5.3.4 Graph kernel regularization

In *manifold regularization* methods with *graph kernels*, the smoothness constraint (5.3.16) is written in the form:

$$\mathcal{S}(\mathbf{f}) = \|\mathbf{f}\|_{\mathcal{H}} = \mathbf{f}^T \mathbf{K} \mathbf{f}, \tag{5.3.37}$$

where $\mathbf{K}$ is a *kernel matrix* that is associated with the *Reproducing Kernel Hilbert Space* (RKHS) $\mathcal{H}$. A kernel matrix is equivalent to the *Gram matrix*. Graph kernels capture the local and global structure of the data space. A function $\mathbf{K}$ is considered to be a kernel function if it is symmetric and positive semi-definite. The exponentiation of any symmetric matrix $\mathbf{H}$ results in a symmetric and positive semi-definite matrix $\mathbf{K}$:

$$\mathbf{K} = e^{\beta \mathbf{H}} = \lim_{n \to \infty} \left(1 + \frac{\beta \mathbf{H}}{n} \right)^n, \tag{5.3.38}$$

which can be used to define an exponential family of kernel functions, where $\mathbf{H}$ is the so-called the generator and β is a bandwidth parameter [KL02]. The exponential kernel $\mathbf{K}$ has the property that if the generator $\mathbf{H}$ represents the local structure of the data space, then $\mathbf{K}$ represents the global structure of the data space. Let us consider an unweighted, undirected graph G and an exponential kernel function with a generator of the form:

$$\mathbf{H} = \begin{cases} 1, & \text{node } i \text{ connected to node } j \\ -d_i, & i=j \\ 0, & \text{otherwise}, \end{cases} \tag{5.3.39}$$

where d_i is the degree of node i. It can be observed that matrix $\mathbf{H}$ is the negative of the Laplacian matrix of graph G. By differentiating equation (5.3.38) with respect to β the following differential equation is obtained:

$$\frac{d\mathbf{K}}{d\beta} = \mathbf{H}\mathbf{K} = -\mathbf{L}\mathbf{K}, \tag{5.3.40}$$

which is the *heat equation* on graph G that is subsequently described in Section 5.4. The resulting kernels are called *diffusion* or *heat* kernels.

Kernel families on graphs can be alternatively derived from the spectral analysis of the normalized graph Laplacian matrix [SK03]. The eigenvectors $\mathbf{v}$ and eigenvalues λ of the normalized graph Laplacian matrix $\mathcal{L}$ contain information about the graph partitions, which renders them an important tool for graph clustering. A class of regularization functionals on graphs can be defined [SK03] to be of the form:

$$\langle \mathbf{f}, \mathbf{f} \rangle_{\mathcal{H}} = \langle \mathbf{f}, r(\mathcal{L})\mathbf{f} \rangle, \tag{5.3.41}$$

where

$$r(\mathcal{L}) := \sum_i r(\lambda_i)\mathbf{v}_i\mathbf{v}_i^T \tag{5.3.42}$$

and $\langle \cdot \rangle_{\mathcal{H}}$ is the inner vector product in the RKHS $\mathcal{H}$ with kernel:

$$\mathbf{K} = \sum_i r^{-1}(\lambda_i)\mathbf{v}_i\mathbf{v}_i^T. \tag{5.3.43}$$

All kernel functions are derived from (5.3.43) with a proper choice of spectral transform $r(\lambda)$ of the Laplacian matrix. For example, the diffusion kernel in equation (5.3.38) is obtained for $r(\lambda) = \exp(\sigma^2/\lambda)$ and the regularized Laplacian [ZLG03]:

$$\mathbf{K} = \mathbf{L} + \mathbf{I}/\sigma^2 \tag{5.3.44}$$

is obtained for $r(\lambda) = \lambda + 1/\sigma^2$.

In [BNS06], a general manifold regularization framework is proposed, where the objective function takes the form:

$$\frac{1}{n_l}\sum_{i=1}^{n_l} V(\mathbf{x}_i, y_i, \mathbf{f}) + \gamma_A\|\mathbf{f}\|_{\mathcal{H}}^2 + \gamma_I\|\mathbf{f}\|_I^2, \tag{5.3.45}$$

where γ_A and γ_I are regularization parameters. The first term is the general form of the cost function on the labeled data, $\|\mathbf{f}\|_{\mathcal{H}}^2$ is a regularization term in the RKHS of the kernel $\mathbf{K}$ and $\|\mathbf{f}\|_I^2$ is a regularization term of the geometry of the probability distribution. *Laplacian Regularized Least Squares* (LapRLS) and *Laplacian Support Vector Machines* (LapSVM) [GCCVMMC08] can be regarded as special cases of manifold regularization.

5.3.5 Inductive label inference

Unlike transductive label propagation methods, which are applied on a specific dataset, inductive methods learn a global representation from the available data and can be applied to unseen data samples. An early work on *inductive semi-supervised learning* can be found in [YTZ04]. The proposed method is based on the regularization framework introduced in [ZBL+04] and adopts RBF basis functions. Another inductive method is introduced in [ZL05], which combines harmonic mixture models with graph-based semi-supervised learning.

In [BDR06], a common framework for different label propagation algorithms is proposed, where a quadratic cost function is minimized. The closed-form solution of the aforementioned cost function is obtained by solving a linear system of size equal to the number of the data samples. This cost criterion provides an extension of the different label propagation algorithms to the inductive setting. Assuming that all the available data samples $\mathcal{X} = \{\mathbf{x}_1, \ldots, \mathbf{x}_N\}$ have been assigned labels $\hat{\mathbf{y}} = \{\hat{y}_1, \ldots, \hat{y}_N\}$ by a label propagation method, the aim is to determine the label of a new data sample $\mathbf{x}$. A new data sample $\mathbf{x}$ is incorporated in the graph, with a new weight matrix $\mathbf{W_x}$. The objective is the minimization of the following cost function:

$$c + \mu\left(\sum_j W_{\mathbf{x}}(\mathbf{x}, \mathbf{x}_j)(\hat{y} - \hat{y}_j)^2 + \epsilon\hat{y}^2\right), \quad \mu, \epsilon > 0, \tag{5.3.46}$$

with respect to the new label $\hat{y}$, where c is a constant. By setting the first derivative with respect to $\hat{y}$ to zero, the minimum of (5.3.46) is calculated as:

$$\hat{y} = \frac{\sum_j W_{\mathbf{x}}(\mathbf{x}, \mathbf{x}_j)\hat{y}_j}{W_{\mathbf{x}}(\mathbf{x}, \mathbf{x}_j) + \epsilon}. \tag{5.3.47}$$

In case the weight matrix $W_{\mathbf{x}}$ is constructed using the k-Nearest Neighbors, then (5.3.47) becomes equivalent to k-NN classification. If $W_{\mathbf{x}}$ is calculated using the Gaussian kernel, then (5.3.47) is equivalent to the *Nadaraya-Watson kernel regression* [Bie87].

5.3.6   Label propagation on data with multiple representations

The methods reviewed so far assume a single data representation. However, in many real world scenarios, multimedia data can be represented in multiple feature spaces. As an example, in the case of 3D video data, each video (node) can be represented by feature vectors describing color or disparity (depth). In such cases, a separate similarity graph can be constructed for each of these representations. The information from the multiple data representations can be fused in two ways. Fusion can take place during graph construction (*early fusion*), e.g., the feature vectors of each representation can be concatenated into a single feature vector. Alternatively, fusion can be performed at the decision level (*late fusion*), e.g., by combining the classification results derived from separate classification schemes for each representation. Late fusion is also referred to as *multi-modal fusion* or *multi-modality learning* [WHH+09].

In an early study [JCST01] convex combinations of independent kernels were adopted:

$$K(\mathbf{x}_1, \mathbf{x}_2) = \alpha K(\mathbf{x}_1, \mathbf{x}_2) + (1 - \alpha)K(\mathbf{x}_1, \mathbf{x}_2), \ \ 0 \le \alpha \le 1. \tag{5.3.48}$$

Such kernels are considered independent if they come from independent data representations. Similar approaches can be found in [AHP05, TSS05, SN05], where a convex combination of graph Laplacians is employed.

In [THL+05], the multi-modality learning problem is approached by extending the regularization framework proposed in [ZBL+04] to the case of multiple graphs. Each feature type (modality) is represented by an individual graph. Two alternative schemes, a linear and a sequential one, are proposed for fusing information from different modalities. In more detail, assuming two feature types, each data sample is associated with two feature vectors $\mathbf{x}_i = \{\mathbf{x}_i^a, \mathbf{x}_i^b\}$, $i = 1, ..., N$, corresponding to each of the modalities a,b. For the first feature type, the weight matrix $\mathbf{W}_a$ of size $N \times N$ is constructed by measuring the similarity between data samples in the modality a. Furthermore, the degree matrix $\mathbf{D}_a$, as well as the normalized weight matrix $\mathbf{S}_a = (\mathbf{D}_a)^{-1/2}\mathbf{W}_a(\mathbf{D}_a)^{-1/2}$ can be calculated. Similarly, matrices $\mathbf{W}_b, \mathbf{D}_b$ and $\mathbf{S}_b$ are defined for the second modality b.

In the linear fusion scheme, the smoothness constraints introduced by $\mathbf{S}_a, \mathbf{S}_b$, as well as the manifold constraint imposed by $\mathbf{Y}$ are fused simultaneously through a weighted sum. In this case, the objective function results directly from (5.3.17) and is expressed as:

$$\begin{aligned}
\mathcal{Q}(\mathbf{F}) = &\mu \sum_{i,j=1}^{N} W_{aij} \left\| \frac{\mathbf{f}_i}{\sqrt{D_{aii}}} - \frac{\mathbf{f}_j}{\sqrt{D_{ajj}}} \right\|^2 + \eta \sum_{i,j=1}^{N} W_{bij} \left\| \frac{\mathbf{f}_i}{\sqrt{D_{bii}}} - \frac{\mathbf{f}_j}{\sqrt{D_{bjj}}} \right\|^2 + \\
&\epsilon \sum_{i=1}^{N} \|\mathbf{f}_i - \mathbf{y}_i\|^2,
\end{aligned} \tag{5.3.49}$$

where $\mu, \eta, \epsilon \in (0, 1)$ and $\mu + \eta + \epsilon = 1$ are weights that control the trade-off among the three constraint terms. The above equation can be also written in the following, more compact form:

$$\mathcal{Q}(\mathbf{F}) = \text{tr}\left(\mu\mathbf{F}^T(\mathbf{I} - \mathbf{S}_a)\mathbf{F} + \eta\mathbf{F}^T(\mathbf{I} - \mathbf{S}_b)\mathbf{F} + \epsilon(\mathbf{F} - \mathbf{Y})^T(\mathbf{F} - \mathbf{Y})\right). \quad (5.3.50)$$

Solving the above optimization problem leads to:

$$\tilde{\mathbf{F}} = (1 - \mu - \eta)(\mathbf{I} - \mu\mathbf{S}^a - \eta\mathbf{S}^b)^{-1}\mathbf{Y}. \quad (5.3.51)$$

In the sequential fusion scheme, the constraints are fused sequentially, leading to a minimization framework solved in two steps:

$$\tilde{\mathbf{F}}_1 = \arg\min_{\mathbf{F}} \text{ tr}\left(\mu\mathbf{F}^T(\mathbf{I} - \mathbf{S}^a)\mathbf{F} + (1 - \mu)(\mathbf{F} - \mathbf{Y})^T(\mathbf{F} - \mathbf{Y})\right) \quad (5.3.52)$$

$$\tilde{\mathbf{F}}_2 = \arg\min_{\mathbf{F}} \text{ tr}\left(\eta\mathbf{F}^T(\mathbf{I} - \mathbf{S}^b)\mathbf{F} + (1 - \eta)(\mathbf{F} - \tilde{\mathbf{F}}_1)^T(\mathbf{F} - \tilde{\mathbf{F}}_1)\right), \quad (5.3.53)$$

where the regularization parameters $0 < \mu, \eta < 1$ regulate the trade-off between the two constraints in each minimization equation. In the first step, an optimal $\tilde{\mathbf{F}}_1$ for the constraints $\mathbf{S}_a, \mathbf{Y}$ is found, while the second step determines an optimal $\tilde{\mathbf{F}}_2$, under the constraints imposed by $\mathbf{S}_b$ and $\tilde{\mathbf{F}}_1$. The classification decision is finally taken according to the values of $\tilde{\mathbf{F}}_2$. The solution to the sequential fusion scheme is calculated by:

$$\tilde{\mathbf{F}}_2 = (1 - \mu)(1 - \eta)(\mathbf{I} - \eta\mathbf{S}^b)^{-1}(\mathbf{I} - \mu\mathbf{S}^a)^{-1}\mathbf{Y}. \quad (5.3.54)$$

Similar approaches to [THL+05] can be also found in [WHY+07] and [WHH+09].

In [ZB07], a different approach to the multiple representations problem is introduced. The proposed algorithm associates each graph with a Markov chain, similar to [ZHS05], and constructs Markov mixture models. The method proposed in [XWTQ07], exploits information from 2D images, as well as from 3D points reconstructed from multiple view images, in order to perform multiple view image segmentation. More precisely, three separate graphs are constructed: a similarity graph between 3D point coordinates, a 2D colour similarity graph and finally, a graph that measures the patch histogram similarity between two joint points. Joint points are vectors containing the 3D coordinates of a point and its corresponding patches in all the images. Subsequently, a single graph, which represents the similarity between two joint points, is constructed, by summing the three graphs.

In [ZNP13], two methods for person identity label propagation in facial images obtained from stereo videos are introduced. The proposed methods combine information from the left and the right channel of a stereo video and are based on the LNP algorithm, proposed in [WZ06]. Initially, a complete graph is constructed for each channel and the corresponding weight matrices $\mathbf{W}_{left}$ and $\mathbf{W}_{right}$ are calculated, based on the mutual information (MI) between 2-D hue and saturation histograms, extracted from the facial images. Subsequently, using the LNP algorithm on the k-nearest neighbors, a sparse graph is constructed from the aforementioned complete graphs, with weight matrices $\mathbf{W}_{Lknn}$ and $\mathbf{W}_{Rknn}$, for the left and right channel respectively. The first proposed method applies label propagation to the left and right channels separately, using the LNP algorithm, thus producing two classification matrices $\mathbf{F}^L$ and $\mathbf{F}^R$. Each stereo facial image is then assigned the label that corresponds to the maximum column of the matrix:

$$\mathbf{F}_{ij}^{max} = \max(\mathbf{F}_{ij}^L, \mathbf{F}_{ij}^R). \quad (5.3.55)$$

In the second method [ZNP13], label propagation is performed on the average graph weight matrix $\mathbf{W}_S$ of the left and right channels:

$$\mathbf{W}_S = \frac{1}{2}\mathbf{W}_{Lknn} + \frac{1}{2}\mathbf{W}_{Rknn}. \quad (5.3.56)$$

In a later study [ZTNP14], another novel method is introduced for propagating personal identity labels across facial images extracted from stereo videos. The proposed algorithm calculates a projection for each data representation, using a technique based on *Locality Preserving Projections* (LPP) [HN04], which incorporates side information in the form of pairwise similarity and dissimilarity constraints. Subsequently, label propagation is applied using both an early and a late fusion scheme. In the early fusion case, the final data representation is calculated as a linear combination of the projections of the different data representations. Label inference can then be performed using a method similar to the one proposed in [ZBL+04]. Late fusion is performed jointly on all different representations, by expressing the regularization framework proposed in [ZBL+04] as a weighted sum of multiple objective functions.

5.3.7 Label propagation on hypergraphs

The label propagation methods reviewed so far consider pairwise relationships between data samples. Nevertheless, real-world applications usually exhibit more complex relationships, which involve an arbitrary number of data samples. For example, in an application, the data may be characterized by multiple labels. In such cases, typical pairwise relationships represented by graphs are not sufficient to capture the relationships between the samples, which results in suboptimal solutions. In order to efficiently represent the complex relationships between data samples, hypergraphs can be adopted.

Let us assume a weighted hypergraph $\mathcal{H} = (\mathcal{E}, \mathcal{V}, w)$ and an initially labeled subset of nodes $\mathcal{S} \subset \mathcal{V}$, with labels $\mathcal{L} = \{l_j, \ j = 1, ..., L\}$. Label propagation algorithms in hypergraphs assign labels to the unlabeled nodes, so that nodes belonging to the same hyperedge are constrained to be assigned the same label. In [ZHS07], a hypergraph transductive label propagation method was introduced that utilizes hypergraph clustering. Given a classification function $\mathbf{f} : \mathcal{V} \to \mathbb{R}^{|\mathcal{V}|}$, the classification decision is given by a framework of the form:

$$\arg \min_{\mathbf{f}} \{R_{emp}(\mathbf{f}) + \mu \Omega(\mathbf{f})\}, \qquad (5.3.57)$$

where $R_{emp}(\mathbf{f})$ is an empirical loss term, $\Omega(\mathbf{f})$ is the clustering objective function, and μ is a regularization parameter.

In [CJ04] and [Tsu05], two information regularization frameworks for hypergraph label propagation are presented. The methods employ label probability distributions, instead of deterministic labels. The framework in [CJ04] minimizes the mixture-type information regularizer (m-regularizer), while the framework in [Tsu05] minimizes the exponential-type information regularizer (e-regularizer), which is the dual of m-regularizer. The advantage of e-regularizer over m-regularizer is that it has a closed form solution.

The method proposed in [SJY08] performs multiple label propagation using hypergraph spectral learning. It is based on the property that the hypergraph spectrum captures the correlation among labels. Furthermore, an approximate hypergraph spectral learning framework is introduced that is suitable for handling large scale multi-label propagation problems. It is based on the approximation of the hypergraph Laplacian matrix $\mathbf{L} \in \mathbb{R}^{N \times N}$ by $\mathbf{L} = \mathbf{H}\mathbf{H}^T$, where $\mathbf{H} \in \mathbb{R}^{N \times L}$ has orthonormal columns.

Another multi-label propagation method, called *Rank-HLapSVM*, is presented in [CZW+09]. It is based on hypergraph normalization. Its objective is the minimization of the ranking loss, while retaining a large margin. It incorporates the hypergraph Laplacian regularizer $\text{tr}\{\mathbf{F}^T\mathbf{LF}\}$ in the objective function of *Ranking-SVM* [EW01]:

$$\min_{\mathbf{F}} \frac{1}{2}\sum_{i=1}^{L}\|\mathbf{w}_i\|^2 + \frac{1}{2}\lambda\text{tr}\{\mathbf{F}^T\mathbf{LF}\} + C\sum_{i=1}^{N}\frac{1}{|y_i||\bar{y}_i|}\sum_{(p,q)\in y_i\times\bar{y}_i}\xi_{ipq}, \tag{5.3.58}$$

$$\text{s.t. } \langle\mathbf{w}_p - \mathbf{w}_q, x_i\rangle \geq 1 - \xi_{ipq}, \ (p,q)\in y_i\times\bar{y}_i, \ \xi_{ipq}\geq 0,$$

where $y_i \subset \mathcal{L}$ is a subset of labels, $\bar{y}_i \subset \mathcal{L}$ is its complementary set and ξ_{ipq} are *slack variables*. In [WWS+09], multi-label propagation with multiple hypergraphs was employed for music style classification. The method integrates three information sources: audio signals, music style correlations and music tag information/correlations. The multiple hypergraphs are combined in a single hypergraph that models the correlations between different modalities. This is done by constructing a hyperedge for each category that contains all the nodes that are relevant to the same category. Subsequently, the method performs hypergraph Laplacian regularization of the form $\text{tr}\{\mathbf{F}^T\mathbf{LF}\}$, similar to the single graph case described in subsection 5.3.3. This regularization ensures that the label assignment function $\mathbf{F} \in \mathbb{R}^{N\times L}$ is smooth on the hypergraph nodes. Hypergraph Laplacian regularization for semi-supervised label propagation is also applied in [DY08, THK09]. In [DY08], a random walk interpretation of hypergraph Laplacian regularization is also presented, as well as the extension of the normalized and the ratio cut (presented in subsection 5.3.3) to hypergraphs.

5.3.8 Label propagation initialization

The initialization of label propagation methods, i.e., the selection of the data samples that are manually initialized with labels, is crucial to the classification performance of the propagation algorithm [ZTNP14]. It has been observed from experiments that the classification accuracy of label propagation methods differs significantly for random initializations of the labeled data set $\mathcal{X}_L$. The problem of initializing label propagation methods is addressed in [ZTNP14]. The method is based on the intuition that the classification accuracy is high when it starts from the most representative class samples and from samples that lie to the border between classes. The initialization method in [ZTNP14] follows an iterative procedure for inserting data samples partially to the initially labeled data set $\mathcal{L}$. According to [ZTNP14], the most representative class samples are the ones that correspond to cluster centers while the data that lie to the border between classes are the ones in which the two highest label scores are similar. The method was employed for person identity labels on facial images extracted from stereo videos.

At the first step of the method, a clustering algorithm is employed on the data in order to extract information for the data structure and select the most representative data samples in each cluster. The most representative data sample of a cluster is considered the sample with the highest similarity to all other samples in the cluster, or equivalently, the sample with the highest within-cluster degree centrality d_i:

$$d_i = \sum_{j\in\mathcal{N}_c} W_{ij}, \tag{5.3.59}$$

where $\mathcal{N}_c$ the set of data that belong to cluster c and $\mathbf{W}$ the similarity graph. It can be observed that the number of data that enter the set $\mathcal{X}_L$ is equal to the number of the data clusters. Then, the initial state matrix $\mathbf{Y}$ is computed and label propagation is performed on the data set according to the decision rule:

$$l(\mathbf{x}_i) = \arg\max_{j} F_{ij}, \tag{5.3.60}$$

where $\mathbf{F}$ is computed by minimizing the regularization framework (5.3.17) introduced in [ZBL$^+$04].

The values in matrix $\mathbf{F}$ are an indication of the "certainty" with which the nodes are assigned a label. More specifically, nodes in which the two highest F_{ij} values are very close to each other are less likely to be assigned the correct label since they lie in a "border" region between two classes. Label assignment to these nodes is considered to be "uncertain". Therefore, the nodes which were assigned a label with the least certainty form the next set of nodes that will be manually labeled and inserted in the labeled nodes set $\mathcal{X}_L$. Then, the initial state matrix $\mathbf{Y}$ is updated, to include the newly inserted labeled samples and label propagation is performed again according to (5.3.18) and (5.3.60). The procedure is repeated and the labeled set $\mathcal{X}_L$ is enriched with a predetermined number of nodes at the time with the least label assignment certainty, until the initially labeled dataset $\mathcal{X}_L$ cardinality is a determined percentage of the overall data number. Usually, the procedure is repeated until $5 - 10\%$ of the data enter the set $\mathcal{X}_L$.

5.3.9 Applications in digital media

The increasing spread of digital media acquisition devices, e.g. cameras and video/audio recorders, has led to a radical expansion of the data residing in users' collections. Furthermore, a vast amount of multimedia data is created, accessed and processed on the Web: on-line multimedia sharing communities, such as Flickr, Picassa and YouTube, as well as on-line social networks, such as Facebook, Google+ and Twitter, allow users to share, rank and annotate multimedia objects. Since the aforementioned websites have grown in popularity over the last years, a huge amount of on-line multimedia data is available.

Label propagation for annotation of video can be performed at pixel level [CC10, BGC10, VG12]. Labels of initially labeled pixels in a small set of video frames are propagated to pixels in the remaining video frames. In this case, a pixel label refers to the semantic concept of the entity it belongs to (e.g., an object depicted in a video frame). The aforementioned process results in intra video frame segmentation [CC11]. Furthermore, label propagation can be applied at a video frame level, i.e., instead of single pixels, labels may characterize entire video frames [QHR$^+$07, THW$^+$09, WHTH09], video snippets [ZXZ$^+$12], or moving/still regions, e.g. persons that appear in the video frames [CCC11]. In these cases, graph nodes represent video frames, video snippets or regions of interest (ROIs) that enclose an object or a person, respectively.

In image annotation methods, label propagation can be also performed in both pixel and image level. The objective of pixel level methods is mainly to automate the image segmentation process, rather than to propagate labels across images. These methods assume that the labels of a small subset of pixels, called seeds, is initially known. Labels of seed pixels are subsequently propagated to the remaining, unlabeled image pixels [Gra06, WWL07, RLF12, KGF12, YWY$^+$13, ML13]. In methods applied at image level, each image is represented by global feature descriptors and is treated as a graph node [GMVS09, CMYC10, HOSS13, HLL$^+$14]. Several methods combine image- and word-based graph learning, by incorporating textual information derived from labels [LLL$^+$09, LHWH13]. As a result, these methods achieve better consistency between image similarity and label similarity. Furthermore, instead of propagating each label individually, several recently proposed multi-label propagation methods spread different labels simultaneously, taking the relationships among the labels into account [CMYC10, BNMY11, LHWH13].

Label propagation methods in digital media are governed by the same rules and assumptions as any graph-based label propagation method. Therefore, any of the graph construction and label inference methods presented in Sections 5.2 and 5.3 can be applied to the case of digital media data. Moreover, the intrinsic characteristics of digital media data can also be exploited in the label propagation process. Throughout Sections 5.2 and 5.3 several methods have been presented that exploit this kind of information for improving the propagation performance.

In many cases, the dataset labels contain "noise," which negatively affects the performance of the label propagation algorithms. Noise may result from incorrect or incomplete annotations provided by users or from poor performance of the label propagation algorithm. In order to deal with the label noise existing in image datasets and to provide labels of improved quality, several methods, referred to as retagging or tag ranking algorithms, have been proposed [WJZZ06, LHY+09, THY+11, LYHZ11, TLLZ14].

Finally, label propagation algorithms find application in *recommendation systems* [SK09]. Recommendation methods that exploit information derived from user ratings and preferences are usually referred to as *collaborative filtering* methods. In addition, content-based recommendation algorithms utilize information related to the multimedia content itself, derived either from metadata or by data feature extraction. Recommendation methods in large video collections perform label propagation on multiple video graphs, and the recommendation results are personalized for each user, [LLH+07, THQ+07, BSS+08]. They exploit co-view information represented by co-view graphs, where graph nodes correspond to videos, while edge weights represent the number of users that have watched the two videos. Furthermore, recommendation, apart from multimedia objects, may concern groups of users with similar interests to a user. In the method proposed in [YJHL11], visual content of images along with text annotations are employed by a label propagation scheme, to recommend user groups according to users' personal photo collections.

5.4 Diffusion processes

Diffusion and label propagation are two closely related research fields, since label propagation can be regarded as an information diffusion process over a graph. In the following subsections, the notions of diffusion in physics, in sociology, as well as in social media are briefly discussed.

5.4.1 Diffusion in physics

Diffusion in physics is used to describe the flow of energy or mass within a medium, which is common in a bundle of physical processes referred to as *transport phenomena*, such as molecular diffusion and heat transfer [AF67].

Molecular diffusion refers to the motion of liquid or gas molecules, due to thermal energy dissipation. The rate of diffusion depends upon the temperature, the viscosity of the fluid, as well as the particle mass. The molecules move from a region with high concentration to one with a lower concentration, which results in a gradual mixing of the material. When the concentrations of the compartments of the material being mixed become equal, *diffusive equilibrium* is reached. Molecular diffusion is described by *Fick's second law*:

$$\frac{\partial \phi}{\partial t} = D\left(\frac{\partial^2 \phi}{\partial x^2} + \frac{\partial^2 \phi}{\partial y^2} + \frac{\partial^2 \phi}{\partial z^2}\right), \tag{5.4.1}$$

where $\phi(x, y, z, t)$ denotes the concentration over the axes x, y and z over time t, while D is the diffusion coefficient that controls the rate of diffusion.

Heat conduction refers to the transfer of energy within and between bodies, as a result of a temperature gradient. Heat flows spontaneously from bodies (or body parts) of higher temperatures to bodies of lower temperatures, until a thermal equilibrium state is reached, when the bodies have the same temperature. The evolution of temperature $T(x, y, z, t)$ in a homogeneous, finite, three-dimensional body is described by the *heat equation*:

$$\frac{\partial T}{\partial t} = \gamma \left(\frac{\partial^2 T}{\partial x^2} + \frac{\partial^2 T}{\partial y^2} + \frac{\partial^2 T}{\partial z^2} \right). \tag{5.4.2}$$

In the above equation, $T(x, y, z, t)$ is the spatio-temporal temperature diffusion over x, y, z and t, while γ denotes *heat diffusivity*, which is the quotient of thermal conductivity κ and heat capacity c. As already mentioned (subsection 5.3.4), the heat equation on a graph G, is associated with the graph Laplacian [CY00, KL02]:

$$\frac{d\mathbf{K}_t}{dt} = -\mathbf{L}\mathbf{K}_t, \tag{5.4.3}$$

where $\mathbf{L}$ is the graph Laplacian and $\mathbf{K}_t$ the heat kernel (equation (5.3.38)), which satisfies the initial condition $\mathbf{K}_0 = \mathbf{I}$. The heat equation on graphs has been adopted in the construction of models of influence [MYLK08, ZL13, BLL$^+$14], as well as in label propagation [WKL11].

5.4.2 Diffusion in sociology

The way information and ideas are communicated within social groups has long been a topic of research in sociology. The *Diffusion of Innovations* theory [Rog95] defines diffusion as the procedure through which an innovation, such as a new idea or product, is spread and adopted over time by the members of a social system. A social system is defined as "a set of interrelated units that are engaged in joint problem-solving to accomplish a common goal." These units can be individual persons, groups of people, corporations or subsystems. The social system constitutes the context of innovation diffusion. Its characteristics determine the *rate of adoption*, i.e., the number of social units that adopt the innovation in a certain time period. The following parameters are decisive in spreading and adopting an innovation: compatibility, complexity, trialability and observability to those within the social system.

The degree of eagerness of a social unit to adopt an innovation characterizes its inovativeness. The social units are divided into five categories, according to their inovativeness: *innovators*, *early adopters*, *early majority*, *late majority*, who are the first 2.5%, next 13.5%, next 34%, next 34%, respectively, of the social units that adopt an innovation and, finally, *laggards*, who are the last 16% of the social units that adopt an innovation. This theory, especially the concept of the adopters categories, has been used in several studies of information diffusion in social networks [KKT03, KKT05, MYLK08].

5.4.3 Diffusion in social media

The evolution of the Internet over the last decade has created new forms of social communities and interactions. The expansion of on-line social network services, such as Facebook, Google+ and Twitter, is closely related to the emergence of Web 2.0, that offers the users the possibility to share information and interact. The development and the growing popularity of such networks is indicative of the transition from older on-line communities where users participated according to their interests (for example business, music, technology), to generic social networking services, where users have the opportunity to communicate to each other, as well as to contribute and share all kind of information.

FIGURE 5.4.1: Information cascades in a social network.

Today, on-line social networks play a key role in information spread and have radically influenced the way ideas, news and trends are communicated among individuals. Information cascades in a network of individuals occur as information or behaviors spread from node to node through the network. For example, a user A in a social network may read an article and share it with user B, who may in turn share it with his/her friends and so on. Such an example of information cascades is illustrated in Figure 5.4.1, where nodes correspond to individuals and arrows indicate the information flow from one individual to another.

Understanding the underlying mechanisms of information diffusion in on-line social networks can be useful in various cases, such as tracking the evolution of specific topics [LBK09], preventing misinformation [NYTE12] and optimizing marketing campaigns [KKT03]. As a consequence, the study of information diffusion cascades in social networks has become an active research field in the last years.

5.5   Social network diffusion models

Modeling the way information is spread among the individuals in a social network is crucial to a diversity of applications, such as viral marketing or emergency management. For this purpose, several approaches have been proposed. In *epidemic* diffusion models, information diffusion is compared to the spread of an infectious disease within a population. Furthermore, *game theoretical* approaches have been employed, where social interactions and individual's behaviors are modeled using concepts from *game theory*. Finally, the commonly used *threshold* and *cascade* models are based on relationships of influence between individuals.

Apart from exploring the ways in which information is spread within a network, several studies are concerned with the problem of *influence maximization*. This consists of discovering the most influential individuals (nodes) of the network, such that, when information is disseminated through these nodes, maximal information spread is achieved. Influence maximization can be of great importance in marketing applications, where the aim is to reach the greatest possible number of customers in a target group.

In the following subsections, we shall discuss the most important models that have been proposed for the study of information diffusion in social networks, as well as for influence maximization. Finally, other applications of information diffusion, apart from influence maximization, are also reviewed.

5.5.1   Game theoretical diffusion models

Social interactions can be modeled using concepts adopted from *game theory* [vNM44]. Individuals are regarded as players, participating in a game (the social interaction), where each individual may choose among a number of strategies, corresponding to different behaviors. Each available choice or *strategy* is assigned a *payoff*, according to a payoff function. This payoff depends not only on player's choice, but also on choices made by the other players. In classic game theory, the players select the choices that lead to the maximization of their payoff. Usually, each player is more likely to interact with certain other players, for example with his/her friends in a real-world scenario, than interacting with all other players. The local nature of such interactions is captured by local interaction games, studied in [Blu93, Ell93, Mor00].

Classic game theory assumes that the players' choices are driven by rationality. However, this fact does not reflect real world situations, because people do not always make decisions according to strictly rational criteria. Furthermore, in real-world situations, individuals tend to alter their behavior, as a result of the experience gained through the interactions and through observation of the behaviors of other people. These aspects of human behavior can be better captured using concepts from evolution game theory [Smi72], which studies the dynamics of the behavior of large populations of players that repeatedly play a game, subjected to evolutionary changes. The player behaviors adjust over time, as a result of experience gained through repetitions.

5.5.2   Epidemic diffusion models

The spread of information through individuals in a social network bears similarities to the transmission of infectious diseases through the population. As a result, various models used in *epidemiology* have been adopted in information diffusion studies. Epidemic models have been developed in order to study how infectious diseases break out and spread over a population [KM27, Bai57, PSV01]. These models describe the disease cycle in a host, using different terms to characterize each stage; for example, when the host is susceptible to the disease (S), becomes infected/infectious (I), or has recovered/is no longer infectious (R). In the *Susceptible-Infected-Recovered* (SIR) epidemic model, a person is initially susceptible to the disease, then becomes infected, and finally recovers. In the recovered stage, the person becomes immune to the disease and is no longer able to infect others. On the contrary, in the *Susceptible-Infected-Recovered-Susceptible* (SIRS) epidemic model, the recovered individuals become susceptible to the disease again. According to the *Susceptible-Infected-Susceptible* (SIS) model, the persons do not become immune to the disease after the infection, but they return directly to the susceptible stage. Finally, *Susceptible-Infected* (SI) models are used to describe fatal diseases, where the infected individual never returns to the recovered or susceptible stages.

In a social interaction context, the notions of susceptibility, infection, and recovery can be adopted to describe the relationship between a person and the mode of being spread (e.g., information). For example, a susceptible person does not yet know the information, an infected person gets to know and can transmit the information, while a recovered person forgets it or has lost interest in it. In [WYH⁺13], two epidemiology-driven models are proposed, which extend the SI and the SIR models respectively, for the analysis of information diffusion in online social networks.

The suitability of the SIR model for large real-world networks is investigated in [BLT12]. [OR13] explores how the structure of the underlying network affects the diffusion process, using the SIR model. In [YQZC12] a SIR model is used to study the information spread taking place in a conjoint framework, including conventional communication, as well as interaction in online social networks. Another work adopting a SIR model can be found in [WSC11], where the spread of violent or extremist topics in social media is explored. The population is divided into three different classes: the susceptible class consists of users who are interested in a topic and read relevant posts. The infected users are those who write posts or answer to threads, once they have read posts associated to a topic. Finally, the recovered class consists of authors whose posts become uninfective. In [XL10], the SIS model was used to simulate the diffusion process in a synthetic Barabási-Albert network and in Facebook, in order to explore their differences. The proposed method takes into consideration the decaying infectiousness of a topic, as well as the generation of new sub-topics during diffusion. Another method utilizing the SIS model for social network simulation is proposed in [SKM09], to study the problem of discovering influential nodes in a network.

5.5.3   Threshold diffusion models

Threshold diffusion models were originally proposed by Granovetter [Gra78] and Schelling [Sch78] and assume that the adoption of an innovation by an individual depends on the number of individuals that have already adopted it. A social network can be modeled by a graph $G = (\mathcal{V}, \mathcal{E})$, where nodes represent individuals and edges their relationships of influence. The edge weights express the probability that a node influences its neighbor. Furthermore, each node is assigned a threshold value. A node is considered to be active or inactive, if it has or has not yet adopted the innovation, respectively. Each node is activated when the fraction of its active neighbors surpasses its threshold value.

The *Linear Threshold* (LT) diffusion model [Gra78] has received much attention. According to this model, each edge is assigned a weight w_{ij} representing the probability that the node i is influenced by the node j, with $\sum_{j \in \mathcal{N}_i} w_{ij} \leq 1$. Additionally, each node i is assigned a threshold value θ_i, that represents the fraction of its neighbors required to have adopted the innovation, so that i adopts it. The Linear Threshold model assumes that a subset of nodes $\mathcal{S} \subset \mathcal{V}$ are initially active. The diffusion process subsequently unfolds at discrete time steps as follows: at time t, the nodes that were activated at $t-1$ remain active, while an inactive node i is activated if:

$$\sum_{j \text{ active neighbor of } i} w_{ij} \geq \theta_i . \tag{5.5.1}$$

Given the threshold values θ_i, the LT diffusion process is deterministic. However, this hypothesis can be lifted [KKT03], if the thresholds are selected uniformly at random in the range $[0, 1]$. In [KKT03], a *General Threshold* (GT) model is introduced, which constitutes an extension of the LT model. In this model, the threshold θ_i of a node i is substituted by a monotone activation function f_i, defined on the set of its neighboring active nodes, which takes values in $[0, 1]$. Taking the aforementioned function into account and following the diffusion procedure discussed above, a node i becomes active at time step t, if $f_i(\mathcal{N}) \geq \theta_i$, where $\mathcal{N}$ denotes the set of its neighbors that are active at time $t - 1$. Threshold θ_i is uniformly chosen at random in the interval $[0, 1]$.

A Linear Threshold model is also adopted in [Wat02], where *global cascades* on random networks are studied. *Global cascades* are cascades that occur infrequently, are triggered by a small seed set of nodes and affect a large portion of the network. The emergence of a global cascade and its governing mechanism have large complexity and vary from system to system. Global cascades can be observed in social and economic systems, as well as in cascades occurring upon failures in physical infrastructure networks. In [Wat02], global cascades are modeled using a sparse random network of interacting individuals, whose decisions are determined by their neighbors' decisions, in accordance to a threshold rule. Initially, every individual is in state 0 (inactive). At time $t = 0$, a small fraction of nodes in state 1 (active) perturbs the graph. At subsequent time steps, the nodes update their states, according to the threshold rule, in a random order. It is assumed that every node in the random graph can be adjacent to at most one seed member, an approximation that is exact only in an infinite network. As consequence, the seed can grow only if at least one of its neighbors has a threshold θ, such that $\theta \leq 1/k$. These nodes are called vulnerable. The success of a cascade depends more on the number and the connectivity of the vulnerable nodes and less on the number of seeds. For a global cascade to occur, it is required that the largest connected vulnerable cluster occupies a finite fraction of an infinite network.

In [AOY11], a stochastic linear threshold model is proposed, that extends the LT model by capturing path dependence, which refers to the fact that the outcome of the diffusion may depend heavily on minor shocks or insignificant events. In contrast to the LT model, in the proposed model, individuals do not necessarily adopt the innovation, if the fraction of their active neighbors surpasses its threshold value. Alternatively, there is a non-zero probability that the individual rejects the adoption. Therefore, at time step t, if the fraction of the active neighbors of a node i is greater than a threshold value, node i will consider the adoption of the innovation and subsequently decide whether to adopt it or not. The outcome of this decision is regarded as a Bernoulli trial, with a parameter $p \in [0, 1]$ denoting the likelihood that the node adopts the innovation, conditioned upon the consideration.

The behavior of the LT model, when applied on a multi-layered social network was studied in [MKJ13]. Multi-layered social networks (MSNs), also referred to as multiplex networks, are networks consisting of more than one layer, each of them representing a different type of interaction between individuals, thus providing a more natural and realistic model of communication in a social network. The nodes are shared between the layers, while their linking varies according to the relationships between the nodes, given the type of interaction corresponding to the layer. In [MKJ13], the authors study how diffusion in the multi-layered network is influenced by the layer number and type, using the LT model.

Finally, a generalization of the LT model is introduced in [PBS10]. In contrast to other diffusion methods, the proposed algorithm considers multiple cascades on the graph, while the nodes can switch between them. Specifically, the proposed model assumes that K different cascades propagate in a graph $G = (\mathcal{V}, \mathcal{E})$. Each node can be either activated by one of the cascades or be inactive, thus being in one of $K + 1$ states. The cascades are simulated using a stochastic graph coloring process, which is proven to be equivalent to a rapidly mixing Markov chain.

5.5.4   Cascade diffusion models

Similarly to threshold models, *cascade models* rely on the hypothesis that the decisions of individuals are influenced by the decisions of other people they interact with. Again, cascade models can be represented with graphs, where the nodes are characterized as active, when they have adopted the innovation, and inactive otherwise. A commonly used model is the *Independent Cascade* (IC) model [GLM01]. According to this model, each node i that becomes active at the (discrete) time instance t is given a single chance to activate each of its currently inactive neighbors j with a probability p_{ji}. Node j will become activated at time $t + 1$, if the attempt of i, or of any other neighbor of j that was also activated at time t, is successful. Once node i has made an attempt to activate its inactive neighbors, it remains active, but it cannot continue activating its neighboring nodes.

A generalization of the IC model is presented in [KKT03]. Unlike the IC model, the proposed *General Cascade* (GC) model takes into account previous attempts for activating of a node performed by its neighbors. More specifically, the probability that the node i activates a neighbor j is calculated by an incremental function $p_j(i, \mathcal{S}) \in [0, 1]$, where $\mathcal{S}$ represents the set of active neighbors of j that have already attempted to activate j. It should be noted that, similar to the IC model, at time step t, all the active neighbors of j attempt to activate it arbitrarily. Therefore, the order in which the attempts take place does not affect the result of the diffusion process. Additionally, if a node i activates node j, further contaminations (activations) by the rest of the neighbors at the same time have no effect. As a result, it is unknown in these models, who contaminated whom. The algorithm proposed in [GRLK10] raises this assumption, by considering the diffusion process in the continuous time domain. Notions are introduced, such as the *hit time* t_j, describing the time when a node j is contaminated by a specific contagion c initiated at some node and the *incubation time*, that is, the time it takes for a contaminated node v to contaminate its neighbors after the hit time. Then the problem of tracing the paths of diffusion in the network is explored.

According to the Independent Cascade (IC) model, each newly activated node is given exactly one chance to activate its inactive neighbors. However, in real-world scenarios, this assumption does not hold, since social interactions are history-dependent. For example, a consumer is usually exposed multiple times to the same advertisement, before he/she decides to buy a product. This fact is taken into consideration in the *History Sensitive Cascade Model* (HSCM) proposed in [Zha09], which is a modification of the IC model. The HSCM model allows the newly activated nodes more than one chance to influence their neighbors.

5.5.5   Influence maximization

In many applications, the objective of information diffusion is to influence the largest possible population within a social network. Examples of such cases include marketing strategies for reaching a wide public to sell products to, pre-election campaigns, or the promotion of certain concepts or ideas, in general. The maximization of influence problem seeks to answer the following question: which set of individuals should the diffusion process start from, in order to obtain the maximum spread? In terms of label propagation on multimedia data, this is related to determining the initial set of labeled data that optimize propagation [ZNP13].

In an early study [DR01], influence maximization is studied from a viral marketing perspective. *Viral marketing* aims at increasing product sales, by exploiting the influence between customers. Instead of treating all the customers in the same way, a company may seek a suitable group of customers, who are the most influential, to promote a marketing plan to. As opposed to the intrinsic value of a customer, measured by the expected profit from directly selling to him/her, the notion of the network value of a customer is introduced, describing the expected profit from sales to other individuals that indirectly result from selling to a particular customer. This is a consequence of the influence one customer has on other customers, and the influence that those customers have on others and so on.

The network value of a customer depends on a number of factors. First, high network connectivity plays an important role. However, successful product marketing also depends on the positive or negative rating of the customer. Second, the influence between customers must be asymmetric. A customer with high network value should influence others much more than others influence him. Third, the network value of a customer is not related just to his/her immediate neighbors, but these neighbors can, in turn, influence other people. This implies that, even if a customer is not highly connected, he may have a strong network value, if his/her neighbors are highly connected. In [DR01] the market is modeled as a probabilistic model of interaction, utilizing a Markov random field. Heuristics are used for choosing customers with a large overall effect on the network. A method is proposed to infer the necessary influence data. A later work [RD02] extends the ideas presented in [DR01], by adopting a linear model to represent the influence between the nodes, which significantly simplifies the computations.

Another approach to influencing maximization is introduced in [KKT03], by finding a set of k initially active nodes that yield the largest expected cascade. In contrast to [DR01], which adopted descriptive models for node interaction, operational models are used in [KKT03], such as the Linear Threshold and Independent Cascade models discussed earlier, that represent step-by-step adoption dynamics. It is proven that the optimization problem of discovering the nodes that maximize the influence is NP-hard. Therefore, the influence maximization problem is addressed by a greedy hill-climbing algorithm, developed in a general framework based on submodular functions. In later work [KKT05], it is proven that the maximization problem can be extended to a very general cascade model, named *Decreasing Cascade* model. According to the Independent Cascade model, the probability $p_j(i)$ that a node j is activated by an active neighbor i is independent of any previous failed attempts conducted by other nodes. Conversely, in the General Cascade model, the influence probabilities depend on previous attempts for node activation by its active neighbors. In this case, probabilities are expressed by an increasing function $p_j(i, \mathcal{S}) \in [0, 1]$, where $\mathcal{S}$ represents the set of active neighbors of j that have already attempted to activate it. According to the Decreasing Cascade model, the functions $p_j(i, \mathcal{S})$ are considered to be non-increasing in $\mathcal{S}$, which means:

$$p_j(i, \mathcal{S}) \geq p_j(i, \mathcal{T}) \quad \text{for} \quad \mathcal{S} \subseteq \mathcal{T}, \tag{5.5.2}$$

where $\mathcal{T}$ denotes the set of nodes. The Decreasing Cascade model assumes that the probability of active node i activating an inactive node j decreases if other nodes have already attempted to activate it, since node j is more "marketing saturated."

In [LKG+07], a problem similar to influence maximization is studied, namely that of *outbreak detection*. This constitutes selecting a set of nodes in a network, such that the detection of the spread of contamination or information is achieved as quickly as possible. The method proposed in [LKG+07] takes into consideration the submodularity property of the influence maximization objective, in order to reduce the number of evaluations of the influence spread function and to obtain solutions that are close to optimal ones. The proposed algorithm, called *Cost-Effective Lazy Forward selection* (CELF) performs 700 times faster than the simple greedy algorithm of [KKT03]. In [CWY09] and later in [CWW10], a new scalable and tunable heuristic is proposed, that is capable of handling influence maximization in large-scale social networks, which outperforms the algorithms of [KKT03] and [LKG+07]. Other influence maximization methods that optimize the CELF algorithm, can be found in [GLL11a] and [GLL11b].

In [CYZ10], a method for scalable influence maximization under the Linear Threshold model is proposed, which adopts *Directed Acyclic Graphs* (DAGs) in order to propagate influence in linear time. The proposed algorithm for influence maximization, called a *Local Directed Acyclic Graph* (LDAG), constructs a local acyclic graph for each node i. Subsequently, influence is propagated from the seed set to node i only through its local DAG, according to the Linear Threshold model.

In [KS06], two new diffusion models are proposed, based on the Independent Cascade model and is studied the problem of influence maximization under these models. According to the first one, called the *Shortest-Path Model* (SPM), each node is activated only through the shortest paths from the seed set, i.e., the nodes that are initially active. More specifically, denoting the seed set by $\mathcal{A}$ and the graph distance from node j to node i by $d(j,i)$, the distance between $\mathcal{A}$ and node i is defined as $d(\mathcal{A},i) = \min_{j \in \mathcal{A}} d(j,i)$. Therefore, in the Shortest-Path model, each node can become active only at time step $t = d(\mathcal{A},i)$. In the second model introduced in [KS06], an extension to the SPM model called the SP1M model is proposed, where each node has a chance to be activated only at time steps $t = d(\mathcal{A},i)$ or $t = d(\mathcal{A},i) + 1$.

In [VK11], the spread of influence is studied, by adopting the Linear Threshold model. Given an initial activation set, recursive expressions for the expected number of the finally influenced nodes are derived, which can be interpreted using acyclic path probabilities in discrete Markov chains. Furthermore, the aforementioned study is applied on different network topologies, such as star, ring, and mesh to derive the optimal initial set for these networks.

Another problem closely related to influence maximization is the diffusion of multiple, competing innovations within a network. A real-world scenario reflecting this case is the competition between companies selling products of the same type, in a quest to dominate the market. In [BKS07], a model is introduced which modifies the IC model, so as to simulate multiple competing innovations. In the proposed model the notion of activation time is introduced, in order to take into account the multiple activation attempts on a node, taking place at the same time step. Therefore, an inactive node i adopts the innovation being spread by the first neighboring node that managed to activate it. In [KOW08], the diffusion of competing rumors was studied, using a game theoretical approach. Other game theoretical approaches to the competing diffusion processes problem can be also found in [AFPT10, TAM12, SA12, GK12].

Although in applications such as viral marketing campaigns the objective is to promote a product or an idea to as many individuals as possible, other applications aim at minimizing the influence spread. Such an example is the limitation of misinformation. This problem bears similarities to the diffusion of competing innovations in a network, where the "good" information competes against misinformation. The techniques used in these scenarios are similar to those of influence maximization. The problem of limitation of misinformation can be defined as discovering a small set of nodes, such that if "good" information is disseminated to them, the effect of the misinformation is minimized, i.e., it reaches the smallest possible number of individuals [BAEA11, NYTE12]. In [BAEA11], a *Multi-Campaign Independent Cascade Model* (MCICM) and a *Campaign-Oblivious Independent Cascade Model* (COICM) are introduced, based on the Independent Cascade model, which describe the diffusion of two competing campaigns (a "good" and a "bad" one), that take place simultaneously in a network. In contrast to the Independent Cascade model, the proposed methods consider the order in which activation attempts occur, when more than two active neighbors try to activate a node i at the same time step. If multiple nodes try to activate an inactive node at the same time step, the "good" campaign is prioritized over the "bad". In the MCICM model, each newly activated node is given a single chance to activate its inactive neighbor i in one of the campaigns, with different probabilities for the two campaigns. In the COICM model, which is similar to the one proposed in [BKS07], the probabilities that a node is activated are the same for the two campaigns. Finally, in [KG13], evolutionary game theory is employed in order to study the spread of misinformation in online social networks.

5.5.6   Cross-Media information diffusion

As already mentioned, online social media play a key role in the spread of trends, news, and opinions. Furthermore, upon the occurrence of an important event, for example elections, natural disasters, or sports events, covered by conventional broadcast audiovisual (AV) media, increased user activity in social media is observed. Such activity reflects people's immediate reactions to the broadcasted event, as well as the process of information/concept diffusion from broadcast media to social media. As an example, during the live broadcasting of a football game on television, relevent tweets emerge in Twitter, while increased tweeting activity takes place upon highlights of the game. By monitoring and analyzing this process, trending and shifting in public opinion can be discovered. This kind of information can be useful in various scenarios, such as in policy making, journalistic investigation, and market analysis.

Jointly broadcasted AV content analysis and social media analysis are relatively new research areas. Work in this field mainly targets the semantic description of large-scale broadcasting events, such as political debates or speeches, through social media analysis, primarily tweets published by viewers, while watching the event [SKC09, DNKS10, HIF13, SCA14]. To this end, statistical tweet analysis is performed, in order to detect the major trends and changes in the semantic structure and content of the event. Moreover, textual tweet analysis is performed, in order to extract the semantic concepts of the event, as well as the sentiments caused to the audience. The aforementioned works employ hashtags and time information for relating tweets with specific video timestamps. Furthermore, several methods have been developed for describing broadcasted AV content, by combining transcribed text and visual metadata [CMC05, IDF$^+$05, GWZ$^+$13, Xu14]. Information describing social network activity during a broadcasted event (e.g., textual/visual information, likes or re-tweets), can be combined with broadcasted AV content description, as shown in Figure 5.5.1. In this way, e.g., the tweet number can be mapped on the audiovisual content timeline, together with actor/speaker timelines. Moreover, audiovisual and social media content description can be combined in a joint rich AV and social media content descriptor.

5.5.7 Other applications of information diffusion

Diffusion methods find application in collaborative filtering algorithms. These methods are commonly employed by recommendation systems, in order to make predictions about the interests of a user, by collecting information regarding the preferences and tastes of other users, expressed through ratings [TH01]. Recommendation systems focus on algorithms for matching users based on their preferences, and weighting the interests of users with similar taste to produce a recommendation for the information seeker. In [SZZZ10], a recommendation model, which uses ternary relationships among users, objects and tags is proposed. A new measure of user similarity is introduced, which incorporates user preferences of both collected objects and used tags. This similarity measure is calculated using a diffusion-based process.

Furthermore, information diffusion methods are adopted for community detection. In [AHH11] and [HAHH12] game theory is adopted for the detection of overlapping communities, i.e., communities whose nodes may belong to more than one community simultaneously. Community formation is considered as an iterative game, where players try to join communities consisting of individuals that share similar interests with them. In [BBM13], another method for community detection is proposed, which is based on information cascades in a social graph.

Another field of study of information diffusion is the citation network. In [STA09], citation networks of publications in computer science are studied from an information diffusion perspective. The structural features of the information paths through these networks, as well as their impact on the information flow are analyzed. Additionally, variations in information diffusion for specific subsets of citation networks are investigated: books versus conference and journal articles, as well as coverage of different computer science domains,

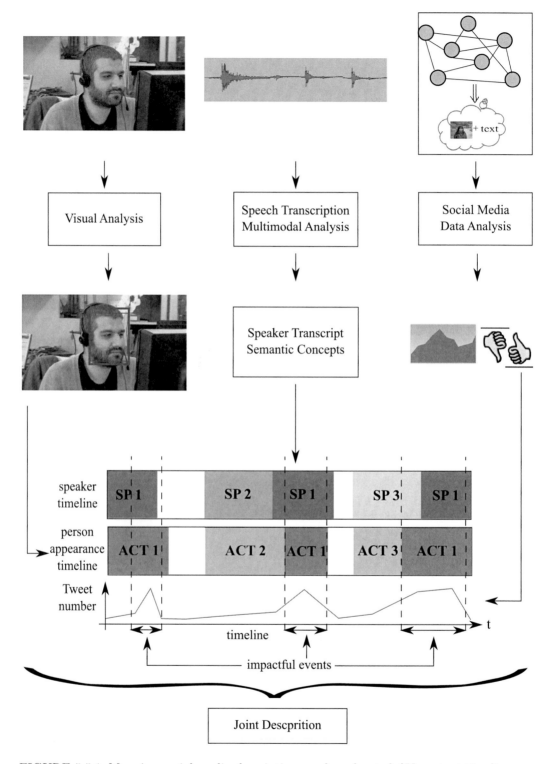

FIGURE 5.5.1: Mapping social media descriptions on broadcasted AV content timeline.

and different time periods. The basic consideration is that many citations are evidence of information flow from one article (and its authors), to another one.

Finally, information diffusion methods find application in emergency situation management, where mass media function as channels for information spread. In the last decade, Internet has played a dominant role in the diffusion of news, especially in cases of emergency. In [YLL09], challenges for information diffusion in the area of emergency events management are discussed. An overview of diffusion models, such as threshold and cascade models is provided. Additionally, four types of emergency situations are investigated using the aforementioned diffusion models: natural disasters, technological disasters, public health incidents, and security incidents. In [HTW+12], a diffusion model is employed in order to study information cascades on Twitter, which occur as a response to an actual crisis event and its accompanying alerts or warning messages from emergency managers. The types of information exchanged during a crisis situation are defined. Furthermore, the way messages spread among the users on Twitter, including the kinds of information cascades or patterns are observed. These patterns provide knowledge on information flow through the network. The proposed method can reveal properties of the diffusion process during the emergency event, offering emergency managers more effective ways to facilitate the spreading through social media or to impede the spread of information. In [ZXP+11], diffusion of information regarding disaster preparation in Twitter was studied. User's re-tweeting behavior was analyzed by studying the factors that might affect their decisions, such as content and network influences, as well as time-related factors.

5.6 Conclusions

In this chapter we discussed graph-based label propagation, with a focus on multimedia applications. Label propagation is a powerful semi-supervised classification method. In contrast to the classical supervised classification methods, label propagation methods require only a few data with known labels from which label inference commences. The basic steps of all the label propagation methods were reviewed, namely graph construction and label inference. Graph construction methods were grouped into three categories, according to the method employed for calculating the edge weights. A common property of all graph construction methods is that they employ local information. This way, the constructed graphs are more robust to noisy data and outliers. Label inference methods are divided into groups, according to the graph type employed (one or more graphs, hypergraphs, etc.) and the label inference rule (iterative, manifold regularization, etc.). Label inference methods employ both local and holistic data information for robust propagation performance. Since label propagation constitutes a diffusion process, several topics regarding information diffusion in social networks have been also discussed. Particular emphasis was placed on the review of the most important models for information diffusion, as well as on the influence maximization problem.

Label propagation methods have been successfully employed in several multimedia-related applications, such as annotation of multimedia content and recommendation systems. Nevertheless, several limitations need to be resolved, such as the insufficiency of the available labeled data, the curse of dimensionality, as well as the choice of appropriate distance measures and graph construction techniques for the data.

Regarding information diffusion, certain challenges have yet to be overcome, such as the selection of the right diffusion model for representing real-world networks and the development of models that reflect realistic properties of social interactions.

Bibliography

[AF67]   M. Alonso and E. J. Finn. *Fundamental University Physics*. Addison-Wesley Publishing, 1967.

[AFPT10]   N. Alon, M. Feldman, A. D. Procaccia, and M. Tennenholtz. A Note on Competitive Diffusion Through Social Networks. *Information Processing Letters*, 110(6):221–225, 2010.

[AHH11]   H. Alvari, S. Hashemi, and A. Hamzeh. Detecting Overlapping Communities in Social Networks by Game Theory and Structural Equivalence Concept. In *Proc. 3rd International Conference on Artificial Intelligence and Computational Intelligence*, volume II, pages 620–630, 2011.

[AHP05]   A. Argyriou, M. Herbster, and M. Pontil. Combining Graph Laplacians for Semi-Supervised Learning. In *Proc. Advances in Neural Information Processing Systems 18*, pages 67–74, 2005.

[AOY11]   D. Acemoglu, A. Ozdaglar, and E. Yildiz. Diffusion of innovations in social networks. In *Proc. 50th IEEE Conference on Decision and Control and European Control Conference*, pages 2329–2334, 2011.

[Azr07]   A. Azran. The Rendezvous Algorithm: Multiclass Semi-supervised Learning with Markov Random Walks. In *Proc. 24th International Conference on Machine Learning*, pages 49–56, 2007.

[BAEA11]   C. Budak, D. Agrawal, and A. El Abbadi. Limiting the Spread of Misinformation in Social Networks. In *Proc. 20th International Conference on World Wide Web*, pages 665–674, 2011.

[Bai57]   N. T. J. Bailey. *The mathematical theory of epidemics*. Griffin, 1957.

[BBC^{+}94] R. Barrett, M. Berry, T. F. Chan, J. Demmel, J. Donato, J. Dongarra, V. Eijkhout, R. Pozo, C. Romine, and H. V. der Vorst. *Templates for the Solution of Linear Systems: Building Blocks for Iterative Methods*. SIAM, 2nd edition, 1994.

[BBM04]   M. Bilenko, S. Basu, and R. Mooney. Integrating constraints and metric learning in semi-supervised clustering. In *Proc. 21st International Conference on Machine Learning*, pages 11–18, 2004.

[BBM13]   N. Barbieri, F. Bonchi, and G. Manco. Cascade-based Community Detection. In *Proc. 6th ACM International Conference on Web Search and Data Mining*, pages 33–42, 2013.

[BC01]   A. Blum and S. Chawla. Learning from Labeled and Unlabeled Data using Graph Mincuts. In *Proc. 18th International Conference on Machine Learning*, pages 19–26, 2001.

[BDR06]   Y. Bengio, O. Delalleau, and N. L. Roux. Label Propagation and Quadratic Criterion. In *Semi-Supervised Learning*, pages 193–216. MIT Press, 2006.

[Ben80]   J. L. Bentley. Multidimensional Divide-and-Conquer. *Communications of the ACM*, 23(4):214–229, 1980.

[Ber99]   D. P. Bertsekas. *Nonlinear Programming*. Athena Scientific, 2nd edition, 1999.

[BG05]   I. Borg and P. J. F. Groenen. *Modern Multidimensional Scaling*. Springer, 2005.

[BGC10]   V. Badrinarayanan, F. Galasso, and R. Cipolla. Label propagation in video sequences. In *Proc. IEEE Conference on Computer Vision and Pattern Recognition*, pages 3265–3272, 2010.

[Bie87]   H. J. Bierens. Kernel estimators of regression functions. In *Advances in Econometrics: 5th World Congress of the Econometric Society*, volume 1, pages 99–144, 1987.

[BKS07]   S. Bharathi, D. Kempe, and M. Salek. Competitive Influence Maximization in Social Networks. In *Proc. 3rd International Conference on Internet and Network Economics*, pages 306–311, 2007.

[BLL$^+$14] S. Bourigault, C. Lagnier, S. Lamprier, L. Denoyer, and P. Gallinari. Learning Social Network Embeddings for Predicting Information Diffusion. In *Proc. 7th ACM International Conference on Web Search and Data Mining*, pages 393–402, 2014.

[BLRR04]   A. Blum, J. Lafferty, M. R. Rwebangira, and R. Reddy. Semi-supervised learning using randomized mincuts. In *Proc. 21st International Conference on Machine Learning*, pages 13–20, 2004.

[BLT12]   D. F. Bernardes, M. Latapy, and F. Tarissan. Relevance of SIR Model for Real-world Spreading Phenomena: Experiments on a Large-scale P2P System. In *Proc. IEEE/ACM International Conference on Advances in Social Networks Analysis and Mining*, pages 327–334, 2012.

[Blu93]   L. E. Blume. The Statistical Mechanics of Strategic Interaction. *Games and Economic Behavior*, 5:387–424, 1993.

[BMN04]   M. Belkin, I. Matveeva, and P. Niyogi. Regularization and semi-supervised learning on large graphs. In *Learning Theory, Lecture Notes in Computer Science*, volume 3120, pages 624–638. Springer, 2004.

[BN03]   M. Belkin and P. Niyogi. Laplacian eigenmaps for dimensionality reduction and data representation. *Neural computation*, 15(6):1373–1396, 2003.

[BNMY11]   B. K. Bao, B. Ni, Y. Mu, and S. Yan. Efficient Region-Aware Large Graph Construction Towards Scalable Multi-Label Propagation. *Pattern Recognition*, 44(3):598–606, 2011.

[BNS06]   M. Belkin, P. Niyogi, and V. Sindhwani. Manifold Regularization: A Geometric Framework for Learning from Labeled and Unlabeled Examples. *Journal of Machine Learning Research*, 7:2399–2434, 2006.

[Bro97]   A. Z. Broder. On the resemblance and containment of documents. In *Proc. Compression and Complexity of Sequences*, pages 21–29, 1997.

[BSS⁺08]   S. Baluja, R. Seth, D. Sivakumar, Y. Jing, J. Yagnik, S. Kumar, D. Ravichandran, and M. Aly. Video suggestion and discovery for YouTube: taking random walks through the view graph. In *Proc. 17th international conference on World Wide Web*, pages 895–904, 2008.

[CC10]   A. Y. C. Chen and J. J. Corso. Propagating multi-class pixel labels throughout video frames. In *Proc. Western New York Image Processing Workshop*, pages 14–17, 2010.

[CC11]   A. Y. C. Chen and J. J. Corso. Temporally consistent multi-class video-object segmentation with the Video Graph-Shifts algorithm. In *Proc. IEEE Workshop on Applications of Computer Vision*, pages 614–621, 2011.

[CCC11]   D. Coppi, S. Calderara, and R. Cucchiara. People appearance tracing in video by spectral graph transduction. In *Proc. IEEE International Conference on Computer Vision Workshops*, pages 920–927, 2011.

[CFS09]   J. Chen, H. Fang, and Y. Saad. Fast Approximate kNN Graph Construction for High Dimensional Data via Recursive Lanczos Bisection. *Journal of Machine Learning Research*, 10:1989–2012, 2009.

[CGK⁺97]   C. S. Chekuri, A. V. Goldberg, D. R. Karger, M. S. Levine, and C. Stein. Experimental Study of Minimum Cut Algorithms. In *Proc. 8th Annual ACM-SIAM Symposium on Discrete Algorithms*, pages 324–333, 1997.

[CJ04]   A. Corduneanu and T. Jaakkola. Distributed information regularization on graphs. *Proc. Neural Information Proccessing Systems*, pages 297–304, 2004.

[CMC05]   S. F. Chang, R. Manmatha, and T. S. Chua. Combining text and audio-visual features in video indexing. In *Proc. IEEE International Conference on Acoustics, Speech, and Signal Processing*, volume 5, pages 1005–1008, 2005.

[CMYC10]   X. Chen, Y. Mu, S. Yan, and T. S. Chua. Efficient Large-scale Image Annotation by Probabilistic Collaborative Multi-label Propagation. In *Proc. International Conference on Multimedia*, pages 35–44, 2010.

[CTLC12]   X. Chen, Z. Tong, H. Liu, and D. Cai. Metric learning with two-dimensional smoothness for visual analysis. In *Proc. IEEE Conference on Computer Vision and Pattern Recognition*, pages 2533–2538, 2012.

[CWW10]   W. Chen, C. Wang, and Y. Wang. Scalable Influence Maximization for Prevalent Viral Marketing in Large-scale Social Networks. In *Proc. 16th ACM SIGKDD International Conference on Knowledge Discovery and Data Mining*, pages 1029–1038, 2010.

[CWY09]   W. Chen, Y. Wang, and S. Yang. Efficient Influence Maximization in Social Networks. In *Proc. 15th ACM SIGKDD International Conference on Knowledge Discovery and Data Mining*, pages 199–208, 2009.

[CY00]   F. Chung and Y. Yau. Discrete Green's Functions. *Journal of Combinatorial Theory, Series A*, 91(12):191 – 214, 2000.

[CYZ10]   W. Chen, Y. Yuan, and L. Zhang. Scalable Influence Maximization in Social Networks Under the Linear Threshold Model. In *Proc. IEEE International Conference on Data Mining*, pages 88–97, 2010.

[CZW⁺09]   G. Chen, J. Zhang, F. Wang, C. Zhang, and Y. Gao. Efficient multi-label classification with hypergraph regularization. In *Proc. IEEE Conference on Computer Vision and Pattern Recognition*, pages 1658–1665, 2009.

[DD08]   J. V. Davis and I. S. Dhillon. Structured Metric Learning for High Dimensional Problems. In *Proc. 14th ACM SIGKDD International Conference on Knowledge Discovery and Data Mining*, pages 195–203, 2008.

[DG08]   J. Dean and S. Ghemawat. MapReduce: Simplified Data Processing on Large Clusters. *Communications of the ACM*, 51(1):107–113, 2008.

[DKJ⁺07]   J. V. Davis, B. Kulis, P. Jain, S. Sra, and I. S. Dhillon. Information-theoretic metric learning. In *Proc. 24th International Conference on Machine Learning*, pages 209–216, 2007.

[DKS09]   S. I. Daitch, J. A. Kelner, and D. A. Spielman. Fitting a graph to vector data. In *Proc. 26th Annual International Conference on Machine Learning*, pages 201–208, 2009.

[DML11]   W. Dong, C. Moses, and K. Li. Efficient K-nearest Neighbor Graph Construction for Generic Similarity Measures. In *Proc. 20th International Conference on World Wide Web*, pages 577–586, 2011.

[DNKS10]   N. Diakopoulos, M. Naaman, and F. Kivran-Swaine. Diamonds in the rough: Social media visual analytics for journalistic inquiry. In *Proc. IEEE Symposium on Visual Analytics Science and Technology*, pages 115–122, Oct 2010.

[DR01]   P. Domingos and M. Richardson. Mining the Network Value of Customers. In *Proc. 7th ACM SIGKDD International Conference on Knowledge Discovery and Data Mining*, pages 57–66, 2001.

[DTC10]   P. S. Dhillon, P. P. Talukdar, and K. Crammer. *Inference Driven Metric Learning (IDML) for Graph Construction*. Technical Report (CIS), University of Pennsylvania, 2010.

[DY08]   L. Ding and A. Yilmaz. Image segmentation as learning on hypergraphs. In *Proc. 7th International Conference on Machine Learning and Applications*, pages 247–252, 2008.

[Ell93]   G. Ellison. Learning, Local Interaction and Coordination. *Econometrica*, 61(5):1047–1071, 1993.

[EW01]   A. Elisseeff and J. Weston. A kernel method for multi-labelled classification. *Proc. Advances in Neural Information Processing Systems*, 14:681–687, 2001.

[Fod02]   I. Fodor. *A Survey of Dimension Reduction Techniques*. Technical Report, Center for Applied Scientific Computing, Lawrence Livermore National Laboratory, 2002.

[GCCVMMC08] L. Gomez-Chova, G. Camps-Valls, J. Munoz-Mari, and J. Calpe. Semisupervised Image Classification With Laplacian Support Vector Machines. *IEEE Geoscience and Remote Sensing Letters*, 5(3):336 –340, 2008.

[GIM99] A. Gionis, P. Indyk, and R. Motwani. Similarity Search in High Dimensions via Hashing. In *Proc. 25th International Conference on Very Large Data Bases*, pages 518–529, 1999.

[GK12] S. Goyal and M. Kearns. Competitive Contagion in Networks. In *Proc. 44th Annual ACM Symposium on Theory of Computing*, pages 759–774, 2012.

[GLL11a] A. Goyal, W. Lu, and L. V. S. Lakshmanan. CELF++: Optimizing the Greedy Algorithm for Influence Maximization in Social Networks. In *Proc. 20th International Conference Companion on World Wide Web*, pages 47–48, 2011.

[GLL11b] A. Goyal, W. Lu, and L. V. S. Lakshmanan. SIMPATH: An Efficient Algorithm for Influence Maximization Under the Linear Threshold Model. In *Proc. IEEE 11th International Conference on Data Mining*, pages 211–220, 2011.

[GLM01] J. Goldenberg, B. Libai, and E. Muller. Talk of the Network: A Complex Systems Look at the Underlying Process of Word-of-Mouth. *Marketing Letters*, 12(3):211–223, 2001.

[GMVS09] M. Guillaumin, T. Mensink, J. Verbeek, and C. Schmid. TagProp: Discriminative metric learning in nearest neighbor models for image auto-annotation. In *Proc. IEEE 12th International Conference on Computer Vision*, pages 309–316, 2009.

[Gra78] M. Granovetter. Threshold models of collective behavior. *American Journal of Sociology*, 83(6):1420–1433, 1978.

[Gra06] L. Grady. Random Walks for Image Segmentation. *IEEE Transactions on Pattern Analysis and Machine Intelligence*, 28(11):1768–1783, 2006.

[GRHS04] J. Goldberger, S. Roweis, G. Hinton, and R. Salakhutdinov. Neighbourhood components analysis. In *Proc. Advances in Neural Information Processing Systems*, volume 17, pages 513–520, 2004.

[GRLK10] M. Gomez Rodriguez, J. Leskovec, and A. Krause. Inferring Networks of Diffusion and Influence. In *Proc. 16th ACM SIGKDD International Conference on Knowledge Discovery and Data Mining*, pages 1019–1028, 2010.

[GWZ+13] Y. Gao, M. Wang, Z. J. Zha, J. Shen, X. Li, and X. Wu. Visual-Textual Joint Relevance Learning for Tag-Based Social Image Search. *IEEE Transactions on Image Processing*, 22(1):363–376, 2013.

[HAHH12] A. Hajibagheri, H. Alvari, A. Hamzeh, and S. Hashemi. Community Detection in Social Networks Using Information Diffusion. In *Proc. IEEE/ACM International Conference on Advances in Social Networks Analysis and Mining*, pages 702–703, 2012.

[HGO+10] K. Heath, N. Gelfand, M. Ovsjanikov, M. Aanjaneya, and L. Guibas. Image webs: Computing and exploiting connectivity in image collections. In *Proc. IEEE Conference on Computer Vision and Pattern Recognition*, pages 3432–3439, 2010.

[HHA+06] R. A. Heckemann, J. V. Hajnal, P. Aljabar, D. Rueckert, and A. Hammers. Automatic anatomical brain MRI segmentation combining label propagation and decision fusion. *NeuroImage*, 33(1):115 – 126, 2006.

[HIF13] S. Huron, P. Isenberg, and J. D. Fekete. PolemicTweet: Video Annotation and Analysis through Tagged Tweets. In *Proc. of the IFIP Conference on Human-Computer Interaction*, pages 135–152, 2013.

[HK92] L. Hagen and A. B. Kahng. New spectral methods for ratio cut partitioning and clustering. *IEEE Transactions on Computer-Aided Design of Integrated Circuits and Systems*, 11(9):1074 –1085, 1992.

[HK10] T. Hwang and R. Kuang. A heterogeneous label propagation algorithm for disease gene discovery. In *Proc. SIAM International Conference on Data Mining*, pages 583–594, 2010.

[HLC08] S. C. H. Hoi, W. Liu, and S. F. Chang. Semi-supervised distance metric learning for collaborative image retrieval. In *Proc. IEEE Conference on Computer Vision and Pattern Recognition*, pages 1–7, 2008.

[HLL+14] L. Huang, Y. Liu, X. Liu, X. Wang, and B. Lang. Graph-based active semi-supervised learning: A new perspective for relieving multi-class annotation labor. In *Proc. IEEE International Conference on Multimedia and Expo*, pages 1–6, 2014.

[HN04] X. He and P. Niyogi. Locality preserving projections. In *Proc. Advances in Neural Information Processing Systems 16*, pages 153–160, 2004.

[HOSS13] M. E. Houle, V. Oria, S. Satoh, and J. Sun. Annotation Propagation in Image Databases Using Similarity Graphs. *ACM Trans. Multimedia Comput. Commun. Appl.*, 10(1):7:1–7:21, 2013.

[HTW+12] C. Hui, Y. Tyshchuk, W. A. Wallace, M. Magdon-Ismail, and M. Goldberg. Information Cascades in Social Media in Response to a Crisis: A Preliminary Model and a Case Study. In *Proc. 21st International Conference Companion on World Wide Web*, pages 653–656, 2012.

[HWLH12] L. C. Hsieh, G. L. Wu, W. Y. Lee, and W. Hsu. Two-stage sparse graph construction using MinHash on MapReduce. In *Proc. IEEE International Conference on Acoustics, Speech and Signal Processing*, pages 1013–1016, 2012.

[IDF+05] G. Iyengar, P. Duygulu, S. Feng, P. Ircing, S. P. Khudanpur, D. Klakow, M. R. Krause, R. Manmatha, H. J. Nock, D. Petkova, B. Pytlik, and P. Virga. Joint Visual-Text Modeling for Automatic Retrieval of Multimedia Documents. In *Proc. 13th Annual ACM International Conference on Multimedia*, pages 21–30, 2005.

[JCST01] T. Joachims, N. Cristianini, and J. Shawe-Taylor. Composite Kernels for Hypertext Categorisation. In *Proc. International Conference on Machine Learning*, pages 250–257, 2001.

[Joa03]   T. Joachims. Transductive learning via spectral graph partitioning. In *Proc. International Conference on Machine Learning*, pages 290–297, 2003.

[Jol02]   I. T. Jolliffe. *Principal Component Analysis*. Springer, 2nd edition, 2002.

[JSWZHZ09]   R. Jin, S. Shijun Wang, and Z. H. Zhi-Hua Zhou. Learning a distance metric from multi-instance multi-label data. In *Proc. IEEE Conference on Computer Vision and Pattern Recognition*, pages 896–902, 2009.

[JWC09]   T. Jebara, J. Wang, and S. Chang. Graph construction and b-matching for semi-supervised learning. In *Proc. 26th Annual International Conference on Machine Learning*, pages 441–448, 2009.

[KG13]   K. K. Kumar and G. Geethakumari. Information diffusion model for spread of misinformation in online social networks. In *Proc. International Conference on Advances in Computing, Communications and Informatics*, pages 1172–1177, 2013.

[KGF12]   D. Kuettel, M. Guillaumin, and V. Ferrari. Segmentation Propagation in Imagenet. In *Proc. 12th European Conference on Computer Vision*, volume VII, pages 459–473, 2012.

[KKT03]   D. Kempe, J. Kleinberg, and E. Tardos. Maximizing the spread of influence through a social network. In *Proc. 9th ACM SIGKDD International Conference on Knowledge Discovery and Data Mining*, pages 137–146, 2003.

[KKT05]   D. Kempe, J. Kleinberg, and E. Tardos. Influential Nodes in a Diffusion Model for Social Networks. In *Proc. 32nd International Conference on Automata, Languages and Programming*, pages 1127–1138, 2005.

[KL02]   R. I. Kondor and J. Lafferty. Diffusion kernels on graphs and other discrete structures. In *Proc. 19th International Conference on Machine Learning*, pages 315–322, 2002.

[KM27]   W. O. Kermack and A. McKendrick. A Contribution to the Mathematical Theory of Epidemics. *Proc. of the Royal Society of London. Series A, Containing Papers of a Mathematical and Physical Character*, 115(772):700–721, 1927.

[KOW08]   J. Kostka, Y. Oswald, and R. Wattenhofer. Word of mouth: Rumor dissemination in social networks. In *Structural Information and Communication Complexity, Lecture Notes in Computer Science*, volume 5058, pages 185–196. Springer, 2008.

[KS06]   M. Kimura and K. Saito. Tractable Models for Information Diffusion in Social Networks. In *Proc. 10th European Conference on Principle and Practice of Knowledge Discovery in Databases*, pages 259–271, 2006.

[KTT⁺12]   K. I. Kim, J. Tompkin, M. Theobald, J. Kautz, and C. Theobalt. Match Graph Construction for Large Image Databases. In *Proc. 12th European Conference on Computer Vision*, volume I, pages 272–285, 2012.

[Lan50]   C. Lanczos. An iterative method for the solution of the eigenvalue problem of linear differential and integral. *Journal of Research of the National Bureau of Standards*, 45:255–282, 1950.

[LBK09]   J. Leskovec, L. Backstrom, and J. Kleinberg. Meme-Tracking and the Dynamics of the News Cycle. In *Proc. 15th ACM SIGKDD International Conference on Knowledge Discovery and Data Mining*, pages 497–506, 2009.

[LHC10]   W. Liu, J. He, and S. Chang. Large graph construction for scalable semi-supervised learning. In *Proc. International Conference on Machine Learning*, pages 679–686, 2010.

[LHWH13]   W. Y. Lee, L. C. Hsieh, G. L. Wu, and W. Hsu. Graph-based Semi-supervised Learning with Multi-Modality Propagation for Large-Scale Image Datasets. *Journal of Visual Communication and Image Representation*, 24(3):295–302, 2013.

[LHY$^+$09]   D. Liu, X. S. Hua, L. Yang, M. Wang, and H. Zhang. Tag Ranking. In *Proc. 18th International Conference on World Wide Web*, pages 351–360, 2009.

[LK03]   S. Letovsky and S. Kasif. Predicting protein function from protein/protein interaction data: a probabilistic approach. *Bioinformatics*, 19:197–204, 2003.

[LKG$^+$07]   J. Leskovec, A. Krause, C. Guestrin, C. Faloutsos, J. VanBriesen, and N. Glance. Cost-effective Outbreak Detection in Networks. In *Proc. 13th ACM SIGKDD International Conference on Knowledge Discovery and Data Mining*, pages 420–429, 2007.

[LLH$^+$07]   J. Liu, W. Lai, X. Hua, Y. Huang, and S. Li. Video search re-ranking via multi-graph propagation. In *Proc. 15th International Conference on Multimedia*, pages 208–217, 2007.

[LLL$^+$09]   J. Liu, M. Li, Q. Liu, H. Lu, and S. Ma. Image Annotation via Graph Learning. *Pattern Recognition*, 42(2):218–228, 2009.

[LMT$^+$10]   W. Liu, S. Ma, D. Tao, J. Liu, and P. Liu. Semi-Supervised Sparse Metric Learning Using Alternating Linearization Optimization. In *Proc. 16th ACM SIGKDD International Conference on Knowledge Discovery and Data Mining*, pages 1139–1148, 2010.

[LPSL10]   D. Lee, J. Park, J. Shim, and S. Lee. An Efficient Similarity Join Algorithm with Cosine Similarity Predicate. In *Database and Expert Systems Applications, Lecture Notes in Computer Science*, volume 6262, pages 422–436. Springer, 2010.

[LV07]   J. A. Lee and M. Verleysen. *Nonlinear Dimensionality Reduction*. Springer, 2007.

[LYHZ11]   D. Liu, S. Yan, X. S. Hua, and H. J. Zhang. Image Retagging Using Collaborative Tag Propagation. *IEEE Transactions on Multimedia*, 13(4):702–712, 2011.

[LYZZ08]   J. Long, J. Yin, W. Zhao, and E. Zhu. Graph-Based Active Learning Based on Label Propagation. In V. Torra and Y. Narukawa, editors, *Modeling Decisions for Artificial Intelligence, Lecture Notes in Computer Science*, volume 5285, pages 179–190. Springer, 2008.

[MKJ13]   R. Michalski, P. Kazienko, and J. Jankowski. Convince a Dozen More and Succeed – The Influence in Multi-layered Social Networks. In *Proc. International Conference on Signal-Image Technology Internet-Based Systems*, pages 499–505, 2013.

[ML13]   T. Ma and L. J. Latecki. Graph Transduction Learning with Connectivity Constraints with Application to Multiple Foreground Cosegmentation. In *Proc. IEEE Conference on Computer Vision and Pattern Recognition*, pages 1955–1962, 2013.

[Mor00]   S. Morris. Contagion. *Review of Economic Studies*, 67:57–78, 2000.

[MYLK08]   H. Ma, H. Yang, M. R. Lyu, and I. King. Mining Social Networks Using Heat Diffusion Processes for Marketing Candidates Selection. In *Proc. 17th ACM Conference on Information and Knowledge Management*, pages 233–242, 2008.

[NG08]   N. Nguyen and Y. Guo. Metric learning: A support vector approach. *Machine Learning and Knowledge Discovery in Databases*, pages 125–136, 2008.

[NJT05]   Z. Y. Niu, D. H. Ji, and C. L. Tan. Word sense disambiguation using label propagation based semi-supervised learning. In *Proc. 43rd Annual Meeting on Association for Computational Linguistics*, pages 395–402, 2005.

[NYTE12]   N. P. Nguyen, G. Yan, M. T. Thai, and S. Eidenbenz. Containment of Misinformation Spread in Online Social Networks. In *Proc. 4th Annual ACM Web Science Conference*, pages 213–222, 2012.

[OC12]   M. Orbach and K. Crammer. Graph-Based Transduction with Confidence. In *Proc. European Conference on Machine Learning and Knowledge Discovery in Databases*, volume II, pages 323–338, 2012.

[OR13]   M. Opuszko and J. Ruhland. Impact of the Network Structure on the SIR Model Spreading Phenomena in Online Networks. In *Proc. The 8th International Multi-Conference on Computing in the Global Information Technology*, pages 22–28, 2013.

[PBS10]   N. Pathak, A. Banerjee, and J. Srivastava. A Generalized Linear Threshold Model for Multiple Cascades. In *Proc. IEEE 10th International Conference on Data Mining*, pages 965–970, 2010.

[PPLJ14]   Y. Park, S. Park, S. Lee, and W. Jung. Greedy filtering: A scalable algorithm for k-nearest neighbor graph construction. In *Database Systems for Advanced Applications, Lecture Notes in Computer Science*, volume 8421, pages 327–341. Springer, 2014.

[PSV01]   R. Pastor-Satorras and A. Vespignani. Epidemic Dynamics and Endemic States in Complex Networks. *Physical Review E*, page 066117, 2001.

[PSZ11]   J. Philbin, J. Sivic, and A. Zisserman. Geometric Latent Dirichlet Allocation on a Matching Graph for Large-Scale Image Datasets. *International Journal of Computer Vision*, 95(2):138–153, 2011.

[QHR+07]   G. J. Qi, X. S. Hua, Y. Rui, J. Tang, T. Mei, and H. J. Zhang. Correlative multi-label video annotation. In *Proc. 15th International Conference on Multimedia*, pages 17–26, 2007.

[QTZ+09]   G. J. Qi, J. Tang, Z. J. Zha, T. S. Chua, and H. J. Zhang. An Efficient Sparse Metric Learning in High-dimensional Space via L1-Penalized Log-Determinant Regularization. In *Proc. 26th Annual International Conference on Machine Learning*, pages 841–848, 2009.

[RD02]   M. Richardson and P. Domingos. Mining knowledge-sharing sites for viral marketing. In *Proc. 8th ACM SIGKDD International Conference on Knowledge Discovery and Data Mining*, pages 61–70, 2002.

[RH05]   H. Rue and L. Held. *Gaussian Markov Random Fields: Theory and Applications*. Chapman & Hall, 2005.

[RLF12]   M. Rubinstein, C. Liu, and W. T. Freeman. Annotation Propagation in Large Image Databases via Dense Image Correspondence. In *Lecture Notes in Computer Science*, volume 7574, pages 85–99. Springer, 2012.

[Rog95]   E. Rogers. *The Diffusion of Innovation*. Free Press, 1995.

[RR09]   D. Rao and D. Ravichandran. Semi-supervised polarity lexicon induction. In *Proc. 12th Conference of the European Chapter of the Association for Computational Linguistics*, pages 675–682, 2009.

[RS00]   S. T. Roweis and L. K. Saul. Nonlinear Dimensionality Reduction by Locally Linear Embedding. *Science*, 290(5500):2323–2326, 2000.

[SA12]   S. Simon and K. Apt. Choosing Products in Social Networks. In *Internet and Network Economics, Lecture Notes in Computer Science*, volume 7695, pages 100–113. Springer, 2012.

[SB11]   A. Subramanya and J. Bilmes. Semi-Supervised Learning with Measure Propagation. *Journal of Machine Learning Research*, 12:3311–3370, 2011.

[SCA14]   P. Sinha, A. D. Choudhury, and A. K. Agrawal. Sentiment analysis of Wimbledon tweets. In *Proc. ACM WWW Microposts workshop*, 2014.

[Sch78]   T. Schelling. *Micromotives and Macrobehavior*. Norton, 1978.

[SJ02]   M. Szummer and T. Jaakkola. Partially labeled classification with Markov random walks. In *Proc. Advances in Neural Information Processing Systems*, pages 945–952, 2002.

[SJY08]   L. Sun, S. Ji, and J. Ye. Hypergraph spectral learning for multi-label classification. In *Proc. 14th ACM SIGKDD International Conference on Knowledge Discovery and Data Mining*, pages 668–676, 2008.

[SK03]   A. Smola and R. Kondor. Kernels and Regularization on Graphs. In *Proc. Conference on Learning Theory and Kernel Machines*, pages 144–158, 2003.

[SK09]   X. Su and T. M. Khoshgoftaar. A survey of Collaborative Filtering Techniques. *Advances in Artificial Intelligence*, 2009:4:2–4:2, 2009.

[SKC09]   D. A. Shamma, L. Kennedy, and E. F. Churchill. Tweet the Debates: Understanding Community Annotation of Uncollected Sources. In *Proc. 1st SIGMM Workshop on Social Media*, pages 3–10, 2009.

[SKM09]   K. Saito, M. Kimura, and H. Motoda. Discovering Influential Nodes for SIS Models in Social Networks. In *Discovery Science, Lecture Notes in Computer Science*, volume 5808, pages 302–316. Springer, 2009.

[Smi72]   J. M. Smith. *On evolution*. Edinburgh University Press, 1972.

[SN05]   V. Sindhwani and P. Niyogi. A co-regularized approach to semi-supervised learning with multiple views. In *Proc. ICML Workshop on Learning with Multiple Views*, 2005.

[SR03]   L. K. Saul and S. T. Roweis. Think globally, fit locally: unsupervised learning of low dimensional manifolds. *Journal of Machine Learning Research*, 4:119–155, 2003.

[SSSN04]   S. Shalev-Shwartz, Y. Singer, and A. Y. Ng. Online and Batch Learning of Pseudo-Metrics. In *Proc. 21st International Conference on Machine Learning*, pages 94–101, 2004.

[SSUB11]   M. Speriosu, N. Sudan, S. Upadhyay, and J. Baldridge. Twitter polarity classification with label propagation over lexical links and the follower graph. In *Proc. 1st Workshop on Unsupervised Learning in NLP*, pages 53–63, 2011.

[STA09]   X. Shi, B. Tseng, and L. Adamic. Information diffusion in computer science citation networks. In *Proc. International Conference on Weblogs and Social Media*, 2009.

[SZZZ10]   M. S. Shang, Z. K. Zhang, T. Zhoub, and Y. C. Zhang. Collaborative filtering with diffusion-based similarity on tripartite graphs. *Physica A*, 389(6):1259–1264, 2010.

[Tal09]   P. P. Talukdar. *Topics in Graph Construction for Semi-Supervised Learning*. Technical Report, University of Pennsylvania, 2009.

[TAM12]   V. Tzoumas, C. Amanatidis, and E. Markakis. A Game-Theoretic Analysis of a Competitive Diffusion Process over Social Networks. In *Internet and Network Economics, Lecture Notes in Computer Science*, volume 7695, pages 1–14. Springer, 2012.

[TC09]   P. P. Talukdar and K. Crammer. New Regularized Algorithms for Transductive Learning. In *Proc. European Conference on Machine Learning and Knowledge Discovery in Databases: Part II*, pages 442–457, 2009.

[TDSL00]   J. B. Tenenbaum, V. De Silva, and J. C. Langford. A global geometric framework for nonlinear dimensionality reduction. *Science*, 290(5500):2319–2323, 2000.

[TH01]   L. Terveen and W. Hill. Beyond Recommender Systems: Helping People Help Each Other. In *HCI In The New Millennium, Jack Carroll ed.* Addison-Wesley, 2001.

[THK09]   Z. Tian, T. Hwang, and R. Kuang. A hypergraph-based learning algorithm for classifying gene expression and arrayCGH data with prior knowledge. *Bioinformatics*, 25(21):2831–2838, 2009.

[THL+05]   H. Tong, J. He, M. Li, C. Zhang, and W. Y. Ma. Graph based multi-modality learning. In *Proc. 13th Annual ACM International Conference on Multimedia*, pages 862–871, 2005.

[THQ+07]   J. Tang, X. S. Hua, G.-J. Qi, T. Mei, and X. Wu. Anisotropic Manifold Ranking for Video Annotation. In *Proc. IEEE International Conference on Multimedia and Expo*, pages 492–495, 2007.

[THQ+08]   J. Tang, X.-S. Hua, G.-J. Qi, Y. Song, and X. Wu. Video Annotation Based on Kernel Linear Neighborhood Propagation. *IEEE Transactions on Multimedia*, 10(4):620 –628, 2008.

[THW+09]   J. Tang, X.-S. Hua, M. Wang, Z. Gu, G.-J. Qi, and X. Wu. Correlative Linear Neighborhood Propagation for Video Annotation. *IEEE Transactions on Systems, Man, and Cybernetics, Part B: Cybernetics*, 39(2):409 –416, 2009.

[THY+11]   J. Tang, R. Hong, S. Yan, T. S. Chua, G.-J. Qi, and R. Jain. Image annotation by kNN-sparse graph-based label propagation over noisily tagged web images. *ACM Transactions on Intelligent Systems and Technology*, 2(2):14:1–14:15, 2011.

[TLLZ14]   J. Tang, M. Li, Z. Li, and C. Zhao. Tag ranking based on salient region graph propagation. *Multimedia Systems*, pages 1–9, 2014.

[TSS05]   K. Tsuda, H. Shin, and B. Schölkopf. Fast protein classification with multiple networks. *Bioinformatics*, 21(2):59–65, 2005.

[Tsu05]   K. Tsuda. Propagating distributions on a hypergraph by dual information regularization. In *Proc. 22nd International Conference on Machine Learning*, pages 920–927, 2005.

[TYH+09]   J. Tang, S. Yan, R. Hong, G. J. Qi, and T. S. Chua. Inferring semantic concepts from community-contributed images and noisy tags. In *Proc. 17th ACM International Conference on Multimedia*, pages 223–232, 2009.

[VG12]   S. Vijayanarasimhan and K. Grauman. Active frame selection for label propagation in videos. In *Proc. 12th European conference on Computer Vision*, volume V, pages 496–509, 2012.

[VK11]   S. Venkatramanan and A. Kumar. Information dissemination in socially aware networks under the linear threshold model. In *National Conference on Communications*, pages 1–5, 2011.

[vNM44]   J. von Neumann and O. Morgenstern. *Theory of Games and Economic Behavior*. Princeton University Press, 1944.

[Wat02]   D. J. Watts. A simple model of global cascades on random networks. In *Proc. of the National Academy of Sciences*, pages 5766–5771, 2002.

[WHH+09]   M. Wang, X. S. Hua, R. Hong, J. Tang, G. J. Qi, and Y. Song. Unified video annotation via multigraph learning. *IEEE Transactions on Circuits and Systems for Video Technology*, 19(5):733–746, 2009.

[WHTH09]   M. Wang, X. S. Hua, J. Tang, and R. Hong. Beyond Distance Measurement: Constructing Neighborhood Similarity for Video Annotation. *IEEE Transactions on Multimedia*, 11(3):465–476, 2009.

[WHY+07]   M. Wang, X. S. Hua, X. Yuan, Y. Song, and L. R. Dai. Optimizing multigraph learning: towards a unified video annotation scheme. In *Proc. 15th International Conference on Multimedia*, pages 862–871, 2007.

[WJC08]   J. Wang, T. Jebara, and S. F. Chang. Graph Transduction via Alternating Minimization. In *Proc. 25th International Conference on Machine Learning*, pages 1144–1151, 2008.

[WJZZ06]   C. Wang, F. Jing, L. Zhang, and H. J. Zhang. Image Annotation Refinement Using Random Walk with Restarts. In *Proc. 14th Annual ACM International Conference on Multimedia*, pages 647–650, 2006.

[WKL11]   D. Wang, I. King, and K. S. Leung. "Like Attracts Like!"– A Social Recommendation Framework Through Label Propagation. In *Proc. Workshop on Social Web Search and Mining: Content Analysis Under Crisis*, 2011.

[WLI+05]   J. Weston, C. Leslie, E. Ie, D. Zhou, and A. Elisseeff. Semi-supervised protein classification using cluster kernels. *Bioinformatics*, 21:3241–3247, 2005.

[WS09]   K. Q. Weinberger and L. K. Saul. Distance Metric Learning for Large Margin Nearest Neighbor Classification. *Journal of Machine Learning Research*, 10:207–244, 2009.

[WSC11]   J. Woo, J. Son, and H. Chen. An SIR model for violent topic diffusion in social media. In *Proc. IEEE International Conference on Intelligence and Security Informatics*, pages 15–19, 2011.

[WWL07]   F. Wang, X. Wang, and T. Li. Efficient label propagation for interactive image segmentation. In *Proc. 6th International Conference on Machine Learning and Applications*, pages 136–141, 2007.

[WWS+09]   F. Wang, X. Wang, B. Shao, T. Li, and M. Ogihara. Tag integrated multi-label music style classification with hypergraph. In *Proc. 10th International Society for Music Information Retrieval*, pages 363–368, 2009.

[WWZ+09]   J. Wang, F. Wang, C. Zhang, H. C. Shen, and L. Quan. Linear Neighborhood Propagation and Its Applications. *IEEE Transactions on Pattern Analysis and Machine Intelligence*, 31(9):1600–1615, 2009.

[WWZ+12]   J. Wang, J. Wang, G. Zeng, Z. Tu, R. Gan, and S. Li. Scalable k-NN graph construction for visual descriptors. In *Proc. IEEE Conference on Computer Vision and Pattern Recognition*, pages 1106–1113, June 2012.

[WYG+09]   J. Wright, A. Y. Yang, A. Ganesh, S. S. Sastry, and Y. Ma. Robust Face Recognition via Sparse Representation. *IEEE Transactions on Pattern Analysis and Machine Intelligence*, 31(2):210–227, 2009.

[WYH+13]   Z. Wei, Y. Yanqing, T. Hanlin, D. Qiwei, and L. Taowei. Information Diffusion Model Based on Social Network. In *Proc. International Conference of Modern Computer Science and Applications, Advances in Intelligent Systems and Computing*, volume 191, pages 145–150. Springer, 2013.

[WZ06]   F. Wang and C. Zhang. Label propagation through linear neighborhoods. In *Proc. 23rd International Conference on Machine Learning*, pages 985–992, 2006.

[XL10]   B. Xu and L. Liu. Information diffusion through online social networks. In *Proc. IEEE International Conference on Emergency Management and Management Sciences*, pages 53–56, 2010.

[XNJR02]   E. P. Xing, A. Y. Ng, M. I. Jordan, and S. Russell. Distance Metric Learning, with Application to Clustering with Side-Information. In *Proc. Advances in Neural Information Processing Systems 15*, volume 15, pages 505–512, 2002.

[Xu14]   J. Xu. Joint visual and textual mining on social media. In *Proc. IEEE International Conference on Data Mining Workshop*, pages 1189–1190, Dec 2014.

[XWTQ07]   J. Xiao, J. Wang, P. Tan, and L. Quan. Joint affinity propagation for multiple view segmentation. In *Proc. IEEE 11th International Conference on Computer Vision*, pages 1–7, 2007.

[YJHL11]   J. Yu, X. Jin, J. Han, and J. Luo. Collection-Based Sparse Label Propagation and Its Application on Social Group Suggestion from Photos. *ACM Transactions on Intelligent Systems and Technology*, 2(2):12:1–12:21, 2011.

[YJZ+06]   L. Yang, D. Ji, G. Zhou, Y. Nie, and G. Xiao. Document re-ranking using cluster validation and label propagation. In *Proc. 15th ACM International Conference on Information and Knowledge Management*, pages 690–697, 2006.

[YLL09]   W. You, L. Liu, and C. Lv. Towards Emergency Management: Review on Diffusion Models and Their Applications. In *Proc. International Symposium on Emergency Management*, pages 633–636, 2009.

[YQZC12]   O. Yagan, D. Qian, J. Zhang, and D. Cochran. Information diffusion in overlaying social-physical networks. In *Proc. 48th Annual Conference in Information Sciences and Systems*, pages 1–6, 2012.

[YR06]   L. Yang and J. Rong. *Distance Metric Learning: A Comprehensive Survey*. Technical Report, Michigan State University, 2006.

[YTZ04]   K. Yu, V. Tresp, and D. Zhou. *Semi-supervised Induction with Basis Functions*. Technical Report No. 141, Max Planck Institute for Biological Cybernetics, 2004.

[YWY+13]   Z. Yong, L. Weishi, Z. Yang, Z. Gang, Q. Dongxiang, Z. Qi, H. Ying, W. Haifeng, H. Xiaobo, and H. Jiaming. Brain MRI Segmentation with Label Propagation. *International Journal of Emerging Trends & Technology in Computer Science*, 2(5):158–163, 2013.

[YWZ12]   W. Yang, K. Wang, and W. Zuo. Fast neighborhood component analysis. *Neurocomputing*, 83:31 – 37, 2012.

[ZB07]   D. Zhou and C. J. C. Burges. Spectral clustering and transductive learning with multiple views. In *Proc. 24th International Conference on Machine Learning*, pages 1159–1166, 2007.

[ZBL+04]   D. Zhou, O. Bousquet, T. N. Lal, J. Weston, and B. Schölkopf. Learning with local and global consistency. In *Proc. Advances in Neural Information Processing Systems 16*, pages 321–328, 2004.

[ZG02]   X. Zhu and Z. Ghahramani. Learning from Labeled and Unlabeled Data with Label Propagation. Technical report, School of CS, CMU, 2002.

[ZGL03]   X. Zhu, Z. Ghahramani, and J. Lafferty. Semi-Supervised Learning Using Gaussian Fields and Harmonic Functions. In *Proc. 20th International Conference on Machine Learning*, pages 912–919, 2003.

[Zha09]   Y. Zhang. A deterministic model for history sensitive cascade in diffusion networks. In *Proc. IEEE International Conference on Systems, Man and Cybernetics*, pages 1977–1982, 2009.

[ZHGL13]   Y. M. Zhang, K. Huang, G. Geng, and C. L. Liu. Fast kNN Graph Construction with Locality Sensitive Hashing. In *Machine Learning and Knowledge Discovery in Databases, Lecture Notes in Computer Science*, volume 8189, pages 660–674. Springer, 2013.

[ZHS05]   D. Zhou, J. Huang, and B. Schölkopf. Learning from labeled and unlabeled data on a directed graph. In *Proc. 22nd International Conference on Machine Learning*, pages 1036–1043, 2005.

[ZHS07]   D. Zhou, J. Huang, and B. Schölkopf. Learning with hypergraphs: Clustering, classification, and embedding. In *Proc. Advances in Neural Information Processing Systems*, volume 19, page 1601, 2007.

[ZK09]   G. D. Zhou and F. Kong. Global learning of noun phrase anaphoricity in coreference resolution via label propagation. In *Proc. Conference on Empirical Methods in Natural Language Processing*, volume 2, pages 978–986, 2009.

[ZL05]   X. Zhu and J. Lafferty. Harmonic Mixtures: Combining Mixture Models and Graph-Based Methods for Inductive and Scalable Semi-Supervised Learning. In *Proc. 22nd International Conference on Machine Learning*, pages 1052–1059, 2005.

[ZL13]   Y. Zhou and L. Liu. Social Influence Based Clustering of Heterogeneous Information Networks. In *Proc. 19th ACM SIGKDD International Conference on Knowledge Discovery and Data Mining*, 2013.

[ZLG03]   X. Zhu, J. Lafferty, and Z. Ghahramani. Semi-Supervised Learning: From Gaussian Fields to Gaussian Processes. Technical report, School of CS, CMU, 2003.

[ZLP13]   G. Zhou, Z. Lu, and Y. Peng. L1-graph construction using structured sparsity. *Neurocomputing*, 120(0):441–452, 2013.

[ZNP13]   O. Zoidi, N. Nikolaidis, and I. Pitas. Exploiting clustering and stereo information in label propagation on facial images. In *Proc. IEEE Workshop on Computational Intelligence in Biometrics and Identity Management*, 2013.

[ZS04]   D. Zhou and B. Schölkopf. Learning from labeled and unlabeled data using random walks. In *Proc. 26th DAGM Symposium on Pattern Recognition*, pages 237–244, 2004.

[ZTNP14]   O. Zoidi, A. Tefas, N. Nikolaidis, and I. Pitas. Person Identity Label Propagation in Stereo Videos. *IEEE Transactions on Multimedia*, 16(5):1358–1368, 2014.

[ZXP⁺11]   J. Zhu, F. Xiong, D. Piao, Y. Liu, and Y. Zhang. Statistically Modeling the Effectiveness of Disaster Information in Social Media. In *Proc. IEEE Global Humanitarian Technology Conference*, pages 431–436, 2011.

[ZXZ⁺12]   T. Zhang, C. Xu, G. Zhu, S. Liu, and H. Lu. A Generic Framework for Video Annotation via Semi-Supervised Learning. *IEEE Transactions on Multimedia*, 14(4):1206–1219, 2012.

Chapter 6

Graph-Based Pattern Classification and Dimensionality Reduction

Alexandros Iosifidis and Ioannis Pitas

Aristotle University of Thessaloniki, Greece

6.1 Introduction

Digital media, like images, videos, etc., play an important role in social networks. For example, users in social networks are able to post images depicting themselves and/or some of their friends in a location, while other users may see some of these images and rate them. Such rates can be used to describe connections between users or between users and locations, etc. Thus, digital media content analysis is of particular interest in social networks. Digital media representations are usually high-dimensional. For example, a facial image of size 200×150 pixels can be represented by a 30000-dimensional vector obtained by using each pixel coordinate as a different dimension. Therefore, significant efforts have been devoted to deriving low-dimensional data representations that retain properties of interest of the data, like pair-wise distances, data dispersion, and class discrimination [YXZ$^+$07, RR13]. Such low-dimensional data representations are essential in order to reduce the computational cost and the physical memory used to store data, especially in social network applications, where the cardinality of the data sets is enormous. In addition, machine learning methods that are able to classify digital media data are required, in order to proceed with automatic data categorization toward decision making and/or recommending appropriate digital media content to users [KZP06].

Graphs have been proven to be a powerful representation of data relations and have been used in many pattern recognition and machine learning methods, in order to express geometric data relations [WHL05]. It should be noted here that graphs have also been used for data representation [KTI03, VBGS06, ZTP07b, RB09a, CMC$^+$09, BR11, GDIHK11]. However, in this chapter, we shall focus our attention on methods employing graph structures in order to express data relationships. Such relationships may be computed either for vectorial data, or for graph-based data representations (e.g., similarity graphs) [NB05, JH05, VSKB10]. For the latter case, several works have been proposed that can be used in order to "embed" a set of graph data representations in vectorial spaces [WHL05, PD05, EWH07, RNB07, RB09b, LRLB13, BPRB13, LRLB13]. An additional reason for exploiting graph based techniques is that this approach can be used to directly capture information appearing in social media/network structures due to the natural relationship between social networks and graphs, as discussed in Chapter 1.

In this chapter, we describe some of the most widely used methods for data dimensionality reduction and classification that assume an underlying graph structure. For simplicity, we focus on methods exploiting one undirected graph structure, while methods exploiting directed graph structures, [SP11, ZHS05, JM07, ZZY$^+$08] and hyper-graphs [LHM11, PF12], also exist. We divide them into three categories, i.e., unsupervised, supervised and semi-supervised, based on their requirement for manual (user-derived) annotation.

6.2 Notations

In this section we introduce notations that will be used in the entire chapter. Let us assume that a set of digital media data, e.g., a set of N facial images, has been preprocessed so that each sample $i = 1, \ldots, N$ is represented by a D-dimensional vector $\mathbf{x}_i \in \mathbb{R}^D$. Let us also define a *similarity measure* $s_{ij} = s(\mathbf{x}_i, \mathbf{x}_j)$ that is used to measure the similarity between two vectors $\mathbf{x}_i$ and $\mathbf{x}_j$. Any similarity measure providing non-negative values (usually $0 \leq s_{ij} \leq 1$) can be used to this end. The most widely adopted choice is the *heat kernel* (also known as the *diffusion kernel* or *RBF kernel*) [KL02], defined as follows:

$$s(\mathbf{x}_i, \mathbf{x}_j) = \exp\left(-\frac{\|\mathbf{x}_i - \mathbf{x}_j\|_2^2}{2\sigma^2}\right), \qquad (6.2.1)$$

where $\|\cdot\|_2$ denotes the L_2 norm of a vector and σ is a parameter used in order to scale the Euclidean distance between $\mathbf{x}_i$ and $\mathbf{x}_j$. It can be easily proven that $0 \leq s(\mathbf{x}_i, \mathbf{x}_j) \leq 1$. By using $w_{ij} = s(\mathbf{x}_i, \mathbf{x}_j)$, $i = 1, \ldots, N$, $j = 1, \ldots, N$, we can form the graph weight matrix $\mathbf{W} \in \mathbb{R}^{N \times N}$. That is, we can assume that the vectors $\mathbf{x}_i$ are embedded in a graph $\mathcal{G} = (\mathcal{V}, \mathcal{E}, \mathbf{W})$, where $\mathcal{V} = \{\mathbf{x}_i\}_{i=1}^N$ denotes the graph vertex set and $\mathcal{E}$ the set of edges connecting $\mathbf{x}_i$. w_{ij} denotes the weight value of the edge connecting the graph vertices $\mathbf{x}_i$ and $\mathbf{x}_j$. Such a toy graph is shown in Figure 6.2.1, where graph nodes are facial images and graph edges' weights denote image similarities. In the case where $\mathbf{W}$ denotes only connections between

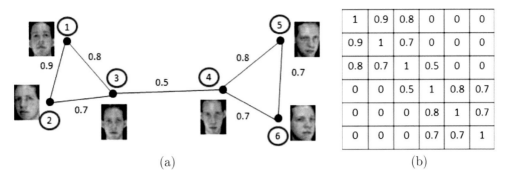

1	0.9	0.8	0	0	0
0.9	1	0.7	0	0	0
0.8	0.7	1	0.5	0	0
0	0	0.5	1	0.8	0.7
0	0	0	0.8	1	0.7
0	0	0	0.7	0.7	1

(a)   (b)

FIGURE 6.2.1: a) Facial image graph, b) similarity matrix.

the graph vertices, i.e., $W_{ij} = \{0, 1\}$, $\mathbf{W}$ is identical to the graph *adjacency matrix* $\mathbf{A}$ defined in Chapter 2.

$\mathcal{G}$ may be either fully connected, i.e., w_{ij} are known for every $i = 1, \ldots, N, j = 1, \ldots, N$, or not. In the first case, the relationships between a sample and all the remaining ones can be exploited. However, as will be described in the following sections, there are cases where we are interested in graph structures that describe the relationships between neighboring graph vertices only. Such graph structures can be obtained either by using an appropriate scale value σ in equation (6.2.1), or by using a node neighborhood definition. Two of the most widely used neighborhood definitions are used in the k-NN and the ϵ-ball graphs, respectively [RS00]. The k-NN graph exploits only the graph weights of a node i that correspond to its k nearest neighboring nodes. That is, the distances between the node i and the remaining $N - 1$ nodes are calculated, sorted in a ascending orders and the weights that correspond to the nodes providing the k smallest distance values are included in $\mathbf{W}$. The ϵ-ball graph exploits the weights w_{ij} that satisfy $\|\mathbf{x}_i - \mathbf{x}_j\|_2 \leq \epsilon$. From the definition of these graph structures, it can be seen that the ϵ-ball graph corresponds to a symmetric weight matrix, i.e. $\mathbf{W} = \mathbf{W}^T$, while the k-NN graph may not correspond to a symmetric weight matrix. A symmetric weight matrix $\mathbf{W}$ for the k-NN graph can be obtained by using the graph weights that correspond to nodes that are k nearest neighbors of each other at the same time. Appropriate k and ϵ parameter values can be obtained by applying a trial and error, or cross-validation procedure, where the optimal k, ϵ values are determined out of a set of predefined values based e.g., on classification performance. It should be noted here that the k-NN graph is usually preferred, as it adapts to the dataset properties, while an appropriate value of ϵ should be defined for each dataset independently.

In the case of *unsupervised learning*, no class labels are used. It is assumed that the only available information is the geometric data relationship encoded in the graph $\mathcal{G}$. This information can be exploited to partition the graph vertex set into smaller clusters (*graph clustering*) [Sch07, NdC11]. It can also be used to embed graph vertices in a low-dimensional feature space, while preserving geometrical data properties [YXZ$^+$07]. In the case of *supervised learning*, each graph vertex $\mathbf{x}_i$ is accompanied by a class label $c_i \in \mathcal{C}, \mathcal{C} = \{\mathcal{C}_1, \ldots, \mathcal{C}_C\}$, e.g., the ID of the person depicted in facial image i. This information can be exploited in training, to learn a mapping $\mathbf{x}_i \overset{f(\mathbf{x}_i)}{\rightarrow} c_i$. Once learned, this mapping can be used to map a new test sample $\mathbf{x}_t$ (not belonging to the training set $\mathcal{V}$) to one of the classes in $\mathcal{C}$. In the case of binary or multi-class classification, the class label set is usually defined as $\mathcal{C} = \{-1, 1\}$, or $\mathcal{C} = \{1, \ldots, C\}$, respectively. In *regression* problems, we typically have $c_i \in \mathbb{R}$. Supervised approaches can also be employed in order to embed the graph vertices in a low-dimensional feature space, while optimizing some class discrimination criteria defined over the graph vertices [YXZ$^+$07].

In *semi-supervised learning*, it is assumed that some of the graph vertices are accompanied by a class label, while the remaining ones are not. That is, out of the N node vectors $\mathbf{x}_i$, $i = 1, \ldots, l, l+1, \ldots, N$ in total, we have only l labeled ones c_i, $i = 1, \ldots, l$. We assume that the first l graph vertices are labeled and the remaining $u = N - l$ ones are not (usually, $u \gg l$). Semi-supervised classification approaches can be divided into two categories, namely *transductive* and *inductive* ones. In the first case, the objective is the exploitation of the geometric data relationships encoded in the graph $\mathcal{G}$ and the label information available for the labeled vertices, in order to assign labels to the unlabeled vertices (data items). This problem is usually referred to as label propagation and is presented in Chapter 5. Inductive methods exploit the geometric data relationships encoded in the graph $\mathcal{G}$ and the label information available for the labeled vertices, in order to learn a mapping $\mathbf{x}_i \overset{f(\mathbf{x}_i)}{\to} c_i$ that can be used in order to map a new test sample $\mathbf{x}_t$ (not belonging to $\mathcal{V}$) to one of the classes in $\mathcal{C}$. Clearly, after learning $f(\cdot)$, it can be used to map the unlabeled vertices $\mathbf{x}_i$, $i = l + 1, \ldots, N$, to one of the classes in $\mathcal{C}$. In this sense, transduction is a special case of the inductive approach. Semi-supervised dimensionality reduction approaches can also be employed to embed the graph vertices in a low-dimensional feature space optimizing a certain class discrimination criterion defined over the labeled graph vertices and preserving geometrical data properties.

6.3 Unsupervised Methods

In this section, we describe unsupervised dimensionality reduction methods exploiting graph structures. Methods related to graph-based clustering are presented in Chapter 3. The objective in unsupervised dimensionality reduction methods is the determination of an appropriate data mapping process that maps the (usually) high-dimensional data $\mathbf{x}_i \in \mathbb{R}^D$ to a reduced-dimensionality feature space $\mathbb{R}^d$, typically $d \ll D$, which preserves some interesting properties of the original data $\mathbf{x}_i$, $i = 1, \ldots, N$. Such properties may be the pairwise distances between data (Multidimensional Scaling (MDS) [Kru64a, Kru64b]), the dataset variance used in Principal Component Analysis (PCA) [HW10] or the local geometric data structure employed in Locally Linear Embedding (LLE) [SR03].

6.3.1 Locality Preserving Projections

Locality Preserving Projections (LPP) [HN04] finds a low-dimensional embedding of the original data $\mathbf{x}_i \in \mathbb{R}^D$, so that nearby samples in the high-dimensional space $\mathbb{R}^D$ remain placed nearby and are similarly co-located with respect to one another in the low-dimensional space $\mathbb{R}^d$. First, it finds the K nearest neighbors of each sample $\mathbf{x}_i$ based on its Euclidean distances from the remaining samples $j \neq i$ that form the neighborhood of vertex i. Then it constructs a neighborhood graph $\mathcal{G}(\mathcal{V}, \mathcal{E})$ such that $\mathcal{E}$ contains the edges e_{ij} connecting neighboring graph vertices in $\mathcal{V} = \{\mathbf{x}_1, \ldots, \mathbf{x}_N\}$. The graph weight matrix $\mathbf{W} \in \mathbb{R}^{N \times N}$ is subsequently defined as follows:

$$W_{ij} = \begin{cases} 1, & i \in \mathcal{N}_j \text{ or } j \in \mathcal{N}_i \\ 0, & \text{otherwise,} \end{cases} \tag{6.3.1}$$

where $\mathcal{N}_i$ denotes the neighborhood of $\mathbf{x}_i$. The global embedding $\mathbf{Y} \in \mathbb{R}^{d \times N}$ of the samples $\mathbf{x}_i$ is given by:

$$\mathbf{P} = \arg\max_{\mathbf{P}^T\mathbf{P}=\mathbf{I}} \sum_{i=1}^{N} \sum_{j=1}^{N} W_{ij} \|\mathbf{P}^T\mathbf{x}_i - \mathbf{P}^T\mathbf{x}_j\|_2^2 = \arg\max_{\mathbf{P}^T\mathbf{P}=\mathbf{I}} tr\left(\mathbf{P}^T\mathbf{X}\mathbf{L}\mathbf{X}^T\mathbf{P}\right), \quad (6.3.2)$$

where $\mathbf{L} = \mathbf{D} - \mathbf{W}$ is the graph Laplacian matrix and $\mathbf{D} \in \mathbb{R}^{N \times N}$ is a diagonal matrix, whose elements are equal to $D_{ii} = \sum_{j=1}^{N} W_{ij}$. The solution of the optimization problem (6.3.2) is found by performing eigenanalysis of the matrix $\mathbf{B} = \mathbf{X}\mathbf{L}\mathbf{X}^T$ and retaining the eigenvectors corresponding to the d smallest eigenvalues. After the calculation of $\mathbf{P}$, the data $\mathbf{x}_i$ are mapped to the reduced-dimensionality feature space $\mathbf{R}^d$ by applying $\mathbf{y}_i = \mathbf{P}^T\mathbf{x}_i$.

6.3.2 Locally Linear Embedding

Locally Linear Embedding (LLE) [SR03] finds a low-dimensional embedding of the original data $\mathbf{x}_i \in \mathbb{R}^D$, where nearby samples in the high-dimensional space $\mathbb{R}^D$ remain placed nearby and are similarly co-located, with respect to one another, in the low-dimensional space $\mathbb{R}^d$. However, this is done in a different way than in LPP. The main difference is that LLE employs a weighted graph, while LPP employs a unweighted graph. In its simplest formulation, LLE defines the K-nearest neighbors of each sample $\mathbf{x}_i$, based on its Euclidean distances from the remaining samples $j \neq i$ and constructs a neighborhood graph $\mathcal{G}(\mathcal{V}, \mathcal{E})$, such that $\mathcal{E}$ contains the edges e_{ij} connecting neighboring graph vertices in $\mathcal{V} = \{\mathbf{x}_1, \ldots, \mathbf{x}_N\}$.

Subsequently, a local fitting step is performed. That is, each sample $\mathbf{x}_i$ is approximated by its neighbors $\mathbf{x}_j$, $j \in \mathcal{N}_i$ according to fitting weights w_{ij} by solving for:

$$\min_{\sum_{j \in \mathcal{N}_i} w_{ij}=1} \|\mathbf{x}_i - \sum_{j \in \mathcal{N}_i} w_{ij}(\mathbf{x}_j - \mathbf{x}_i)\|_2^2. \quad (6.3.3)$$

This can be done by solving a regularized least squares problem. That is, we create the Gram matrix $\mathbf{C}^i \in \mathbb{R}^K$ having its elements equal to $C_{jk}^i = (\mathbf{x}_j - \mathbf{x}_i)^T(\mathbf{x}_k - \mathbf{x}_i)$ and obtain the weight vector $\mathbf{w}_i$ by solving for $\mathbf{w}_i = \left(\mathbf{C}^i - \mu\mathbf{I}\right)^{-1}\mathbf{1}$, where $\mu \geq 0$ is a regularization parameter and $\mathbf{1} \in \mathbb{R}^K$ is a vector of ones. $\mathbf{w}_i$ is subsequently normalized to have unit l_1 norm.

After the calculation of the weight vectors $\mathbf{w}_i$, $i = 1, \ldots, N$, we define the weight matrix $\mathbf{W} \in \mathbb{R}^{N \times N}$ having its elements equal to:

$$W_{ij} = \begin{cases} w_{ij}, & j \in \mathcal{N}_i \\ 0, & \text{otherwise.} \end{cases} \quad (6.3.4)$$

The global embedding $\mathbf{Y} \in \mathbb{R}^{d \times N}$ of the samples $\mathbf{x}_i$ is found by minimizing:

$$\min_{\mathbf{Y}} \sum_{i=1}^{N} \|\mathbf{y}_i - \sum_{j=1}^{N} W_{ij}\mathbf{y}_j\|_2^2 = tr\left(\mathbf{Y}(\mathbf{I} - \mathbf{W})^T(\mathbf{I} - \mathbf{W})\mathbf{Y}^T\right). \quad (6.3.5)$$

The optimization problem (6.3.5) is solved by performing eigenanalysis of the positive semi-definite matrix $\mathbf{B} = (\mathbf{I} - \mathbf{W})^T(\mathbf{I} - \mathbf{W})$. Subsequently, the eigenvalues $\lambda_1, \ldots, \lambda_N$ are sorted in ascending order and the eigenvectors $\mathbf{v}_1, \ldots, \mathbf{v}_d$ corresponding to the smallest eigenvalues $\lambda_1, \ldots, \lambda_d$ are retained. The optimal embedding is finally given by $\mathbf{Y} = \left[\frac{1}{\sqrt{\lambda_1}}\mathbf{v}_1, \ldots, \frac{1}{\sqrt{\lambda_d}}\mathbf{v}_d\right]^T$. An advantage of LLE, when compared to LPP, is that additional local geometric information can be exploited by exploiting a weighted graph, rather than the unweighted graph used in LPP.

6.3.3 ISOMAP

ISOMAP [TS00] determines a low-dimensional embedding of the original data $\mathbf{x}_i \in \mathbb{R}^D$, so that the pairwise geodesic distances between the data are preserved in the low-dimensional space $\mathbb{R}^d$. ISOMAP constructs a neighborhood graph $\mathcal{G}(\mathcal{V}, \mathcal{E})$, such that $\mathcal{E}$ contains the edges e_{ij} connecting neighboring graph vertices i, j in $\mathcal{V} = \{\mathbf{x}_1, \ldots, \mathbf{x}_N\}$, if $j \in \mathcal{N}_i$. Then, the elements of the graph weight matrix $\mathbf{W}$ are set to $W_{ij} = \|\mathbf{x}_i - \mathbf{x}_j\|_2$. Subsequently, the shortest path distances:

$$d_{ij} = \min_{\mathcal{P}=\{\mathbf{x}_i,\ldots,\mathbf{x}_j\}} \left(\|\mathbf{x}_i - \mathbf{x}_{t_1}\|_2 + \cdots + \|\mathbf{x}_{t_{k-1}} - \mathbf{x}_j\|_2 \right) \qquad (6.3.6)$$

are computed, denoting the length of the graph shortest path $\mathcal{P} = \{\mathbf{x}_i, \ldots, \mathbf{x}_{p_1}, \ldots, \mathbf{x}_{p_{k-1}}, \ldots, \mathbf{x}_j\}$ connecting the vertices i and j and spanning k edges [Ata98].

Subsequently, the standard MDS algorithm [FC11] is applied by using the matrix $\mathbf{D} \in \mathbb{R}^{N \times N}$, where $D_{ij} = d_{ij}^2$. That is, eigenanalysis is performed on the symmetric matrix $\mathbf{B} = -\frac{1}{2}\mathbf{H}\mathbf{D}\mathbf{H}^T$, where $\mathbf{H} = \mathbf{I} - \frac{1}{N}\mathbf{1}\mathbf{1}^T$. Subsequently, the eigenvalues $\lambda_1, \ldots, \lambda_N$ are sorted in descending order $\lambda_1 \geq \lambda_2 \geq \cdots \geq \lambda_N$ and the top d eigenvectors $\mathbf{v}_1, \ldots, \mathbf{v}_d$ corresponding to the d biggest eigenvalues are retained. The optimal embedding is finally given by $\mathbf{Y} = \mathbf{\Lambda}^{\frac{1}{2}}[\mathbf{v}_1, \ldots, \mathbf{v}_d]^T$. Similarly to LLE, an advantage of ISOMAP, when compared to LPP is that, by exploiting a weighted graph, additional local geometric information can be exploited. However, ISOMAP requires additional computational cost in order to calculate the shortest path distances d_{ij}.

6.3.4 Laplacian Embedding

Laplacian Eigenmaps (LE) [BN03] compute a low-dimensional embedding of the original data $\mathbf{x}_i \in \mathbb{R}^D$, with the property that nearby samples in the high-dimensional space $\mathbb{R}^D$ remain placed nearby in the low-dimensional space $\mathbb{R}^d$. LE constructs a neighboring graph $\mathcal{G}(\mathcal{V}, \mathcal{E})$ such that $\mathcal{E}$ contains the edges connecting neighboring graph vertices in $\mathcal{V} = \{\mathbf{x}_1, \ldots, \mathbf{x}_N\}$ and the elements of the graph weight matrix $\mathbf{W}$ are set to $W_{ij} = \exp\left(-\frac{\|\mathbf{x}_i-\mathbf{x}_j\|_2^2}{2\sigma^2}\right)$. Subsequently, the optimal embedding is given by minimizing:

$$\min_{\mathbf{Y}} \sum_{i=1}^{N} \sum_{j=1}^{N} \|\mathbf{y}_i - \mathbf{y}_j\|_2^2 W_{ij} = \min_{\mathbf{Y}} tr\left(\mathbf{Y}^T \mathbf{L} \mathbf{Y}\right), \qquad (6.3.7)$$

where $tr(\cdot)$ denotes the trace of a matrix, $\mathbf{L} = \mathbf{D} - \mathbf{W}$ is the graph Laplacian matrix and $\mathbf{D} \in \mathbb{R}^{N \times N}$ is a diagonal matrix having its elements equal to $D_{ii} = \sum_{j=1}^{N} W_{ij}$. In order to remove any arbitrary scaling factor in the embedding, the constraint $\mathbf{Y}^T \mathbf{D} \mathbf{Y} = \mathbf{I}$ is set, leading to the following optimization problem:

$$\min_{\mathbf{Y}^T \mathbf{D} \mathbf{Y} = \mathbf{I}} tr\left(\mathbf{Y}^T \mathbf{L} \mathbf{Y}\right). \qquad (6.3.8)$$

The objective function (6.3.8) is minimized by solving the generalized eigendecomposition problem $\mathbf{L}\mathbf{y} = \lambda \mathbf{D}\mathbf{y}$ and retaining the eigenvectors corresponding to the smallest eigenvalues. By exploiting a complete weighted graph structure, it can exploit both local and global geometric information, depending on the value of the parameter σ. This is an advantage in the cases where a smooth low-dimensional embedding is searched for.

6.3.5 Diffusion Maps

Diffusion Maps [CL06] are a method for the analysis of the geometry of general datasets, based on the definition of Markov chains [Nor98]. For a fixed value ϵ, the isotropic diffusion kernel can be defined:

$$k_\epsilon(\mathbf{x}_i, \mathbf{x}_j) = \exp\left(-\frac{\|\mathbf{x}_i - \mathbf{x}_j\|_2^2}{4\epsilon}\right). \tag{6.3.9}$$

Assuming that the transition probability between the vertices $\mathbf{x}_i$ and $\mathbf{x}_j$ is proportional to $k_\epsilon(\mathbf{x}_i, \mathbf{x}_j)$, the Markov matrix $\mathbf{M} \in \mathbb{R}^{N \times N}$ can be constructed, which has elements equal to:

$$M_{ij} = \frac{k_\epsilon(\mathbf{x}_i, \mathbf{x}_j)}{p_\epsilon(\mathbf{x}_i)}, \tag{6.3.10}$$

where $p_\epsilon(\mathbf{x}_i)$ is a normalization constant given by:

$$p_\epsilon(\mathbf{x}_i) = \sum_{i=1}^{N} k_\epsilon(\mathbf{x}_i, \mathbf{x}_j). \tag{6.3.11}$$

For large values of ϵ, $\mathbf{M}$ is fully connected. Thus, it has eigenvalues $\lambda_1 > \lambda_2 \geq \cdots \geq \lambda_N$ with $\lambda_1 = 1$ and right eigenvectors $\mathbf{u}_l$, $l = 1, \ldots, N$. The diffusion distance at time t is defined as follows:

$$D_t^2(\mathbf{x}_i, \mathbf{x}_j) = \sum_k \left(p(\mathbf{x}_k, t|\mathbf{x}_i) - p(\mathbf{x}_k, t|\mathbf{x}_j)\right)^2 w(\mathbf{x}_k), \tag{6.3.12}$$

where $p(\mathbf{x}_k, t|\mathbf{x}_i)$ is the probability that the random walk is located at $\mathbf{x}_k$ at time t, given a starting location $\mathbf{x}_i$ at time $t = 0$. Using a weight function $w(\mathbf{x}_k) = \frac{1}{p_\epsilon(\mathbf{x}_k)}$ [NLCK05], we obtain:

$$D_t^2(\mathbf{x}_i, \mathbf{x}_j) = \sum_l \lambda_l^{2t} \left(u_l(i) - u_l(j)\right)^2, \tag{6.3.13}$$

where $u_l(i)$ denotes the i-th element of $\mathbf{u}_l$. The diffusion map at time t of $\mathbf{x}_i$ is defined, as in the normalized graph Laplacian case [SM97, Wei99, NJW02], by:

$$\mathbf{y}_i^t = [\lambda_1^t u_1(i), \lambda_2^t u_2(i), \ldots, \lambda_d^t u_d(i)]^T. \tag{6.3.14}$$

6.4 Supervised Methods

As has been previously mentioned, supervised methods employ label information that is available for graph vertices, in order to determine graph node vector mappings to a low-dimensional feature space while optimizing certain class discrimination criteria that are defined over the labeled graph vertices. In this section, we first describe some of the most widely employed supervised dimensionality reduction methods that assume an underlying graph structure. Their graph embedding formulation follows. Subsequently, we describe supervised classification methods that exploit criteria used in supervised dimensionality reduction.

6.4.1 Linear Discriminant Analysis

Linear Discriminant Analysis (LDA) [DHS00] determines a projection subspace $\mathbb{R}^d$ for data projection, where the samples belonging to the same class come close to each another, while samples belonging to different classes distance themselves as far as possible. This is achieved by minimizing the within-class data scatter and maximizing the between-class data scatter that are expressed by the following matrices:

$$\mathbf{S}_w = \sum_{c=1}^{C} \sum_{\mathbf{x}_i \in \mathcal{C}_c} (\mathbf{x}_i - \boldsymbol{\mu}_c)(\mathbf{x}_i - \boldsymbol{\mu}_c)^T \qquad (6.4.1)$$

$$\mathbf{S}_b = \sum_{c=1}^{C} N_c (\boldsymbol{\mu}_c - \boldsymbol{\mu})(\boldsymbol{\mu}_c - \boldsymbol{\mu})^T, \qquad (6.4.2)$$

respectively, where N_c denotes the number of samples belonging to class c, $\boldsymbol{\mu}_c = \frac{1}{N_c} \sum_{\mathbf{x}_i \in \mathcal{C}_c} \mathbf{x}_i$ is the arithmetic mean vector of class c and $\boldsymbol{\mu} = \frac{1}{N} \sum_{i=1}^{N} \mathbf{x}_i$ is the arithmetic mean vector of the entire training dataset.

Although it has not been proposed as a graph-based method, LDA can be defined by using graph notation. Let us assume that the training samples $\mathbf{x}_i$ are employed in order to construct a graph $\mathcal{G}(\mathcal{V}, \mathcal{E})$ [YXZ+07]. Two graph weight matrices $\mathbf{W}_w$ and $\mathbf{W}_b$ are defined, having elements equal to:

$$W_{w,ij} = \begin{cases} \frac{1}{N_{c_i}}, & c_j = c_i \\ 0, & \text{otherwise} \end{cases} \qquad (6.4.3)$$

$$W_{b,ij} = \begin{cases} \frac{1}{N_{c_i}} - \frac{1}{N}, & c_j = c_i \\ -\frac{1}{N}, & \text{otherwise.} \end{cases} \qquad (6.4.4)$$

By using $\mathbf{W}_w$ and $\mathbf{W}_b$, the within-class and between-class scatter matrices are described by:

$$\mathbf{S}_w = \mathbf{V}(\mathbf{D}_w - \mathbf{W}_w)\mathbf{V}^T \qquad (6.4.5)$$

$$\mathbf{S}_b = \mathbf{V}(\mathbf{D}_b - \mathbf{W}_b)\mathbf{V}^T, \qquad (6.4.6)$$

where $\mathbf{D}_w$, $\mathbf{D}_b$ are diagonal matrices having their elements equal to $D_{w,ii} = \sum_{j=1}^{N} W_{w,ij}$ and $D_{b,ii} = \sum_{j=1}^{N} W_{b,ij}$, respectively.

The optimal data projection matrix $\mathbf{P} \in \mathbb{R}^{D \times d}$ can be subsequently obtained by solving the *trace ratio* optimization problem defined by [WYX+07, JNZ09]:

$$\mathbf{P} = \arg\max_{\mathbf{P}} \frac{tr(\mathbf{P}^T \mathbf{S}_b \mathbf{P})}{tr(\mathbf{P}^T \mathbf{S}_w \mathbf{P})}. \qquad (6.4.7)$$

Since the trace ratio problem does not have a direct closed-form globally optimal solution, it is conventionally approximated by solving the *ratio trace* problem defined by:

$$\mathbf{P} = \arg\max_{\mathbf{P}} tr\left((\mathbf{P}^T \mathbf{S}_w \mathbf{P})^{-1} (\mathbf{P}^T \mathbf{S}_b \mathbf{P}) \right), \qquad (6.4.8)$$

which is equivalent to the optimization problem:

$$\mathbf{S}_w \mathbf{p} = \lambda \mathbf{S}_b \mathbf{p}, \quad \lambda \neq 0 \qquad (6.4.9)$$

and can be solved by performing the eigenanalysis of matrix $\mathbf{S} = \mathbf{S}_b^{-1} \mathbf{S}_w$, if $\mathbf{S}_b$ is invertible, or $\mathbf{S} = \mathbf{S}_w^{-1} \mathbf{S}_b$, if $\mathbf{S}_w$ is invertible.

Although the original trace ratio problem (6.4.7) does not have a closed form solution, it has been shown that it can be converted to an equivalent *trace difference* problem having the form [WYX+07, JNZ09]:

$$\mathbf{P} = \arg\max_{\mathbf{P}^T \mathbf{P} = \mathbf{I}} trace\left(\mathbf{P}^T (\mathbf{S}_b - \lambda \mathbf{S}_w) \mathbf{P} \right), \qquad (6.4.10)$$

which is solved by applying an efficient constrained maximization algorithm based on the Newton-Raphson method [WYX$^+$07, JNZ09]. After the calculation of $\mathbf{P}$, the data $\mathbf{x}_i$ are mapped to the reduced-dimensionality feature space $\mathbf{R}^d$ by applying $\mathbf{y}_i = \mathbf{P}^T \mathbf{x}_i$.

While LDA has been shown to provide satisfactory performance in a variety of applications, its underlying assumption is the use of unimodal classes following normal distributions. However, this assumption is rather restrictive. In order to take into account class multimodality, Subclass Discriminant Analysis (SDA) [ZM06] and Clustering-based Discriminant Analysis (CDA) [CH03] have been proposed. In these approaches, it is assumed that each class is formed by several unimodal subclasses, each following a normal distribution. Therefore, the scatter matrices (6.4.1), (6.4.2) are modified accordingly. In addition, in order to overcome the class normality assumption, the adoption of optimized class representations has been proposed in [ITP13b]. It has been shown that both of these approaches outperform LDA in the cases where the assumptions of LDA are not met.

6.4.2 Marginal Fisher Analysis

In *Marginal Fisher Analysis* (MFA) [YXZ$^+$07], the definition of the intra-class and between-class relationships are described by following a local approach. It is assumed that the data are embedded in an intrinsic graph $\mathcal{G}(\mathcal{V}, \mathcal{E}, \mathbf{W})$ and a penalty graph $\mathcal{G}^{(p)}(\mathcal{V}, \mathcal{E}, \mathbf{W}^{(p)})$, where the matrix $\mathbf{W}$ expresses the local relationships between the data belonging to the same class and the matrix $\mathbf{W}^{(p)}$ expresses the penalty weights used to increase inter-class discrimination. $\mathbf{W}$, $\mathbf{W}^{(p)}$ are defined by:

$$W_{ij} = \begin{cases} 1, & c_i = c_j \text{ and } j \in \mathcal{N}_i \\ 1, & c_i = c_j \text{ and } i \in \mathcal{N}_j \\ 0, & \text{otherwise} \end{cases} \tag{6.4.11}$$

$$W_{ij}^{(p)} = \begin{cases} 1, & c_i \neq c_j \text{ and } j \in \mathcal{N}_i \\ 1, & c_i \neq c_j \text{ and } i \in \mathcal{N}_j \\ 0, & \text{otherwise.} \end{cases} \tag{6.4.12}$$

Similarly to LDA, the optimal data projection matrix $\mathbf{P}$ is obtained by solving for:

$$\mathbf{P} = \arg\min_{\mathbf{P}} \frac{tr\left(\mathbf{P}^T \mathbf{X} (\mathbf{D} - \mathbf{W}) \mathbf{X}^T \mathbf{P}\right)}{tr\left(\mathbf{P}^T \mathbf{X} (\mathbf{D}^p - \mathbf{W}^{(p)}) \mathbf{X}^T \mathbf{P}\right)}, \tag{6.4.13}$$

where $\mathbf{D}$, $\mathbf{D}^p$ are diagonal matrices having their elements equal to $D_{ii} = \sum_{j=1}^{N} W_{ij}$ and $D_{ii}^p = \sum_{j=1}^{N} W_{ij}^{(p)}$, respectively. $\mathbf{P}$ can be calculated by solving the generalized eigenanalysis problem $\mathbf{X}(\mathbf{D} - \mathbf{W})\mathbf{X}^T \mathbf{q} = \lambda \mathbf{X}(\mathbf{D}^p - \mathbf{W}^{(p)})\mathbf{X}^T \mathbf{q}$.

After the calculation of $\mathbf{P}$, the data $\mathbf{x}_i$ are mapped to the reduced-dimensionality feature space $\mathbf{R}^d$ by applying $\mathbf{y}_i = \mathbf{P}^T \mathbf{x}_i$. MFA can exploit local intra-class and inter-class information by employing local graph structures. This is advantageous in cases where the assumptions of LDA are not met [YXZ$^+$07].

6.4.3 Local Fisher Discriminant Analysis

Local Fisher Discriminant Analysis (LFDA) [Sug07] defines the within-class and between-class relationships, by using graph relationships. It is assumed that the data are embedded in an intrinsic graph $\mathcal{G}(\mathcal{V}, \mathcal{E}, \tilde{\mathbf{W}}^{(w)})$ and a penalty graph $\mathcal{G}^{(b)}(\mathcal{V}, \mathcal{E}, \tilde{\mathbf{W}}^{(b)})$, where matrix $\tilde{\mathbf{W}}^{(w)}$ expresses local relationships between data belonging to the same class, while the matrix $\tilde{\mathbf{W}}^{(b)}$ expresses local relationships between data placed at the borders of different classes. $\tilde{\mathbf{W}}^{(w)}$, $\tilde{\mathbf{W}}^{(b)}$ are defined by:

$$W_{ij}^{(w)} = \begin{cases} \frac{s_{ij}}{N_{c_i}}, & c_j = c_i \\ 0, & \text{otherwise} \end{cases} \tag{6.4.14}$$

$$W_{ij}^{(b)} = \begin{cases} s_{ij}\left(\frac{1}{N} - \frac{1}{N_{c_i}}\right), & c_j = c_i \\ \frac{1}{N}, & \text{otherwise,} \end{cases} \tag{6.4.15}$$

where s_{ij} is a measure of similarity between $\mathbf{x}_i$ and $\mathbf{x}_j$, such as the heat kernel function (6.2.1). The optimal data projection matrix $\mathbf{P}$ is obtained by maximizing the objective function:

$$\mathbf{P} = \arg\max_{\mathbf{P}} \, tr\left(\left(\mathbf{P}^T\tilde{\mathbf{S}}^{(w)}\mathbf{P}\right)^{-1}\left(\mathbf{P}^T\tilde{\mathbf{S}}^{(b)}\mathbf{P}\right)\right), \tag{6.4.16}$$

where the matrices $\tilde{\mathbf{S}}^{(w)}$, $\tilde{\mathbf{S}}^{(b)}$ are defined as follows:

$$\tilde{\mathbf{S}}^{(w)} = \frac{1}{2}\sum_{i=1}^{N}\sum_{j=1}^{N}\tilde{W}_{ij}^{(w)}\left(\mathbf{x}_i - \mathbf{x}_j\right)\left(\mathbf{x}_i - \mathbf{x}_j\right)^T \tag{6.4.17}$$

$$\tilde{\mathbf{S}}^{(b)} = \frac{1}{2}\sum_{i=1}^{N}\sum_{j=1}^{N}\tilde{W}_{ij}^{(b)}\left(\mathbf{x}_i - \mathbf{x}_j\right)\left(\mathbf{x}_i - \mathbf{x}_j\right)^T. \tag{6.4.18}$$

$\mathbf{P}$ can be calculated by solving the generalized eigenanalysis problem $\tilde{\mathbf{S}}^{(w)}\mathbf{q} = \lambda\tilde{\mathbf{S}}^{(b)}\mathbf{q}$.

After the calculation of $\mathbf{P}$, the data $\mathbf{x}_i$ are mapped to the reduced-dimensionality feature space $\mathbf{R}^d$ by applying $\mathbf{y}_i = \mathbf{P}^T\mathbf{x}_i$. By employing weighted local graph structures, LFDA can exploit local intra-class and inter-class information using additional geometric information, when compared to MFA. Sometimes, this is advantageous in terms of classification performance [Sug07].

6.4.4 Graph Embedding

It has been shown in [YXZ+07] that a wide range of (linear and non-linear) dimensionality reduction criteria can be described from a graph embedding point of view. Let $\mathcal{G}(\mathcal{V}, \mathcal{E})$ be an undirected weighted graph, where we assume that the training samples $\mathbf{x}_i$ reside on graph vertices and $\mathbf{W} \in \mathbb{R}^{N \times N}$ is the corresponding graph weight (or adjacency) matrix. The diagonal matrix $\mathbf{D} \in \mathbb{R}^{N \times N}$ and the graph Laplacian matrix $\mathbf{L} \in \mathbb{R}^{N \times N}$ are defined by $D_{ii} = \sum_i \mathbf{W}_{ij}$, $i = 1, ..., N$ and $\mathbf{L} = \mathbf{D} - \mathbf{W}$, respectively. The graph Laplacian matrix $\mathbf{L}$ can be employed, in order to describe criteria exploited in several subspace learning techniques, like LDA, ISOMAP, LLE, LE, etc. Let us denote by $\mathbf{L}_X$ the graph Laplacian matrix describing a certain criterion X. Then, the criterion X can be modelled using a matrix of the form:

$$\mathbf{S}_X = \mathbf{X}\mathbf{L}_X\mathbf{X}^T. \tag{6.4.19}$$

For example, the total scatter matrix employed in Principal Component Analysis (PCA) [DHS00] can be expressed as follows:

$$\mathbf{S}_T = \mathbf{X}\mathbf{L}_T\mathbf{X}^T. \tag{6.4.20}$$

The within-class and between-class scatter matrices employed in LDA have the form:

$$\mathbf{S}_w = \mathbf{X}\mathbf{L}_w\mathbf{X}^T \tag{6.4.21}$$

$$\mathbf{S}_b = N\mathbf{S}_T - \mathbf{S}_w. \tag{6.4.22}$$

The corresponding graph Laplacian matrices $\mathbf{L}_T$, $\mathbf{L}_w$ are given by:

$$\mathbf{L}_T = \mathbf{I} - \frac{1}{N}\mathbf{e}\mathbf{e}^T \tag{6.4.23}$$

$$\mathbf{L}_w = \mathbf{I} - \sum_{c=1}^{C} \frac{1}{N_c}\mathbf{e}^c\mathbf{e}^{cT}, \tag{6.4.24}$$

where $\mathbf{e} \in \mathbb{R}^N$ is a vector of ones, $\mathbf{I} \in \mathbb{R}^{N \times N}$ is the identity matrix and $\mathbf{e}^c \in \mathbb{R}^N$ is a vector with $e_j^c = 1$, if $c_j = C$ and $e_l^c = 0$ otherwise. After the calculation of the graph Laplacian matrices describing the criteria of interest, the data projection matrix can be obtained by minimization:

$$\mathbf{P} = \operatorname*{arg\,min}_{tr(\mathbf{P}^T\mathbf{X}\mathbf{L}^p\mathbf{X}^T\mathbf{P})} \ tr\left(\mathbf{P}^T\mathbf{X}\mathbf{L}\mathbf{X}^T\mathbf{P}\right), \tag{6.4.25}$$

where $\mathbf{L}$ is the graph Laplacian matrix describing the criterion to be minimized and $\mathbf{L}^p$ describes the penalty criterion (to be maximized).

6.4.5 Minimum Class Variance Extreme Learning Machine

Minimum Class Variance Extreme Learning Machine (MCVELM) [ITP13a] is an algorithm for Single-hidden Layer Feedforward Neural (SLFN) network training that exploits nonlinear data relationships. For a classification problem involving vectors $\mathbf{x}_i \in \mathbb{R}^D$, each belonging to one of the C classes, the network should consist of D input, L hidden and C output neurons. Usually, the number L of hidden layer neurons is much greater than the number C of action classes that are involved in the classification problem, i.e., $L \gg C$ [HZS04, HZDZ12].

The network target vectors $\mathbf{t}_i = [t_{i1}, ..., t_{iC}]^T$, each corresponding to a vector $\mathbf{x}_i$, are set to $t_{ik} = 1$, when $c_i = k$, and to $t_{ik} = -1$ otherwise. The network input weights $\mathbf{W}_{in} \in \mathbb{R}^{D \times L}$ and the hidden layer bias values $\mathbf{b} \in \mathbb{R}^L$ are randomly assigned, while the network output weights $\mathbf{W}_{out} \in \mathbb{R}^{L \times C}$ are analytically calculated. It has been shown that almost any nonlinear piecewise continuous activation functions $\Phi(\cdot)$ can be used for the calculation of the network hidden layer outputs, like the sigmoid, sine, Gaussian, hard-limiting, and Radial Basis Functions (RBF), Fourier series, etc [HCS06, HZDZ12]. By using $\mathbf{W}_{in}$ and $\mathbf{b}$, each training vector $\mathbf{x}_i$ is mapped to a vector $\boldsymbol{\phi}_i$, corresponding to the network hidden layer output.

The network output weights are subsequently determined by solving the following optimization problem:

$$\min_{\mathbf{W}_{out}} \frac{1}{2}\|\mathbf{S}_w^{\frac{1}{2}}\mathbf{W}_{out}\|_F^2 + \frac{\gamma}{2}\sum_{i=1}^{N}\|\boldsymbol{\xi}_i\|_2^2, \tag{6.4.26}$$

subject to the constraints:

$$\mathbf{W}_{out}^T\boldsymbol{\phi}_i = \mathbf{t}_i - \boldsymbol{\xi}_i, \ \ i = 1, ..., N, \tag{6.4.27}$$

where $\boldsymbol{\xi}_i \in \mathbb{R}^C$ is the error vector corresponding to training vector $\mathbf{x}_i$ and γ is a parameter denoting the importance of the training error in the optimization problem. $\mathbf{S}_w$ is the within-class variance matrix used in LDA (6.4.5) to describe class dispersions. When the classes are multimodal in the feature space determined by the network hidden layer outputs, the intra-class dispersion can be described accordingly [ITP13a]. An extension of the above-described formulation that exploits the total scatter of the data (6.4.20) has also been proposed in [ITP14b].

By substituting (6.4.27) in the optimization problem (6.4.26) and solving for $\frac{\vartheta L_P}{\vartheta \mathbf{W}_{out}} = 0$, $\mathbf{W}_{out}$ is given by:

$$\mathbf{W}_{out} = \left(\mathbf{\Phi}\mathbf{\Phi}^T + \frac{1}{\gamma}\mathbf{S}_w \right)^{-1} \mathbf{\Phi}\mathbf{T}^T. \qquad (6.4.28)$$

An extension of the algorithm that exploits local class information defined on a neighborhood graph for $\mathbf{W}_{out}$ calculation has been proposed in [ITP14a], where the following optimization problem is solved:

$$\min_{\mathbf{W}_{out}} \left(\frac{1}{2}\|\mathbf{W}_{out}\|_F^2 + \frac{\gamma}{2}\sum_{i=1}^{N}\|\boldsymbol{\xi}_i\|_2^2 + \frac{\lambda}{2}tr\left(\mathbf{W}_{out}^T(\mathbf{\Phi}\mathbf{L}\mathbf{\Phi}^T)\mathbf{W}_{out} \right) \right), \qquad (6.4.29)$$

$$s.t. \ \mathbf{W}_{out}^T\boldsymbol{\phi}_i = \mathbf{t}_i - \boldsymbol{\xi}_i, \ \ i = 1, ..., N. \qquad (6.4.30)$$

In (6.4.29), $tr(\cdot)$ is the trace operator and $\mathbf{L}$ is a graph Laplacian matrix calculated by using a class neighboring graph on $\boldsymbol{\phi}_i$. The network output weights $\mathbf{W}_{out}$ are given by:

$$\mathbf{W}_{out} = \left(\mathbf{\Phi}\left(\mathbf{I} + \frac{\lambda}{\gamma}\mathbf{L} \right)\mathbf{\Phi}^T + \frac{1}{\gamma}\mathbf{I} \right)^{-1} \mathbf{\Phi}\mathbf{T}^T. \qquad (6.4.31)$$

ELM-based classification schemes have been found to be both efficient and effective [HCS06, HZDZ12]. When compared to standard neural network training approaches, like the Back-Propagation [RHW86] and the Levenberg-Marquardt [HM94] algorithms, ELM requires lower human supervision and leads to faster network training. ELM algorithms exploiting class variance criteria are able to enhance its classification performance [ITP13a, ITP14a, ITP14b].

6.4.6   Minimum Class Variance Support Vector Machine

Minimum Class Variance SVM (MCVSVM) [TKP01, ZTP07c] exploits statistical class properties in the SVM optimization process. Specifically, it modifies the regularizer of the SVM formulation, in order to exploit the intra-class dispersion of the training data used in LDA (6.4.5), leading to the following SVM formulation:

$$\min_{\mathbf{w},b} \frac{1}{2}\mathbf{w}^T\mathbf{S}_w\mathbf{w} + \gamma\sum_{i=1}^{N}\xi_i, \ \ \mathbf{w}^T\mathbf{S}_w\mathbf{w} \geq 0, \qquad (6.4.32)$$

subject to the constraints:

$$c_i\left(\mathbf{w}^T\mathbf{x}_i + b \right) \geq 1 - \xi_i, \ \ \ \xi_i \geq 0, \ \ \ i = 1, ..., N. \qquad (6.4.33)$$

Non-linear decision functions have been determined by exploiting the kernel trick [SS01, BNS06]. In the cases of multi-modal classes, i.e., classes formed by multiple subclasses, the intra-class dispersion can be described accordingly [OT12, GMK12]. MCVSVM, exploits both information relating to the support vectors and the class compactness by incorporating the within-class variance of the training data in the SVM optimization problem. In several cases, this leads to enhanced classification performance [OT12, GMK12].

6.4.7   Graph Embedded Support Vector Machines

Graph Embedded SVM [AT12] (GESVM) extends MCVSVM, in order to incorporate geometrical criteria of the data used in the Graph Embedding framework [YXZ+07]. It assumes that the data have been embedded in a graph $\mathcal{G}(\mathcal{V}, \mathcal{E})$, with a graph weight matrix $\mathbf{W}$ expressing geometric data relationships. The following regularizer is incorporated in the SVM formulation:

$$J(\mathbf{w}) = \mathbf{w}^T \left(\mathbf{XLX}^T \right) \mathbf{w}, \tag{6.4.34}$$

leading to the following SVM optimization problem:

$$\min_{\mathbf{w},b} \frac{1}{2} \|\mathbf{w}\|_2^2 + \gamma \sum_{i=1}^{N} \xi_i + \frac{\lambda}{2} \mathbf{w}^T \left(\mathbf{XLX}^T \right) \mathbf{w}, \tag{6.4.35}$$

subject to the constraints:

$$c_i \left(\mathbf{w}^T \mathbf{x}_i + b \right) \geq 1 - \xi_i, \quad \xi_i \geq 0, \quad i = 1, \ldots, N. \tag{6.4.36}$$

In the above, γ, λ are the two regularization parameters relating to the training error of the SVM classifier and the trade-off between the two regularization terms, respectively. Nonlinear decision functions can been determined by exploiting the kernel trick [SS01, BNS06]. Depending on the structure of the graph employed by the GESVM, global or local class information is incorporated in the SVM optimization problem. In several cases, this leads to enhanced classification performance.

6.5 Semi-Supervised Methods

Semi-Supervised methods employ labeling information that is available for some graph vertices and the geometric data relationships that are encoded in the graph, in order to either determine a data mapping to a low-dimensional feature space, or to learn a mapping $\mathbf{x}_i \overset{f(\mathbf{x}_i)}{\rightarrow} c_i$ that can be used in order to map a new test sample $\mathbf{x}_t$ (not belonging to the graph) to one of the classes in $\mathcal{C}$. In this section, we first describe an extension of the supervised dimensionality reduction methods that is able to exploit both labeled and unlabeled data. Subsequently, we describe semi-supervised classification methods that exploit geometric information provided by the graph structure, in order to enhance classification performance.

Before starting the description of these methods, we briefly present the assumptions imposed on the structure of the underlying data distribution. Semi-supervised methods make use of at least one of the following assumptions [CSZ06]:

1. Smoothness assumption. Data which are close to each other are more likely to belong to the same class.

2. Cluster assumption. The data tend to form discrete clusters. Data in each cluster are more likely to belong to the same class.

3. Manifold assumption. The data lie on a manifold of much lower dimensionality than that of the original data domain $\mathbb{R}^D$.

From the above, the first assumption indicates the need of simple decision boundaries, residing in low data-density regions. The second assumption indicates that classes may be multimodal, i.e., classes may be formed by multiple clusters. Finally, the third assumption is useful in cases where the input space $\mathbb{R}^D$ is high-dimensional, in order to reduce dimensionality to avoid the so-called *curse of dimensionality* [Mem00].

6.5.1 Semi-Supervised Discriminant Analysis

Semi-Supervised Discriminant Analysis (SDA) [CHH07] aims at finding a low-dimensional feature space by using discriminant information inferred from the labeled data and geometrical information inferred both from the labeled and unlabeled data. It is assumed that the data are embedded in a graph $\mathcal{G}(\mathcal{V}, \mathcal{E})$, with a graph weight matrix $\mathbf{W}$ expressing the local relationships of both the labeled and unlabeled graph nodes. $\mathbf{W}$ is defined by:

$$W_{ij} = \begin{cases} w_{ij}, & i \in \mathcal{N}_j \text{ and } j \in \mathcal{N}_i \\ 0, & \text{otherwise.} \end{cases} \tag{6.5.1}$$

The geometrical data structure is expressed by incorporating the following regularizer:

$$J(\mathbf{P}) = \mathbf{P}^T \mathbf{X} \mathbf{L} \mathbf{X}^T \mathbf{P}, \tag{6.5.2}$$

in the LDA optimization problem. By using $J(\mathbf{P})$, the optimal data projection is given by:

$$\mathbf{P} = \arg\max_{\mathbf{P}} \frac{tr(\mathbf{P}^T \mathbf{S}_b \mathbf{P})}{tr(\mathbf{P}^T (\mathbf{S}_T + \alpha \mathbf{X} \mathbf{L} \mathbf{X}^T) \mathbf{P})}, \tag{6.5.3}$$

where α is a regularization parameter determining the importance of the geometrical data structure in the optimization problem and $\mathbf{S}_T$ is the total scatter matrix calculated by using both the labeled and the unlabeled data. Furthermore, an extension of SDA that exploits global and local geometrical data structure by incorporating regularization terms related to PCA and LPP has been described in [ZLPW13]. $\mathbf{P}$ can be calculated by solving the generalized eigenanalysis problem $\mathbf{S}_b \mathbf{q} = \lambda (\mathbf{S}_T + \alpha \mathbf{X} \mathbf{L} \mathbf{X}^T) \mathbf{q}$.

After the calculation of $\mathbf{P}$, the data $\mathbf{x}_i$ are mapped to the reduced-dimensionality feature space $\mathbf{R}^d$ by applying $\mathbf{y}_i = \mathbf{P}^T \mathbf{x}_i$.

6.5.2 Laplacian Support Vector Machine

Laplacian Support Vector Machine (LapSVM) has been proposed within the context of *Manifold Regularization* [BNS06, MB11]. It incorporates information relating to the geometrical structure of both the labeled and unlabeled data in the SVM optimization process. In its simplest form, it assumes that the data have been embedded in a graph $\mathcal{G}(\mathcal{V}, \mathcal{E})$, with a graph weight matrix $\mathbf{W}$ expressing local relationships of both the labeled and unlabeled data. The k-nearest neighbor graph based on the heat kernel is usually employed. The following regularizer is incorporated in the SVM formulation:

$$J(\mathbf{w}) = \mathbf{w}^T \left(\mathbf{X} \mathbf{L} \mathbf{X}^T \right) \mathbf{w}, \tag{6.5.4}$$

leading to the following SVM optimization problem:

$$\min_{\mathbf{w}, b} \|\mathbf{w}\|_2^2 + c_1 \sum_{i=1}^{N} \xi_i + \frac{c_2}{N^2} \mathbf{w}^T \left(\mathbf{X} \mathbf{L} \mathbf{X}^T \right) \mathbf{w}, \tag{6.5.5}$$

where $\mathbf{L}$ is the graph Laplacian matrix calculated by using both the labeled and unlabeled data. This optimization is subject to the constraints:

$$c_i \left(\mathbf{w}^T \mathbf{x}_i + b \right) \geq 1 - \xi_i, \quad \xi_i \geq 0, \quad i = 1, \ldots, N. \tag{6.5.6}$$

In the above, c_1, c_2 are the regularization parameters relating the training error and the Laplacian regularization. The normalized graph Laplacian:

$$\tilde{\mathbf{L}} = \mathbf{D}^{-\frac{1}{2}} \mathbf{L} \mathbf{D}^{-\frac{1}{2}} \tag{6.5.7}$$

can also be used. Non-linear decision functions can been determined by exploiting the kernel trick [SS01, BNS06]. Instead of using regularization based on similarity metrics, regularizers exploiting sparsity constraints can be used [FGQZ11]. An extension of the above described formulation has also been proposed for semi-supervised feature selection in [XKLJ10].

6.5.3 Semi-Supervised Extreme Learning Machine

Semi-Supervised Extreme Learning Machine (S-ELM) [LCLZ11] is an algorithm for SLFN network training that can incorporate geometric information related to both labeled and unlabeled training data in ELM training. As in the MCVELM algorithm described in Section 6.4.5, the SELM input weights $\mathbf{W}_{in} \in \mathbb{R}^{D \times L}$ and the hidden layer bias values $\mathbf{b} \in \mathbb{R}^{L}$ are randomly assigned, while the network output weights $\mathbf{W}_{out} \in \mathbb{R}^{L \times C}$ are analytically calculated. The network target vectors $\mathbf{t}_i \in \mathbb{R}^{C}$ are set to $a)$ $t_{ij} = 1$ for vectors belonging to class j, i.e., $c_i = j$, $b)$ $t_{ij} = -1$ for vectors not belonging to class j, i.e., and $c_i \neq j$ and $c)$ $t_{ij} = 0$, for unlabeled vectors.

Let us denote by $\boldsymbol{\phi}_i$, $i = 1, \ldots, l, \ldots, N$ the hidden layer outputs for the entire training set. SELM solves the following optimization problem for the calculation of the network output weights:

$$\mathbf{W}_{out} = \underset{\mathbf{W}_{out}}{\arg\min} \|\mathbf{W}_{out}^T \boldsymbol{\Phi} - \mathbf{T}\|_F,$$

$$s.t. \quad \sum_{i=1}^{N} \sum_{j=1}^{N} w_{ij} \left(\mathbf{W}_{out}^T \boldsymbol{\phi}_i - \mathbf{W}_{out}^T \boldsymbol{\phi}_j\right)^2 = 0, \tag{6.5.8}$$

where $\boldsymbol{\Phi} = [\boldsymbol{\phi}_1, \ldots, \boldsymbol{\phi}_N]$ is a matrix containing the network hidden layer outputs, $\mathbf{T} \in \mathbb{R}^{C \times N}$ is a matrix containing the network target vectors $\mathbf{t}_i$ and w_{ij} is a value denoting the similarity between $\boldsymbol{\phi}_i$ and $\boldsymbol{\phi}_j$ in the feature space determined by the network hidden layer outputs. That is, in SELM, it is assumed that the data representations in the so-called ELM space are embedded in a graph $\mathcal{G}(\mathcal{V}, \mathcal{E})$, where $\mathcal{V}$ contains the nodes $\{\boldsymbol{\phi}_1, \ldots, \boldsymbol{\phi}_N\}$ and the correspoinding graph weight matrix $\mathbf{W}$ expresses the similarity of the training data in the ELM space according to a metric, usually the heat kernel function. The calculation of the data similarity values w_{in} in the ELM space $\mathbb{R}^L$, rather than in the input space $\mathbb{R}^D$, has the advantage that nonlinear data relationships can be better exploited.

By solving (6.5.8), $\mathbf{W}_{out}$ is given by:

$$\mathbf{W}_{out} = \left(\left(\mathbf{J} + \lambda \mathbf{L}^T\right) \boldsymbol{\Phi}\right)^{\dagger} \mathbf{J} \mathbf{T}^T. \tag{6.5.9}$$

The diagonal matrix $\mathbf{J} = diag(1, 1, \ldots, 0, 0)$ has the first l diagonal entries equal to 1 and the rest equal to 0. $\mathbf{L}$ is the Graph Laplacian matrix [BNS06] encoding the similarity between the training vectors and $\mathbf{A}^{\dagger} = \left(\mathbf{A}\mathbf{A}^T\right)^{-1} \mathbf{A}$ is the Moore-Penrose pseudoinverse of $\mathbf{A}^T$. A regularized version of the SELM algorithm has been proposed in [HSGW14]. In addition, a regularized version of the SELM algorithm that also exploits discriminan criteria has been proposed in [ITP14c]. Semi-supervised learning using ELM-approaches has been shown to be effective, since it has been shown that it can outperform related approaches [HSGW14, ITP14c], like LapSVM.

6.6 Applications

In this section, we discuss some problems, where graph-based pattern recognition and machine learning techniques have been applied, focusing on areas relating to social media analysis.

Recent advances in technological equipment, like digital cameras, smart-phones, etc., have led to an explosive increase of the captured digital media, e.g., images and videos. As expected, most of these data are acquired in order to describe human presence and activity and are exploited, e.g., for monitoring (visual surveillance and security) or for personal/social use and entertainment. Images play an important role in many applications, including social media. For example, social network users may share images or rate the images of their friends, or tag them in their images [WGLF10, DY11]. Two tasks that are of particular interest in social media tagging and annotation propagation are image segmentation and face recognition [Pit00]. In the first case, the objective is to automatically partition an image in homogeneous regions. This is an important pre-processing step that facilitates other image processing tasks, e.g., face/object recognition. Graph-based techniques have been found to be effective in this task, since, by exploiting graph structures, both the quantitative and topological relationships of the image regions can be exploited [SM00, FH04, KH09, ZFFX10, JLFW10, JT11].

Face recognition assigns a person ID label to a facial image depicting this person. Graph-based techniques have been employed for this task in both the image representation and classification phases. In the first case, graph-based representations combined with elastic graph matching have been found to be effective [WFKM97, TKP01, TKP02, ZTP07a, ZTP07b]. In the latter case, machine learning techniques that exploit an underlying graph structure can be employed to determine a low-dimensional manifold where the facial images reside [RS00, TS00, MK01, BN03, YXZ+07], or to classify the facial images [BNS06, ZTP07c, AT12, HZDZ12].

Video content analysis, especially human action recognition from videos, has been a very active research field nowadays. Several graph-based approaches have been proposed to this end, including spatio-temporal graphs [BT11, GZSRC11, CWSL12, AMA14] and generative Delaunay graphs [TH06]. Methods exploiting graph structures have also been employed in human action classification [YZLZ10, DFHP10, DFHP10, TCFL12, ITP13a, ITP14c].

Recently, social media, i.e., images and videos, have been exploited for automatic social behaviour inference [WCW09, EPL09, GCR09, YLPR09, CZG09, DY10, WGLF10]. This approach exploits the fact that the people depicted in social media images or videos often share social relationships, e.g., are members of the same family [WGLF10], colleagues in work or school, athletes in a football game [PY10], opponents in a wrangle [LW09], etc. Furthermore, the social relationship between the depicted persons influences their relative position and appearance in the image/video. Thus, social media analysis can be exploited to automatically infer such social relationships in movies, personal collections, sports, surveillance, etc. An analysis that has been shown to provide promising results describes such relationships by using graph structures, in which graph nodes represent the different actors/persons and graph edges express the corresponding social relationships. Such relationships may be considered to be static, i.e., not changing in time [WCW09, YLPR09], or evolving [YCZ+09, DY11].

6.7   Conclusions

In this chapter, some of the most widely used methods for dimensionality reduction and classification that assume an underlying graph structure have been described. The methods have been divided into three categories, i.e. unsupervised, supervised and semi-supervised, based on their requirement of manual (user-derived) annotation. Their application in social media analysis has also been discussed by reviewing recent work in different applications.

Bibliography

[AMA14]   B. N. Aoun, M. Mejdoub, and C. B. Amar. Graph-based approach for human action recognition using spatio-temporal features. *Journal of Visual Communication and Image Representation*, 25(2):329–338, 2014.

[AT12]   G. Arvanitidis and A. Tefas. Exploiting graph embedding in support vector machines. In *Proc. IEEE International Workshop on Machine Learning for Signal Processing*, 2012.

[Ata98]   M. J. Atallah. *Basic Graph Algorithms In Algorithms and Theory of Computation Handbook*. CRC Press, 1998.

[BN03]   M. Belkin and P. Niyogi. Laplacian eigenmaps for dimensionality reduction and data representation. *Neural Computation*, 15(6):1373–1396, 2003.

[BNS06]   M. Belkin, P. Niyogi, and V. Sindhwani. Manifold regularization: A geometric framework for learning from labeled and unlabeled examples. *Journal of Machine Learning Research*, 7:2399–2434, 2006.

[BPRB13] E. Z. Borzeshi, M. Piccardi, K. Riesen, and H. Bunke. Discriminative prototype selection methods for graph embedding. *Pattern Recognition*, 46(6):1648–1657, 2013.

[BR11]   H. Bunke and K. Riesen. Recent advances in graph-based pattern recognition with applications in document analysis. *Pattern Recognition*, 44(5):1057–1067, 2011.

[BT11]   W. Brendel and S. Todorovic. Learning spatiotemporal graphs of human activities. In *Proc. International Conference on Computer Vision*, 2011.

[CH03]   X. W. Chen and T. Huang. Facial expression recognition: A clustering based approach. *Pattern Recognition Letters*, 24(9):1295–1302, 2003.

[CHH07]   D. Cai, X. He, and J. Han. Semi-supervised discriminant analysis. In *Proc. International Conference on Computer Vision*, pages 1–7, 2007.

[CL06]   R. R. Coifman and S. Lafon. Diffusion maps. *Applied and Computational Harmonic Analysis*, 21(1):5–30, 2006.

[CMC$^+$09] T. S. Caetano, J. J. McAuley, L. Cheng, Q. V. Le, and A. J. Smola. Learning graph matching. *IEEE Transactions on Pattern Analysis and Machine Intelligence*, 31(6):1048–1058, 2009.

[CSZ06]   O. Chapelle, B. Scholkopf, and A. Zien. *Semi-supervised learning*. MIT Press, 2006.

[CWSL12] O. Celiktutan, C. Wolf, B. Sankur, and E. Lombardi. Real-time exact graph matching with application in human action recognition. In *Proc. International Conference on Human Behavior Understanding*, 2012.

[CZG09]   J. Chen, O. Zaiane, and R. Goebel. Detecting communities in social networks using max-min modularity. *SIAM International Conference on Data Mining*, 2009.

[DFHP10] L. Ding, Q. Fan, J. Hsiao, and S. Pankanti. Graph-based event detection from realistic videos using weak feature correspondence. In *Proc. IEEE International Conference on Accoustics, Speech, and Signal Processing*, 2010.

[DHS00] R. O. Duda, P. E. Hart, and D. G. Stork. *Pattern Classification*, 2nd ed. Wiley, 2000.

[DY10] L. Ding and A. Yilmaz. Learning relations among movie characters: A social network perspective. In *Proc. European Conference on Computer Vision*, 2010.

[DY11] L. Ding and A. Yilmaz. Inferring social relations from visual concepts. In *Proc. International Conference on Computer Vision*, 2011.

[EPL09] N. Eagle, A. Pentland, and D. Lazer. Inferring friendship network structure by using mobile phone data. In *Proc. of the National Academy of Sciences*, volume 106, pages 15274–15278, 2009.

[EWH07] D. Emms, R. Wilson, and E. Hancock. Graph embedding using quantum commute times. In *Graph-Based Representations in Pattern Recognition*. Springer, 2007.

[FC11] S. L. France and J. D. Carroll. Two-way multidimensional scaling: A review. *IEEE Transactions on Systems, Man and Cybernetics - Part C: Applications and Reviews*, 41(5):644–661, 2011.

[FGQZ11] M. Fan, N. Gu, H. Qiao, and B. Zhang. Sparse regularization for semi-supervised classification. *Pattern Recognition*, 44:1777–1784, 2011.

[FH04] P. F. Felzenszwalb and D. P. Huttenlocher. Efficient graph-based image segmentation. *International Journal of Computer Vision*, 59(2):167–181, 2004.

[GCR09] W. Ge, R. Collins, and B. Ruback. Automatically detecting the small group structure of a crowd. *Applications of Computer Vision*, 2009.

[GDIHK11] R. Gonzalez-Diaz, A. Ion, M. I. Ham, and W. G. Kropatsch. Invariant representative cocycles of cohomology generators using irregular graph pyramids. *Computer Vision and Image Understanding*, 115(7):1011–1022, 2011.

[GMK12] N. Gkalelis, V. Mezaris, and I. Kompatsiaris. Linear subclass support vector machines. *IEEE Signal Processing Letters*, 19(9):575–5784, 2012.

[GZSRC11] U. Gaur, Y. Zhu, A. Song, and A. Roy-Chowdhury. A string of feature graphs model for recognition of complex activities in natural videos. In *Proc. International Conference on Computer Vision*, 2011.

[HCS06] G. B. Huang, L. Chen, and C. K. Siew. Universal approximation using incremental constructive feedforward networks with random hidden nodes. *IEEE Transactions on Neural Networks*, 17(4):879–892, 2006.

[HM94] M. T. Hagan and M. B. Menhaj. Training feedforward networks with the Marquardt algorithm. *IEEE Transactions on Neural Networks*, 5(6):989–993, 1994.

[HN04] X. He and P. Niyogi. Locality preserving projections. In *Proc. Neural Information Processing Systems Conference*, 2004.

[HSGW14]   G. Huang, S. Song, J. N. D. Gupta, and C. Wu. Semi-supervised and un-supervised extreme learning machines. *IEEE Transactions on Cybernetics*, 44(12):2405–2417, 2014.

[HW10]   Abdi. H. and L. J. Williams. Principal component analysis. *Wiley Interdisciplinary Reviews: Computational Statistics*, 2:433–459, 2010.

[HZDZ12]   G. B. Huang, H. Zhou, X. Ding, and R. Zhang. Extreme learning machine for regression and multiclass classification. *IEEE Transactions on Systems, Man, and Cybernetics, Part B: Cybernetics*, 42(2):513–529, 2012.

[HZS04]   G. B. Huang, Q. Y. Zhu, and C. K. Siew. Extreme learning machine: a new learning scheme of feedforward neural networks. In *Proc. IEEE International Joint Conference on Neural Networks*, 2004.

[ITP13a]   A. Iosifidis, A. Tefas, and I. Pitas. Minimum class variance extreme learning machine for human action recognition. *IEEE Transactions on Circuits and Systems for Video Technology*, 23(11):1968–1979, 2013.

[ITP13b]   A. Iosifidis, A. Tefas, and I. Pitas. On the optimal class representation in linear discriminant analysis. *IEEE Transactions on Neural Networks and Learning Systems*, 24(9):1491–1497, 2013.

[ITP14a]   A. Iosifidis, A. Tefas, and I. Pitas. Exploiting local class information in extreme learning machine. In *Proc. International Conference on Neural Computation Theory and Applications*, 2014.

[ITP14b]   A. Iosifidis, A. Tefas, and I. Pitas. Minimum variance extreme learning machine for human action recognition. In *Proc. IEEE International Conference on Acoustics, Speech and Signal Processing*, 2014.

[ITP14c]   A. Iosifidis, A. Tefas, and I. Pitas. Semi-supervised classification of human actions based on neural networks. In *Proc. IEEE International Conference on Pattern Recognition*, 2014.

[JH05]   D. Justice and A. Hero. A binary linear programming formulation of the graph edit distance. *IEEE Transactions on on Pattern Analysis and Machine Intelligence*, 28(8):1200–1214, 2005.

[JLFW10]   M. Jiang, C. Li, J. Feng, and L. Wang. Segmentation via NCuts and lossy minimum description length: a unified approach. In *Proc. Asian Conference on Computer Vision*, 2010.

[JM07]   J. Johns and S. Mahadevan. Constructing basis functions from directed graphs for value function approximation. In *Proc. International Conference on Machine Learning*, 2007.

[JNZ09]   Y. Jia, F. Nie, and C. Zhang. Trace ratio problem revisited. *IEEE Transactions on Neural Networks*, 20(4):729–735, 2009.

[JT11]   S. Jouili and S. Tabbone. Towards performance evaluation of graph-based representation. In *Proc. International Conference on Graph-based Representations in Pattern Recognition*, 2011.

[KH09]   H. Kim and A. Hilton. Graph-based foreground extraction in extended color space. In *Proc. IEEE International Conference on Image Processing*, 2009.

[KL02]   R. I. Kondor and J. D. Lafferty. Diffusion kernels on graphs and other discrete input spaces. In *Proc. International Conference on Machine Learning*, 2002.

[Kru64a]   J. B. Kruskal. Multidimensional scaling by optimizing goodness of fit to a nonmetric hypothesis. *Psychometrika*, 29:1–27, 1964.

[Kru64b]   J. B. Kruskal. Nonmetric multidimensional scaling: A numerical method. *Psychometrika*, 29:115–129, 1964.

[KTI03]   H. Kashima, K. Tsuda, and A. Inokuchi. Marginalized kernels between labeled graphs. In *Proc. International Conference on Machine Learning*, 2003.

[KZP06]   S. B. Kotsiantis, I. D. Zaharakis, and P. E. Pintelas. Machine learning: a review of classification and combining techniques. *Artificial Intelligence Review*, 26(3):159–190, 2006.

[LCLZ11]   J. Liu, Y. Cheng, M. Liu, and Z. Zhao. Semi-supervised ELM with application in sparse calibrated location estimation. *Neurocomputing*, 74:2566–2572, 2011.

[LHM11]   Q. Liu, Y. Huang, and D. N. Metaxas. Hypergraph with sampling for image retrieval. *Pattern Recognition*, 44(10):2255–2262, 2011.

[LRLB13]   M. M. Luqman, J. Y. Ramel, J. Llados, and T. Brouard. Fuzzy multilevel graph embedding. *Pattern Recognition*, 46:551–565, 2013.

[LW09]   J. Lin and W. Wang. Weakly-supervised violence detection in movies with audio and video based co-training. *Advances in Multimedia Information Processing*, 2009.

[MB11]   S. Melacci and M. Belkin. Laplacian support vector machines trained in the primal. *Journal of Machine Learning Research*, 12:1149–1184, 2011.

[Mem00]   A. Memoire. High-dimensional data analysis: The curses and blessings of dimensionality. In *Proc. American Mathematical Society Conference Math Challenges of the 21st Century*, 2000.

[MK01]   A. M. Martinez and A. C. Kak. PCA versus LDA. *IEEE Transactions on Pattern Analysis and Machine Intelligence*, 23(2):228–233, 2001.

[NB05]   M. Neuhaus and H. Bunke. Self-organizing maps for learning the edit costs in graph matching. *IEEE Transactions on Systems, Man, and Cybernetics, Part B: Cybernetics*, 35(3):503–514, 2005.

[NdC11]   M. Nascimento and A. de Carvalho. Spectral methods for graph clustering - a survey. *European Journal of Operational Research*, 211(2):221–231, 2011.

[NJW02]   A. Y. Ng, M. I. Jordan, and Y. Weiss. On spectral clustering: Analysis and an algorithm. *Advances in Neural Information Processing Systems*, 2002.

[NLCK05]   B. Nadler, S. Lafon, R. R. Coifman, and I. G. Kevrekidis. Diffusion maps, spectral clustering and eigenfunctions of Fokker-Planck operators. *Advances in Neural Information Processing Systems*, 2005.

[Nor98]   J. R. Norris. *Markov chains*. Cambridge University Press, 1998.

[OT12]   G. Orphanidis and A. Tefas. Exploiting subclass information in support vector machines. In *Proc. IEEE International Conference on Pattern Recognition*, 2012.

[PD05]   E. Pekalska and R. Duin. *The dissimilarity representation for pattern recognition: foundations and applications*. World Scientific Publishing, 2005.

[PF12]   L. Pu and B. Faltings. Hypergraph learning with hyperedge expansion. In *Proc. European Conference on Machine Learning and Knowledge Discovery in Databases*, 2012.

[Pit00]   I. Pitas. *Digital Image Processing Algorithms and Applications*. Wiley, 2000.

[PY10]   K. J. Park and A. Yilmaz. Social network approach to analysis of soccer games. In *Proc. IEEE International Conference on Pattern Recognition*, 2010.

[RB09a]   K. Riesen and H. Bunke. Approximate graph edit distance computation by means of bipartite graph matching. *Image and Vision Computing*, 27(7):950–959, 2009.

[RB09b]   K. Riesen and H. Bunke. Graph classification by means of Lipschitz embedding. *IEEE Transactions on Systems, Man, and Cybernetics, Part B: Cybernetics*, 39(6):1472–1483, 2009.

[RHW86]   D. E. Rumelhart, G. E. Hinton, and R. J. Williams. Learning representations by back-propagating errors. *Nature*, 323(6088):533–536, 1986.

[RNB07]   K. Riesen, M. Neuhaus, and H. Bunke. Graph embedding in vector spaces by means of prototype selection. In *Proc. International Conference on Graph-based Representations in Pattern Recognition*, 2007.

[RR13]   A. Romey and S. Ruggieri. A multidisciplinary survey on discrimination analysis. *The Knowledge Engineering Review*, FirstView:1–57, 2013.

[RS00]   S. T. Roweis and L. K. Saul. Nonlinear dimensionality reduction by locally linear embedding. *Science*, 290(5500):2323–2326, 2000.

[Sch07]   S. E. Schaeffer. Graph clustering. *Computer Science Review*, 1(1):27–64, 2007.

[SM97]   J Shi and J. Malik. Normalized cuts and image segmentation. In *Proc. IEEE Conference on Computer Vision and Pattern Recognition*, 1997.

[SM00]   J. Shi and J. Malik. Normalized cuts and image segmentation. *IEEE Transactions on Pattern Analysis and Machine Intelligence*, 22(8):888–905, 2000.

[SP11]   V. Satuluri and S. Parthasarathy. Symmetrizations for clustering directed graphs. In *Proc. International Conference on Extending Database Technology*, 2011.

[SR03]   L. K. Saul and S. T. Roweis. Think globally, fit locally: Unsupervised learning of low dimensional manifolds. *Journal of Machine Learning Research*, 4:119–155, 2003.

[SS01]   B. Scholkopf and A. J. Smola. *Learning with kernels*. MIT Press, 2001.

[Sug07]   M. Sugiyama. Dimensionality reduction of multimodal labeled data by local fisher discriminant analysis. *Journal of Machine Learning Research*, 8:1027–1061, 2007.

[TCFL12]   C. C. Tseng, J. C. Chen, C. H. Fang, and J. J. J. Line. Human action recognition based on graph-embedded spatio-temporal subspace. *Pattern Recognition*, 45(10):3611–3624, 2012.

[TH06]   A. Torsello and E. R. Hancock. Learning shape-classes using a mixture of tree-unions. *IEEE Transactions on Pattern Analysis and Machine Intelligence*, 28(6):954–967, 2006.

[TKP01]   A. Tefas, C. Kotropoulos, and I. Pitas. Using support vector machines to enhance the performance of elastic graph matching for frontal face authentication. *IEEE Transactions on Pattern Analysis and Machine Intelligence*, 23(7):735–746, 2001.

[TKP02]   A. Tefas, C. Kotropoulos, and I. Pitas. Face verification using elastic graph matching based on morphological signal decomposition. *Signal Processing*, 82(6):833–851, 2002.

[TS00]   J. Tenenbaum and J. C. Silva, V. andLangford. A global geometric framework for nonlinear dimensionality reduction. *Science*, 290:2319–2323, 2000.

[VBGS06]   S. V. N. Vishwanathan, K. M. Borgwardt, O. Guttman, and A. J. Smola. Kernel extrapolation. *Neurocomputing*, 69(7):721–729, 2006.

[VSKB10]   S. V. N. Vishwanathan, N. N. Schraudolph, R. Kondor, and K. M. Borgwardt. Graph kernels. *Journal of Machine Learning Research*, 11(4):1201–1242, 2010.

[WCW09]   C. Y. Weng, W. T. Chu, and J. L. Wu. Rolenet: Movie analysis from the perspective of social networks. *IEEE Transactions on Multimedia*, 11(2):256–271, 2009.

[Wei99]   Y. Weis. Segmentation using eigenvectors: A unifying view. In *Proc. IEEE Conference on Computer Vision and Pattern Recognition*, 1999.

[WFKM97]   L Wiskott, J. M. Fellous, N. Kuiger, and C. Malsburg. Face recognition by elastic bunch graph matching. *IEEE Transactions on Pattern Analysis and Machine Intelligence*, 19(7):775–779, 1997.

[WGLF10]   G. Weang, A. Gallagher, J. Luo, and D. Forshyth. Seeing people in social context: Recognizing people and social relationships. *European Conference on Computer Vision*, 2010.

[WHL05]   R. Wilson, E. Hancock, and B. Luo. Pattern vectors from algebraic graph theory. *IEEE Transactions on Pattern Analysis and Machine Intelligence*, 27(7):1112–1124, 2005.

[WYX+07]   H. Wang, S. Yan, D. Xu, X. Tang, and T. Huang. Trace ratio vs. ratio trace for dimensionality reduction. In *Proc. IEEE Conference on Computer Vision and Pattern Recognition*, 2007.

[XKLJ10]   Z. Xu, I. Kin, M. R. T. Lyu, and R. Jin. Discriminative semi-supervised feature selection via manifold regularization. *IEEE Transactions on Neural Networks*, 21(7):1033–1047, 2010.

[YCZ+09]   T. Yang, Y. Chi, S. Zhu, Y. Gong, and R. Jin. A Bayesian approach toward finding communities and their evolutions in dynamic social networks. In *Proc. SIAM International Conference on Data Mining*, 2009.

[YLPR09]   T. Yu, S. N. Lim, K. Patwardhan, and N. Rkahnstoever. Monitoring, recognizing and discovering social networks. In *Proc. IEEE Conference on Computer Vision and Pattern Recognition*, 2009.

[YXZ+07]   S. Yan, D. Xu, B. Zhang, H. J. Zhang, Q. Yang, and S. Lin. Graph embedding and extensions: A general framework for dimensionality reduction. *IEEE Transactions on Pattern Analysis and Machine Intelligence*, 29(1):40–50, 2007.

[YZLZ10]   Y. Yuan, H. Zheng, Z. Li, and D. Zhang. Video action recognition with spatiotemporal graph embedding and spline modeling. In *Proc. International Conference on Acoustics, Speech and Signal Processing*, 2010.

[ZFFX10]   B. Zhao, L. Fei-Fei, and E. P. Xing. Image segmentation with topic random field. In *Proc. European Conference on Computer Vision*, 2010.

[ZHS05]   D. Zhou, J. Huang, and B. Scholkopf. Learning from labeled and unlabeled data on a directed graph. In *Proc. International Conference on Machine Learning*, 2005.

[ZLPW13]   X. Zhao, X. Li, C. Pang, and S. Wang. Human action recognition based on semi-supervised discriminant analysis with global constraint. *Neurocomputing*, 105:45–50, 2013.

[ZM06]   Z. Zhu and A. Martnez. Subclass discriminant analysis. *IEEE Transactions on Pattern Analysis and Machine Intelligence*, 28(8):1274–1286, 2006.

[ZTP07a]   S. Zafeiriou, A. Tefas, and I. Pitas. The discriminant elastic graph matching algorithm applied to frontal face verification. *Pattern Recognition*, 40(10):2798–2810, 2007.

[ZTP07b]   S. Zafeiriou, A. Tefas, and I. Pitas. Learning discriminant person-specific facial models using expandable graphs. *IEEE Transactions on Information Forensics and Security*, 2(1):55–68, 2007.

[ZTP07c]   S. Zafeiriou, A. Tefas, and I. Pitas. Minimum class variance support vector machines. *IEEE Transactions on Image Processing*, 16(10):2551–2564, 2007.

[ZZY+08]   D. Zhou, S. Zhu, K. Yu, X. Song, B. L. Tseng, H. Zha, and C. L. Giles. Learning multiple graphs for document recommendations. In *Proc. International Conference on World Wide Web*, 2008.

Chapter 7

Matrix and Tensor Factorization with Recommender System Applications

Panagiotis Symeonidis

Aristotle University of Thessaloniki, Greece

7.1 Introduction

Representing data in lower dimensional spaces has been extensively used in many disciplines, such as natural language and image processing, data mining and information retrieval [DDF+90]. Recommender systems deal with challenging issues such as scalability, noise, and sparsity, and thus, matrix and tensor factorization techniques appear as an interesting tool to be exploited. Symeonidis et al. [SNPM06, Sym07], for example, used SVD for the prediction of items/ratings in recommender systems. They assumed that there is only a small number of factors influencing the users' preferences, and that a user's preference for an item is determined by how each factor applies to the user and the item. More recently, due to the Netflix challenge, research on matrix factorization methods, a class of latent factor models, gained renewed momentum in the recommender systems literature, given that many of the best performing methods used on the challenge were based on matrix factorization techniques [Kor08, SM08, Kor09]. Please note that the Netflix challenge was a competition for the best recommender system algorithm to predict user ratings for movies. The competition was held by Netflix (http://www.netflixprize.com/), an on-line DVD-rental service. In this chapter we describe matrix and tensor factorization techniques (i.e., SVD on matrices and HOSVD on tensors) in recommender systems and social tagging systems, respectively. In addition, we present a real-world recommender system for Location-Based Social Networks, which employs tensor decomposition techniques.

While technology is developed fast, data become larger as well, and as the size of data grows so does the difficulty of processing it. One way to view and process data easily is as a matrix, which is a rectangular array of numbers. However, matrices suffer from big data as well. The dimensions of matrices keep on growing fast and this fact makes analysts' jobs more difficult. The problem of dimensionality reduction appears when data are in fact of a higher dimension than being manageable. Dimensionality reduction attempts to reduce the dimensionality of data to a manageable size, while keeping as much of the original important information as possible.

The "information overload" problem affects our everyday experience while we are searching for knowledge on a topic. To overcome this problem, we often rely on suggestions from others who have more experience of the topic. In Web, this is attained with the usage of *Collaborative Filtering* (CF), which provides recommendations based on the suggestions of users who have similar preferences. Since CF is able to capture the particular preferences of a user, it has become one of the most popular methods in recommender systems. The classic CF (i.e., user-based CF and item-based CF) methods are also known as memory-based methods and constitute the first subgroup of the categorization of CF systems. Memory-based methods first load the rating matrix into the main memory and afterwards provide recommendations based on the relationship between a user-item pair and the rest of the rating matrix.

The second main category of CF algorithms is known as model-based algorithms, which recommend by first developing a model of user ratings for items. These methods fit a parameterized model to the given rating matrix and provide recommendations based on the fitted model. It has been shown that model-based algorithms can efficiently handle scalability and improve accuracy of recommendations in large datasets. Model-based approaches can combine the effectiveness of the nearest-neighbor CF algorithms in terms of accuracy, with the efficiency in terms of execution time.

SVD is a technique that has been extensively used in informational retrieval. It detects latent relationships between documents and terms. In CF, SVD can be used to form user trends from individual preferences, by detecting latent relationships between users and items. Therefore, with SVD, a higher level representation of the original user-item matrix is produced, which presents a three-fold advantage: (i) it contains the main trends of user preferences, (ii) noise is removed, (iii) it is much more condensed than the original matrix, thus it favors scalability.

In the following, we describe matrix and tensor factorization techniques in Sections 7.2 and 7.3, respectively. Finally, in Section 7.4, we present a real-world recommender system, which is based on HOSVD.

7.2 Singular Value Decomposition on Matrices for Recommender Systems

A well-known latent factor model for matrix decomposition is singular value decomposition. The singular value decomposition (SVD) [Str06] of a matrix $\mathbf{A}_{I_1 \times I_2}$ can be written as a product of three matrices, as shown in equation (7.2.1):

$$\mathbf{A}_{I_1 \times I_2} = \mathbf{U}_{I_1 \times I_1} \cdot \mathbf{S}_{I_1 \times I_2} \cdot \mathbf{V}_{I_2 \times I_2}^T, \qquad (7.2.1)$$

where $\mathbf{U}$ is the matrix with the left singular vectors of $\mathbf{A}$, $\mathbf{V}^T$ is the transpose of the matrix V with the right singular vectors of $\mathbf{A}$, and $\mathbf{S}$ is the diagonal matrix of ordered singular values of $\mathbf{A}$. Please note that the singular values determined by the factorization of equation (7.2.1) are unique and satisfy $\sigma_1 \geq \sigma_2 \geq \sigma_3 \geq \cdots \geq \sigma_{I_2} \geq 0$.

By preserving only the largest $c < \min\{I_1, I_2\}$ singular values of $\mathbf{S}$, SVD results in matrix $\hat{\mathbf{A}}$, which is an approximation of $\mathbf{A}$. In information retrieval, this technique is used by LSI [FDD+88], to deal with the latent semantic associations of terms in texts and to reveal the major trends in $\mathbf{A}$.

To perform the SVD over a user-item matrix $\mathbf{A}$, we tune the value of parameter c, of singular values (i.e., dimensions) with the objective to reveal the major trends. The tuning of c, is determined by the rank of matrix $\mathbf{A}$. A rule of thumb, for defining parameter c is to compute the sum of elements in the main diagonal of the $\mathbf{S}$ matrix (also known as the nuclear norm). Next, we preserve a sufficient percentage of this sum for the creation of an approximation of the original matrix $\mathbf{A}$. If we have the allowance to use less information percentage with similar results, we just have to reduce the value of c and sum the corresponding elements of the main diagonal of the $\mathbf{S}$ matrix. Therefore, a c-dimensional space is created and each of the c dimensions corresponds to a distinctive rating trend. Next, given the current ratings of the target user u, we enter a pseudo-user vector in the c-dimensional space. Finally, we find the k nearest neighbors of the pseudo-user vector in the c-dimensional space and apply either user- or item-based similarity to compute the top-N recommended items. Conclusively, the provided recommendations consider the existence of user rating trends, as the similarities are computed in the reduced c-dimensional space, where dimensions correspond to trends.

To simplify the discussion, we will use the running example illustrated in Figure 7.2.1, where $I_1, \ldots, I_4$ are items and $U_1, \ldots, U_4$ are users. As shown, the example data set is divided into a training set and a test set. The null cells(no rating) are presented as zeros.

	I_1	I_2	I_3	I_4
U_1	4	1	1	4
U_2	1	4	2	0
U_3	2	1	4	5

	I_1	I_2	I_3	I_4
U_4	1	4	1	0

(a)   (b)

FIGURE 7.2.1: (a) Training Set (3×4), (b) Test Set (1×4).

7.2.1   Applying the SVD and Preserving the Largest Singular Values

Initially, we apply the SVD to a $n \times m$ matrix $\mathbf{A}$ (i.e., the training data of our running example) that produces the decomposition shown in Equation (7.2.2). The matrices of our running example are shown in Figure 7.2.2.

$$\mathbf{A}_{n \times m} = \mathbf{U}_{n \times n} \cdot \mathbf{S}_{n \times m} \cdot \mathbf{V}^{T}_{m \times m}. \qquad (7.2.2)$$

4	1	1	4
1	4	2	0
2	1	4	5

$\mathbf{A}_{n \times m}$

−0.61	0.28	−0.74
−0.29	−0.95	−0.12
−0.74	0.14	0.66

$\mathbf{U}_{n \times n}$

8.87	0	0	0
0	4.01	0	0
0	0	2.51	0

$\mathbf{S}_{n \times m}$

−0.47	−0.28	−0.47	−0.69
0.11	−0.85	−0.27	0.45
−0.71	−0.23	0.66	0.13
−0.52	0.39	−0.53	0.55

$\mathbf{V}^{T}_{m \times m}$

FIGURE 7.2.2: Example of: $\mathbf{A}_{n \times m}$ (initial matrix $\mathbf{A}$), $\mathbf{U}_{n \times m}$ (left singular vectors of $\mathbf{A}$), $\mathbf{S}_{n \times m}$ (singular values of $\mathbf{A}$), $\mathbf{V}^{T}_{m \times m}$ (right singular vectors of $\mathbf{A}$).

It is possible to reduce the $n \times m$ matrix $\mathbf{S}$ to have only c largest singular values. Then, the reconstructed matrix is the closest rank-c approximation of the initial matrix $\mathbf{A}$, as it is shown in equation (7.2.3) and Figure 7.2.3:

$$\mathbf{A}^{*}_{n \times m} = \mathbf{U}_{n \times c} \cdot \mathbf{S}_{c \times c} \cdot \mathbf{V}^{T}_{c \times m}. \qquad (7.2.3)$$

2.69	0.57	2.22	4.25
0.78	3.93	2.21	0.04
3.17	1.38	2.92	4.78

$\mathbf{A}^{*}_{n \times i}$

−0.61	0.28
−0.29	−0.95
−0.74	0.14

$\mathbf{U}_{n \times c}$

8.87	0
0	4.01

$\mathbf{S}_{c \times c}$

−0.47	−0.28	−0.47	−0.69
0.11	−0.85	−0.27	0.45

$\mathbf{V}^{T}_{c \times m}$

FIGURE 7.2.3: Example of: $\mathbf{A}^{*}_{n \times m}$ (approximation matrix of $\mathbf{A}$), $\mathbf{U}_{n \times c}$ (left singular vectors of $\mathbf{A}^{*}$), $\mathbf{S}_{c \times c}$ (singular values of $\mathbf{A}^{*}$), $\mathbf{V}^{T}_{c \times m}$ (right singular vectors of $\mathbf{A}^{*}$).

We tune the number c of singular values (i.e., dimensions) with the objective to reveal the major trends. The tuning of c is determined by the information percentage that is preserved compared to the original matrix. Therefore, a c-dimensional space is created and each of the c dimensions corresponds to a distinctive rating trend. We have to notice that in the running example we create a 2-dimensional space using 83,7% of the total information of the matrix (12,88/15,39). Please note that the number 15,39 is the sum of elements in the main diagonal of $\mathbf{S}_{c \times c}$ (singular values of $\mathbf{A}^{*}$).

7.2.2   Generating the Neighborhood of Users/Items

Having reduced the dimensional representation of the original space, we form the neighborhoods of users/items in that space. Please note that the original space consists of two subspaces:

- range of ($\mathbf{A}$) whose $\mathbf{U}$ (see Figure 7.2.3) is an orthonormal basis. This vector space is the column space of $\mathbf{A}$, referring to users,

- range of ($\mathbf{A}^T$) whose $\mathbf{V}$ (see Figure 7.2.3) is an orthonormal basis. This vector space is the row space of $\mathbf{A}$, referring to items.

In particular, there are two subspaces: The first is the range of $\mathbf{A}$, whose matrix $\mathbf{U}_{n \times c}$ is its orthonormal basis. This vector space is the column space of $\mathbf{A}$ and refers to users. The second is the range of $\mathbf{A}^T$, whose matrix $\mathbf{V}_{m \times c}$ is its orthonormal basis. This vector space is the row space of $\mathbf{A}$ and refers to items. A user-based approach relies on the predicted value of a rating that a user gives on an item I_j. This value is computed as an aggregation of the ratings of the user neighborhood (i.e., similar users) on this particular item. Whereas, an item-based approach takes into consideration only the user-item rating matrix (e.g., a user rated a movie with a rate of 3).

For the user-based approach, we find the k nearest neighbors of pseudo user vector in the c-dimensional space. The similarities between training and test users can be based on Cosine Similarity. First, we compute the matrix $\mathbf{U}_{n \times c} \cdot \mathbf{S}_{c \times c}$ and then we perform vector similarity among rows. This $n \times c$ matrix is the c-dimensional representation for the n users. For the item based approach, we find the k nearest neighbors of item vector in the c-dimensional space. First, we compute the matrix $\mathbf{S}_{c \times c} \cdot \mathbf{V}_{c \times m}^T$ and then we determine the vector similarity among columns. This $c \times m$ matrix is the c-dimensional representation for the m items.

7.2.3   Generating the Recommendation List

The most often used technique for the generation of the top-N list of recommended items, is the one that counts the frequency of each item inside the found neighborhood, and recommends the N most frequent ones. Henceforth, this technique is denoted as Most-Frequent item recommendation (MF). Based on MF, we sort (in descending order) the items according to their frequency in the found neighborhood of the target user, and recommend the first N of them.

As another method, someone could use the predicted values for each item to rank them. This ranking criterion, denoted as Highest Predicted Rated item recommendation (HPR), is influenced by the the Mean Absolute Error (MAE)[1] between the predicted and the real preferences of a user for an item. HPR opts for recommending the items that are more likely to receive a higher rating. Notice that HPR shows poor performance for the classic CF algorithms. However, it has very good results when it is used in combination with SVD. The reason is that in the latter it is based only on the major trends of users.

As another method, we can sum the positive rates of the items in the neighborhood, instead of just counting their frequency. This method is denoted as the Highest Sum of Rates item recommendation (HSR). The top-N list consists of the N items with the highest sum. The intuition behind HSR is that it takes into account both the frequency (as MF) and the actual ratings, because it wants to favor items that appear most frequently in the neighborhood and have the best ratings. Assume, for example, an item I_j that has just a

---

[1]$MAE = \frac{1}{n} \sum_{i=1}^{n} |f_i - y_i|$: The Mean Absolute Error (MAE) is the average of the absolute errors $e_i = f_i - y_i$, where f_i is the prediction and y_i the true value.

smaller frequency than an item I_k. If I_j is rated much higher than I_k, then HSR will prefer it over I_k, whereas MF will favor I_k.

7.2.4   Inserting a Test User in the c-dimensional Space

Related work [SKKR00] has studied SVD on CF considering the test data as apriori known. It is evident that for a user-based approach, the test data should be considered as unknown in the c-dimensional space. Thus, a specialized insertion process should be used. Given the current ratings of the test user u, we enter a pseudo-user vector in the c-dimensional space using equation (7.2.4) [FDD$^+$88]. In the current example,we insert U_4 into the 2-dimensional space, as it is shown in Figure 7.2.4:

$$\mathbf{u}_{new} = \mathbf{u} \cdot \mathbf{V}_{m \times c} \cdot \mathbf{S}_{c \times c}^{-1} \tag{7.2.4}$$

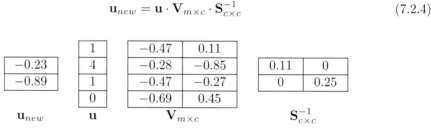

FIGURE 7.2.4: Example of: $\mathbf{u}_{new}$ (inserted new user vector), $\mathbf{u}$ (user vector), $\mathbf{V}_{m \times c}$ (two right singular vectors of V), $\mathbf{S}_{c \times c}^{-1}$ (two singular values of inverse $\mathbf{S}$).

In equation (7.2.4), $\mathbf{u}_{new}$ denotes the mapped ratings of the test user u, whereas $\mathbf{V}_{m \times c}$ and $\mathbf{S}_{c \times c}^{-1}$ are matrices derived from the SVD. This $\mathbf{u}_{new}$ vector should be added at the end of the $\mathbf{U}_{n \times c}$ matrix, which is shown in Figure 7.2.3. Notice that the inserted vector values of test user U_4 are very similar to these of U_2 after the insertion, as shown in Figure 7.2.5. This is reasonable, because these two users have similar ratings, as shown in Figure 7.2.1a and Figure 7.2.1b.

7.2.5   Other Factorization Methods

There are many methods to decompose a matrix, in order to deal with a high-dimensional data set. Principal-Component Analysis or simply PCA is a data mining technique that replaces the high-dimensional original data by its projection onto the most important axes. It is a simple method, which is based on eigenvalues and eigenvectors of a matrix and it is quite effective.

Another useful decomposition method is UV decomposition. UV is an instance of SVD decomposition and its philosophy is that the original matrix is actually the product of two long "thin" matrices, $\mathbf{U}$ and $\mathbf{V}$. UV's most often problem is overfitting. To address this problem, we can extend UV with L_2 regularization, which is also known as Tikhonov regularization [Tik63]. That is, since the basic idea of the UV decomposition is to minimize

−0.61	0.28
−0.29	**−0.95**
−0.74	0.14
−0.23	**−0.89**

FIGURE 7.2.5: The new $\mathbf{U}_{n+1,c}$ matrix containing the new user ($\mathbf{u}_{new}$) that we have added.

an element-wise loss on the elements of the predicted/approximation matrix by optimizing the square loss, we can extend it with L_2 regularization terms. After the application of a regularized optimization criterion the possible overfitting can be reduced.

Another widely known method in dimensionality reduction and data analysis is *Non-Negative Matrix Factorization* (NMF). The non-negative matrix factorization, also known as non-negative matrix approximation, is a group of algorithms in multivariate analysis and linear algebra where a matrix **A** is factorized into (usually) two matrices **U** and **V**, with the property that all three matrices have no negative elements. This non-negativity makes the resulting matrices easier to inspect. Since the problem is not exactly solvable in general, it is commonly approximated numerically [BBL+06].

Assume that $\mathbf{a}_i, \ldots, \mathbf{a}_N$ are N non-negative input vectors and we organize them as the columns of non-negative data matrix **A**. Non-negative matrix factorization seeks a small set of K non-negative representative vectors $\mathbf{v}_i, \ldots, \mathbf{v}_K$ that can be non-negatively combined to approximate the input vectors $\mathbf{a}_i$:

$$\mathbf{A} \approx \mathbf{U} \times \mathbf{V} \tag{7.2.5}$$

and

$$\mathbf{a}_i \approx \sum_{k=1}^{K} \mathbf{u}_{kn} \mathbf{v}_k, \quad 1 \leq n \leq N, \tag{7.2.6}$$

where the combining coefficients $\mathbf{u}_{kn}$ are restricted to be non-negative [DS05]. There are several ways in which the **U** and **V** may be found. The main equation we presented before (see equation (7.2.5)) is the most popular method used to find **U** and **V** matrices.

Another method used for the decomposition of a matrix is the CUR decomposition [MD09], which presents good properties over SVD in some cases. As it is already described, SVD is able to reduce dimensions without losing approximation accuracy. However, many times in high-dimensional datasets, the produced SVD matrices tend to be very dense, which makes their process a big challenge. In contrast, CUR decomposition confronts this problem as it decomposes the original matrix into two sparse matrices **C**, **R** and only one dense matrix **U**, whose size is quite small and doesn't much affect the time complexity of the method. Another difference between SVD and CUR decompositions is that CUR gives an exact decomposition no matter how large the rank of the original matrix is, whereas in SVD the k largest singular values will be at least as many as the rank of the original matrix.

7.3 Higher Order Singular Value Decomposition (HOSVD) on Tensors

HOSVD is a generalization of singular value decomposition and has been successfully applied in several areas. In this section, we summarize the HOSVD procedure, apply HOSVD for recommendations in *Social Tagging Systems* (STSs), combine HOSVD with other methods and study the limitations of HOSVD.

7.3.1 From SVD to HOSVD

Formally, a *tensor* is a multi-dimensional matrix. A N-order tensor $\mathcal{A}$ is denoted as $\mathcal{A} \in \mathbb{R}^{I_1 \times \cdots \times I_N}$, with elements $a_{i_1, \ldots, i_N}$. The high-order singular value decomposition [LMV00] generalizes the SVD computation to tensors. To apply HOSVD on a 3-order tensor $\mathcal{A}$, three *matrix unfolding* operations are defined as follows [LMV00]: $\mathbf{A}_1 \in \mathbb{R}^{I_1 \times (I_2 I_3)}$, $\mathbf{A}_2 \in \mathbb{R}^{I_2 \times (I_1 I_3)}$, $\mathbf{A}_3 \in \mathbb{R}^{(I_3 I_1) \times I_2}$. $\mathbf{A}_1, \mathbf{A}_2, \mathbf{A}_3$ are called the mode-1, mode-2, mode-3 matrix unfolding of $\mathcal{A}$, respectively. Please note that we define as "matrix unfolding" of a given tensor the matrix representations of that tensor in which all the column vectors are stacked one after the other. The unfoldings of $\mathcal{A}$ in the three modes are illustrated in Figure 7.3.1.

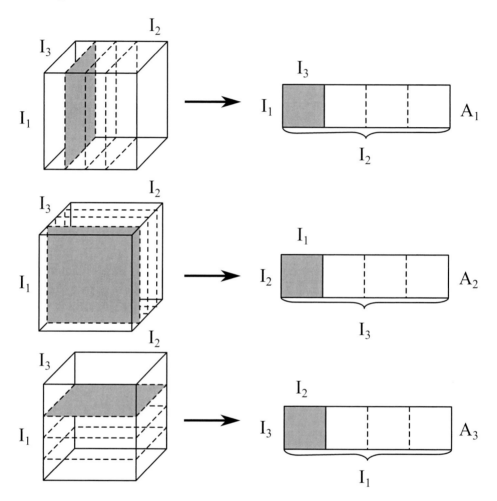

FIGURE 7.3.1: Visualization of the three unfoldings of a 3-order tensor.

In the following, we will present an example of tensor decomposition adopted from [LMV00]:

Example 7.3.1. *Define a tensor* $\mathcal{A} \in \mathbb{R}^{3 \times 2 \times 3}$ *by* $a_{1,1,1} = a_{1,1,2} = a_{2,1,1} = -a_{2,1,2} = 1, a_{2,1,3} = a_{3,1,1} = a_{3,1,3} = a_{1,2,1} = a_{1,2,2} = a_{2,2,1} = -a_{2,2,2} = 2, a_{2,2,3} = a_{3,2,1} = a_{3,2,3} = 4, a_{1,1,3} = a_{3,1,2} = a_{1,2,3} = a_{3,2,2} = 0$. *The tensor and its mode-1 matrix unfolding* $\mathbf{A}_1 \in \mathbb{R}^{I_1 \times I_2 \times I_3}$ *are illustrated in Figure 7.3.2.*

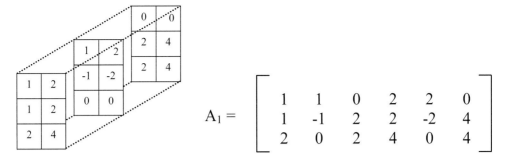

$$A_1 = \begin{bmatrix} 1 & 1 & 0 & 2 & 2 & 0 \\ 1 & -1 & 2 & 2 & -2 & 4 \\ 2 & 0 & 2 & 4 & 0 & 4 \end{bmatrix}$$

FIGURE 7.3.2: Visualization of tensor $\mathcal{A} \in \mathbb{R}^{3 \times 2 \times 3}$ and its mode-1 matrix unfolding.

Next, we define the mode-n product of a N-order tensor $\mathcal{A} \in \mathbb{R}^{I_1 \times \cdots \times I_N}$ by a matrix $\mathbf{U} \in \mathbb{R}^{J_n \times I_n}$, which is denoted as $\mathcal{A} \times_n \mathbf{U}$. The result of the mode-n product is a $(I_1 \times I_2 \times \cdots \times I_{n-1} \times J_n \times I_{n+1} \times \cdots \times I_N)$-tensor, the entries of which are defined as follows:

$$(\mathcal{A} \times_n \mathbf{U})_{i_1 i_2 \ldots i_{n-1} j_n i_{n+1} \ldots i_N} = \sum_{i_n} a_{i_1 i_2 \ldots i_{n-1} i_n i_{n+1} \ldots i_N} u_{j_n, i_n}. \tag{7.3.1}$$

Since we focus on 3-order tensors, $n \in \{1, 2, 3\}$, we use mode-1, mode-2, and mode-3 products. In terms of mode-n products, SVD on a regular two-dimensional matrix (i.e., 2-order tensor), can be rewritten as follows [LMV00]:

$$\mathbf{F} = \mathbf{S} \times_1 \mathbf{U}^{(1)} \times_2 \mathbf{U}^{(2)}, \tag{7.3.2}$$

where $\mathbf{U}^{(1)} = [\mathbf{u}_1^{(1)} \mathbf{u}_2^{(1)} \ldots \mathbf{u}_{I_1}^{(1)}]$ is a *unitary* $(I_1 \times I_1)$-matrix and $\mathbf{U}^{(2)} = [\mathbf{u}_1^{(2)} \mathbf{u}_2^{(2)} \ldots \mathbf{u}_{I_2}^{(2)}]$ is a unitary $(I_2 \times I_2)$-matrix. Please note that a $n \times n$ matrix $\mathbf{U}$ is said to be unitary if its column vectors form an orthonormal set in the complex inner product space $\mathbb{C}^n$. That is, $\mathbf{U}^T \mathbf{U} = \mathbf{I}_n$. Also, $\mathbf{S}$ is a $(I_1 \times I_2)$-matrix with the properties of:

i. pseudo-diagonality: $\mathbf{S} = \mathrm{diag}(\sigma_1, \sigma_2, \ldots, \sigma_{\min\{I_1, I_2\}})$

ii. ordering: $\sigma_1 \geq \sigma_2 \geq \cdots \geq \sigma_{\min\{I_1, I_2\}} \geq 0$.

By extending this form of SVD, HOSVD of a 3-order tensor $\mathcal{A}$ can be written as follows [LMV00]:

$$\mathcal{A} = \mathbf{S} \times_1 \mathbf{U}^{(1)} \times_2 \mathbf{U}^{(2)} \times_3 \mathbf{U}^{(3)}, \tag{7.3.3}$$

where $\mathbf{U}^{(1)}$, $\mathbf{U}^{(2)}$, and $\mathbf{U}^{(3)}$ contain the orthonormal vectors (called the mode-1, mode-2 and mode-3 singular vectors, respectively) spanning the column space of the $\mathbf{A}_1, \mathbf{A}_2, \mathbf{A}_3$ matrix unfoldings. $\mathbf{S}$ is called core tensor and has the property of "all orthogonality." Please note that "all orthogonality" means that the different "horizontal matrices" of $\mathbf{S}$ (the first index i_1 is kept fixed, while the two other indices, i_2 and i_3, are free) are mutually orthogonal with respect to the scalar product of matrices (i.e., the sum of the products of the corresponding entries vanishes); at the same time, the different "frontal" matrices (i_2 fixed) and the different "vertical" matrices (i_3 fixed) should be mutually orthogonal as well. For more information, see [LMV00]. This decomposition also refers to a general factorization model known as Tucker decomposition [Tuc66].

In the following, we provide a solid description of the tensor reduction method with an outline of the algorithm for the case of Social Tagging Systems, where we have three participatory entities (user, item, tag). In particular, we provide details on how HOSVD is applied to tensors and how item/tag recommendation is performed, based on the detected latent associations.

The tensor reduction approach initially constructs a tensor based on usage data triplets $\{u, i, t\}$ of user, item and tag. The motivation is to use all three objects that interact inside a social tagging system. Consequently, we proceed to the unfolding of $\mathcal{A}$, where we build three new matrices. Then, we apply SVD in each new matrix. Finally, we build the core tensor $\mathcal{S}$ and the resulting tensor $\hat{\mathcal{A}}$. The six steps of the HOSVD approach are summarized as follows:

- *Step 1*: The initial tensor $\mathcal{A}$ construction, which is based on usage data triplets (user, item, tag).

- *Step 2*: The matrix unfoldings of tensor $\mathcal{A}$, where we matricize the tensor in all three modes, creating three new matrices (one for each mode).

- *Step 3*: The application of SVD in all three new matrices, where we keep the c-most important singular values for each matrix.

- *Step 4*: The construction of the core tensor $\mathcal{S}$, that reduces the dimensionality. (see equation (7.3.2))

- *Step 5*: The construction of the $\hat{\mathcal{A}}$ tensor, that is an approximation of tensor $\mathcal{A}$. (see equation (7.3.3))

- *Step 6*: Based on the weights of the elements of the reconstructed tensor $\hat{\mathcal{A}}$, we recommend item/tag to the target user u.

Steps $1 - 5$ build a model and can be performed off-line. The recommendation in Step 6 is performed on-line, i.e., each time we have to recommend a item/tag to a user, based on the built model.

7.3.2   HOSVD for Recommendations in Social Tagging Systems

In this subsection, we elaborate on how HOSVD can be employed for computing recommendations in STSs, and present an example on how one can recommend items according to the detected latent associations. Although we illustrate only the recommendation of items, once the approximation $\hat{\mathcal{A}}$ is computed, the recommendation of users or tags is straightforward [SNM10].

The ternary relation of users, items, and tags can be represented as a third-order tensor $\mathcal{A}$, such that tensor factorization techniques can be employed in order to exploit the underlying latent semantic structure in $\mathcal{A}$. While the idea of computing low rank tensor approximations has already been used for many different purposes [LMV00, SH05, SZL$^+$05, WA08, CWZ07, KS08], just recently it has been applied for the problem of recommendations in STS. The basic idea is to cast the recommendation problem as a third-order tensor completion problem — completing the non-observed entries in $\mathcal{A}$.

Formally, a social tagging system is defined as a relational structure $\mathbb{F} := (\mathcal{U}, \mathcal{I}, \mathcal{T}, \mathcal{Y})$ in which:

- $\mathcal{U}$, $\mathcal{I}$, and $\mathcal{T}$ are disjoint non-empty finite sets, whose elements are called users, items, and tags, respectively, and

- $\mathcal{Y}$ is the set of observed ternary relations between them, i.e., $\mathcal{Y} \subseteq \mathcal{U} \times \mathcal{I} \times \mathcal{T}$, whose elements are called tag assignments.

- A post corresponds to the set of tag assignments of a user for a given item, i.e., a triple $(u, i, \mathcal{T}_{u,i})$ with $u \in \mathcal{U}$, $i \in \mathcal{I}$, and a non-empty set $\mathcal{T}_{u,i} := \{t \in \mathcal{T} \mid (u, i, t) \in \mathcal{Y}\}$.

In the following we present several approaches for recommending in STS based on tensor factorization. $\mathcal{Y}$ which represents the ternary relation of users, items and tags can be depicted by the binary tensor $\mathcal{A} = (a_{u,i,t}) \in \mathbb{R}^{|\mathcal{U}| \times |\mathcal{I}| \times |\mathcal{T}|}$ where 1 indicates observed tag assignments and 0 missing values, i.e.,

$$a_{u,i,t} := \begin{cases} 1, & (u, i, t) \in \mathcal{Y} \\ 0, & \text{else.} \end{cases}$$

Now, we express the tensor decomposition as

$$\hat{\mathcal{A}} := \hat{\mathcal{C}} \times_u \hat{\mathbf{U}} \times_i \hat{\mathbf{I}} \times_t \hat{\mathbf{T}}, \qquad (7.3.4)$$

where $\hat{\mathbf{U}}$, $\hat{\mathbf{I}}$, and $\hat{\mathbf{T}}$ are low-rank feature matrices representing a mode (i.e., user, items, and tags, respectively) in terms of its small number of latent dimensions $k_{\mathbf{U}}$, $k_{\mathbf{I}}$, $k_{\mathbf{T}}$, and $\hat{\mathcal{C}} \in \mathbb{R}^{k_{\mathbf{U}} \times k_{\mathbf{I}} \times k_{\mathbf{T}}}$ is the core tensor representing interactions between the latent factors. The model parameters are represented by the quadruple $\hat{\theta} := (\hat{\mathcal{C}}, \hat{\mathbf{U}}, \hat{\mathbf{I}}, \hat{\mathbf{T}})$ (see Figure 7.3.3). The basic idea of the HOSVD algorithm is to minimize an element-wise loss on the elements of $\hat{\mathcal{A}}$ by optimizing the square loss, i.e.,

$$\arg \min_{\hat{\theta}} \sum_{(u,i,t) \in Y} (a_{u,i,t} - \hat{a}_{u,i,t})^2.$$

After the parameters are optimized, predictions can be done as follows:

$$\hat{a}(u, i, t) := \sum_{\tilde{u}=1}^{k_{\mathbf{U}}} \sum_{\tilde{i}=1}^{k_{\mathbf{I}}} \sum_{\tilde{t}=1}^{k_{\mathbf{T}}} \hat{c}_{\tilde{u},\tilde{i},\tilde{t}} \cdot \hat{u}_{u,\tilde{u}} \cdot \hat{i}_{i,\tilde{i}} \cdot \hat{t}_{t,\tilde{t}}, \qquad (7.3.5)$$

where $\hat{\mathbf{U}} = [\hat{u}_{u,\tilde{u}}]_{\tilde{u}=1,\dots,k_{\mathbf{U}}}^{u=1,\dots,\mathbf{U}}, \hat{\mathbf{I}} = [\hat{i}_{i,\tilde{i}}]_{\tilde{i}=1,\dots,k_{\mathbf{I}}}^{i=1,\dots,\mathbf{I}}, \hat{\mathbf{T}} = [\hat{t}_{t,\tilde{t}}]_{\tilde{t}=1,\dots,k_{\mathbf{T}}}^{t=1,\dots,\mathbf{T}}$ and indices over the feature dimension of a feature matrix are marked with a tilde, and elements of a feature matrix are marked with a hat (i.e., $\hat{t}_{t,\tilde{t}}$).

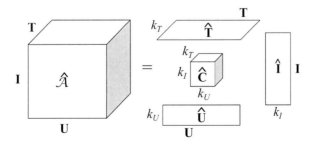

FIGURE 7.3.3: Tensor decomposition in STS. Figure adapted from [RMNST09].

Example 7.3.2. *The HOSVD algorithm takes $\mathcal{A}$ as input and outputs the reconstructed tensor $\hat{\mathcal{A}}$. $\hat{\mathcal{A}}$ measures the strength of associations between users, items, and tags. Each element of $\hat{\mathcal{A}}$ can be represented by a quadruplet $\{u, i, t, p\}$, where p measures the likeliness that user u will tag item i with tag t. Therefore, items can be recommended to u according to their weights associated with the $\{u, t\}$ pair.*

In this subsection, in order to illustrate how HOSVD for item recommendation works, we apply HOSVD to a toy example. As illustrated in Figure 7.3.4, three users tagged three different items (web links). In Figure 7.3.4, the part of an arrow line (sequence of arrows with the same annotation) between a user and an item represents that the user tagged the corresponding item, and the part between an item and a tag indicates that the user tagged this item with the corresponding tag. Thus, the annotated numbers on the arrow lines gives the correspondence among the three types of objects. For example, user u_1 tagged item i_1 with tag "BMW," denoted as t_1. The remaining tags are "Jaguar," denoted as t_2, "CAT," denoted as t_3.

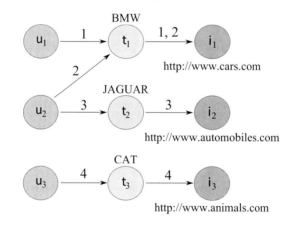

FIGURE 7.3.4: Usage data of the running example.

From Figure 7.3.4, we can see that users u_1 and u_2 have common interests in cars, while user u_3 is interested in cats. A 3-order tensor $\mathcal{A} \in \mathbb{R}^{3 \times 3 \times 3}$, can be constructed from the usage data. We use the co-occurrence frequency (denoted as weights) of each triplet user, item, and tag as the elements of tensor $\mathcal{A}$, which are given in Table 7.3.1. Note that all associated weights are initialized to 1. Figure 7.3.5 presents the tensor construction of our running example.

After performing the tensor reduction analysis, we can get the reconstructed tensor of $\hat{\mathcal{A}}$, which is presented in Table 7.3.2, whereas Figure 7.3.6 depicts the contents of $\hat{\mathcal{A}}$

TABLE 7.3.1: Associations of the running example.

Arrow Line	User	Item	Tag	Weight
1	u_1	i_1	t_1	1
2	u_2	i_1	t_1	1
3	u_2	i_2	t_2	1
4	u_3	i_3	t_3	1

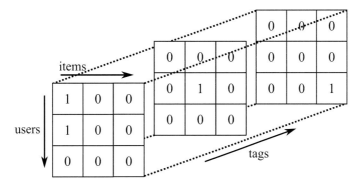

FIGURE 7.3.5: The tensor construction of our running example.

TABLE 7.3.2: Associations derived on the running example.

Arrow Line	User	Item	Tag	Weight
1	u_1	i_1	t_1	0.72
2	u_2	i_1	t_1	1.17
3	u_2	i_2	t_2	0.72
4	u_3	i_3	t_3	1
5	**u_1**	**i_2**	**t_2**	**0.44**

graphically (the weights are omitted). As shown in Table 7.3.2 and Figure 7.3.6, the output of the tensor reduction algorithm for the running example is interesting, because a new association among these objects is revealed. The new association is between u_1, i_2, and t_2. This association is represented with the last (bold faced) row in Table 7.3.2 and with the dashed arrow line in Figure 7.3.6).

If we have to recommend to u_1 an item for tag t_2, then there is no direct indication for this task in the original tensor $\mathcal{A}$. However, we see that in Table 7.3.2 the element of $\hat{\mathcal{A}}$ associated with (u_1, i_2, r_2) is 0.44, whereas for u_1 there is no other element associating other tags with i_2. Thus, we recommend item i_2 to user u_1, who used tag t_2. For the current example, the resulting $\hat{\mathcal{A}}$ tensor is presented in Figure 7.3.7.

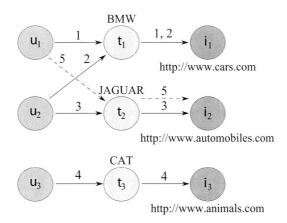

FIGURE 7.3.6: Illustration of the tensor reduction algorithm output for the running example.

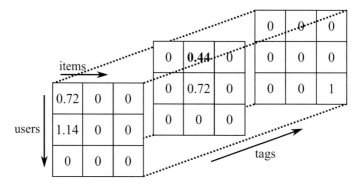

FIGURE 7.3.7: The resulting $\hat{\mathcal{A}}$ tensor for the running example.

The resulting recommendation is reasonable, because u_1 is interested in cars rather than cats. That is, the tensor reduction approach is able to capture the latent associations among the multi-type data objects: user, items, and tags. The associations can then be used to improve the item recommendation procedure.

7.3.3   Handling the Sparsity Problem

Sparsity is a severe problem in 3-dimensional data, and it can affect the outcome of SVD. To address this problem, instead of SVD we can apply kernel-SVD [CST04, CSS06] in the three unfolded matrices. Kernel-SVD is the application of SVD in the Kernel-defined feature space. Smoothing with kernel SVD is also applied by Symeonidis et al. in [SNM10].

For each unfolding $\mathbf{A}_i$ ($1 \leq i \leq 3$) we have to non-linearly map its contents to a higher dimensional space using a mapping function ϕ. Therefore, from each $\mathbf{A}_i$ matrix we can derive an $\mathbf{F}_i$ matrix, where each element a_{xy} of $\mathbf{A}_i$ is mapped to the corresponding element f_{xy} of $\mathbf{F}_i$, i.e., $f_{xy} = \phi(a_{xy})$. Next, we can apply SVD and decompose each $\mathbf{F}_i$ as follows:

$$\mathbf{F}_i = \mathbf{U}^{(i)}\mathbf{S}^{(i)}(\mathbf{V}^{(i)})^T. \qquad (7.3.6)$$

The resulting $\mathbf{U}^{(i)}$ matrices are then used to construct the core tensor.

Nevertheless, to avoid the explicit computation of $\mathbf{F}_i$, all computations must be done in the form of inner products. In particular, as we are interested to compute only the matrices with the left-singular vectors, for each mode i we can define a matrix $\mathbf{B}_i$ as follows:

$$\mathbf{B}_i = \mathbf{F}_i\mathbf{F}_i^T. \qquad (7.3.7)$$

As $\mathbf{B}_i$ is computed using inner products from $\mathbf{F}_i$, we can substitute the computation of inner products with the results of a kernel function. This technique is called the "kernel trick" [CST04] and avoids the explicit (and expensive) computation of $\mathbf{F}_i$. As each $\mathbf{U}^{(i)}$ and $\mathbf{V}^{(i)}$ are orthogonal and each $\mathbf{S}^{(i)}$ is diagonal, it easily follows from equations (7.3.6) and (7.3.7) that:

$$\mathbf{B}_i = (\mathbf{U}^{(i)}\mathbf{S}^{(i)}(\mathbf{V}^{(i)})^T)(\mathbf{U}^{(i)}\mathbf{S}^{(i)}(\mathbf{V}^{(i)})^T)^T = \mathbf{U}^{(i)}(\mathbf{S}^{(i)})^2(\mathbf{V}^{(i)})^T. \qquad (7.3.8)$$

Therefore, each required $\mathbf{U}^{(i)}$ matrix can be computed by diagonalizing each $\mathbf{B}_i$ matrix (which is square) and taking its eigen-vectors.

Regarding the kernel function, in our experiments we use the Gaussian kernel $K(x, y) = e^{-\frac{||x-y||^2}{c}}$, which is commonly used in many applications of kernel SVD. As Gaussian Kernel parameter c, we use the estimate for standard deviation in each matrix unfolding.

7.3.4 Inserting New Users, Tags, or Items

As new users, tags, or items are being introduced to the system, the tensor $\hat{\mathcal{A}}$, which provides the recommendations, has to be updated. The most demanding operation is the updating of the SVD of the corresponding mode in equations (7.3.6) and (7.3.8). As we would like to avoid the costly batch re-computation of the corresponding SVD, we can consider incremental solutions [Bur02, SKR02]. Depending on the size of the update (i.e., number of new users, tags, or items), different techniques have been followed in related research. For small update sizes we can consider the *folding-in* technique [FDD+88, SKR02], whereas for larger update sizes we can consider Incremental SVD techniques [Bur02]. Both techniques are described next [SNM10].

Update by folding-in. Given a new user, we first compute the new 1-mode matrix unfolding $\mathbf{A}_1$. It is easy to see that the entries of the new user result to the appending of new row in $\mathbf{A}_1$. This is exemplified in Figure 7.3.8. Figure 7.3.8a shows the insertion of a new user in the tensor of the current example. The new values are presented in the last (fourth) row of each frontal slice of the tensor. Notice that to ease the presentation, the new user's tags and items are identical with those of user U_2.

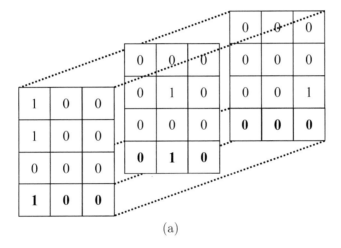

(a)

$$A_1 = \begin{bmatrix} 1 & 0 & 0 & 0 & 0 & 0 & 0 & 0 & 0 \\ 1 & 0 & 0 & 0 & 1 & 0 & 0 & 0 & 0 \\ 1 & 0 & 0 & 0 & 0 & 0 & 0 & 0 & 1 \\ \mathbf{1} & \mathbf{0} & \mathbf{0} & \mathbf{0} & \mathbf{1} & \mathbf{0} & \mathbf{0} & \mathbf{0} & \mathbf{0} \end{bmatrix}$$

(b)

FIGURE 7.3.8: Example of folding in a new user: a) the insertion of a new user into the tensor, b) the new 1-mode unfolded matrix $\mathbf{A}_1$.

Let $\mathbf{u}$ denote the new row that is appended to $\mathbf{A}_1$. Figure 7.3.8b presents the new $\mathbf{A}_1$, i.e., the 1-mode unfolded matrix, where it is shown that the contents of $\mathbf{u}$ (highlighted in bold) have been appended as a new row on the end of $\mathbf{A}_1$.

Since $\mathbf{A}_1$ changed, we have to compute its SVD. To avoid batch SVD recomputation, we can use the existing basis $\mathbf{U}_{c1}^{(1)}$ of left singular vectors, to project the row $\mathbf{u}$ onto the the reduced $c1$-dimensional space of users in the $\mathbf{A}_1$ matrix. This projection is called folding-in and is computed by using the following equation (7.3.9) [FDD^{+}88]:

$$\mathbf{u}_{new} = \mathbf{u} \cdot \mathbf{V}_{c1}^{(1)} \cdot (\mathbf{S}_{c1}^{(1)})^{-1}. \qquad (7.3.9)$$

In equation (7.3.9), $\mathbf{u}_{new}$ denotes the mapped row, which will be appended to $\mathbf{U}_{c1}^{(1)}$, whereas $\mathbf{V}_{c1}^{(1)}$ and $(\mathbf{S}_{c1}^{(1)})^{-1}$ are the dimensionally reduced matrices derived when SVD was originally applied to $\mathbf{A}_1$, i.e., before the insertion of the new user. In the current example, the computation of $\mathbf{u}_{new}$ is described in Figure 7.3.9.

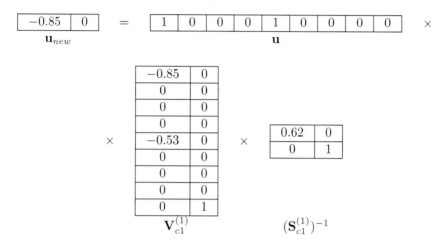

FIGURE 7.3.9: The result of folding-in for the current example.

The $\mathbf{u}_{new}$ vector should be appended in the end of the $\mathbf{U}_{c1}^{(1)}$ matrix. For the current example, appending should be done to the previously $\mathbf{U}_{c1}^{(1)}$ matrix. Notice that in the example, $\mathbf{u}_{new}$ is identical to the second column of the transpose of $\mathbf{U}_{c1}^{(1)}$, as shown in Figure 7.3.8a. The reason is that the new user has identical tags and items with user U_2 and we mapped them on the same space (recall that the folding-in technique maintains the same space computed originally by SVD).

Finally, to update tensor $\hat{\mathcal{A}}$, we have to perform the products given in equation (7.3.6). Notice that only $\mathbf{U}^{(1)c1}$ has been modified in this equation. Thus, to optimize the insertion of new users, as mode products are interchangeable, we can perform this product as $\left[\mathbf{S} \times_2 \mathbf{U}_{c2}^{(2)} \times_3 \mathbf{U}_{c3}^{(3)}\right] \times_1 \mathbf{U}_{c1}^{(1)}$, where the left factor (inside the brackets), which is unchanged, can be pre-stored so as to avoid its re-computation. For the current example, the resulting $\hat{\mathcal{A}}$ tensor is shown in Figure 7.3.10.

An analogous insertion procedure can be followed for the insertion of a new item or tag. For a new item insertion, we have to apply equation (7.3.9) on the 2-mode matrix unfolding of tensor $\mathcal{A}$, while for a new tag we apply equation (7.3.9) on the 3-mode matrix unfolding of tensor $\mathcal{A}$.

Update by Incremental SVD. Folding-in incrementally updates SVD but the resulting model is not a perfect SVD model, because the space is not orthogonal [SKR02]. When the update size is not big, loss of orthogonality may not be a severe problem in practice. Nevertheless, for larger update sizes the loss of orthogonality may result in an inaccurate SVD model. In this case, we need to incrementally update SVD so as to ensure orthogonality. This can be attained in several ways. Next we describe the approach proposed by Brand [Bur02].

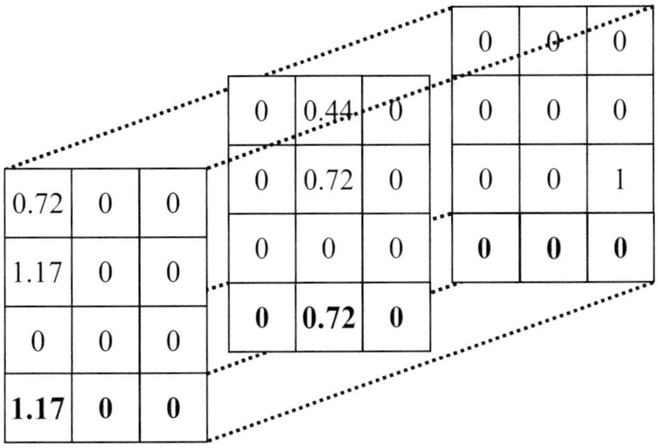

0.72	0	0
1.17	0	0
0	0	0
1.17	0	0

0	0.44	0
0	0.72	0
0	0	0
0	0.72	0

0	0	0
0	0	0
0	0	1
0	0	0

FIGURE 7.3.10: The resulting $\hat{\mathcal{A}}$ tensor of the running example after the insertion of the new user.

Let $\mathbf{M}_{p \times q}$ be a matrix, upon we which apply SVD and maintain the first r singular values, i.e.:

$$\mathbf{M}_{p \times q} = \mathbf{U}_{p \times r} \mathbf{S}_{r \times r} \mathbf{V}^T_{r \times q}. \tag{7.3.10}$$

Assume that each column of matrix $\mathbf{C}_{p \times c}$ contains the additional elements. Let $\mathbf{L} \overset{\triangle}{=} \mathbf{U} \backslash \mathbf{C} = \mathbf{U}^T \mathbf{C}$ be the projection of $\mathbf{C}$ onto the orthogonal basis of $\mathbf{U}$. Let also $\mathbf{H} = (\mathbf{I} - \mathbf{U}\mathbf{U}^T)\mathbf{C} = \mathbf{C} - \mathbf{U}\mathbf{L}$ be the component of $\mathbf{C}$ orthogonal to the subspace spanned by $\mathbf{U}$ ($\mathbf{I}$ is the identity matrix). Finally, let $\mathbf{J}$ be an orthogonal basis of $\mathbf{H}$ and let $\mathbf{K} = \mathbf{J} \backslash \mathbf{H} = \mathbf{J}^T \mathbf{H}$ be the projection of $\mathbf{C}$ onto the subspace orthogonal to $\mathbf{U}$. Consider the following identity:

$$[\mathbf{U}\ \mathbf{J}] \begin{bmatrix} \mathbf{S} & \mathbf{L} \\ \mathbf{0} & \mathbf{K} \end{bmatrix} \begin{bmatrix} \mathbf{V} & \mathbf{0} \\ \mathbf{0} & \mathbf{I} \end{bmatrix}^T = [\mathbf{U}(\mathbf{I} - \mathbf{U}\mathbf{U}^T)\mathbf{C}/\mathbf{K}] \begin{bmatrix} \mathbf{S} & \mathbf{U}^T\mathbf{C} \\ \mathbf{0} & \mathbf{K} \end{bmatrix} \begin{bmatrix} \mathbf{V} & \mathbf{0} \\ \mathbf{0} & \mathbf{I} \end{bmatrix}^T =$$
$$[\mathbf{U}\mathbf{S}\mathbf{V}^T\ \mathbf{C}] = [\mathbf{M}\ \mathbf{C}]. \tag{7.3.11}$$

Like an SVD, the left and right matrixes in the product are unitary and orthogonal. The middle matrix, denoted as $\mathbf{Q}$, should be diagonal. To incrementally update the SVD, $\mathbf{Q}$ must be diagonalized. If we apply SVD on $\mathbf{Q}$ we get:

$$\mathbf{Q} = \mathbf{U}'\mathbf{S}'(\mathbf{V}')^T. \tag{7.3.12}$$

Additionally, define $\mathbf{U}'', \mathbf{S}'', \mathbf{V}''$ as follows:

$$\mathbf{U}'' = [\mathbf{U}\ \mathbf{J}]\mathbf{U}', \quad \mathbf{S}'' = \mathbf{S}', \quad \mathbf{V}'' = \begin{bmatrix} \mathbf{V} & \mathbf{0} \\ \mathbf{0} & \mathbf{I} \end{bmatrix} \mathbf{V}'. \tag{7.3.13}$$

Then, the updated SVD of matrix $[\mathbf{M}\ \mathbf{C}]$ is:

$$[\mathbf{M}\ \mathbf{C}] = [\mathbf{U}\mathbf{S}\mathbf{V}^T\ \mathbf{C}] = \mathbf{U}''\mathbf{S}''(\mathbf{V}'')^T. \tag{7.3.14}$$

This incremental update procedure takes $O((p+q)r^2 + pc^2)$ time.

Returning to the application of incremental update for new users, items, or tags, as described in the subsection on updating by folding in in the current section, in each case we obtain a result with a number of new rows that are appended at the end of the unfolded matrix of the corresponding mode. Therefore, we need an incremental SVD procedure in the case where we add new rows, whereas the aforementioned method works in the case where we add new columns. In this case, we simply swap $\mathbf{U}$ for $\mathbf{V}$ and $\mathbf{U}''$ for $\mathbf{V}''$.

7.3.5   Other Scalable Factorization Models

The HOSVD approach has an important drawback. That is, its runtime complexity is cubic in the size of the latent dimensions. This can be seen in equation (7.3.5), where three nested sums have to be calculated just for predicting a single (user, item, tag)-triple. There are several approaches to improve the efficiency of HOSVD [DM07, Tur07, KS08].

The limitation in runtime of HOSVD stems from its model which is the Tucker Decomposition. In the following, we will discuss a second factorization model (i. e., PARAFAC) that has been proposed for tag recommendation. We investigate its model assumptions, complexity and its relation with HOSVD.

The underlying tensor factorization model of HOSVD is the *Tucker Decomposition* (TD) [Tuc66]. As noted before, for tag recommendation, the model reads:

$$\hat{\mathcal{A}} := \hat{\mathbf{C}} \times_u \hat{\mathbf{U}} \times_i \hat{\mathbf{I}} \times_t \hat{\mathbf{T}}, \tag{7.3.15}$$

or equivalently

$$\hat{a}_{u,i,t} = \sum_{\tilde{u}=1}^{k_U} \sum_{\tilde{i}=1}^{k_I} \sum_{\tilde{t}=1}^{k_T} \hat{c}_{\tilde{u},\tilde{i},\tilde{t}} \cdot \hat{u}_{u,\tilde{u}} \cdot \hat{i}_{i,\tilde{i}} \cdot \hat{t}_{t,\tilde{t}} \tag{7.3.16}$$

The reason for the cubic complexity (i. e., $O(k^3)$ with $k := \min(k_U, k_I, k_T)$) of TD is the core tensor.

The *Parallel Factor Analysis* (PARAFAC) [Har70] model or *canonical decomposition* [CC70] reduces the complexity of the TD model by assuming a diagonal core tensor.

$$c_{\tilde{u},\tilde{i},\tilde{t}} \stackrel{!}{=} \begin{cases} 1, & \text{if } \tilde{u} = \tilde{i} = \tilde{t} \\ 0, & \text{else,} \end{cases} \tag{7.3.17}$$

which allows us to rewrite the model equation:

$$\hat{a}_{u,i,t} = \sum_{f=1}^{k} \hat{u}_{u,f} \cdot \hat{i}_{i,f} \cdot \hat{t}_{t,f}. \tag{7.3.18}$$

In contrast to TD, the model equation of PARAFAC can be computed in $O(k)$. In total, the model parameters $\hat{\theta}$ of the PARAFAC model are $\hat{\mathbf{U}} \in \mathbb{R}^{|\mathbf{U}| \times k}, \hat{\mathbf{I}} \in \mathbb{R}^{|\mathbf{I}| \times k}, \hat{\mathbf{T}} \in \mathbb{R}^{|\mathbf{T}| \times k}$. The assumption of a diagonal core tensor is a restriction of the TD model.

A graphical representation of Tucker Decomposition (TD) and Parallel Factor Analysis (PARAFAC) shown in Figure 7.3.11. It can be seen that any PARAFAC model can be expressed by a TD model (with a diagonal core tensor).

Let $\mathcal{M}$ be the set of models that can be represented by a model class. In [Ren10] it is shown that for the tag recommendation:

$$\mathcal{M}^{\text{PARAFAC}} \subset \mathcal{M}^{\text{TD}}. \tag{7.3.19}$$

This means that any PARAFAC model can be expressed with a TD model, but there are

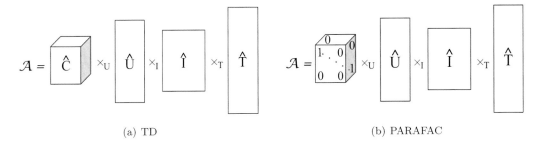

(a) TD   (b) PARAFAC

FIGURE 7.3.11: Relationship between Tucker Decomposition and Parallel Factor Analysis (PARAFAC).

TD models that cannot be represented by a PARAFAC model. In [RST10, Ren10] it was pointed out that this does not mean that TD is guaranteed to have a higher prediction quality than PARAFAC. On the contrary, because all the model parameters are estimated from limited data, restricting the expressiveness of a model can lead to higher prediction quality if the restriction is in line with the true parameters.

7.4 A Real Geo-Social System-Based on HOSVD

This section presents a real-world recommender system for Location-Based Social Networks (LBSNs). Our GeoSocialRec website allows us to test, evaluate and compare different recommendation styles in an online setting, where the users of GeoSocialRec actually receive recommendations during their check-in process.

The GeoSocialRec recommender system consists of several components. The system's architecture is illustrated in Figure 7.4.1, where three main sub-systems are described: (i) the website, (ii) the Database Profiles and (iii) the Recommendation Engine. In the following sections, we describe each subsystem of GeoSocialRec in detail.

7.4.1 GeoSocialRec Website

The GeoSocialRec system uses a website [2] to interact with the users. The website consists of four subsystems: (i) the friend recommendation, (ii) the location recommendation, (iii) the activity recommendation, and (iv) the check-in subsystem. The friend recommendation subsystem is responsible for evaluating incoming data from the Recommendation Engine of GeoSocialRec and providing updated friend recommendations. We provide friend, location, and activity recommendations, where new and updated location and activity recommendations presented to the user as new check-ins are stored in the Database profiles. Finally, the check-in subsystem is responsible for passing the data inserted by the users to the respective Database profiles.

Figure 7.4.2 presents a scenario where the GeoSocialRec system recommends four possible friends to the target user. As shown, the first table recommends 2 possible friends, who are connected to him with 2-hop paths. The results are ordered based on the second

---

[2]http://delab.csd.auth.gr/geosocialrec

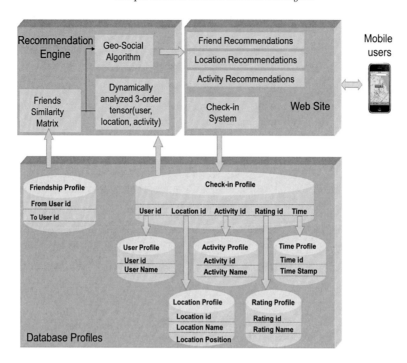

FIGURE 7.4.1: Components of the Geo-social recommender system.

to last column of the table, which indicates the number of common friends that the target user shares with each possible friend.

As also shown in Figure 7.4.2, Anastasia is the top recommendation because she shares 3 common friends with the target user. The common friends are then presented in the last column of the table. The second table contains two users, who are connected to the target user via 3-hop paths. The last column of the second table indicates the number of found paths that connect the target user with the recommended friends. Manolis is now the top recommendation, because he is connected to the target user via three 3-hop paths. It is obvious that the second explanation style is more analytical and detailed, since users can see, in a transparent way, the paths that connect them with the recommended friends.

Figure 7.4.3a shows a location recommendation, while Figure 7.4.3b depicts an activity recommendation. As shown in Figure 7.4.3a, the target user can provide to the system the activity she wants to do and the place she is (i.e., Bar in Athens). Then, the system provides a map with bar places (i.e., place A, place B, place C, etc.) along with a table, where these places are ranked based on the number of users' check-ins and their average rating. As shown in Figure 7.4.3a, the top recommended Bar is Mojo (i.e., place A), which is visited 3 times (by the target user's friends) and is rated highly (i.e., 5 stars). Regarding the activity recommendation, as shown in Figure 7.4.3b, the user selects a nearby city (i.e., Thessaloniki) and the system provides activities that she could perform. In this case, the top recommended activity is sightseeing at the White Tower of Thessaloniki, because it is visited 14 times and has an average rating of 4.36.

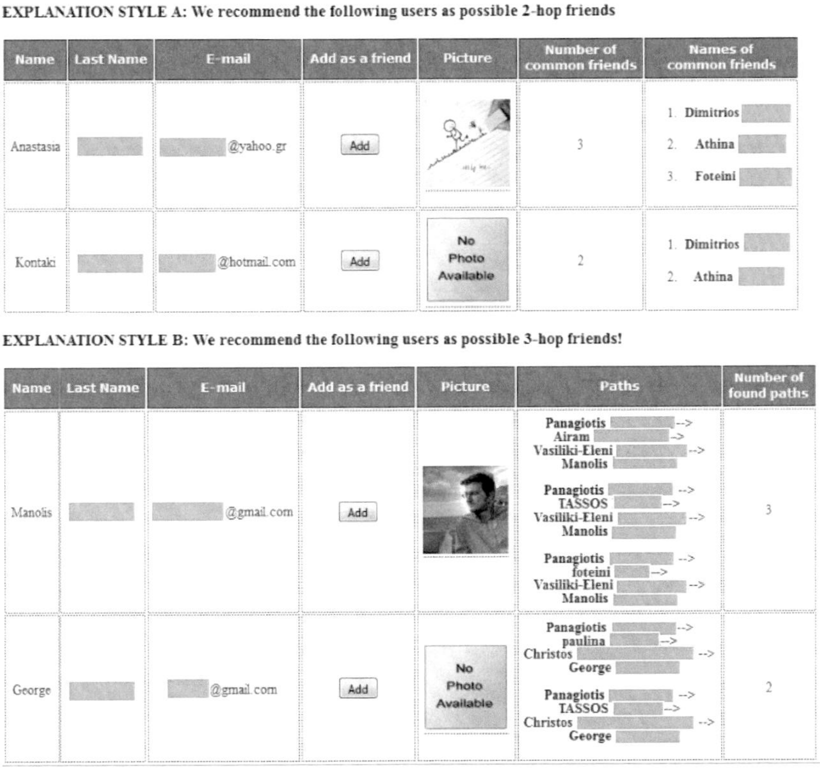

EXPLANATION STYLE A: We recommend the following users as possible 2-hop friends

Name	Last Name	E-mail	Add as a friend	Picture	Number of common friends	Names of common friends
Anastasia		@yahoo.gr	Add		3	1. Dimitrios 2. Athina 3. Foteini
Kontaki		@hotmail.com	Add	No Photo Available	2	1. Dimitrios 2. Athina

EXPLANATION STYLE B: We recommend the following users as possible 3-hop friends!

Name	Last Name	E-mail	Add as a friend	Picture	Paths	Number of found paths
Manolis		@gmail.com	Add		Panagiotis --> Airam --> Vasiliki-Eleni --> Manolis Panagiotis --> TASSOS --> Vasiliki-Eleni --> Manolis Panagiotis --> foteini --> Vasiliki-Eleni --> Manolis	3
George		@gmail.com	Add	No Photo Available	Panagiotis --> paulina --> Christos --> George Panagiotis --> TASSOS --> Christos --> George	2

FIGURE 7.4.2: Friend recommendations provided by the GeoSocialRec system.

7.4.2 GeoSocialRec Database and Recommendation Engine

The database that supports the GeoSocialRec system is a MySQL(v.5.5.8) [3] database. MySQL is an established Database Management System (DBMS), which is widely used in on-line, dynamic, database driven websites.

The database profile sub-system contains five profiles where data about the users, locations, activities and their corresponding ratings are stored. As shown in Figure 7.4.1, these data are received by the Check-In profile and along with the Friendship profile, they provide the input for the Recommendation Engine sub-system. Each table field represents the respective data that is collected by the Check-In profile. User-id, Location-id and Activity-id refer to specific ids given to users, locations, and activities respectively.

The recommendation engine is responsible for collecting the data from the database and producing the recommendations, which will then be displayed on the website. As shown in Figure 7.4.1, the recommendation engine constructs a friends-similarity matrix based on FriendLink [PSM11] algorithm. The average geographical distances (*in kilometers*) between users' check-ins are used as link weights. To obtain the weights, we calculate the average distance between all pairs of Points Of Interests (POIs) that two users have checked in. The recommendation engine also produces a dynamically analyzed 3-order tensor, which is first constructed by the HOSVD algorithm and is then updated using incremental methods [Bra02, SKKR02], both of which are explained previously in Sections 7.2.4 and 7.3.4, respectively.

---

[3]http://www.mysql.com

We recommend the following Point(s) of Interest
for the Activity: Bar
in the city of: Athens

EXPLANATION STYLE A:

We recommend the following POI's (Point of Interest) based on total Check-ins!

A/A	Point Of Interest	POI Address	Explanation Style A: Total Check-Ins	Average Rating from style A	Go To
A	Mojo	Παπαδιαμαντοπούλου 8, Ζωγράφου 157 71, Ελλάδα	3	5.0000	Move!
B	A for Athens	Ερμού 82, Αθήνα 105 55, Ελλάδα	3	3.6667	Move!
C	Rox Box	Ειρήνης 2-10, Φιλαδέλφεια Χαλκηδόνα 143 41, Ελλάδα	2	4.5000	Move!
D	Holy Spirit	Λαοδίκης 18, Γλυφάδα 166 74, Ελλάδα	2	4.0000	Move!
E	Mo Better	Καλέττη 28-42, Αθήνα, Ελλάδα	2	4.0000	Move!
F	Allo Bar	Thoukydidou 9-13, Chalandri 15232, Greece	2	2.0000	Move!

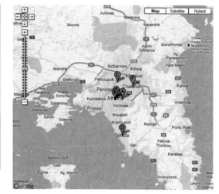

We recommend the following Activities
in the city of: Thessaloniki

EXPLANATION STYLE A:

We recommend the following activities based on total Check-ins!

A/A	Activity	Point Of Interest	POI Address	Explanation Style A: Total Check-Ins	Average Rating from style A	Go To
A	Sight-seeing	White Tower	Nikis Avenue-- Paralia Thessalonikis	14	4.3571	Move!
B	Education	Aristotle University of Thessaloniki	Egnatia & Kondriktonos-- Aristotle Campus	13	4.2308	Move!
C	Sight-seeing	Aristotelous Square	Aristotle Square- -City Center	11	4.1818	Move!
D	Museums	Archaeological Museum of Thessaloniki	Ανδρόνικου, Θεσσαλονίκη 54621, Ελλάς--	8	4.0000	Move!
E	Cinema	Video Land	Ιθάκης 63-71, Εύοσμο 56224, Ελλάς--	8	2.6250	Move!
F	Bar	BRISTOL	George Papandreou 24-- Poseidonio	7	4.2857	Move!

(a)

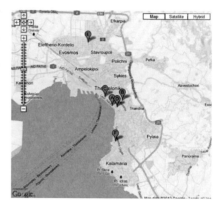

(b)

FIGURE 7.4.3: Location and activity recommendations made by the Geo-Social recommender system.

7.4.3 Experiments

In this section, we study the performance of the FriendLink [PSM11] algorithm and the HOSVD method in terms of friend, location, and activity recommendations. To evaluate the aforementioned recommendations we have chosen two real data sets. The first one, denoted as the GeoSocialRec data set, is extracted from the GeoSocialRec site [4]. It consists of 102 users, 46 locations and 18 activities. The second data set, denoted as UCLAF [ZCZ+10], consists of 164 users, 168 locations and 5 different types of activities, including "Food and Drink," "Shopping," "Movies and Shows," "Sports and Exercise," and "Tourism and Amusement."

---

[4]http://delab.csd.auth.gr/~symeon

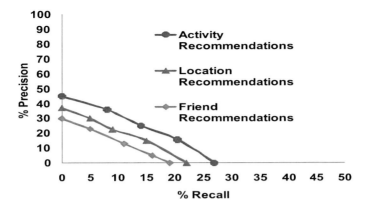

FIGURE 7.4.4: Precision-recall diagram of HOSVD and FriendLink for activity, location and friend recommendations on the GeoSocialRec data set.

The numbers c_1, c_2, and c_3 of left singular vectors of matrices $U^{(1)}$, $U^{(2)}$, $U^{(3)}$ for HOSVD, after appropriate tuning, are set to 25, 12, and 8 for the GeoSocialRec dataset, and to 40, 35, and 5 for the UCLAF data set. Due to lack of space we do not present experiments for the tuning of c_1, c_2, and c_3 parameters. The core tensor dimensions are fixed, based on the aforementioned c_1, c_2, and c_3 values.

We perform 4-fold cross validation and the default size of the training set is 75% (we pick, for each user, 75% of his check-ins and friends randomly). The task of all three recommendation types (i.e., friend, location, activity) is to predict the friends/locations/activities of the 25% remaining user check-ins and friends, respectively. As performance measures we use precision and recall, which are standard in such scenarios.

Next, we study the accuracy performance of HOSVD in terms of precision and recall. This reveals the robustness of HOSVD in attaining high recall with minimal losses in terms of precision. We examine the top-N ranked list, which is recommended to a test user, starting from the top friend/location/activity. In this situation, the recall and precision vary as we proceed with the examination of the top-N list. In Figure 7.4.4, we plot a precision versus recall curve.

As it can be seen, the HOSVD approach presents high accuracy. The reason is that we exploit altogether the information that concerns the three entities (friends, locations, and activities) and thus, we are able to provide accurate location/activity recommendations. Notice that activity recommendations are more accurate than location recommendations. A possible explanation could be the fact that the number of locations is bigger than the number of activities. That is, it is easier to accurately predict an activity than a location. Notice that for the task of friend recommendation, the performance of Friendlink is not so high. The main reason is data sparsity. In particular, the friendship network has an average nodes' degree equal to 2.7 and an average shortest distance between nodes of 4.7, which means that the friendship network can not be considered a "small world" network and friend recommendations can not be so accurate.

For the UCLAF data set, as shown in Figure 7.4.5, the HOSVD algorithm attains analogous results. Notice that the recall for the activity recommendations, reaches 100% because the total number of activities is 5. Moreover, notice that in this diagram, we do not present results for the friend recommendation task, since there is no friendship network in the corresponding UCLAF data set.

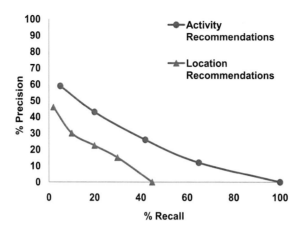

FIGURE 7.4.5: Precision-recall diagram of HOSVD for activity and location recommendations on the UCLAF data set.

7.5   Conclusion

In this chapter, we described matrix and tensor factorization techniques in recommender systems and social tagging systems, respectively. In addition, we presented a real-world recommender system for location-based social networks, which employs tensor decomposition techniques. As shown, matrix and tensor decompositions are suitable for scenarios in which the data is extremely large, very sparse, and too noisy, since the reduced representation of the data can be interpreted as a de-noisified approximation of the "true" data.

Bibliography

[BBL⁺06]   M. W. Berry, M. Browne, A. N. Langville, V. P. Pauca, and R. J. Plemmons. Algorithms and applications for approximate nonnegative matrix factorization. In *Computational Statistics and Data Analysis*, pages 155–173, 2006.

[Bra02]   M. Brand. Incremental singular value decomposition of uncertain data with missing values. In *Proc. of the 7th European Conference on Computer Vision*, pages 707–720, Copenhagen, Denmark, 2002.

[Bur02]   R. Burke. Hybrid recommender systems: Survey and experiments. *User Modeling and User-Adapted Interaction*, 12(4):331–370, 2002.

[CC70]   J. Carroll and J. Chang. Analysis of individual differences in multidimensional scaling via an n-way generalization of "Eckart-Young decomposition." *Psychometrika*, 35:283–319, 1970.

[CSS06]   T. Chin, K. Schindler, and D. Suter. Incremental kernel SVD for face recognition with image sets. In *Proc. of the 7th International Conference on Automatic Face and Gesture Recognition*, pages 461–466, 2006.

[CST04]   N. Cristianini and J. Shawe-Taylor. *Kernel methods for pattern analysis*. Cambridge University Press, 2004.

[CWZ07]   S. Chen, F. Wang, and C. Zhang. Simultaneous heterogeneous data clustering based on higher order relationships. In *Proc. of the 7th IEEE International Conference on Data Mining Workshops*, pages 387–392, 2007.

[DDF+90] S. Deerwester, S. T. Dumais, G. W. Furnas, T. K. Landauer, and R. Harshman. Indexing by latent semantic analysis. *Journal of the American Society for Information Science*, 41(6):391–407, 1990.

[DM07]   P. Drineas and M. W. Mahoney. A randomized algorithm for a tensor-based generalization of the SVD. *Linear Algebra and Its Applications*, 420(2–3):553–571, 2007.

[DS05]   I. S. Dhillon and S. Sra. Generalized nonnegative matrix approximations with Bregman divergences. In *Proc. Advances in Neural Information Processing Systems*, pages 283–290, 2005.

[FDD+88] G. W. Furnas, S. Deerwester, S. T. Dumais, T. K. Landauer, R. A. Harshman, L. A. Streeter, and K. E. Lochbaum. Information retrieval using a singular value decomposition model of latent semantic structure. In *Proc. of the 11th Annual International ACM SIGIR Conference on Research and Development in Information Retrieval*, pages 465–480, 1988.

[Har70]   R. A. Harshman. Foundations of the PARAFAC procedure: models and conditions for an "explanatory" multimodal factor analysis. *UCLA Working Papers in Phonetics*, 16:1–84, 1970.

[Kor08]   Y. Koren. Factorization meets the neighborhood: a multifaceted collaborative filtering model. In *Proc. of the 14th ACM SIGKDD International Conference on Knowledge Discovery and Data Mining*, pages 426–434, 2008.

[Kor09]   Y. Koren. Collaborative filtering with temporal dynamics. In *Proc. of the 15th ACM SIGKDD International Conference on Knowledge Discovery and Data Mining*, pages 447–456, 2009.

[KS08]   T. G. Kolda and J. Sun. Scalable tensor decompositions for multi-aspect data mining. In *Proc. of the 8th IEEE International Conference on Data Mining*, pages 363–372, 2008.

[LMV00]   L. D. Lathauwer, B. D. Moor, and J. Vandewalle. A multilinear singular value decomposition. *SIAM Journal on Matrix Analysis and Applications*, 21(4):1253–1278, 2000.

[MD09]   M. W. Mahoney and P. Drineas. CUR matrix decompositions for improved data analysis. *Proceedings of the National Academy of Sciences*, 106(3):697–702, 2009.

[PSM11]   A. Papadimitriou, P. Symeonidis, and Y. Manolopoulos. Friendlink: Link prediction in social networks via bounded local path traversal. In *Proc. of the 3rd Conference on Computational Aspects of Social Networks*, 2011.

[Ren10]   S. Rendle. *Context-Aware Ranking with Factorization Models*. Springer, 1st edition, November 2010.

[RMNST09] S. Rendle, L. B. Marinho, A. Nanopoulos, and L. Schimdt-Thieme. Learning optimal ranking with tensor factorization for tag recommendation. In *Proc. of the 15th ACM SIGKDD International Conference on Knowledge Discovery and Data Mining*, pages 727–736, 2009.

[RST10] S. Rendle and L. Schmidt-Thieme. Pairwise interaction tensor factorization for personalized tag recommendation. In *Proc. of the 3rd ACM International Conference on Web Search and Data Mining*, 2010.

[SH05] A. Shashua and T. Hazan. Non-negative tensor factorization with applications to statistics and computer vision. In *Proc. of the 22nd International Conference on Machine Learning*, pages 792–799, 2005.

[SKKR00] B. Sarwar, G. Karypis, J. Konstan, and J. Riedl. Application of dimensionality reduction in recommender system - a case study. In *Proc. of the ACM SIGKDD Workshop on Web Mining for E-Commerce - Challenges and Opportunities*, Boston, MA, 2000.

[SKKR02] B. Sarwar, G. Karypis, J. Konstan, and J. Riedl. Incremental singular value decomposition algorithms for highly scalable recommender systems. In *Proc. of the 5th International Conference on Computer and Information Technology*, pages 27–28, Dhaka, Bangladesh, 2002.

[SKR02] B. Sarwar, J. Konstan, and J. Riedl. Incremental singular value decomposition algorithms for highly scalable recommender systems. In *Proc. International Conference on Computer and Information Science*, 2002.

[SM08] R. Salakhutdinov and A. Mnih. Bayesian probabilistic matrix factorization using Markov chain Monte Carlo. In *Proc. of the 25th International Conference on Machine Learning*, pages 880–887, 2008.

[SNM10] P. Symeonidis, A. Nanopoulos, and Y. Manolopoulos. A unified framework for providing recommendations in social tagging systems based on ternary semantic analysis. *IEEE Transactions on Knowledge and Data Engineering*, 22(2), 2010.

[SNPM06] P. Symeonidis, A. Nanopoulos, A. Papadopoulos, and Y. Manolopoulos. Collaborative filtering based on users trends. In *Proc. of the 30th Conference of the German Classification Society*, 2006.

[Str06] G. Strang. *Linear Algebra and Its Applications*. Brooks Cole, 2006.

[Sym07] P. Symeonidis. Content-based dimensionality reduction for recommender systems. In *Proc. of the 31st Conference of the German Classification Society*, Freiburg, 2007.

[SZL+05] J.-T. Sun, H.-J. Zeng, H. Liu, Y. Lu, and Z. Chen. CubeSVD: a novel approach to personalized web search. In *Proc. of the 14th International Conference on World Wide Web*, pages 382–390, 2005.

[Tik63] A. Tikhonov. Solution of incorrectly formulated problems and the regularization method. In *Soviet Math. Doklady*, volume 4, pages 1035–1038, 1963.

[Tuc66] L. Tucker. Some mathematical notes on three-mode factor analysis. *Psychometrika*, pages 279–311, 1966.

[Tur07]   P. Turney. Empirical evaluation of four tensor decomposition algorithms. *Technical Report (NRC/ERB-1152)*, 2007.

[WA08]   H. Wang and N. Ahuja. A tensor approximation approach to dimensionality reduction. *International Journal of Computer Vision*, 76:217–229, 2008.

[ZCZ⁺10]   V. Zheng, B. Cao, Y. Zheng, X. Xie, and Q. Yang. Collaborative filtering meets mobile recommendation: A user-centered approach. In *Proc. of the 24th AAAI Conference on Artificial Intelligence*, pages 236–241, Atlanta, GA, 2010.

Chapter 8

Multimedia Social Search Based on Hypergraph Learning

Constantine Kotropoulos

Aristotle University of Thessaloniki, Greece

8.1 Introduction

Multimedia social search refers to a class of problems, such as tagging, retrieval, and recommendation, where both multimedia content and social context (i.e., information distilled from the web that is correlated to the content) are exploited. *Ranking* is the heart of multimedia social search. The motivation behind multimedia social search is that retrieval or recommendation based only on content yields frequently unsatisfactory results due to the well known semantic gap. Thanks to online social sharing sites (e.g., Flickr [http://www.flickr.com], Lastfm [http://www.last.fm], etc.) multimedia come with additional metadata, including, ownership, tags, geo-location, etc. Such metadata offer rich complementary information worth exploiting. A new transductive learning framework has recently emerged that addresses the aforementioned problems as ranking on *hypergraphs* by jointly analyzing the content and its associated context defined by the metadata in a unified manner.

Hypergraphs generalize the concept of graphs by allowing their edges, called *hyperedges* hereafter, to connect more than two vertices. This way, the hypergraphs can capture more complex relations, i.e., three-way or higher-order ones. Hypergraphs have been used in various domains, such as, databases [Fag83], data mining [HKKM98], biology [KHT06], or to model complex networks [ERV06], to mention a few, well in advance of their use in multimedia social search. The roots of hypergraphs are traced back in mathematics [BM73, Ber89]. The goal of this chapter is two-fold: (1) To provide a comprehensive, but not superficial, self-contained survey of the theory related to hypergraphs, promoting the coherence of knowledge; (2) To describe sample applications in high-order web link analysis, object recognition, music/image recommendation and tagging, image search, and tourism recommendation. This way the chapter might be useful to senior undergraduate students, graduate students, and Ph.D. candidates as well as scientists, engineers, and practitioners.

Starting with the basic definitions related to hypergraphs, given in Section 8.2, the fundamentals of the most studied member of the family of hypergraphs, the so-called κ-*uniform hypergraph*, are summarized in Section 8.2.1. Emphasis is given to the spectrum of the κ-uniform hypergraphs. κ-uniform hypergraphs are important for two reasons. First, they are described in terms of an adjacency tensor, a degree tensor, and a Laplacian tensor. Accordingly, tensor decompositions can be used for detecting communities in the web, as is demonstrated in Section 8.6.1. Hereafter, we shall refrain from using the term tensor and shall prefer the more precise term *hypermatrix*. Moreover, feature matching, which is strongly related to content-based image retrieval, can be cast as a uniform hypergraph matching problem, as is demonstrated in Section 8.6.2. Second, uniform hypergraph clustering can be cast into a non-cooperative multi-player game, where the notion of a cluster is equivalent to a classical *game-theoretic equilibrium* concept, as is detailed in Section 8.3. The Nash equilibria of the clustering game correspond to strict local maxima of a *homogeneous polynomial*, a subject that goes back to the seminal *Baum-Eagon theorem*.

Section 8.4 elaborates spectral clustering for *arbitrary hypergraphs*. *Spectral hypergraph clustering* extends the spectral clustering originally applied to graphs. In particular, a hypergraph is transformed into a graph, whose edge weights are mapped from the weights of the original hypergraph. Two transformations are studied, namely the click expansion and the star expansion. The latter is shown to be closely related to the *hypergraph normalized cut criterion* that generalizes the normalized cut method applied to graphs. This is tantamount to the eigen-decomposition of a positive semidefinite matrix, the so-called *hypergraph Laplacian*.

Hypergraphs were applied to exploit visual-duplicates for video re-ranking using a random walk algorithm [TNW08]. A probabilistic hypergraph was proposed to describe both the higher order grouping information and the affinity relation among the vertices within hyperedges for image retrieval [HLZM10]. Recently, Kapoor et al. [KSS13] proposed node degree centrality metrics for weighted hypergraphs. Tramasco et al. evaluated various semantic and structural hypotheses for academic team formation [TCR10]. Going beyond collaborative filtering that is heavily based on user ratings, hypergraphs were employed for music recommendation [BTC+10]. Indeed, a hypergraph models the various objects (i.e., users, user groups, tags, tracks, albums, or artists) and the relations among them, such as friendship relations, membership relations, listening relations, tagging relations on tracks, tagging relations on albums, tagging relations on artists, track-album inclusion relations, album-artist inclusion relations, and similarities between the tracks. Figure 8.1.1 shows various hyperedges, capturing information related to music recommendation. For example, e_1 models a listening relation, which employs a user and a music recording; e_2 and e_3 capture tagging relations, which embrace a user, a music recording, and a tag; e_4 models content

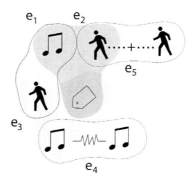

FIGURE 8.1.1: Various hyperedges that model information useful for music recommendation.

similarity; and e_5 captures the friends of a particular user. Section 8.5 elaborates hypergraph ranking. In Section 8.5.1, it is shown that recommendation can be cast as an optimization of an objective function made up by a smoothness constraint, which guarantees that vertices with the same value in the ranking vector are strongly connected plus an ℓ_2 norm measuring the difference between the obtained ranking scores and the pre-given query vector. The former constraint is related to the spectral clustering of an arbitrary hypergraph, while the latter one guarantees that the hypergraph vertices are ranked based on their relevance to the given query. By properly structuring the query vector, recommendation, tagging, or retrieval could also be addressed. Besides the construction of the query vector, open questions in ranking on hypergraphs are how hyperedges are generated, how group sparsity or other structural group penalties can be enforced, how the hyperedge weights can be learnt. Hyperedge weight learning helps us to assess how the various subsets of vertices affect the recommendation or tagging quality for music recordings and images. The associated constrained optimization problems are thoroughly studied in Sections 8.5.1 and 8.5.2 as well as in the applications to music recommendation and tagging (cf. Section 8.6.3), image recommendation, retrieval, and tagging demonstrated in Sections 8.6.4-8.6.6.

Section 8.7 deals with the so-called *Big Data* case. *Randomized matrix/hypermatrix factorizations* are surveyed, because they can cope efficiently with big data. Such techniques complement the discussions on adaptive latent models, incremental spectral clustering, and incremental tensor adaptation as well as their parallel and distributed implementations in Chapter 11. Figure 8.1.2 depicts the dependencies among the various sections of the chapter.

Throughout the chapter, calligraphic uppercase letters are reserved for graphs, and hypergraphs (e.g., $\mathcal{G}$, $\mathcal{H}$). Sets appear as blackboard bold uppercase letters (e.g., $\mathbb{V}$), hypermatrices are denoted by boldface Euler script calligraphic letters (e.g., $\boldsymbol{\mathcal{A}}$), matrices are indicated by uppercase boldface letters (e.g., $\mathbf{A}$), and vectors are denoted by lowercase boldface letters (e.g., $\mathbf{a}$). The elements of all the aforementioned mathematical structures are denoted by lowercase letters indexed by one or more indices. For example, the elements of matrix $\mathbf{A}$ are denoted as $a_{i_1 i_2}$. Occasionally, the elements of $\mathbf{A}$ are indicated as $A(i_1, i_2)$ or $[\mathbf{A}]_{i_1 i_2}$. In addition, $\mathbf{a}_i$ is a shorthand notation for the i-th column of $\mathbf{A}$, while $\mathbf{a}^{(j)}$ refers to the j-th row of $\mathbf{A}$. $\mathbb{R}$, $\mathbb{Z}$, $\mathbb{C}$ denote the fields of real, integer, and complex numbers, respectively, unless it is defined otherwise.

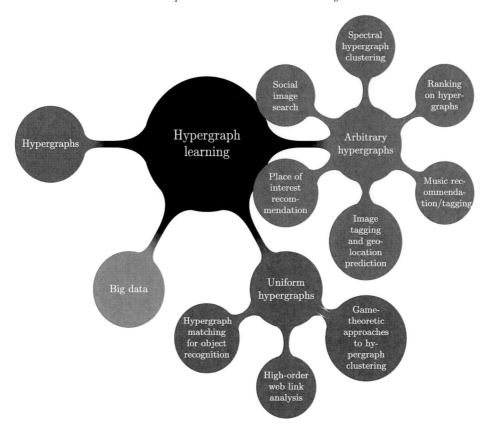

FIGURE 8.1.2: Roadmap of this chapter.

8.2 Hypergraphs

Let $\mathbb{V} = \{v_1, v_2, \ldots, v_N\}$ be a finite set of cardinality $|\mathbb{V}| = N$. The elements v_i, $i = 1, 2, \ldots, N$ are called *vertices*. A *hypergraph* on $\mathbb{V}$, $\mathcal{H}(\mathbb{V}, \mathbb{E})$, is formally defined as a family of nonempty subsets of $\mathbb{V}$, called *hyperedges*, whose union yields $\mathbb{V}$ [Ber89]. To simplify notation, let us refer to this collection by $\mathbb{E} = (e_1, e_2, \ldots, e_M)$. Let $\mathbb{P}(\mathbb{V})$ be the *power set* of $\mathbb{V}$, i.e., the set composed of all subsets of $\mathbb{V}$. Its cardinality is 2^N. It is evident that $\mathbb{E} \subseteq \mathbb{P}(\mathbb{V})$. Moreover, the number of hyperedges could be gargantuan even for small N.

The aforementioned definition of hypergraph does not allow for repeated vertices within a hyperedge (often called hyperloops). To include hyperloops, the definition has to be expanded by employing the notion of *multiset*. A multiset is a generalization of a set in which the members are allowed to appear more than once. Accordingly, a *multi-hypergraph* $\mathcal{H}$ is a pair $(\mathbb{V}, \mathbb{E})$, where $\mathbb{E}$ is a set of multisets of $\mathbb{V}$ [PZ13].

Given two hypergraphs $\mathcal{H} = (\mathbb{V}, \mathbb{E})$ and $\mathcal{H}' = (\mathbb{V}', \mathbb{E}')$, if $\mathbb{V}' \subseteq \mathbb{V}$ and $\mathbb{E}' \subseteq \mathbb{E}$, then $\mathcal{H}'$ is said to be a *subgraph* of $\mathcal{H}$. A set of vertices $\mathbb{S} \subseteq \mathbb{V}$ is said to *induce* the subgraph $\mathcal{H}[\mathbb{S}] = (\mathbb{S}, \mathbb{E} \cap \mathbb{P}(\mathbb{S}))$. Subgraphs formally describe communities in a hypergraph.

A vertex $v \in \mathbb{V}$ and a hyperedge $e \in \mathbb{E}$ are called *incident*, if $v \in e$. Two vertices v_i and v_k are connected by a *hyperpath*, if there is an alternating sequence of distinct vertices and hyperedges $v_1, e_1, v_2, e_2, \ldots, e_{k-1}, v_k$, such that $\{v_i, v_{i+1}\} \subseteq e_i$ for $1 \leq i \leq k - 1$. A

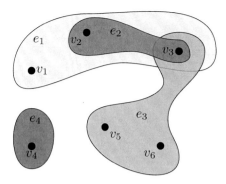

FIGURE 8.2.1: A hypergraph.

hypergraph is *connected*, if there is a hyperpath for every pair of vertices [ZHS06, BTC+10]. A hypergraph may be drawn as a set of points representing the vertices. The order of vertices incident to a hyperedge is irrelevant. Hyperedges are represented by closed curves enclosing their elements, as shown in Figure 8.2.1.

A hypergraph can be defined by its *incidence matrix* $\mathbf{H} \in \mathbb{R}^{N \times M}$ whose element $h(v, e)$ equals 1 if $v \in e$ and 0 otherwise. Strictly speaking, the aforementioned definition yields the so-called *unsigned incidence matrix* in graph theory. The incidence matrix of the hypergraph shown in Figure 8.2.1 is given by

$$
\mathbf{H} = \begin{array}{c} v_1 \\ v_2 \\ v_3 \\ v_4 \\ v_5 \\ v_6 \end{array} \overset{\begin{array}{cccc} e_1 & e_2 & e_3 & e_4 \end{array}}{\left[\begin{array}{cccc} 1 & 0 & 0 & 0 \\ 1 & 1 & 0 & 0 \\ 1 & 1 & 1 & 0 \\ 0 & 0 & 0 & 1 \\ 0 & 0 & 1 & 0 \\ 0 & 0 & 1 & 0 \end{array} \right]} . \tag{8.2.1}
$$

The *dual of a hypergraph* with incidence matrix $\mathbf{H}$ is also a hypergraph with incidence matrix $\mathbf{H}^T$, where the superscript T denotes matrix transposition. Accordingly, the vertices of the dual hypergraph correspond to the edges of the original hypergraph and its edges $e_i^* = \{e_j : v_i \in e_j\}$. If two vertices v_i and v_k are adjacent in the original hypergraph, then their corresponding edges e_i^* and e_k^* in the dual hypergraph are adjacent. If two edges e_j and e_l are adjacent in the original hypergraph, then they will correspond to adjacent vertices in the dual hypergraph. The dual of the hypergraph in Figure 8.2.1 is shown in Figure 8.2.2.

The cardinality of e defines the *degree of the hyperedge*, i.e., $\delta(e) = |e|$. Using the definition of the incidence matrix, the degree of a hyperedge is given by

$$
\delta(e) = \sum_{v \in \mathbb{V}} h(v, e). \tag{8.2.2}
$$

Frequently, each hyperedge is assigned a positive real weight $w(e)$. The hypergraph is then called *weighted* and is denoted as $\mathcal{H} = (\mathbb{V}, \mathbb{E}, w)$. The *degree of a vertex* is defined as:

$$
\delta(v) = \sum_{e \in \mathbb{E} | v \in e} w(e) = \sum_{e \in \mathbb{E}} h(v, e) \, w(e). \tag{8.2.3}
$$

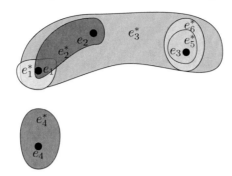

FIGURE 8.2.2: The dual of the hypergraph in Figure 8.2.1.

In the following, we shall use the diagonal matrices $\mathbf{D}_e \in \mathbb{R}^{M \times M}$ and $\mathbf{D}_v \in \mathbb{R}^{N \times N}$ having the hyperedge and vertex degrees in their main diagonal, respectively. Similarly, we define the diagonal matrix $\mathbf{W} \in \mathbb{R}^{M \times M}$ containing the hyperedge weights in its main diagonal. The *adjacency matrix* $\mathbf{A} \in \mathbb{R}^{N \times N}$ of a hypergraph is defined as:

$$\mathbf{A} = \mathbf{H} \, \mathbf{W} \, \mathbf{H}^T - \mathbf{D}_v. \tag{8.2.4}$$

The *rank* of a hypergraph is defined as $\max_{e \in \mathbb{E}} \delta(e)$, while the *anti-rank* is defined as $\min_{e \in \mathbb{E}} \delta(e)$. If the rank of a hypergraph equals its anti-rank, then the hypergraph is called *uniform*. When the degree of each hyperedge is constant, say κ, then the hypergraph is called κ-uniform. In undirected κ-uniform hypergraphs, the ordering of vertices in a hyperedge does not matter.

The degree of hyperedges κ is related to the order of the model underlying the objects in κ-uniform hypergraphs. For example, let us elaborate the case when a set of patterns in $\mathbb{R}^D$ should be clustered into K groups. Assume that the patterns lie on K subspaces of dimension ι. The patterns could be associated to the vertices of a κ-uniform hypergraph. To cluster the set of vertices $\mathbb{V}$ into K groups, affinity measures over more than ι vertices should be considered. That is, given a hyperedge $e \subset \mathbb{V}$ with degree $\delta(e) = \kappa > \iota$, a subspace should be fit to e, say, via singular value decomposition (SVD). The fitting error ϑ can be transformed into a hyperedge weight $w(e) = \exp(-\frac{\vartheta^2}{\sigma^2})$. Accordingly, the hyperedge e captures the existence of a subspace with a "degree of confidence" given by its weight $w(e)$. In practice, $\kappa = \iota + 1$.

Hypergraph-based learning techniques are classified into two categories. The first category uses hypermatrices for clustering, such as the high-order extension of spectral methods for graphs. Despite the fact that the methods resorting to hypermatrices are mathematically attractive, they are confined to the κ-uniform hypergraphs. The second category deals with arbitrary hypergraphs, which model mixed higher-order relationships. The latter techniques first approximate a hypergraph via a standard weighted graph and then resort to graph-based clustering and semi-supervised learning.

8.2.1 Uniform hypergraphs

Spectral graph theory has been thoroughly studied and has found a plethora of applications in combinatorics, computer science, social sciences, operations research, and biology [CD12]. Its aim is to establish connections of the graph intrinsic structures (e.g., connectivity, diameters) with the spectra (i.e., the eigenvalues) of various associated matrices, such as the adjacency matrix, the incidence matrix, and the graph Laplacian [Chu97, PZ13]. Several attempts have been made to define the spectra of κ-uniform hypergraphs through the eigenvalues of special nonnegative symmetric hypermatrices, such as the adjacency hypermatrix or the Laplacian hypermatrix.

Formally, hypermatrices are coordinate representations of *tensors*, just as matrices are coordinate representations of linear operators. A tensor is an element of a tensor product of κ vector spaces. The latter can be interpreted as the space of multilinear functionals. Tensors are represented as hypermatrices by choosing a basis for each vector space [Lim14]. Let us confine ourselves to N-dimensional vector spaces $\mathcal{V}_l$, $l = 1, 2, \ldots, \kappa$. For each vector space, the typical choice is the standard basis $\left\{ \boldsymbol{\epsilon}_1^{[l]}, \boldsymbol{\epsilon}_2^{[l]}, \ldots, \boldsymbol{\epsilon}_N^{[l]} \right\}$, where $\boldsymbol{\epsilon}_i^{[l]} \in \mathbb{R}^N$ has its i-th element equal to 1 and all other elements equal to 0. Then:

$$\mathcal{A} = \sum_{i_1=1}^{N} \sum_{i_2=1}^{N} \cdots \sum_{i_\kappa=1}^{N} a_{i_1 i_2 \ldots i_\kappa} \; \boldsymbol{\epsilon}_{i_1}^{[l]} \circ \boldsymbol{\epsilon}_{i_2}^{[l]} \circ \cdots \circ \boldsymbol{\epsilon}_{i_\kappa}^{[l]}, \tag{8.2.5}$$

where $\circ$ denotes the (Segre) outer product among vectors [Lim14]. That is, all the information about tensor $\mathcal{A}$ is captured by the coordinates $a_{i_1 i_2 \ldots i_\kappa}$, which are gathered in a κ-order N-dimensional hypermatrix $\boldsymbol{A}$. Accordingly, a κ-order N-dimensional hypermatrix $\boldsymbol{A}$ is a collection of N^κ elements.

For notation simplicity, let us denote the set of hypergraph vertices as $\mathbb{V} = \{1, 2, \ldots, N\}$, so that we deal only with the indices $i_l \in \mathbb{V}$, $l = 1, 2, \ldots, \kappa$. For a κ-uniform hypergraph $\mathcal{H}$ of N vertices, the (normalized) *adjacency hypermatrix* $\boldsymbol{A}$ is the κ-order N-dimensional hypermatrix with elements [CD12]:

$$a_{i_1 i_2 \ldots i_\kappa} = \begin{cases} \frac{1}{(\kappa-1)!} & \text{if } \{i_1, i_2, \ldots, i_\kappa\} \in \mathbb{E} \\ 0 & \text{otherwise.} \end{cases} \tag{8.2.6}$$

If $\mathcal{H}$ is a multi-hypergraph, then the denominator in (8.2.6) has to be replaced by the binomial coefficient $\binom{\kappa-1}{\nu_1, \nu_2, \ldots, \nu_s}$, where ν_t is the times the unique element i_{ι_t} occurs in the multiset $\{i_2, i_3, \ldots, i_\kappa\}$, $t \leq (\kappa - 1)$.

The hypermatrix $\boldsymbol{A}$ in (8.2.6) is super-symmetric, because the elements associated to the same index sets are the same. That is, $a_{i_1 i_2 \ldots i_\kappa} = a_{i_{\varpi(1)} i_{\varpi(2)} \ldots i_{\varpi(\kappa)}}$ for all the permutations of indices. A κ-order N-dimensional super-symmetric hypermatrix $\boldsymbol{A}$ uniquely defines a *homogenous polynomial* of degree κ in N variables [CD12]:

$$F_{\mathcal{A}}(\mathbf{x}) = \sum_{i_1=1}^{N} \sum_{i_2=1}^{N} \cdots \sum_{i_\kappa=1}^{N} a_{i_1 i_2 \ldots i_\kappa} \; x_{i_1} x_{i_2} \cdots x_{i_\kappa}. \tag{8.2.7}$$

Equation (8.2.7) can be expressed as the scalar product of the hypermatrix $\boldsymbol{A}$ with the κ-order N-dimensional hypermatrix $\boldsymbol{\mathcal{X}}$ with elements $x_{i_1 i_2 \ldots i_\kappa} = x_{i_1} \cdot x_{i_2} \cdots x_{i_\kappa}$ [Lat97], i.e., $F_{\mathcal{A}}(\mathbf{x}) = <\boldsymbol{A}, \boldsymbol{\mathcal{X}}>$. More insight offers the interpretation of (8.2.7) as:

$$F_{\mathcal{A}}(\mathbf{x}) = \kappa \sum_{e \in \mathbb{E}} x^{[e]}, \tag{8.2.8}$$

where $x^{[e]}$ is the monomial $\prod_{i_l \in e} x_{i_l}$. Indeed, there are $\kappa!$ active monomials in (8.2.7) weighted by $\frac{1}{(\kappa-1)!}$ giving rise to the factor κ appearing in (8.2.8).

Let $\overline{\boldsymbol{\mathcal{X}}}_{i_1}$ be the $(\kappa - 1)$ order N-dimensional hypermatrix with elements $x_{i_2 i_3 \ldots i_\kappa} = x_{i_2} \cdot x_{i_3} \cdots x_{i_\kappa}$. The inner product of hypermatrices $\boldsymbol{A}$ and $\overline{\boldsymbol{\mathcal{X}}}_{i_1}$ over the common indices

$i_2, i_3, \ldots, i_\kappa$ yields a vector in $\mathbb{R}^N$ whose i-th element is:

$$(< \mathcal{A}, \overline{\mathcal{X}}_{i_1} >)_i = \sum_{i_2=1}^{N} \sum_{i_3=1}^{N} \cdots \sum_{i_\kappa=1}^{N} a_{i\, i_2 \ldots i_\kappa}\, x_{i_2 i_3 \ldots i_\kappa}. \qquad (8.2.9)$$

A real number λ is called an *eigenvalue* of $\mathcal{A}$, if there is a non-zero vector $\mathbf{x} = (x_1, x_2, \ldots, x_N)^T \in \mathbb{R}^N$, which is called an *eigenvector*, such that:

$$(< \mathcal{A}, \overline{\mathcal{X}}_{i_1} >)_i = \lambda\, x_i^{\kappa-1}, \quad i = 1, 2, \ldots, N. \qquad (8.2.10)$$

The pair $(\lambda, \mathbf{x})$ satisfying (8.2.10) is called an H-eigenpair [PZ13]. The indices of the elements of $\mathbf{x}$ correspond to the vertices of the hypergraph under consideration. The left-hand side of (8.2.9) can be interpreted as:

$$(< \mathcal{A}, \overline{\mathcal{X}}_{i_1} >)_i = \sum_{e \in \overline{\mathbb{E}}_i} x^{[e]}, \qquad (8.2.11)$$

where $\overline{\mathbb{E}}_i = \{e \setminus \{i\} \mid i \in e \in \mathbb{E}\}$ is the set of hyperedges obtained by removing the vertex i from each hyperedge of $\mathcal{H}$ containing i.

Let $\lambda_{\max}$ denote the eigenvalue with largest modulus in (8.2.10). Let us also define the set:

$$\mathbb{S}_{\geq 0} = \left\{ \mathbf{x} \in \mathbb{R}^N \mid \sum_{i=1}^{N} x_i^k = 1 \text{ and } x_i \geq 0 \text{ for } i = 1, 2, \ldots, N \right\}. \qquad (8.2.12)$$

We call support of $\mathbf{x}$, supp($\mathbf{x}$), all the indices of the non-zero elements of $\mathbf{x}$. The following lemmata hold [CD12]:

Lemma 8.2.1 *If* $\mathbf{v} \in \mathbb{S}_{\geq 0}$ *maximizes* $F_{\mathcal{A}}(\mathbf{x})$ *on* $\mathbb{S}_{\geq 0}$, *then* supp($\mathbf{v}$) *induces some collection of connected components of* $\mathcal{H}$.

Lemma 8.2.2 *If* $\mathcal{H}$ *is a connected* κ-*uniform hypergraph, then it has a strictly positive eigenpair* $(\lambda, \mathbf{v})$, *where* λ *is the maximum value of* $F_{\mathcal{A}}(\mathbf{x})$ *admitted at* $\mathbf{x} = \mathbf{v}$ *on* $\mathbb{S}_{\geq 0}$.

Lemma 8.2.3 *Let* $\mathbf{z} = \frac{1}{N^{\frac{1}{\kappa}}} \mathbf{1} \in \mathbb{R}^N$, *where* $\mathbf{1}$ *is a vector of ones.* $F_{\mathcal{A}}(\mathbf{z})$ *equals the average vertex degree* $\overline{\delta}$.

Lemma 8.2.4 *If* $\mathcal{H}$ *is a connected* κ-*order* N-*dimensional uniform hypergraph, then the real eigenvalue* λ *given by Lemma 8.2.2 is the only eigenvalue with a strictly positive eigenvector. If* λ' *is any other eigenvalue of* $\mathcal{H}$, *then* $|\lambda'| \leq \lambda$.

Lemma 8.2.5 *For any non-empty* κ-*order* N-*dimensional uniform hypergraph,* $\lambda_{\max}$ *can be chosen to be a positive real number. If* $\mathcal{H}$ *is connected, then a corresponding eigenvector* $\mathbf{x}$ *can be chosen to be strictly positive.*

Lemma 8.2.6 *If* $\mathcal{H}$ *is a* κ-*order* N-*dimensional uniform hypergraph,* $\lambda_{\max}$ *is bounded as follows:*

$$\overline{\delta} \leq \lambda_{\max} \leq \Delta, \qquad (8.2.13)$$

where Δ *is the maximum vertex degree in* $\mathcal{H}$.

A special case of hypergraphs are the δ-*regular* ones, which have vertices with the same degree, i.e., $\delta(i) = \delta$, $i = 1, 2, \ldots, N$. For δ-regular hypergraphs, $\lambda_{\max} = \delta$ [CQ14]. An example of a regular hypergraph is the *circulant* hypergraph. A κ-order N-dimensional hypergraph $\mathcal{H}$ is called circulant if $e = \{i_1, i_2, \ldots, i_\kappa\} \in \mathbb{E}$ implies that $\tilde{e} = \{i'_1, i'_2, \ldots, i'_\kappa\} \in \mathbb{E}$ for $i'_l = i_l + 1$, $l = 1, 2, \ldots, \kappa$ [CQ14].

Lemma 8.2.7 *If $\mathcal{G}$ is a subgraph of $\mathcal{H}$, then $\lambda_{max}(\mathcal{G}) \leq \lambda_{\max}(\mathcal{H})$.*

For a κ-order N-dimensional uniform hypergraph, the following hypermatrices are defined in $\mathbb{R}^{\overbrace{N \times N \times \ldots \times N}^{\kappa}}$ [CQ14]:

degree hypermatrix denoted by $\mathcal{D}$, i.e., a diagonal hypermatrix having elements

$$d_{i_1 i_2 \ldots i_\kappa} = \begin{cases} \delta(l) & \text{if } i_1 = i_2 = \cdots = i_\kappa = l, \, l = 1, 2, \ldots N \\ 0 & \text{otherwise.} \end{cases} \tag{8.2.14}$$

Laplacian hypermatrix $\mathcal{L} = \mathcal{D} - \mathcal{A}$,

unsigned Laplacian hypermatrix $\mathcal{Q} = \mathcal{D} + \mathcal{A}$.

Like the adjacency hypermatrix, the Laplacian and unsigned Laplacian hypermatrices are super-symmetric. In addition, the adjacency and unsigned Laplacian hypermatrices are nonnegative. If κ is even, the Laplacian and the unsigned Laplacian hypermatrices are positive semi-definite. All the aforementioned hypermatrices possess H-eigenvalues. The smallest H-eigenvalue of the Laplacian hypermatrix is zero with corresponding eigenvector $\mathbf{1} \in \mathbb{R}^N$. Furthermore, if the uniform hypergraph is δ-regular, then the largest H-eigenvalue of the Laplacian and the unsigned Laplacian hypermatrices are δ and 2δ, respectively. For a circulant hypergraph, all the aforementioned hypermatrices are super-symmetric circulant hypermatrices.

Let $||\mathbf{x}||_2$ be the ℓ_2 norm of vector $\mathbf{x}$, $\mathbb{R}_+^N = \{\mathbf{x} = (x_1, x_2, \ldots, x_N)^T \in \mathbb{R}^N \mid x_i \geq 0, \, i = 1, 2, \ldots, N\}$ be the set of nonnegative N-dimensional vectors, and $\mathbb{S}^{N-1}$ be the standard unit sphere in $\mathbb{R}^N$. The maximization of the homogenous polynomial $F_{\mathcal{A}}(\mathbf{x})$ subject to the constraint $||\mathbf{x}||_2^2 = 1$ yields a *Z-eigenpair* $(\lambda, \mathbf{v})$, where λ is real and positive, known as the spectral radius of $\mathcal{A}$, $\varrho(\mathcal{A})$, satisfying [PZ13]:

$$\lambda \overset{\triangle}{=} \varrho(\mathcal{A}) = \max_{\mathbf{x} \in \mathbb{S}^{N-1} \cap \mathbb{R}_+^N} F_{\mathcal{A}}(\mathbf{x}) \tag{8.2.15}$$

and $\mathbf{v} = \underset{\mathbf{x} \in \mathbb{S}^{N-1} \cap \mathbb{R}_+^N}{\arg\max} F_{\mathcal{A}}(\mathbf{x})$. It can be shown that [PZ13]:

$$\frac{1}{N^{\frac{\kappa}{2}}} \sum_{i=1}^{N} \delta(i) \leq \lambda \leq \min\left(\Delta\sqrt{N}, \kappa M\right), \tag{8.2.16}$$

where Δ is the maximum vertex degree in $\mathcal{H}$ and $M = |\mathbb{E}| \leq \binom{N}{\kappa}$ is the cardinality of the set of hyperedges.

8.3 Game-Theoretic approaches to uniform hypergraph clustering

An interesting perspective to the hypergraph clustering (cf. Section 8.4) was proposed in [BP13], extending the prior work to pairwise clustering [PP07]. Instead of partitioning the input objects (i.e., hypergraph vertices), and hence obtaining the clusters, a rigorous notion of a cluster was proposed. In particular, a cluster is defined as a maximally coherent subset of vertices, $\mathbb{V}' \subset \mathbb{V}$, which satisfies both an *internal* criterion (i.e., all vertices in

$\mathbb{V}'$ are highly similar to each other) and an *external* one (i.e., all elements outside $\mathbb{V}'$ are highly dissimilar to those within $\mathbb{V}'$). Such formal definition of cluster was given in terms of the classical *equilibrium* concept from evolutionary game theory. It has been shown that there exists a one-to-one correspondence between the equilibria and the local solutions of a linearly constrained polynomial optimization problem [BP13], following similar lines to [PP07]. This implies that a powerful class of dynamical systems can be exploited to extract the clusters. The just mentioned dynamical systems are based on the well-known Baum-Eagon inequality [BE67], which generalizes the pairwise replicator dynamics [HS98] from evolutionary game theory to higher-order interactions. In addition, there is no need to set a priori the number of clusters.

A κ-order N-dimensional weighted uniform hypergraph $\mathcal{H} = (\mathbb{V}, \mathbb{E}, w)$ is cast into a κ-*player hypergraph clustering game* $\mathcal{G} = (\mathbb{P}, \mathbb{V}, b)$ with a *set of players* $\mathbb{P} = \{1, 2, \ldots, \kappa\}$, *set of pure strategies* available to each player equal to the hypergraph vertex set $\mathbb{V} = \{v_1, v_2, \ldots, v_N\}$, and *payoff function* $b(e)$, which assigns a utility to each hyperedge $e = \{v_1, v_2, \ldots, v_\kappa\} \in \mathbb{E}$ that is proportional to the similarity of the vertices e is incident to, i.e., [BP13]:

$$b(e) = \begin{cases} \frac{1}{\kappa!} w(e) & \text{if } e = \{v_1, v_2, \ldots, v_\kappa\} \in \mathbb{E} \\ 0 & \text{otherwise.} \end{cases} \tag{8.3.1}$$

It is seen that the hyperedges play the role of *strategy profiles*, which are ordered sets of pure strategies (i.e., vertices) played by the different players. In addition, it is assumed that all players share the same set of pure strategies and payoff function. In contrast to classical game theory, the players are not supposed to behave rationally or to have complete knowledge of the game details. The game is played with an evolutionary setting, where κ players are drawn at random from a large population and each player is assumed to play a pre-assigned strategy. Let $\mathbb{D} = \left\{ \mathbf{x} \in \mathbb{R}^N \mid x_i \geq 0, \sum_{i=1}^{N} x_i = 1 \right\}$ be the *standard simplex*. Given a population $\mathbf{x} \in \mathbb{D}$, x_i represents the fraction of players that is programmed to select $v_i \in \mathbb{V}$ from the vertices to be clustered.

If $\mathbf{x}^{[\kappa]} = \underbrace{(\mathbf{x}, \mathbf{x}, \ldots, \mathbf{x})}_{\kappa}$ and $\boldsymbol{\epsilon}_i \in \mathbb{R}^N$ is the vector whose i-th element equals 1 and all other elements equal 0, the following functions are defined [BP13]:

$$u(\mathbf{x}^{[\kappa]}) = \sum_{\{v_1, v_2, \ldots, v_\kappa\} \in \mathbb{V}^\kappa} b(v_1, v_2, \ldots, v_\kappa) \, x^{[e]} \tag{8.3.2}$$

$$u(\mathbf{x}^{[\kappa-1]}, \boldsymbol{\epsilon}_i) = \sum_{\{v_1, v_2, \ldots, v_{\kappa-1}\} \in \mathbb{V}^{\kappa-1}} b(v_1, v_2, \ldots, v_{\kappa-1}, i) \, x^{[\overline{e}_\kappa]},$$
$$\forall v_i \in \mathbb{V} \tag{8.3.3}$$

and $x^{[e]}$ is the monomial $\prod_{j=1}^{\kappa} x_{v_j}$ and $\overline{e}_\kappa = \{v_1, v_2, \ldots, v_{\kappa-1}\}$. The *expected payoff* earned by a player selecting $v_i \in \mathbb{V}$ in a population $\mathbf{x} \in \mathbb{D}$ is given by (8.3.3), which measures the average similarity of v_i with respect to the cluster. The expected payoff over the entire population is given by (8.3.3), because

$$u(\mathbf{x}^{[\kappa]}) = \sum_{v_i \in \mathbb{V}} u(\mathbf{x}^{[\kappa-1]}, \boldsymbol{\epsilon}_i) \, x_{v_i}. \tag{8.3.4}$$

The expected payoff (8.3.3) can be treated as a measure of cluster internal coherence, capturing the average similarity of the vertices forming the cluster.

Intuitively, an evolutionary process reaches a *Nash equilibrium* $\mathbf{x} \in \mathbb{D}$, when every individual in the population obtains the same expected payoff and no strategy can prevail upon the other ones, i.e.:

$$u(\mathbf{x}^{[\kappa-1]}, \boldsymbol{\epsilon}_i) \leq u(\mathbf{x}^{[\kappa]}) \quad \forall v_i \in \mathbb{V}. \tag{8.3.5}$$

A strategy $\mathbf{x} \in \mathbb{D}$ is an *evolutionary stable strategy* (ESS) if and only if for all $\mathbf{y} \in \mathbb{D} \setminus \{\mathbf{x}\}$, $\exists v_i \in \bar{e}_\kappa$ such that both conditions:

$$u\left((\mathbf{y} - \mathbf{x})^{[i+1]}, \mathbf{x}^{[\kappa-1-i]}\right) < 0 \tag{8.3.6}$$

$$u\left((\mathbf{y} - \mathbf{x})^{[l+1]}, \mathbf{x}^{[\kappa-1-l]}\right) = 0, \quad l = 0, 1, \ldots, i-1, \tag{8.3.7}$$

are satisfied [BP13]. The indices of the nonzero elements of $\mathbf{x}$ correspond to the indices of the vertices forming a cluster. That is, the support of $\mathbf{x}$ provides a measure of the degree of membership of its elements.

An ESS of the hypergraph clustering game defines an ESS-cluster in $\mathcal{H}$. If $\mathbf{x} \in \mathbb{D}$ is an ESS-cluster of $\mathcal{H}$ with support supp$(\mathbf{x})$, then [BP13]

$$u(\mathbf{x}^{[\kappa-1]}, \boldsymbol{\epsilon}_i) = u(\mathbf{x}^{[\kappa]}) \quad \forall i \in \text{supp}(\mathbf{x}). \tag{8.3.8}$$

The resulting ESS cluster is $\mathbb{V}' = \{v_i \mid i \in \text{supp}(\mathbf{x})\}$. Moreover, $\mathbb{V}'$ is a *two-cover* of $\mathcal{H}$. That is, for any pair of vertices $\{v_j, v_l\} \in \mathbb{V}'$, there exists an edge $e \in \mathbb{E}$ such that $\{v_j, v_l\} \subseteq e \subset \mathbb{V}'$ [BP13].

The ESS-clusters also satisfy a property of *external coherence*. That is:

$$u(\mathbf{x}^{[\kappa-1]}, \boldsymbol{\epsilon}_i) \leq u(\mathbf{x}^{[\kappa]}) \quad \forall i \notin \text{supp}(\mathbf{x}). \tag{8.3.9}$$

Equation (8.3.9) is a consequence of the fact that $\mathbf{x} \in \mathbb{D}$ is a Nash equilibrium with support supp$(\mathbf{x})$. Whenever one deviates from an ESS-cluster $\mathbf{x} \in \mathbb{D}$, e.g., by adding an external element to its support, the cluster similarity drops, provided that the deviation is not too large.

Bulo and Pelillo [BP13] proved:

Theorem 8.3.1 *Let $\mathcal{H} = (\mathbb{V}, \mathbb{E}, w)$ be a hypergraph clustering problem, $\mathcal{G} = (\mathbb{P}, \mathbb{V}, b)$ be the corresponding clustering game, and $f(\mathbf{x})$ be defined as:*

$$f(\mathbf{x}) = u(\mathbf{x}^{[\kappa]}) = \sum_{e \in \mathbb{E}} b(e) \, x^{[e]}. \tag{8.3.10}$$

Nash equilibria of $\mathcal{G}$ have one-to-one correspondence with the critical points of the maximization of $f(\mathbf{x})$ over $\mathbb{D}$, i.e., they satisfy the first-order necessary Karush-Kuhn-Tucker(KKT) conditions [Kuh76] of this optimization problem. Moreover, the ESS-clusters of $\mathcal{H}$ have one-to-one correspondence with the strict local maximizers of $f(\mathbf{x})$ over $\mathbb{D}$.

The extraction of the ESS-clusters can be cast into finding strict local maximizers of (8.3.10) in $\mathbb{D}$. The function $f(\mathbf{x})$ in (8.3.10) is a homogeneous polynomial in the variables x_i with nonnegative coefficients. Accordingly, it is a special case of the Baum-Eagon theorem [BE67]. In particular, we obtain:

$$\frac{\partial f(\mathbf{x})}{\partial x_i} = \frac{1}{\kappa} u(\mathbf{x}^{[\kappa-1]}, \boldsymbol{\epsilon}_i), \quad i = 1, 2, \ldots, N, \tag{8.3.11}$$

which implies

$$\sum_{l=1}^{N} x_l \frac{\partial f(\mathbf{x})}{\partial x_l} = \frac{1}{k} u(\mathbf{x}^{[\kappa]}). \tag{8.3.12}$$

Accordingly, the *growth transformation* that extracts an ESS-cluster takes the form:

$$x_i(t+1) = \frac{x_i(t) \frac{\partial f(\mathbf{x})}{\partial x_i} |_{x_i(t)}}{\sum_{l=1}^{N} x_l(t) \frac{\partial f(\mathbf{x})}{\partial x_l} |_{x_i(t)}} = x_i(t) \frac{u(\mathbf{x}^{[\kappa-1]}(t), \boldsymbol{\epsilon}_i)}{u(\mathbf{x}^{[\kappa]}(t))}, \quad i = 1, 2, \ldots, N. \tag{8.3.13}$$

During the evolution of (8.3.13), better-than-average strategies, i.e., strategies satisfying $u(\mathbf{x}^{[\kappa-1]}, \boldsymbol{\epsilon}_i) > u(\mathbf{x}^{[\kappa]})$, will spread in the population, while the remaining ones will be eliminated, establishing a Darwinian selection process. Accordingly, (8.3.13) generalizes the classical formulation of natural selection process in a two-player evolutionary game theory, known as *replicator dynamics* [HS98].

The population dynamics are initialized with the barycenter of the simplex $\mathbb{D}$, i.e., $\mathbf{x}(0) = \frac{1}{N}\mathbf{1}$, where $\mathbf{1} \in \mathbb{R}^N$ is a vector of ones, which defines a uniform distribution over the set of vertices in $\mathbb{V}$. By doing so, no particular vertex is favored. The growth transformation (8.3.13) satisfies the invariant property $\mathrm{supp}(\mathbf{x}(t)) \subset \mathrm{supp}(\mathbf{x}(0))$ for $t > 0$. If the numerator in (8.3.13) is positive for all $i \in \mathrm{supp}(\mathbf{x}(0))$, then $\mathrm{supp}(\mathbf{x}(t)) = \mathrm{supp}(\mathbf{x}(0))$ for all finite $t > 0$. That is, the limit point of the trajectory $\mathbf{x}_{\mathrm{opt}} = \lim_{t\to\infty} \mathbf{x}(t)$ only asymptotically satisfies $\mathrm{supp}(\mathbf{x}_{\mathrm{opt}}) \subset \mathrm{supp}(\mathbf{x}(0))$, suggesting the need for thresholding the elements of $\mathbf{x}$ in order to obtain the support of the corresponding ESS-cluster. Having found an ESS-cluster $\mathbb{V}'$ with dynamics (8.3.13), one removes its vertices from $\mathbb{V}$ and repeats the just described procedure for the remaining vertices. That is, one cluster is extracted at a time.

A unified method for clustering from κ-ary affinity relations can be found in [LLY10] that is motivated by the intuitive observation that there may exist $\binom{m}{\kappa}$ possible κ-ary affinity relations for a cluster of m objects capturing the internal coherence conditions. Let $\mathcal{B}$ be the weighted adjacency hypermatrix having elements:

$$b(e) = \begin{cases} w(e) & \text{if } e = \{v_1, v_2, \ldots, v_\kappa\} \in \mathbb{E} \\ 0 & \text{otherwise.} \end{cases} \qquad (8.3.14)$$

For each edge e, there are $\kappa!$ duplicate entries in $\mathcal{B}$. For a subset of N' vertices, $\mathbb{V}' \subseteq \mathbb{V}$, its hyperedge set is denoted as $\mathbb{E}'$. If $\mathbb{V}'$ is a cluster, then the majority of hyperedges in $\mathbb{E}'$ should have large weights. This property can be captured by the sum of all entries in $\mathcal{B}$ that are associated with hyperedges containing only vertices in $\mathbb{V}'$ [LLY10]:

$$S(\mathbb{V}') = \sum_{\{v_1,v_2,\ldots,v_\kappa\}\in\mathbb{V}'} b(v_1, v_2, \ldots, v_\kappa) = \sum_{\{v_1,v_2,\ldots,v_\kappa\}\in\mathbb{V}} b(v_1, v_2, \ldots, v_\kappa)\, y^{[e]}, \qquad (8.3.15)$$

where $y^{[e]}$ is the monomial $\prod_{i=1}^{\kappa} y_{v_i}$ with $y_{v_i} = 1$ if $v_i \in \mathbb{V}'$ and zero otherwise. Since the cardinality of $\mathbb{V}'$ is N', $(N')^\kappa$ terms are added in (8.3.15) yielding an average

$$S_{\mathrm{av}}(\mathbb{V}') = \frac{1}{(N')^\kappa} S(\mathbb{V}') = \sum_{\{v_1,v_2,\ldots,v_\kappa\}\in\mathbb{V}} b(v_1, v_2, \ldots, v_\kappa)\, x^{[e]}, \qquad (8.3.16)$$

where $x^{[e]}$ is the monomial $\prod_{i=1}^{\kappa} x_{v_i}$ with $x_{v_i} = \frac{y_{v_i}}{N'}$ if $v_i \in \mathbb{V}'$ and zero otherwise. That is, the ℓ_1 norm of $\mathbf{x} \in \mathbb{R}^N$ is 1. Accordingly, the clustering problem can be cast as maximization of $S_{\mathrm{av}}(\mathbb{V}')$ [LLY10]. However, neither N' is known nor are known the N' objects to be chosen. To relax this NP-hard problem, the i-th element of $\mathbf{x}$, x_i is allowed to vary in the continuous range $[0, \varepsilon]$, where the constant $\varepsilon \leq 1$:

$$\max f(\mathbf{x}) = \sum_{\{v_1,v_2,\ldots,v_\kappa\}\in\mathbb{V}} b(v_1, v_2, \ldots, v_\kappa)\, x^{[e]}$$

$$\text{s.t. } \mathbf{x} \in \mathbb{D}_\varepsilon = \left\{ \mathbf{x} \in \mathbb{R}^N \mid 0 \leq x_i \leq \varepsilon,\ \textstyle\sum_{i=1}^{N} x_i = 1 \right\}, \qquad (8.3.17)$$

where the acronym s.t. stands for "subject to." The constraint in (8.3.17) enables an intuitive interpretation of x_i as the probability for a cluster to contain the i-th object. Moreover, such a constraint sparsifies the solution. It is seen that the maximization of (8.3.10) in the standard simplex is a special case of (8.3.17) for $\varepsilon = 1$. Setting $\varepsilon < 1$ implies that the probability of choosing each vertex in the game has a known upper bound, which is the prior, while $\varepsilon = 1$ represents a noninformative prior [LLY10]. In fact, ε offers a tool to control the least number of objects in a cluster. The problem (8.3.17) has many local maxima. The large ones correspond to true clusters, while the small ones usually form meaningless subsets. Let us elaborate the constrained optimization problem (8.3.17) by inserting Lagrange multipliers ξ for the equality constraint $\sum_{i=1}^{N} x_i = 1$, $\beta_i > 0$ for the inequality constraints $x_i \geq 0$, and γ_i for the inequality constraints $x_i \leq \varepsilon$, $i = 1, 2, \ldots, N$. The resulting Lagrangian function is:

$$\mathcal{L}(\mathbf{x}, \xi, \boldsymbol{\beta}, \boldsymbol{\gamma}) = f(\mathbf{x}) - \xi\left(\mathbf{1}^T\mathbf{x} - 1\right) + \sum_{i=1}^{N} \beta_i\, x_i + \sum_{i=1}^{N} \gamma_i\,(\varepsilon - x_i). \tag{8.3.18}$$

Let $\bar{e}_\kappa = \{v_1, v_2, \ldots, v_{\kappa-1}\}$. The *reward* at vertex i is defined as [LLY10]:

$$r_i(\mathbf{x}) = \sum_{\bar{e}_\kappa \in \mathbb{V}} b(\bar{e}_\kappa, i) x^{[\bar{e}_\kappa]}. \tag{8.3.19}$$

It can be easily proved that the gradient of $f(\mathbf{x})$ with respect to x_i is proportional to $r_i(x)$, i.e.:

$$\frac{\partial f(\mathbf{x})}{\partial x_i} = \kappa\, r_i(\mathbf{x}). \tag{8.3.20}$$

Any local maximizer $\mathbf{x}_{\mathrm{opt}}$ satisfies the KKT conditions. That is:

$$
\begin{aligned}
\kappa\, r_i(\mathbf{x}_{\mathrm{opt}}) - \xi + \beta_i - \gamma_i &= 0, \; i = 1, 2, \ldots, N \\
\sum_{i=1}^{N} x_{\mathrm{opt},i}\, \beta_i &= 0 \\
\sum_{i=1}^{N} (\varepsilon - x_{\mathrm{opt},i})\, \gamma_i &= 0.
\end{aligned}
\tag{8.3.21}
$$

Since $x_{opt,i}$, β_i, and γ_i are all nonnegative for $i = 1, 2, \ldots, N$, (8.3.21) can be rewritten as [LLY10]:

$$
r_i(\mathbf{x}_{\mathrm{opt}})
\begin{cases}
\leq \frac{\xi}{\kappa} & x_{\mathrm{opt},i} = 0 \\
= \frac{\xi}{\kappa} & 0 < x_{\mathrm{opt},i} < \varepsilon \\
\geq \frac{\xi}{\kappa} & x_{\mathrm{opt},i} = \varepsilon.
\end{cases}
\tag{8.3.22}
$$

Indeed, if $x_{\mathrm{opt},i} > 0$, then $\beta_i = 0$ and if $x_{\mathrm{opt},i} < \varepsilon$, then $\gamma_i = 0$, giving rise to equation in the second line of (8.3.22). If $x_{\mathrm{opt},i} = \varepsilon$, $\gamma_i > 0$ and $\beta_i = 0$. Solving the equation in the first line of (8.3.21) for γ_i and replacing into $\gamma_i \geq 0$ yields the inequality in the third line of (8.3.22). Similarly if $x_{\mathrm{opt},i} = 0$, then $\beta_i \geq 0$ and $\gamma_i = 0$. Solving the equation in the first line of (8.3.21) for β_i and replacing into $\beta_i \geq 0$ yields the inequality in the first line of (8.3.22).

It is seen that the vertices in $\mathbb{V}$ are divided into three disjoint subsets, namely $\mathbb{V}_1(\mathbf{x}) = \{v_i \mid x_i = 0\}$, $\mathbb{V}_2(\mathbf{x}) = \{v_i \mid x_i \in (0, \varepsilon)\}$, and $\mathbb{V}_3(\mathbf{x}) = \{v_i \mid x_i = \varepsilon\}$. Let $\mathbb{V}_d(\mathbf{x}) = \mathbb{V}_2(\mathbf{x}) \cup \mathbb{V}_3(\mathbf{x})$ be the set of non-zero components in $\mathbf{x}$, while $\mathbb{V}_u(\mathbf{x}) = \mathbb{V}_1(\mathbf{x}) \cup \mathbb{V}_2(\mathbf{x})$ be the set of components in $\mathbf{x}$ that are smaller than ε. For any $\mathbf{x}$, if $f(\mathbf{x})$ must increase, then the values from some components in $\mathbb{V}_d(\mathbf{x})$ must decrease and the values of some components in $\mathbb{V}_u(\mathbf{x})$ must increase [LLY10]. If $\mathbf{x}$ is a maximizer of (8.3.17), then $r_i(\mathbf{x}) \leq r_j(\mathbf{x})$, $\forall v_i \in \mathbb{V}_u(\mathbf{x})$ and $\forall v_j \in \mathbb{V}_d(\mathbf{x})$. If $\mathbf{x}$ is not a maximizer, then we should increase x_i and decrease x_j. This is achieved as follows [LLY10]. If $\bar{e}_{\kappa,\kappa-1} = \{v_1, v_2, \ldots, v_{\kappa-2}\}$, define

$$r_{ij}(\mathbf{x}) = \sum_{\bar{e}_{\kappa,\kappa-1}} b(v_1, v_2, \ldots, v_{\kappa-2}, i, j)\, x^{[\bar{e}_{\kappa,\kappa-1}]} \tag{8.3.23}$$

$$\zeta = \begin{cases} \min(x_j, \varepsilon - x_i) & \text{if } r_{ij}(\mathbf{x}) \leq 0 \\ \min(x_j, \varepsilon - x_i, \frac{r_i(\mathbf{x}) - r_j(\mathbf{x})}{2(\kappa-1)\, r_{ij}(\mathbf{x})}) & \text{otherwise} \end{cases} \tag{8.3.24}$$

and update $\mathbf{x}$ according to

$$x_l(t+1) = \begin{cases} x_l(t) + \zeta & \text{if } l = i \\ x_l(t) - \zeta & \text{if } l = j \\ x_l(t) & \text{otherwise.} \end{cases} \tag{8.3.25}$$

Multiple initializations (i.e., priors) are needed to obtain the significant maxima of (8.3.17), which correspond to true clusters in general. Informative priors can be efficiently constructed from the vertices in $\mathbb{E}(v)$, which are connected with vertex v through a hyperedge, by means of Algorithm 8.3.1 [LLY10]. In Algorithm 8.3.1, $[\frac{1}{\varepsilon}]$ represents the smallest

---

Algorithm 8.3.1: Construction of a prior $\mathbf{x}(0)$ containing vertex v.

Input: Hyperedge set $\mathbb{E}(v)$ and ε.
Output: A prior $\mathbf{x}(0)$.

1: Sort the hyperedges $e \in \mathbb{E}(v)$ in descending order according to their weight $w(e)$
2: **for** $e \in |\mathbb{E}(v)|$ **do**
3:     Add all vertices associated with the hyperedge e to $\mathbb{L}$
4:     **if** $|\mathbb{L}| \geq [\frac{1}{\varepsilon}]$ **then**
5:         **break**
6:     **end if**
7: **end for**
8: **for** each vertex $v \in \mathbb{L}$ **do**
9:     Set the corresponding component $x_v(0) = \frac{1}{|\mathbb{L}|}$
10: **end for**

---

integer that is larger than or equal to $\frac{1}{\varepsilon}$. Having an informative prior $\mathbf{x}(0)$, a local maximizer of (8.3.17) can be obtained thanks to Algorithm 8.3.2 [LLY10].

As mentioned previously, the hyperedge degree $\delta(e) = \kappa$ is frequently chosen as model order ι plus 1, i.e, the smallest possible. Even in this case, the number of possible hyperedges in a κ-uniform N dimensional hypergraph, $\binom{N}{\kappa}$, is too large for exhaustive listing. Most of the existing techniques *sample* the set of hyperedges to construct sparse hypergraphs. In addition, by limiting κ to the smallest number, the chance of identifying pure hyperedges, i.e., those containing objects that are likely from the same cluster, is maximized [PCAS14]. Accordingly, computational feasibility is the underlying justification for imposing the aforementioned constraint in the hyperedge degree.

Purkait et al. investigated whether there was any benefit to employing hyperedges of higher degree, i.e., $\kappa > \iota + 1$ for hypergraph clustering and whether one could sample large hyperedges without significant effort [PCAS14]. Sampling strategies based on random cluster models were proposed as a solution.

Shashua et al. [SZH06] cast the clustering problem with high-order relations into a nonnegative factorization problem of the closest hyper-stochastic version of the input affinity hypermatrix, extending the work in [Gov05], where a pairwise similarity matrix was derived

---

Algorithm 8.3.2: Computation of a local maximizer $\mathbf{x}_{\text{opt}}$ given an informative prior $\mathbf{x}(0)$.

Input: Weighted adjacency hypermatrix $\mathcal{B}$ and prior $\mathbf{x}(0)$.
Output: Local maximizer $\mathbf{x}_{\text{opt}}$.

1: $t = 0$
2: **repeat**
3:     Compute the reward $r_i(\mathbf{x}(t))$ for each vertex v_i using (8.3.19)
4:     Compute $\mathbb{V}_1(\mathbf{x}(t)) = \{v_i \mid x_i(t) = 0\}$, $\mathbb{V}_2(\mathbf{x}(t)) = \{v_i \mid x_i(t) \in (0, \varepsilon)\}$, $\mathbb{V}_3(\mathbf{x}(t)) = \{v_i \mid x_i(t) = \varepsilon\}$, $\mathbb{V}_d(\mathbf{x}(t)) = \mathbb{V}_2(\mathbf{x}(t)) \cup \mathbb{V}_3(\mathbf{x}(t))$, and $\mathbb{V}_u(\mathbf{x}(t)) = \mathbb{V}_1(\mathbf{x}(t)) \cup \mathbb{V}_2(\mathbf{x}(t))$
5:     Find the vertex $v_i \in \mathbb{V}_u(\mathbf{x}(t))$ with the largest reward and the vertex $v_j \in \mathbb{V}_d(\mathbf{x}(t))$ with the smallest reward.
6:     Compute ζ using (8.3.24)
7:     Compute $\mathbf{x}(t+1)$ using (8.3.25)
8:     Set $t \leftarrow t + 1$
9: **until** $\mathbf{x}$ is a local maximizer

---

from the factorization of the input affinity hypermatrix and was given as input to standard graph spectral clustering techniques.

8.4 Spectral clustering for arbitrary hypergraphs

Given a similarity measure, clustering aims at organizing a set of objects into groups (i.e., the *clusters*) so that the objects grouped together are similar and, at the same time, dissimilar to other objects assigned to different clusters. Object similarities are typically expressed as pairwise relations. That is, the similarity relation between the objects forms a weighted graph $\mathcal{G}(\mathbb{V}, \mathbb{E}, w)$. Various unsupervised and semi-supervised machine learning techniques have been formulated as operations on this graph. For example, in spectral clustering, the relation between the structural and the spectral properties have been analyzed by matrix theoretic methods that are graph theoretic as well. Indeed, the *Laplacian* of the graph is such a matrix that is used to study the structure of the graph [Chu97]. The *unnormalized Laplacian* (also known as combinatorial Laplacian) is defined as [Chu97]:

$$\mathbf{L} = \mathbf{D}_v - \mathbf{A}, \tag{8.4.1}$$

where $\mathbf{D}_v$ is the diagonal matrix consisting of vertex degrees and $\mathbf{A} \in \mathbb{R}^{|\mathbb{V}| \times |\mathbb{V}|}$ is the adjacency matrix with entry (u, v) equal to the weight of the edge (u, v) if they are connected and 0 otherwise. The adjacency matrix can be expressed in terms of the incidence matrix $\mathbf{H} \in \mathbb{R}^{|\mathbb{V}| \times |\mathbb{E}|}$ and the diagonal matrix of weights $\mathbf{W}$ as:

$$\mathbf{A} = \frac{1}{2} \mathbf{H} \, \mathbf{W} \, \mathbf{H}^T. \tag{8.4.2}$$

Equation (8.4.1) can be rewritten as:

$$\mathbf{L} = \mathbf{D}_v^{\frac{1}{2}} \underbrace{\left(\mathbf{I} - \mathbf{D}_v^{-\frac{1}{2}} \mathbf{A} \, \mathbf{D}_v^{-\frac{1}{2}} \right)}_{\tilde{\mathbf{L}}} \mathbf{D}_v^{\frac{1}{2}}, \tag{8.4.3}$$

where $\tilde{\mathbf{L}}$ is the so-called *normalized Laplacian*. By substituting (8.4.2) into (8.4.1) and the definition of the normalized Laplacian in (8.4.3), we obtain:

$$\mathbf{L} = \frac{1}{2}\left(2\,\mathbf{D}_v - \mathbf{H}\,\mathbf{W}\,\mathbf{H}^T\right) \tag{8.4.4}$$

$$\tilde{\mathbf{L}} = \mathbf{I} - \frac{1}{2}\,\mathbf{D}_v^{-\frac{1}{2}}\,\mathbf{H}\,\mathbf{W}\,\mathbf{H}^T\,\mathbf{D}_v^{-\frac{1}{2}}. \tag{8.4.5}$$

The graph Laplacian is the discrete analog of the *Laplace-Beltrami operator* on compact Riemannian manifolds [Chu97, ABB06]. The generalized eigenvalue problem involving the graph Laplacian was used in the development of spectral clustering algorithms by relaxing the graph partitioning into a continuous optimization problem [AKY99, SM00, NJW01].

However, in several applications, such as face clustering [ALZM$^+$05], perceptual grouping [Gov05], parametric motion segmentation [SZH06], image retrieval [HLZM10], and image categorization [HLL$^+$11], higher-order relations are shown to be more appropriate. To cluster such high-order similarities, we may resort to *spectral hypergraph clustering*, where the vertices are the objects to be clustered and the weighted hyperedges encode the high-order similarities. In the following, we shall not assume that each hyperedge contains exactly κ vertices, as in Section 8.3. Such arbitrary hypergraphs include the undirected, weighted hypergraphs $\mathcal{H}(\mathbb{V}, \mathbb{E}, w)$, a superset containing the undirected, weighted graphs as a special case.

It is worth noting that spectral hypergraph clustering has a long history in Very Large Scale Integration (VLSI) [AK95, KAKS97, ZSC99, KK00]. There, vertices correspond to circuit elements and hyperedges correspond to wiring that may connect more than two elements. Finding the cuts with minimum cost allows one to partition the elements into modules with minimum interconnections. The leading algorithm in VLSI design is based on a two-phase multilevel approach [KK00]. In the first phase, a hierarchy of hypergraphs is constructed, where the hypergraph at each level is a coarser version of the hypergraph at the previous one, according to some measure of homogeneity. In the second phase, starting from a partition at the coarsest level, the algorithm proceeds downward in the hierarchy and each level updates the partition obtained at the previous one in a greedy fashion. Recently, an increasing interest in spectral hypergraph clustering has been observed in various domains, notably computer vision, multimedia information retrieval, machine learning, etc.

In this section, spectral hypergraph clustering is addressed by resorting to graph-based clustering methods. That is, to transform a hypergraph into a graph whose edge-weights are mapped from the weights of the original hypergraph. Two approaches were proposed, namely the *clique expansion* and the *star expansion* [ZSC99]. The clique expansion constructs a graph $\mathcal{G}_x(\mathbb{V}, \mathbb{E}_x, w_x)$ from the hypergraph $\mathcal{H}(\mathbb{V}, \mathbb{E}, w)$ by replacing each hyperedge $e = (v_1, v_2, \ldots, v_\kappa) \in \mathbb{E}$ with an edge for each pair of vertices in the hyperedge $\mathbb{E}_x = \{(v_i, v_j) \mid v_i \in e, v_j \in e, e \in \mathbb{E}\}$. The weight assigned to the resulting edge $w_x(v_i, v_j)$ should minimize the difference between itself and the weight of each hyperedge e incident to both v_i and v_j. Accordingly:

$$w_x(v_i, v_j) = \sum_{e\mathbb{E}\mid v_i \in e,\, v_j \in e} w(e). \tag{8.4.6}$$

The normalized Laplacian of $\mathcal{G}_x$ can be expressed as [ZSC99]:

$$\tilde{\mathbf{L}}_x = \mathbf{I} - \mathbf{C}, \tag{8.4.7}$$

where $\mathbf{C}$ has elements

$$C(i, j) = \begin{cases} 0 & \text{if } \nexists e \in \mathbb{E} \text{ such that } v_i \in e \text{ and } v_j \in e \\ \frac{w_x(v_i, v_j)}{\sqrt{\delta_x(v_i)}\sqrt{\delta_x(v_j)}} & \text{otherwise}, \end{cases} \tag{8.4.8}$$

with:

$$\delta_x(v_i) = \sum_{e \in \mathbb{E}} h(v_i, e) \ (\delta(e) - 1) \ w(e). \tag{8.4.9}$$

The star expansion algorithm constructs a graph $\mathcal{G}_*(\mathbb{V}_*, \mathbb{E}_*, w_*)$ from the hypergraph $\mathcal{H}(\mathbf{V}, \mathbf{E}, w)$ by introducing a new vertex for every hyperedge $e \in \mathbb{E}$. Thus, $\mathbb{V}_* = \mathbb{V} \cup \mathbb{E}$. The resulting edge set is $\mathbb{E}_* = \{(v, e) \mid v \in e, e \in \mathbb{E}\}$. Star expansion assigns the scaled hyperedge weight to each corresponding graph edge:

$$w_*(v, e) = \frac{w(e)}{\delta(e)}. \tag{8.4.10}$$

$\mathcal{G}_*$ is a bipartite graph with vertices corresponding to $\mathbb{E}$ on the one side and vertices corresponding to $\mathbb{V}$ on the other, since there are no edges from $\mathbb{V}$ to $\mathbb{V}$ or from $\mathbb{E}$ to $\mathbb{E}$ [ABB06]. Let us assume that the vertex set $\mathbb{V}_*$ has been ordered such that all elements of $\mathbb{V}$ appear on the top of the elements of $\mathbb{E}$. The adjacency matrix of $\mathcal{G}_*$ having size $(|\mathbb{V}| + |\mathbb{E}|) \times (|\mathbb{V}| + |\mathbb{E}|)$ can be written as [ABB06]:

$$\mathbf{A}_* = \begin{bmatrix} \mathbf{0}_{|\mathbb{V}| \times |\mathbb{V}|} & \mathbf{H} \, \mathbf{W}_* \\ \mathbf{W}_* \, \mathbf{H}^T & \mathbf{0}_{|\mathbb{E}| \times |\mathbb{E}|} \end{bmatrix}, \tag{8.4.11}$$

where $\mathbf{W}_*$ is the properly scaled hyperedge weight matrix. The degrees of the vertices in $\mathcal{G}_*$ are given by:

$$\delta_*(v) = \sum_{e \in \mathbb{E}} h(v, e) \, w_*(v, e), \quad v \in \mathbb{V} \tag{8.4.12}$$

$$\delta_*(e) = \sum_{v \in \mathbb{V}} h(v, e) \, w_*(v, e), \quad e \in \mathbb{E}. \tag{8.4.13}$$

Let $\mathbf{D}_{v*}$ be the $|\mathbb{V}| \times |\mathbb{V}|$ diagonal matrix with elements in the main diagonal $\delta_*(v)$. Similarly let $\mathbf{D}_{e*}$ be the $|\mathbb{E}| \times |\mathbb{E}|$ diagonal matrix with elements in the main diagonal $\delta_*(e)$. The normalized Laplacian of $\mathcal{G}_*$ is written in the form [ABB06]:

$$\begin{aligned}
\tilde{\mathbf{L}}_* &= \begin{bmatrix} \mathbf{I}_{|\mathbb{V}| \times |\mathbb{V}|} & \mathbf{0} \\ \mathbf{0} & \mathbf{I}_{|\mathbb{E}| \times |\mathbb{E}|} \end{bmatrix} - \begin{bmatrix} \mathbf{D}_{v*}^{-\frac{1}{2}} & \mathbf{0} \\ \mathbf{0} & \mathbf{D}_{e*}^{-\frac{1}{2}} \end{bmatrix} \mathbf{A}_* \begin{bmatrix} \mathbf{D}_{v*}^{-\frac{1}{2}} & \mathbf{0} \\ \mathbf{0} & \mathbf{D}_{e*}^{-\frac{1}{2}} \end{bmatrix} \\
&= \begin{bmatrix} \mathbf{I}_{|\mathbb{V}| \times |\mathbb{V}|} & -\mathbf{B} \\ -\mathbf{B}^T & \mathbf{I}_{|\mathbb{E}| \times |\mathbb{E}|} \end{bmatrix},
\end{aligned} \tag{8.4.14}$$

where $\mathbf{B} \in \mathbb{R}^{|\mathbb{V}| \times |\mathbb{E}|}$ is:

$$\mathbf{B} = \mathbf{D}_{v*}^{-\frac{1}{2}} \, \mathbf{H} \, \mathbf{W}_* \, \mathbf{D}_{e*}^{-\frac{1}{2}}. \tag{8.4.15}$$

Any eigenvector $\phi = \left(\phi_v^T \ \phi_e^T\right)^T$ of the normalized Laplacian $\tilde{\mathbf{L}}_*$ with corresponding eigenvalue λ satisfies the identity [ABB06]:

$$\mathbf{B} \, \mathbf{B}^T = (1 - \lambda)^2 \phi_v. \tag{8.4.16}$$

That is, the $|\mathbb{V}|$ elements of the eigenvectors of the normalized Laplacian, which correspond to the vertices of the hypergraph $\mathbb{V}$, are the eigenvectors of the $|\mathbb{V}| \times |\mathbb{V}|$ matrix $\mathbf{B} \, \mathbf{B}^T$. If (8.4.10) is substituted into (8.4.12) and (8.4.13), we obtain:

$$\delta_*(v) = \sum_{e \in \mathbb{E}} h(v, e) \, \frac{w(e)}{\delta(e)}, \quad v \in \mathbb{V} \tag{8.4.17}$$

$$\delta_*(e) = w(e), \quad e \in \mathbb{E}. \tag{8.4.18}$$

It is not difficult to show that the (i, j) element of $\mathbf{B}\,\mathbf{B}^T$ is given by:

$$[\mathbf{B}\,\mathbf{B}^T]_{ij} = \frac{1}{\sqrt{\delta_*(v_i)}\,\sqrt{\delta_*(v_j)}} \sum_{e \in \mathbb{E}} h(v_i, e)\, h(v_j, e)\, \frac{w(e)}{\delta^2(e)}. \qquad (8.4.19)$$

If we substitute:

$$w_*^c = w(e)\,(\delta(e) - 1) \qquad (8.4.20)$$

into (8.4.12) and (8.4.13), the vertex degree of the clique expansion $\mathcal{G}_x$ results [ABB06]:

$$\delta_*^c(v) \;=\; \delta_x(v) \qquad (8.4.21)$$
$$\delta_*^c(e) \;=\; w(e)\,\delta(e)\,(1 - \delta(e)). \qquad (8.4.22)$$

Accordingly, the equivalent bipartite graph matrix $\mathbf{B}_*^c\,(\mathbf{B}_*^c)^T$ for the clique expansion has elements:

$$[\mathbf{B}_*^c\,(\mathbf{B}_*^c)^T]_{i,j} = \sum_{e \in \mathbb{E}} \frac{h(v_i, e)\, h(v_j, e)\, w(e)}{\delta(e)\,(1 - \delta(e))}. \qquad (8.4.23)$$

If instead of (8.4.6), we choose:

$$w_x^c(v_i, v_j) = \frac{1}{\delta(e)\,(\delta(e) - 1)} \sum_{e \in \mathbb{E}} h(v_i, e)\, h(v_j, e)\, w(e), \qquad (8.4.24)$$

the vertex degree for star expansion $\mathcal{G}_*$ (8.4.17) emerges. The equivalent normalized graph Laplacian of $\mathcal{G}_x^c$, $\mathbf{C}_x^c$, for star expansion has elements [ABB06]:

$$[\mathbf{C}_x^c]_{ij} = \frac{1}{\sqrt{\delta_*(v_i)}\,\sqrt{\delta_*(v_j)}} \sum_{e \in \mathbb{E}} \frac{1}{\delta(e)\,(\delta(e) - 1)}\, h(v_i, e)\, h(v_j, e)\, w(e). \qquad (8.4.25)$$

For κ-uniform hypergraphs, all vertex degrees are equal to κ. In this case, the following pairs of matrices coincide up to a multiplicative constant:

- $\mathbf{C}$ with elements (8.4.8) and $\mathbf{B}_*^c\,(\mathbf{B}_*^c)^T$ with elements (8.4.23),

- $\mathbf{C}_x^c$ with elements (8.4.25) and $\mathbf{B}\,\mathbf{B}^T$ with elements (8.4.19).

Thus, the eigenvectors of the normalized Laplacian $\mathbf{B}_*^c\,(\mathbf{B}_*^c)^T$ for the bipartite graph $\mathcal{G}_*^c$ are exactly the eigenvectors of the normalized Laplacian $\mathbf{C}$ of the clique expansion graph $\mathcal{G}_x$. Similarly, the eigenvectors of the clique matrix $\mathbf{C}_x^c$ are exactly the eigenvectors of the normalized Laplacian for star expansion $\mathbf{B}\,\mathbf{B}^T$.

Bolla defined a Laplacian matrix for an unweighted hypergraph (i.e., $\mathbf{W} = \mathbf{I}$) in terms of the vertex degree matrix $\mathbf{D}_v$, the edge degree matrix $\mathbf{D}_e$, and the incidence matrix $\mathbf{H}$, $\mathbf{L}_{\text{Bolla}} = \mathbf{D}_v - \mathbf{H}\,\mathbf{D}_e^{-1}\,\mathbf{H}^T$, and established a link between the spectral properties of $\mathbf{L}_{\text{Bolla}}$ and the minimum cut of the hypergraph [Bol93]. It has been shown that $\mathbf{L}_{\text{Bolla}}$ is equal to the unnormalized Laplacian of the associated clique expansion with weight matrix $\mathbf{W}_{\text{Bolla}} = \mathbf{H}\,\mathbf{D}_e^{-1}\,\mathbf{H}^T$ [ABB06].

Rodriguez has disclosed the same results to those in [Bol93] by transforming the hypergraph into a graph using the clique expansion [Rod03]. In particular, a weighted graph $\mathcal{G}_{\text{Rodriguez}}(\mathbb{V}, \mathbb{E}_x, w_r)$ has been constructed from the unweighted hypergraph $\mathcal{H}(\mathbb{V}, \mathbb{E})$ by replacing each hyperedge by a clique. The weight of the edge is set to the number of edges containing both v_i and v_j, i.e., $w_{\text{Rodriguez}}(v_1, v_2) = |\{e \in \mathbb{E} \mid v_i \in e, \; v_j \in e\}|$. The graph Laplacian of $\mathcal{G}_{\text{Rodriguez}}$ was expressed as $\mathbf{L}_{\text{Rodriguez}} = \mathbf{D}_v^r - \mathbf{H}\,\mathbf{H}^T$, where $\mathbf{D}_v^r$ is the vertex degree of the graph $\mathcal{G}_{\text{Rodriguez}}$. It has been shown that there is a relationship between the spectral properties of $\mathbf{L}_{\text{Rodriguez}}$ and the minimum cut of the original hypergraph. Agarwal et al. proposed the *clique averaging* method and reported better results than the clique expansion method [ALZM$^+$05].

A hypergraph normalized cut criterion has been proposed in [ZHS06] that generalizes the well-known normalized cut method [SM00]. In particular, let $(\mathbb{S}, \overline{\mathbb{S}})$ be a partitioning of the vertices of $\mathbb{V}$ in $\mathcal{H}$ by a cut, where $\mathbb{S} \cup \overline{\mathbb{S}} = \mathbb{V}$. This can be achieved by removing the hyperedges in the *cut set* $\mathbb{E}_c(\mathbb{S}, \overline{\mathbb{S}}) = \{e \in \mathbb{E} \mid e \cap \mathbb{S} \neq \emptyset, \; e \cap \overline{\mathbb{S}} \neq \emptyset\}$, which yields the disjoint sets $\mathbb{S}$ and $\overline{\mathbb{S}}$. The cost of the cut was defined by the following volume [ZHS06]:

$$\text{vol}(\mathbb{S}, \overline{\mathbb{S}}) = \sum_{e \in \mathbb{E}_c(\mathbb{S}, \overline{\mathbb{S}})} w(e) \underbrace{\frac{|e \cap \mathbb{S}| \, |e \cap \overline{\mathbb{S}}|}{\delta(e)}}_{\zeta(e \mid \mathbb{S}, \overline{\mathbb{S}})} \tag{8.4.26}$$

that is interpreted as the sum of weights of the hyperedges, which are cut. The volume of cluster $\mathbb{S}$ was defined as the sum of the degrees of the vertices in $\mathbb{S}$:

$$\text{vol}(\mathbb{S}) = \sum_{v \in \mathbb{S}} \delta(v). \tag{8.4.27}$$

The definition in (8.4.26) results from treating each hyperedge e as a *clique* (i.e., a fully connected subgraph) and assigning the same weight $\frac{w(e)}{\delta(e)}$ to all subedges of the (virtual) subgraph. When a hyperedge e is cut, $|e \cap \mathbb{S}| \, |e \cap \overline{\mathbb{S}}|$ subedges should be cut. Using (8.4.26) and (8.4.27), the normalized cut criterion for partitioning $\mathbb{V}$ into $(\mathbb{S}, \overline{\mathbb{S}})$ was set as:

$$\text{ncut}(\mathbb{S}) = \text{vol}(\mathbb{S}, \overline{\mathbb{S}}) \left(\frac{1}{\text{vol}(\mathbb{S})} + \frac{1}{\text{vol}(\overline{\mathbb{S}})} \right) \tag{8.4.28}$$

and the optimal $\mathbb{S}$, $\mathbb{S}_{\text{opt}}$, was obtained as the solution of the optimization problem:

$$\mathbb{S}_{\text{opt}} = \underset{\emptyset \neq \mathbb{S} \subset \mathbb{V}}{\arg\min} \, \text{ncut}(\mathbb{S}), \tag{8.4.29}$$

which formally expresses the natural requirements of a partition, namely a dense connection among the vertices in the same cluster and a sparse connection between two clusters. For a simple graph, $|e \cap \mathbb{S}| = |e \cap \overline{\mathbb{S}}| = 1$ and $\delta(e) = 2$. Thus, the right-hand side of (8.4.26) reduces to the ordinary graph normalized cut [SM00] up to a factor $\frac{1}{2}$.

Another interpretation worth mentioning is associated with *random walks* [ZHS06] that generalizes the natural random walk on ordinary graphs. Let us associate each hypergraph with a natural random walk with the following transition rule: Given the current vertex $v \in \mathbb{V}$, first choose a hyperedge e among those incident with v with probability proportional to $w(e)$, and then choose a vertex $v' \in e$ uniformly at random. The resulting transition probability matrix $\mathbf{P}$ is given by:

$$\mathbf{P} = \mathbf{D}_v^{-1} \, \mathbf{H} \, \mathbf{W} \, \mathbf{D}_e^{-1} \, \mathbf{H}^T. \tag{8.4.30}$$

Its elements are $p(v, v') = \sum_{e \in \mathbb{E}} w(e) \frac{h(v,e)}{\delta(v)} \frac{h(v',e)}{\delta(e)}$. The stationary distribution $\boldsymbol{\pi}$ of the random walk satisfies the identity:

$$\boldsymbol{\pi}^T = \boldsymbol{\pi}^T \, \mathbf{P}. \tag{8.4.31}$$

It can be verified that $\pi(v) = \frac{\delta(v)}{\text{vol}(\mathbb{V})}$ or in matrix notation:

$$\boldsymbol{\pi}^T = \frac{1}{\text{vol}(\mathbb{V})} \, \boldsymbol{\delta}_v^T, \tag{8.4.32}$$

where $\boldsymbol{\delta}_v = \mathrm{vec}(\boldsymbol{D}_v) = \mathbf{H}\mathbf{W}\mathbf{1}_M$ with vec() being the operation that returns the elements in the main diagonal of the diagonal matrix inside parentheses as a column vector of compatible dimensions and $\mathbf{1}_M$ denoting the $M \times 1$ vector of ones. Indeed if (8.4.32) and (8.4.30) are substituted into (8.4.31), we obtain:

$$
\begin{aligned}
\boldsymbol{\pi}^T \mathbf{P} &= \frac{1}{\mathrm{vol}(\mathbb{V})} \underbrace{\boldsymbol{\delta}_v^T \mathbf{D}_v^{-1}}_{\mathbf{1}_N^T} \mathbf{H}\mathbf{W}\mathbf{D}_e^{-1}\mathbf{H}^T = \frac{1}{\mathrm{vol}(\mathbb{V})} \underbrace{\mathbf{1}_N^T \mathbf{H}}_{\boldsymbol{\delta}_e^T} \mathbf{W}\mathbf{D}_e^{-1}\mathbf{H}^T \\
&= \frac{1}{\mathrm{vol}(\mathbb{V})} \underbrace{\boldsymbol{\delta}_e^T \mathbf{D}_e^{-1}}_{\mathbf{1}_M^T} \mathbf{W}\mathbf{H}^T = \frac{1}{\mathrm{vol}(\mathbb{V})} \mathbf{1}_M^T \mathbf{W}\mathbf{H}^T \\
&= \frac{1}{\mathrm{vol}(\mathbb{V})} \boldsymbol{\delta}_v^T,
\end{aligned}
\tag{8.4.33}
$$

where we used that $\boldsymbol{\delta}_e^T = \mathrm{vec}(\boldsymbol{D}_e) = \mathbf{1}_N^T \mathbf{H}$ and the fact that the diagonal matrices $\mathbf{W}$ and $\mathbf{D}_e^{-1}$ commute in multiplication. The hypergraph normalized cut (8.4.28) can be rewritten as [ZHS06]:

$$
\mathrm{ncut}(\mathbb{S}) = \frac{\mathrm{vol}(\mathbb{S},\bar{\mathbb{S}})}{\mathrm{vol}(\mathbb{V})} \left(\frac{1}{\frac{\mathrm{vol}(\mathbb{S})}{\mathrm{vol}(\mathbb{V})}} + \frac{1}{\frac{\mathrm{vol}(\bar{\mathbb{S}})}{\mathrm{vol}(\mathbb{V})}} \right).
\tag{8.4.34}
$$

Moreover, it is seen that:

$$
\frac{\mathrm{vol}(\mathbb{S})}{\mathrm{vol}(\mathbb{V})} = \sum_{v \in \mathbb{S}} \frac{\delta(v)}{\mathrm{vol}(\mathbb{V})} = \sum_{v \in \mathbb{S}} \pi(v)
\tag{8.4.35}
$$

and similarly:

$$
\frac{\mathrm{vol}(\mathbb{S},\bar{\mathbb{S}})}{\mathrm{vol}(\mathbb{V})} = \sum_{v \in \mathbb{S}} \sum_{v' \in \bar{\mathbb{S}}} \pi(v)\, p(v,v').
\tag{8.4.36}
$$

That is, the ratio $\frac{\mathrm{vol}(\mathbb{S},\bar{\mathbb{S}})}{\mathrm{vol}(\mathbb{V})}$ is the probability with which one sees a jump of the random walk from $\mathbb{S}$ to $\bar{\mathbb{S}}$ under the stationary distribution [ZHS06]. Accordingly, the hypergraph normalized cut criterion can be interpreted as seeking a cut such that the probability with which the random walk crosses different clusters is as small as possible, while the probability with which the random walk stays in the same cluster is as large as possible. This interpretation is consistent with the random walk view of ordinary graphs [MS01].

The optimization problem (8.4.29) in NP-complete and it can be relaxed into a real-valued optimization problem [ZHS06]:

$$
\underset{\mathbf{f} \in \mathbb{R}^N}{\mathrm{argmin}} \frac{1}{2} \sum_{e \in \mathbb{E}} \sum_{\{v,v'\} \subseteq e} \frac{w(e)}{\delta(e)} \left(\frac{f(v)}{\sqrt{\delta(v)}} - \frac{f(v)}{\sqrt{\delta(v)}} \right)^2
$$
$$
\text{s.t. } \|\mathbf{f}\|_2^2 = 1 \text{ and } \sum_{v \in \mathbb{V}} f(v)\sqrt{\delta(v)} = 0.
\tag{8.4.37}
$$

One may coin the $N \times N$ matrix:

$$
\tilde{\mathbf{L}} = \mathbf{I} - \boldsymbol{\Theta}
\tag{8.4.38}
$$

as *hypergraph Laplacian*, where:

$$
\boldsymbol{\Theta} = \mathbf{D}_v^{-\frac{1}{2}} \mathbf{H}\mathbf{W}\mathbf{D}_e^{-1}\mathbf{H}^T \mathbf{D}_v^{-\frac{1}{2}}
\tag{8.4.39}
$$

and $\mathbf{I}$ is the identity matrix of compatible dimensions. Indeed, for an ordinary graph (i.e., $\mathbf{D}_e = 2\mathbf{I}$), we obtain:

$$
\begin{aligned}
\tilde{\mathbf{L}} &= \mathbf{I} - \frac{1}{2} \mathbf{D}_v^{-\frac{1}{2}} \underbrace{\mathbf{H}\,\mathbf{W}\,\mathbf{H}^T}_{\mathbf{D}_v + \mathbf{A}} \mathbf{D}_v^{-\frac{1}{2}} = \mathbf{I} - \frac{1}{2} \mathbf{D}_v^{-\frac{1}{2}} \left(\mathbf{D}_v + \mathbf{A} \right) \mathbf{D}_v^{-\frac{1}{2}} \\
&= \frac{1}{2} \left(\mathbf{I} - \mathbf{D}_v^{-\frac{1}{2}} \mathbf{A} \mathbf{D}_v^{-\frac{1}{2}} \right),
\end{aligned}
\tag{8.4.40}
$$

which is the ordinary graph normalized Laplacian up to a factor $\frac{1}{2}$. It is trivial to show that the double sum in (8.4.37) is simply the quadratic form $2\,\mathbf{f}^T\,\tilde{\mathbf{L}}\,\mathbf{f}$. For $w(e) \geq 0$, the quadratic form is nonnegative, accordingly, $\tilde{\mathbf{L}}$ is positive semi-definite. It can be checked that the smallest eigenvalue of $\tilde{\mathbf{L}}$ is zero and its corresponding eigenvector is $\sqrt{\boldsymbol{\delta}_v} = \mathrm{vec}(\mathbf{D}_v^{\frac{1}{2}})$, where the $\sqrt{}$ is retained in the left-hand side for the easy of linear algebra manipulations following:

$$
\begin{aligned}
\tilde{\mathbf{L}}\sqrt{\boldsymbol{\delta}_v} &= \sqrt{\boldsymbol{\delta}_v} - \boldsymbol{\Theta}\sqrt{\boldsymbol{\delta}_v} = \sqrt{\boldsymbol{\delta}_v} - \mathbf{D}_v^{-\frac{1}{2}}\mathbf{H}\,\mathbf{W}\,\mathbf{D}_e^{-1}\mathbf{H}^T \underbrace{\mathbf{D}_v^{-\frac{1}{2}}\sqrt{\boldsymbol{\delta}_v}}_{\mathbf{1}_N} \\
&= \sqrt{\boldsymbol{\delta}_v} - \mathbf{D}_v^{-\frac{1}{2}}\mathbf{H}\,\mathbf{W}\,\mathbf{D}_e^{-1}\underbrace{\mathbf{H}^T\mathbf{1}_N}_{\boldsymbol{\delta}_e} \\
&= \sqrt{\boldsymbol{\delta}_v} - \mathbf{D}_v^{-\frac{1}{2}}\mathbf{H}\,\mathbf{W}\underbrace{\mathbf{D}_e^{-1}\boldsymbol{\delta}_e}_{\mathbf{1}_M} \\
&= \sqrt{\boldsymbol{\delta}_v} - \mathbf{D}_v^{-\frac{1}{2}}\underbrace{\mathbf{H}\,\mathbf{W}\,\mathbf{1}_M}_{\boldsymbol{\delta}_v} = \sqrt{\boldsymbol{\delta}_v} - \underbrace{\mathbf{D}_v^{-\frac{1}{2}}\boldsymbol{\delta}_v}_{\sqrt{\boldsymbol{\delta}_v}} = 0.
\end{aligned}
\tag{8.4.41}
$$

The second constraint in (8.4.37) indicates that the solution should be orthogonal to the eigenvector $\sqrt{\boldsymbol{\delta}_v}$. Accordingly, the solution of this optimization problem is an eigenvector $\boldsymbol{\phi}_v$ of $\tilde{\mathbf{L}}$ associated with the second smallest eigenvalue (i.e., the smallest nonzero eigenvalue). Consequently, the vertex set $\mathbb{V}$ is split into two disjoint clusters $\mathbb{S} = \{v \in \mathbb{V} \mid \boldsymbol{\phi}_v(v) \geq 0\}$ and $\overline{\mathbb{S}} = \{v \in \mathbb{V} \mid \boldsymbol{\phi}_v(v) < 0\}$.

The normalized hypergraph Laplacian $\tilde{\mathbf{L}}$ in (8.4.38) is related with the Laplacian of a graph with adjacency matrix (8.4.11) with $\mathbf{W}_* = \mathbf{W}$ [ABB06]. The degree matrix for this graph is given by the diagonal matrix:

$$
\mathbf{D}_{\mathrm{Zhou}} = \begin{bmatrix} \mathbf{D}_v & \mathbf{0} \\ \mathbf{0} & \mathbf{W}\,\mathbf{D}_e \end{bmatrix}.
\tag{8.4.42}
$$

The normalized Laplacian for this bipartite graph is given by:

$$
\begin{aligned}
\tilde{\mathbf{L}}_{\mathrm{Zhou}} &= \begin{bmatrix} \mathbf{I}_{|\mathbb{V}|\times|\mathbb{V}|} & \mathbf{0} \\ \mathbf{0} & \mathbf{I}_{|\mathbb{E}|\times|\mathbb{E}|} \end{bmatrix} - \begin{bmatrix} \mathbf{D}_v^{-\frac{1}{2}} & \mathbf{0} \\ \mathbf{0} & \mathbf{D}_e^{-\frac{1}{2}}\mathbf{W}^{-\frac{1}{2}} \end{bmatrix} \\
&\quad \begin{bmatrix} \mathbf{0}_{|\mathbb{V}|\times|\mathbb{V}|} & \mathbf{H}\,\mathbf{W} \\ \mathbf{W}\,\mathbf{H}^T & \mathbf{0}_{|\mathbb{E}|\times|\mathbb{E}|} \end{bmatrix} \begin{bmatrix} \mathbf{D}_v^{-\frac{1}{2}} & \mathbf{0} \\ \mathbf{0} & \mathbf{D}_e^{-\frac{1}{2}}\mathbf{W}^{-\frac{1}{2}} \end{bmatrix} \\
&= \begin{bmatrix} \mathbf{I}_{|\mathbb{V}|\times|\mathbb{V}|} & -\mathbf{B} \\ -\mathbf{B}^T & \mathbf{I}_{|\mathbb{E}|\times|\mathbb{E}|} \end{bmatrix}
\end{aligned}
\tag{8.4.43}
$$

for $\mathbf{B} = \mathbf{D}_v^{-1/2}\mathbf{H}\,\mathbf{W}^{\frac{1}{2}}\mathbf{D}_e^{-\frac{1}{2}}$. Any eigenvector $\boldsymbol{\phi} = \left(\boldsymbol{\phi}_v^T\ \boldsymbol{\phi}_e^T \right)^T$ of the normalized Laplacian

$\tilde{\mathbf{L}}_{\text{Zhou}}$ with corresponding eigenvalue λ' satisfies (8.4.16), where $\mathbf{B}\,\mathbf{B}^T = \boldsymbol{\Theta}$, i.e.:

$$
\boldsymbol{\Theta}\,\boldsymbol{\phi}_v \;=\; (1-\lambda')^2\,\boldsymbol{\phi}_v \;\;\text{or}
$$
$$
\underbrace{(\mathbf{I}-\boldsymbol{\Theta})}_{\tilde{\mathbf{L}}}\,\boldsymbol{\phi}_v \;=\; \underbrace{\left(1-(1-\lambda')^2\right)}_{\lambda}\,\boldsymbol{\phi}_v. \tag{8.4.44}
$$

That is, the $N = |\mathbb{V}|$ elements of the eigenvectors of the normalized Laplacian $\tilde{\mathbf{L}}_{\text{Zhou}}$, which correspond to the vertices of the hypergraph, are the eigenvectors of the $N \times N$ matrix $\tilde{\mathbf{L}} = \mathbf{I} - \boldsymbol{\Theta}$. Since $\lambda = \lambda'\,(2-\lambda') \geq 0$, we infer that $0 \leq \lambda' \leq 2$.

The spectral hypergraph clustering approach can be extended to K-*way* partitioning, i.e., to finding the partition $(\mathbb{V}_1, \mathbb{V}_2, \ldots, \mathbb{V}_K)$ with $\cup_{j=1}^{K}\mathbb{V}_j = \mathbb{V}$ and $\mathbb{V}_j \cap \mathbb{V}'_j = \emptyset$ for $j \neq j'$ and $1 \leq j, j' \leq K$. Zhou et al. [ZHS06] proved the following theorem:

Theorem 8.4.1 *Assume a hypergraph* $\mathcal{H} = (\mathbb{V}, \mathbb{E}, w)$ *with* $|\mathbb{V}| = N$. *Denote* $\lambda_1 \leq \lambda_2 \leq \ldots \leq \lambda_N$ *the eigenvalues of the hypergraph Laplacian* $\tilde{\mathbf{L}}$. *If* $\mathrm{ncut}_K(\mathcal{H}) = \min \mathrm{ncut}(\mathbb{V}_1, \mathbb{V}_2, \ldots, \mathbb{V}_K)$ *is the cost of* K-*way partitioning, then* $\mathrm{ncut}_K(\mathcal{H}) \geq \sum_{j=1}^{K}\lambda_j$.

It is seen that the real-valued optimization problem derived from the relaxation is actually a lower bound of the original combinatorial optimization problem. It is unclear how one may employ multiple eigenvectors simultaneously in order to obtain a K-way partition [ZHS06]. Among the many heuristics proposed in the literature, the most popular works as follows [NJW01].

1. Create matrix $\mathbf{X} = [\boldsymbol{\phi}_1 | \boldsymbol{\phi}_2 | \cdots \boldsymbol{\phi}_K]$ where $\boldsymbol{\phi}_j$, $j = 1, 2, \ldots, K$ are the eigenvectors of $\tilde{\mathbf{L}}$ associated to the K smallest eigenvalues.

2. Treat the row vectors of $\mathbf{X}$ as representations of the hypergraph vertices in the K-dimensional Euclidean space. They are expected to be well separated.

3. Obtain a good partition by running K-means on them once.

The existence of the multiplier $\zeta(e \mid \mathbb{S}, \overline{\mathbb{S}})$ in (8.4.26) differentiates the normalized cut on hypergraphs from that on ordinary graphs. As mentioned, this multiplier arises from the projection of the hypergraph to a normal graph, where each hyperedge e is replaced by the fully connected subgraph with vertices e to which the weight $\frac{w(e)}{\delta(e)}$ is assigned [ABB06]. The cost (8.4.28) is the cut cost of the projected graph. $\zeta(e \mid \mathbb{S}, \overline{\mathbb{S}})$ can be factored as [PCAS14]:

$$
\zeta(e \mid \mathbb{S}, \overline{\mathbb{S}}) = \eta\,(1-\eta)\,\delta(e), \tag{8.4.45}
$$

where $\eta = \frac{|e \cap \mathbb{S}|}{\delta(e)}$ varies within the range:

$$
\frac{1}{\delta(e)} \leq \eta \leq \frac{\delta(e)-1}{\delta(e)}. \tag{8.4.46}
$$

In [PCAS14], it has been shown empirically that the cost of cutting a hyperedge e is highest, if e is divided into equal halves (i.e., $\eta = 0.5$). Moreover, given the same η, $\zeta(e \mid \mathbb{S}, \overline{\mathbb{S}})$ is always higher, when $\delta(e) = \kappa$ increases (i.e., for larger hyperedges), since the numerator in $\zeta(e \mid \mathbb{S}, \overline{\mathbb{S}})$ increases quadratically with respect to $\delta(e)$, while the denominator increases linearly with respect to $\delta(e)$. Hence, given two hyperedges of the same weight $w(e)$ and the same η, NCut will inherently favor preserving the larger hyperedge and cutting the smaller hyperedge. Intuitively, the larger hyperedges convey more evidence on the existence of a cluster than the smaller ones, even if the model is fitted equally well in both cases. For example, fitting a circle (i.e., a third-order model) to points can be done by clustering either a 4-uniform hypergraph or an 8-uniform hypergraph. However, 4 points cannot constrain a circle, especially if they are spatially close. On the contrary, the 8-uniform hypergraph yields a more accurate fitting. The same applies to all the aforementioned approaches, which although were designed to deal with high-order relations, they are reduced to standard pairwise approaches [ABB06]. However, approximating the high-order relations in terms of pairwise interactions can lead to a substantial information loss. Moreover, the intrinsic bias toward larger hyperedges also affects the clique averaging [ALZM$^+$05] and the max projection [OB12].

Another assumption frequently made is that clustering algorithms assign each object to one class exclusively. Consequently, the clusters are not modeled, but they are obtained as a by-product of the partition of objects into a predetermined number of classes. That is, the resulting clusters are by definition disjoint sets. For example, in graph clustering into 2 classes, the goodness of the partitioning is inversely proportional to the cost of the cut that separates the vertices, which is a function of the weights of those edges having vertices in both clusters [PCAS14]. In several applications, overlapping clusters are more appropriate [HG07]. To address the need for *hypergraph soft clustering*, the constraints imposed by crisp partitions are relaxed so that soft boundaries between the clusters emerge.

Co-occurrence relations, such as co-citation and co-purchase relations, typically involve more than two items. Consequently, they are represented by a hyperedge in a hypergraph. As has been seen, the hyperedges are usually transformed into cliques of edges by clique expansion, star expansion, or normalized hypergraph cut. Let us call the just-mentioned transformations, *vertex expansions*. Another transformation, called *hyperedge expansion* has been proposed in [PF12]. Hyperedge expansion operates on the dual hypergraph, i.e., the hypergraph having as vertices the hyperedges of the original hypergraph. It is first carried out on the hyperedge level and the learning induced by the hyperedges is projected back to the vertices through the adjacency matrix of the dual hypergraph. Formally, a multiclass labeling on the hypergraph $\mathcal{H}(\mathbb{V}, \mathbb{E}, w)$ is defined that associates each vertex $v \in \mathbb{V}$ with a single label $y(v) \in \mathbb{Y}$. For a hyperedge $e \in \mathbb{E}$, $y(e) = \{y(v) \mid v \in e\}$ is the set of labels associated to e. When $|y(e)| > 1$, it is said that the hyperedge e is broken or violated by y [PF12]. Hyperedge expansion finds a set of hyperedges of minimum weight that separates the vertices of the hypergraph into two disjoint subsets by minimizing the cost function:

$$\text{HE} = \min_{y} \sum_{e \in \mathbb{E} \, |y(e)| > 1} w(e), \tag{8.4.47}$$

i.e., the sum of the weights of broken hyperedges. The cost function (8.4.47) was proposed a long time ago [Law73] and was revisited recently in [Fuk10]. However instead of dealing with combinatorial algorithms that minimize (8.4.47), which are efficient only when the number of classes is small, a relaxation of (8.4.47) to a continuous optimization problem is described next.

A directed graph $\hat{\mathcal{G}}$ is constructed by inserting two vertices e^+ and e^- for each hyperedge e in the original hypergraph $\mathcal{H}(\mathbb{V}, \mathbb{E}, w)$. The vertices in $\hat{\mathcal{G}}$ are twice as much as the hyperedges of $\mathcal{H}$. Accordingly, $\hat{\mathcal{G}}$ is associated to the dual hypergraph of $\mathcal{H}$. For a pair of overlapping hyperedges e_1 and e_2 in $\mathcal{H}$, two directed edges (e_1^-, e_2^+) and (e_2^-, e_1^+) are included in $\hat{\mathcal{G}}$ with weights $w(e_1)$ and $w(e_2)$, respectively, where $w(\cdot)$ is the weighting function in $\mathcal{H}$. The adjacency matrix of $\hat{\mathcal{G}}$, $\mathbf{A} \in \mathbb{R}^{2M \times 2M}$, is defined as:

$$\mathbf{A}_{\hat{\mathcal{G}}} = \begin{bmatrix} \mathbf{0} & \mathbf{A}^* \mathbf{W} \\ \mathbf{W} & \mathbf{0} \end{bmatrix}, \tag{8.4.48}$$

where $M = |\mathbb{E}|$, $\mathbf{A}^* \in \mathbf{R}^{M \times M}$ is the adjacency matrix of the dual hypergraph of $\mathcal{H}$ (i.e., the adjacency matrix of hyperedges) with elements:

$$A^*(i,j) = \begin{cases} 1 & \text{if } e_i \cap e_j \neq \emptyset \text{ and } e_i \neq e_j \\ 0 & \text{otherwise} \end{cases} \qquad (8.4.49)$$

and $\mathbf{W} \in \mathbb{R}^{M \times M}$ is the diagonal matrix of the hyperedge weights. Let us sort the rows and columns of $\mathbf{A}_{\hat{\mathcal{G}}}$ in the following order $[e_1^-, e_2^-, \ldots e_M^-, e_1^+, e_2^+, \ldots e_M^+]$. The *out-degree matrix* of $\hat{\mathcal{G}}$ is [PF12]:

$$\mathbf{D}_{\hat{\mathcal{G}}} = \begin{bmatrix} \hat{\mathbf{D}} & \mathbf{0} \\ \mathbf{0} & \mathbf{W} \end{bmatrix}, \qquad (8.4.50)$$

where $\hat{\mathbf{D}} = \text{diag}(\mathbf{1}_M^T \mathbf{W} \mathbf{A}^*)$ is a diagonal matrix of size $M \times M$. Then, the unnormalized *out-degree Laplacian* of $\hat{\mathcal{G}}$ is given by:

$$\mathbf{L}_{\hat{\mathcal{G}}} = \mathbf{D}_{\hat{\mathcal{G}}} - \mathbf{A}_{\hat{\mathcal{G}}} = \begin{bmatrix} \hat{\mathbf{D}} & \mathbf{A}^* \mathbf{W} \\ -\mathbf{W} & \mathbf{W} \end{bmatrix}. \qquad (8.4.51)$$

Let us elaborate a minimum cut in $\hat{\mathcal{G}}$. That is, to seek for a set $\mathbb{S}$, such that $\mathbb{S} + \overline{\mathbb{S}} = \hat{\mathbb{V}}$, where $\hat{\mathbb{V}} = \{e_i^-, e_i^+ \mid e_i \in \mathbb{E}\}$. Let also $\hat{\mathbb{E}} = \{(e_i^-, e_j^+), (e_j^-, e_i^+) \mid e_i \cap e_j \neq \emptyset, \mathbb{E} \ni e_i \neq e_j \in \mathbb{E}\}$. We introduce the vector $\mathbf{f} \in \{\frac{1}{\sqrt{|\mathbb{S}|}}, 0\}^{2M}$ with elements $f(e) = \frac{1}{\sqrt{|\mathbb{S}|}}$ and 0 otherwise. Such a unit ℓ_2 norm vector $\mathbf{f}$ minimizes the cost of the cut:

$$\text{HEcut} = \sum_{\substack{e_1, e_2 \in \hat{\mathbb{V}} \\ (e_1, e_2) \in \hat{\mathbb{E}}}} |\mathbb{S}| \, (f(e_1) - f(e_2)) \, f(e_1), \qquad (8.4.52)$$

where the last term ensures that the edges from $\mathbb{S}$ to $\overline{\mathbb{S}}$ are counted, whenever $f(e_1) = \frac{1}{\sqrt{|\mathbb{S}|}}$ and $f(e_2) = 0$ [PF12]. If $\mathbf{f}$ admits real positive values, then the optimization problem (8.4.52) is relaxed to the minimization of HEcut with respect to $\mathbf{f}$ subject to $\mathbf{f}^T \mathbf{f} = 1$. The solution of this optimization problem is [PF12]:

$$\mathbf{f}^T \underbrace{\left(2\mathbf{D}_{\hat{\mathcal{G}}} - \mathbf{A}_{\hat{\mathcal{G}}} \right)}_{\mathbf{D}_{\hat{\mathcal{G}}} + \mathbf{L}_{\hat{\mathcal{G}}}} = 2 \, \lambda \mathbf{f}^T, \qquad (8.4.53)$$

i.e., $\mathbf{f}$ is a left eigenvector of the matrix in the left-hand side of (8.4.53), which is closely related to $\mathbf{L}_{\hat{\mathcal{G}}}$. Both the matrix in the left-hand side of (8.4.53) and $\mathbf{L}_{\hat{\mathcal{G}}}$ are non symmetric matrices. Under certain conditions, it can be shown the eigenvalues of $\mathbf{L}_{\hat{\mathcal{G}}}$ are non-negative real numbers and the left eigenvectors admit real values [PF12]. Suppose that the $2M \times 1$ left (real) eigenvectors $\hat{\phi}_1, \hat{\phi}_2, \ldots, \hat{\phi}_K$ are associated to the K smallest eigenvalues of $\mathbf{L}_{\hat{\mathcal{G}}}$. To map the hyperedge embedding back to the vertices of the hypergraph $\mathcal{H}$, one has to multiply the $N \times M$ incidence matrix of the hypergraph $\mathbf{H}$ with the $M \times K$ matrix $\left[\hat{\phi}_1^- | \hat{\phi}_2^- | \cdots | \hat{\phi}_K^- \right]$, where $\hat{\phi}_j^-$ is the vector formed by retaining the top M elements.

8.5   Ranking on hypergraphs

For both clustering and semi-supervised learning, the key factor is the *cut functional* (e.g., (8.4.28) or (8.4.52)). The *total variation* on a hypergraph is introduced as the Lovasz extension of the hypergraph cut [HSJR13]. In particular, a family of regularization functionals was proposed, which interpolates between the total variation and the regularization functionals enforcing smoother functions on the hypergraph, such as the quadratic form $\Omega(\mathbf{f}) = \mathbf{f}^T \mathbf{L} \mathbf{f}$, where $\mathbf{L}$ is a proper Laplacian. It is shown that there exists a tighter relaxation of the normalized hypergraph cut. For both clustering and semi-supervised learning, convex optimization problems are solved whose core is proximal mapping.

In the following, we shall refer to vertices via their indices.

Definition 1 *[HSJR13] The total variation $TV_{\mathcal{H}} : \mathbb{R}^{N \times 1} \to \mathbb{R}$ on hypergraph $\mathcal{H}(\mathbb{V}, \mathbb{E}, w)$ is defined as:*

$$\text{TV}_{\mathcal{H}}(\mathbf{f}) = \sum_{e \in \mathbb{E}} w(e) \left(\max_{i \in e} f_i - \min_{j \in e} f_j \right) = \sum_{e \in \mathbb{E}} w(e) \max_{e \ni i, j \in e} |f_i - f_j|. \tag{8.5.1}$$

It is a generalization of the total variation on graphs $\text{TV}_{\mathcal{G}}(\mathbf{f}) = \frac{1}{2} \sum_{i=1}^{N} \sum_{j=1}^{N} w_{ij} |f_i - f_j|$. For nonnegative hyperedge weights $w(e)$, (8.5.1) is a nonnegative combination of convex functions, and thus is also a convex function.

Let $\nabla_e : \mathbb{R}^{N \times 1} \to \mathbb{R}^{N \times N}$ be the difference operator for the hyperedge $e \in \mathbb{E}$ with elements:

$$[\nabla_e(\mathbf{f})]_{ij} = \begin{cases} f_i - f_j & i \in e \text{ and } j \in e \\ 0 & \text{otherwise.} \end{cases} \tag{8.5.2}$$

Then, the total variation (8.5.1) can be expressed as $\text{TV}_{\mathcal{H}}(\mathbf{f}) = \sum_{e \in \mathbb{E}} w(e) \|\nabla_e(\mathbf{f})\|_\infty$, which can be seen as inducing a group sparse structure on the gradient level. This interpretation gives rise to a family of regularization functionals $\Omega_{\mathcal{H},p} : \mathbb{R}^{N \times 1} \to \mathbb{R}$ for a hypergraph $\mathcal{H}(\mathbb{V}, \mathbb{E}, w)$ for $p \geq 1$:

$$\Omega_{\mathcal{H},p}(\mathbf{f}) = \sum_{e \in \mathbb{E}} w(e) \left(\max_{i \in e} f_i - \min_{j \in e} f_j \right)^p. \tag{8.5.3}$$

Accordingly, $\text{TV}_{\mathcal{H}}(\mathbf{f})$ is equivalent to $\Omega_{\mathcal{H},1}(\mathbf{f})$. If $\mathcal{H}$ is a graph and $p \geq 1$, $\Omega_{\mathcal{H},p}(\mathbf{f})$ reduces to a Laplacian regularization:

$$\Omega_{\mathcal{G},p}(\mathbf{f}) = \sum_{i=1}^{N} \sum_{j=1}^{N} w_{ij} |f_i - f_j|^p. \tag{8.5.4}$$

It can be shown that:

$$\Omega_{\mathcal{H},p}(\mathbf{1}_N) = \sum_{\substack{e \in \mathbb{E}: \\ e \cap \mathbb{S} \neq \emptyset \\ e \cap \overline{\mathbb{S}} \neq \emptyset}} w(e) = \text{cut}(\mathbb{S}, \overline{\mathbb{S}}). \tag{8.5.5}$$

8.5.1   Enforcing structural constraints

Let $\mathcal{H}(\mathbb{V}, \mathbb{E}, w)$ denote a hypergraph with set of vertices $\mathbb{V}$ and set of hyperedges $\mathbb{E}$ to which a weight function $w: \mathbb{E} \to \mathbb{R}$ is assigned. The vertex set $\mathbb{V}$ is frequently made by concatenating sets of objects of different type (e.g., images, users, social groups, geo-tags, tags). Hereafter, each vertex subset is referred to as *object group* in order to avoid confusion with social groups. In general, each object group contributes differently to the ranking procedure. The different impact of each object group can be taken into account by proper regularization terms in the ranking procedure, as is detailed next.

The vertices and hyperedges form a $|\mathbb{V}| \times |\mathbb{E}|$ incidence matrix with elements $H(v, e) = 1$, if $v \in e$ and 0 otherwise. Let $\mathbf{D}_v$ denote the vertex degree diagonal matrix of size $|\mathbb{V}| \times |\mathbb{V}|$, $\mathbf{D}_e$ be the hyperedge degree diagonal matrix of size $|\mathbb{E}| \times |\mathbb{E}|$, and $\mathbf{W}$ represent the $|\mathbb{E}| \times |\mathbb{E}|$ diagonal matrix of hyperedge weights. Then, $\tilde{\mathbf{L}} = \mathbf{I} - \mathbf{\Theta}$ is the positive semi-definite hypergraph Laplacian (8.4.38) [ZHS06]. The elements of $\mathbf{\Theta}$ defined in (8.4.39), $\Theta(u, v)$, can be interpreted as a measure of relatedness between the vertices u and v. A real valued ranking vector $\mathbf{f} \in \mathbb{R}^{|\mathbb{V}|}$ that minimizes $\Omega(\mathbf{f}) = \frac{1}{2}\mathbf{f}^T\tilde{\mathbf{L}}\mathbf{f}$ yields a clustering of the hypergraph, where all vertices with the same value in the ranking vector $\mathbf{f}$ are strongly connected [ABB06]. By including the ℓ_2 regularization norm between the ranking vector $\mathbf{f}$ and a query vector $\mathbf{y} \in \mathbb{R}^{|V|}$, a *recommendation problem* was solved [BTC$^+$10]. That is, the function to be minimized is expressed as:

$$\tilde{\Psi}(\mathbf{f}) = \Omega(\mathbf{f}) + \vartheta \, ||\mathbf{f} - \mathbf{y}||_2^2, \tag{8.5.6}$$

where ϑ is a regularization parameter. The ranking vector $\mathbf{f}^* = \underset{\mathbf{f}}{\mathrm{argmin}} \, \tilde{\Psi}(\mathbf{f})$ is [BTC$^+$10]:

$$\mathbf{f}^* = \frac{\vartheta}{1 + \vartheta} \left(\mathbf{I} - \frac{1}{1 + \vartheta}\mathbf{\Theta}\right)^{-1} \mathbf{y}. \tag{8.5.7}$$

A Group Lasso regularizing term is more appropriate than the ℓ_2 norm in this kind of problem [YL06]. In [PK14c], the hypergraph vertices are split into R non-overlapping object groups (images, users, social groups, geo-tags, tags) and different weights γ_r, $r = 1, 2, \ldots, R$ are assigned to each object group, yielding the following objective function to be minimized:

$$\Psi(\mathbf{f}) = \Omega(\mathbf{f}) + \vartheta \sum_{r=1}^{R} \sqrt{\gamma_r \, (\mathbf{f} - \mathbf{y})^T \mathbf{K}_r (\mathbf{f} - \mathbf{y})}. \tag{8.5.8}$$

In (8.5.8), ϑ is also a regularization parameter and $\mathbf{K}_r$ is the $|\mathbb{V}| \times |\mathbb{V}|$ diagonal matrix with elements equal to 1 for the vertices, which belong to the r-th object group. The latter minimization problem is expressed as:

$$\mathbf{f}^* = \underset{\mathbf{f}}{\mathrm{argmin}}\Psi(\mathbf{f}). \tag{8.5.9}$$

Let $\mathbf{x} = \mathbf{f} - \mathbf{y}$. By introducing the auxiliary variable $\mathbf{z} = \mathbf{x}$, the right-hand side of (8.5.9) is rewritten as:

$$\underset{\mathbf{x}}{\mathrm{argmin}} \frac{1}{2}(\mathbf{x} + \mathbf{y})^T \tilde{\mathbf{L}}(\mathbf{x} + \mathbf{y}) + \vartheta \sum_{r=1}^{R} \sqrt{\gamma_r \, \mathbf{z}^T \mathbf{K}_r \mathbf{z}}$$
$$\text{s.t. } \mathbf{z} = \mathbf{x}. \tag{8.5.10}$$

The solution of (8.5.10) is obtained by minimizing the augmented Lagrangian function:

$$\mathcal{L}(\mathbf{x}, \mathbf{z}, \boldsymbol{\lambda}) = \frac{1}{2}(\mathbf{x} + \mathbf{y})^T \tilde{\mathbf{L}}(\mathbf{x} + \mathbf{y}) + \vartheta \sum_{r=1}^{R} \sqrt{\gamma_r \mathbf{z}^T \mathbf{K}_r \mathbf{z}}$$
$$+ \boldsymbol{\lambda}^T (\mathbf{z} - \mathbf{x}) + \frac{\mu}{2}||\mathbf{z} - \mathbf{x}||_2^2, \tag{8.5.11}$$

where $\boldsymbol{\lambda}$ is the vector of the Lagrange multipliers, which is updated at each iteration and μ is a parameter regularizing the violation of the constraint $\mathbf{x} = \mathbf{z}$. (8.5.11) can be solved by the Alternating Directions Method [LLS11], as shown in Algorithm 8.5.1.

Solving for $\mathbf{x}^{t+1}$ in line 3 yields:

$$\mathbf{x}^{t+1} = (\tilde{\mathbf{L}} + \mu^t\mathbf{I})^{-1}(\boldsymbol{\lambda}^t + \mu^t\mathbf{z}^t - \tilde{\mathbf{L}}\mathbf{y}). \tag{8.5.12}$$

Algorithm 8.5.1: Alternating Directions Method

1: Given $\mathbf{x}^t$, $\mathbf{z}^t$ and $\boldsymbol{\lambda}^t$.
2: Set tolerance ϵ and initialize μ^0.
3: $\mathbf{x}^{t+1} \leftarrow \underset{\mathbf{x}}{\operatorname{argmin}} \mathcal{L}(\mathbf{x}, \mathbf{z}^t, \boldsymbol{\lambda}^t)$
4: $\mathbf{z}^{t+1} \leftarrow \underset{\mathbf{z}}{\operatorname{argmin}} \mathcal{L}(\mathbf{x}^{t+1}, \mathbf{z}, \boldsymbol{\lambda}^t)$
5: **if** $\|\mathbf{z} - \mathbf{x}\|_2^2 > \epsilon$ **then**
6: $\quad \boldsymbol{\lambda}^{t+1} \leftarrow \boldsymbol{\lambda}^t + \mu^t(\mathbf{z}^{t+1} - \mathbf{x}^{t+1})$
7: $\quad \mu^{t+1} = \min(1.1\mu^t, 10^6)$
8: **else**
9: $\quad$ return $\mathbf{x}^{t+1}$, $\mathbf{z}^{t+1}$.
10: $\quad \mathbf{f} = \mathbf{x}^{t+1} + \mathbf{y}$
11: **end if**

A careful look at (8.5.12) reveals that matrix inversion is not needed at each iteration. Only one eigen-decomposition is needed. Indeed, let $\mathbf{Q}_t = \tilde{\mathbf{L}} + \mu^t \mathbf{I}$. Then, $\mathbf{Q}_t^{-1} = \frac{1}{\mu^t - \mu^{t-1}} \left[\mathbf{I} + \frac{1}{\mu^t - \mu^{t-1}} \mathbf{Q}_{t-1} \right]^{-1}$. $\mathbf{Q}_0 = \tilde{\mathbf{L}} + \mu^0 \mathbf{I}$ is a symmetric matrix. Therefore, it is diagonalizable: $\mathbf{Q}_0 = \mathbf{U} \boldsymbol{\Lambda}_0 \mathbf{U}^T$, where $\mathbf{U} \mathbf{U}^T = \mathbf{U}^T \mathbf{U} = \mathbf{I}$. It can be easily derived that $\mathbf{Q}_1^{-1} = \mathbf{U} \left[(\mu^1 - \mu^0) \mathbf{I} + \boldsymbol{\Lambda}_0 \right]^{-1} \mathbf{U}^T$, and in general:

$$\mathbf{Q}_t^{-1} = \mathbf{U} \left[(\mu^t - \mu^0) \mathbf{I} + \boldsymbol{\Lambda}_0 \right]^{-1} \mathbf{U}^T. \tag{8.5.13}$$

The minimization problem described in line 4 of Algorithm 8.5.1 is expressed as:

$$\min_{\mathbf{z}} \mu^t \left\{ \frac{\vartheta}{\mu^t} \sum_{r=1}^{R} \sqrt{\gamma_r} \sqrt{\mathbf{z}^T \mathbf{K}_r \mathbf{z}} + \frac{1}{2} \|\mathbf{z} - (\mathbf{x}^{t+1} - \frac{1}{\mu^t} \boldsymbol{\lambda}^t)\|_2^2 \right\}. \tag{8.5.14}$$

By applying the soft-thresholding operator [QS10], we obtain:

$$z_j = \frac{\rho_j}{\|\boldsymbol{\rho}_r\|_2} \max\left(0, \|\boldsymbol{\rho}_r\|_2 - \vartheta \mu^t \frac{1}{\sqrt{\gamma_r}} \right), \tag{8.5.15}$$

where $\rho_j = x_j^{t+1} - \frac{1}{\mu^t} \lambda_j^t$, r is the object group where the j-th element belongs, and $\boldsymbol{\rho}_r$ denotes the segment of $\boldsymbol{\rho}$ corresponding to the r-th object group.

8.5.2 Learning hyperedge weights

In Section 8.5.1, the hyperedges are assigned a fixed weight by the user. Since the aforementioned weights indicate the importance given to each relationship captured by a hyperedge, a more suitable approach is to learn the weights by solving the optimization problem [GWZ+13]:

$$\underset{\mathbf{f}, \mathbf{w}}{\operatorname{argmin}} \ \tilde{\Psi}'(\mathbf{f}, \mathbf{w}) = \mathbf{f}^T \tilde{\mathbf{L}} \mathbf{f} + \vartheta_1 \|\mathbf{f} - \mathbf{y}\|_2^2 + \vartheta_2 \|\mathbf{w}\|_2^2$$

$$\text{s.t. } \mathbf{w}^T \mathbf{1} = 1, \tag{8.5.16}$$

where $\mathbf{w}$ has elements like those in the main diagonal of matrix $\mathbf{W}$. In (8.5.16), ϑ_1 and ϑ_2 are positive regularization parameters. The objective function in (8.5.16) includes the hypergraph smoothing, the discrepancy between the ranking vector $\mathbf{f}$ and the query vector

$\mathbf{y}$, and the hypergraph weight regularizer. An alternating optimization can be employed to solve the optimization problem under study. That is, first $\mathbf{w}$ is kept fixed and we optimize with respect to $\mathbf{f}$ and then the optimal $\mathbf{f}$ is fixed and we optimize with respect to $\mathbf{w}$ in each iteration.

When $\mathbf{w}$ is kept fixed, the solution of (8.5.16) with respect to $\mathbf{f}$ is given by (8.5.7) for $\vartheta = \vartheta_1$. Next, we fix $\mathbf{f} = \mathbf{f}^*$ and we solve the optimization problem:

$$\operatorname*{argmin}_{\mathbf{w}} \mathbf{f}^T \tilde{\mathbf{L}} \mathbf{f} + \vartheta_2 ||\mathbf{w}||_2^2$$

$$\text{s.t. } \mathbf{w}^T \mathbf{1} = 1 \tag{8.5.17}$$

for $\mathbf{w}$. Having ignored the contribution of the vertex degree matrix $\mathbf{D}_v = \operatorname{diag}(\mathbf{H}\,\mathbf{w})$ having as elements in its main diagonal those of the vector $\mathbf{H}\,\mathbf{w}$, the following solution is obtained for the i-th weight [GWZ$^+$13]:

$$w_i = \frac{1}{|\mathbb{E}|} - \frac{\mathbf{f}^T \mathbf{\Gamma} \mathbf{D}_e^{-1} \mathbf{\Gamma}^T \mathbf{f}}{2|\mathbb{E}|\,\vartheta_2} + \frac{\mathbf{f}^T \mathbf{\Gamma}_i \mathbf{\Gamma}_i^T \mathbf{f}}{2\vartheta_2\,D_e(i,i)}, i = 1, 2, \ldots, |\mathbb{E}|, \tag{8.5.18}$$

where $\mathbf{\Gamma} = \mathbf{D}_v^{-\frac{1}{2}} \mathbf{H}$ and $\mathbf{\Gamma}_i$ is the i-th column of $\mathbf{\Gamma}$.

An interesting probabilistic interpretation of the just described method [GWZ$^+$13]. At the optimal $\mathbf{f}$ and $\mathbf{w}$, the posterior probability density function given the samples $\mathbb{X}$ and the query vector $\mathbf{y}$ is maximized, i.e.:

$$\{\mathbf{f}^*, \mathbf{w}^*\} = \operatorname*{argmin}_{\mathbf{f}, \mathbf{w}} p(\mathbf{f}, \mathbf{w}|\mathbb{X}, \mathbf{y}). \tag{8.5.19}$$

If the data and the weights are conditionally independent and the constant term $p(\mathbb{X}, \mathbf{y})$ is ignored, the conditional joint density $p(\mathbf{f}, \mathbf{w}|\mathbb{X}, \mathbf{y})$ is approximated by:

$$p(\mathbf{f}, \mathbf{w}|\mathbb{X}, \mathbf{y}) = p(\mathbf{y}|\mathbb{X}, \mathbf{f}, \mathbf{w})\, p(\mathbf{f}|\mathbb{X}, \mathbf{w})\, p(\mathbf{w}). \tag{8.5.20}$$

Let

$$p(\mathbf{y}|\mathbb{X}, \mathbf{f}, \mathbf{w}) = p(\mathbf{y}|\mathbb{X}, \mathbf{f}) = \frac{1}{Z_1} \exp\left(-\frac{||\mathbf{y} - \mathbf{f}||_2^2}{\frac{1}{\vartheta_1}} \right) \tag{8.5.21}$$

$$p(\mathbf{f}|\mathbb{X}, \mathbf{w}) = \frac{1}{Z_2} \exp\left(-\mathbf{f}^T \tilde{\mathbf{L}}\,\mathbf{f} \right) \tag{8.5.22}$$

$$p(\mathbf{w}) = \frac{1}{Z_3} \exp\left(-\frac{||\mathbf{w} - \frac{1}{|\mathbb{E}|}\mathbf{1}||_2^2}{\frac{1}{\vartheta_2}} \right), \tag{8.5.23}$$

where Z_i, $i = 1, 2, 3$ is a normalizing constant so that the integral of the associated probability density function1 is 1 and $\mathbf{1}$ is a vector of ones of size $|\mathbb{E}| \times 1$. The first two terms in (8.5.20) model the two assumptions that the final ranking vector should not deviate too much for the query vector and the conditional density function of the ranking vector should be smooth on the hypergraph. The term (8.5.23) assigns a Gaussian density function on the hyperedge weights instead of treating them as fixed parameters.

8.6   Applications

To begin with, two applications employing uniform hypergraphs are described in Sections 8.6.1 and 8.6.2. The first one deals with high-order link analysis and the second one deals with object recognition. Next, four applications of ranking on arbitrary hypergraphs are discussed. Section 8.6.3 deals with music recommendation and personalized music tagging. Section 8.6.4 addresses simultaneous image tagging and geo-location prediction. Section 8.6.5 is devoted to social image search exploiting joint visual-textual information. Finally, Section 8.6.6 describes annotation, classification, and tourism recommendation driven by latent semantic analysis. Although the comparisons against the state-of-the-art techniques demonstrate the great potential of the methods described in Sections 8.6.3-8.6.6, the main utility of the details disclosed is to enable the readers to implement and assess the merits of these methods.

8.6.1   High-order web link analysis

The hyperlink structure between hypertexts (i.e., web pages) is the simplest case of explicit, directed links between documents. They mimic the citations between the papers in bibliometrics . There, the papers can be treated as vertices of a graph. Directed links (i.e., edges) exist whenever one paper cites another. Link analysis (e.g., PageRank [BP98], HITS [Kle99]) decomposes a proper matrix, capturing the hyperlink structure. More precisely, let $\mathbf{A} \in \mathbb{R}^{N \times N}$ be the adjacency matrix of the graph with elements:

$$A_{ij} = \begin{cases} 1, & \text{if there exists an edge from vertex } i \text{ to vertex } j \\ 0, & \text{otherwise,} \end{cases} \qquad (8.6.1)$$

where N is the number of nodes. Let $\mathbf{a}^{(i)}$ and $\mathbf{a}_j$ denote the i-th row of $\mathbf{A}$ and its j-th column, respectively. If $\mathbf{1} \in \mathbb{R}^N$ is a vector of N ones, the row normalized adjacency matrix $\tilde{\mathbf{A}}$ has elements $\tilde{A}_{ij} = \frac{1}{\mathbf{1}^T \mathbf{a}^{(i)}}$. PageRank seeks for the eigenvector associated to the dominant eigenvalue of the *Google matrix*, that is a convex combination of a stochastic variant of the row normalized adjacency matrix $\tilde{\mathbf{A}}$ and the matrix $\frac{1}{N} \mathbf{1} \mathbf{v}^T$, where $\mathbf{v} \in \mathbb{R}^N_+$ is the so-called personalization or teleportation vector [LM06]. Starting from a neighborhood, HITS seeks for the dominant eigenvector of either $\mathbf{A}^T \mathbf{A}$ or $\mathbf{A} \mathbf{A}^T$ [LM06]. The former is known as the *authority matrix*, while the latter is called the *hub matrix*. The aforementioned matrices are closely related to the co-citation and co-reference matrices in bibliometrics [DHZS02].

Undirected links may be also established, capturing similarities between titles, abstracts, and keywords or joint authorship. Accordingly, graphs with multiple link types (i.e., hypergraphs) emerge. For example, by exploiting both citation analysis and the just mentioned implicit, derived, links one may obtain a full picture of publication impact. Another closely related example is social network analysis. Social networks frequently include many link types, capturing friendship relations, organizational relations, geographical ones, and so on. High-order web link analysis[KBK05, DKK11] refers to the set of techniques used to analyze such hypergraphs.

As previously mentioned, the hypergraphs are represented by hypermatrices. In particular, the adjacency hypermatrix can be formed by stacking the adjacency matrix for each link type. Let us consider a hypergraph having N nodes capturing K link types. The hypergraph can be represented by a 3rd order adjacency hypermatrix $\boldsymbol{\mathcal{A}}$ of size $N \times N \times K$, where $\boldsymbol{\mathcal{A}}(i, j, k)$ is nonzero if vertex i is connected to vertex j by link type k. Let us denote by $\mathcal{A}(:,:,k)$ the k-th frontal slice of the adjacency hypermatrix $\mathcal{A}$, hereafter. For example, K could be 5 and the slices could be given the following interpretation in the context of bibliometric data [DKK11]:

Abstract similarity. The ij element of the 1st slice $\mathcal{A}(i,j,1)$ is a measure of the similarity of abstracts for documents i and j (e.g., the cosine similarity between the abstracts). Having found the term-abstract matrix $\mathbf{T}$, say by means of the product of term frequency and inverse document frequency, each column of $\mathbf{T}$ is normalized to unit norm, yielding the matrix $\tilde{\mathbf{T}}$ whose j-th column is $\frac{\mathbf{t}_j}{||\mathbf{t}_j||_2}$, where $||\;||_2$ denotes the ℓ_2 norm of a vector. Then, $\mathcal{A}(:,:,1) = \mathbf{T}^T\mathbf{T}$.

Title similarity. The ij element of the 2nd slice $\mathcal{A}(i,j,2)$ is a measure of the similarity of titles for documents i and j and can be computed in a similar way to abstract similarity.

Keyword similarity. The ij element of the 3rd slice $\mathcal{A}(i,j,3)$ is a measure of the similarity between the keywords associated to documents i and j and can be computed in a similar way to abstract similarity.

Author similarity. Let $\mathbf{Q} \in \mathbb{R}^{L \times N}$ be the author-document indicator matrix with element $Q_{li} = 1$ if author l, $l = 1, 2, \ldots, L$, has authored document i, $i = 1, 2, \ldots, N$ and zero otherwise. Let $\tilde{\mathbf{Q}}$ be a properly normalized author-document matrix with elements $\tilde{Q}_{li} = \frac{1}{\sqrt{\mathbf{1}^T\mathbf{q}_i}}$. The denominator is simply the square root of the number of authors of the i-th document. Then, $\mathcal{A}(:,:,4) = \tilde{\mathbf{Q}}^T\tilde{\mathbf{Q}}$.

Citations. The fifth slice captures the citation information, i.e.,

$$\mathcal{A}(i,j,5) = \begin{cases} \xi, & \text{if document } i \text{ cites document } j \\ 0, & \text{otherwise,} \end{cases} \tag{8.6.2}$$

where ξ is an arbitrary number, e.g., $\xi = 2$ [KBK05, DKK11].

Figure 8.6.1 shows pictorially the frontal slices of the adjacency hypermatrix $\mathcal{A}$. The just described choices for the slices of the hypermatrix $\mathcal{A}$ are not unique. In any case, each slice

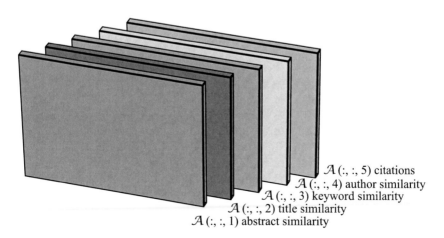

$\mathcal{A}(:,:,5)$ citations
$\mathcal{A}(:,:,4)$ author similarity
$\mathcal{A}(:,:,3)$ keyword similarity
$\mathcal{A}(:,:,2)$ title similarity
$\mathcal{A}(:,:,1)$ abstract similarity

FIGURE 8.6.1: Frontal slices of the adjacency hypermatrix $\mathcal{A}$.

can be treated as the adjacency matrix of a particular graph. All the graphs can be combined into a hypergraph associated to the just described third-order hypermatrix, since all the graphs refer to the same set of vertices. Moreover, the slices can be sparsified by retaining only the elements that exceed a proper threshold, in order to facilitate computations.

The aforementioned discussion can easily be customized to accommodate any metadata associated to images uploaded to Flickr or videos to YouTube. This is straightforward for title, keyword, and author similarity. Other slices can be appended in order to capture user comments or click-through data.

A straightforward procedure is to decompose the adjacency hypermatrix $\mathcal{A}$ of the hypergraph. Various decompositions can be found in the literature [Lat97, CZPA09, KB09, LPV13] The Canonical Decomposition (CANDECOMP)/Parallel Factor Analysis (PARAFAC) or CP for short is a higher order analog of the SVD for matrices. The CP decomposition generates feature vectors comprising all the linkages simultaneously for each vertex of the hypergraph. If CP with R factors is applied to $\mathcal{A}$, then $\mathcal{A}$ is approximated as:

$$\mathcal{A} \approx \sum_{r=1}^{R} \lambda_r \, \mathbf{x}_r \circ \mathbf{y}_r \circ \mathbf{z}_r, \tag{8.6.3}$$

where $\circ$ denotes the outer product of vectors. Each term $\lambda_r \, \mathbf{x}_r \circ \mathbf{y}_r \circ \mathbf{z}_r$ is identified as a rank-one hypermatrix (i.e., factor), that represents a "community" within the data. The number of factors R in (8.6.3) is loosely related to the number of the communities in the data. Figure 8.6.2 demonstrates graphically the CP decomposition (8.6.3). Let $\mathbf{X} \in \mathbb{R}^{N \times R}$, $\mathbf{Y} \in \mathbb{R}^{N \times R}$, and $\mathbf{Z} \in \mathbb{R}^{K \times R}$ be the matrices formed by the column vectors $\mathbf{x}_r$, $\mathbf{y}_r$, and $\mathbf{z}_r$, $r = 1, 2, \ldots, R$.

The elements of the third-order hypermatrix $\mathcal{A} \in \mathbb{R}^{N \times N \times K}$ can be reordered into the following three mode-n matrices for $n = 1, 2, 3$:

$$\begin{aligned}
A_{(1)}(i,p) &= \mathcal{A}(i,j,k) & p &= j + (k-1)\,N \\
A_{(2)}(j,p) &= \mathcal{A}(i,j,k) & p &= i + (k-1)\,N \\
A_{(3)}(k,p) &= \mathcal{A}(i,j,k) & p &= i + (j-1)\,N.
\end{aligned} \tag{8.6.4}$$

To derive the optimal approximation $\hat{\mathcal{A}}$ in (8.6.3), one has to minimize the squared Frobenius norm of the approximation error $\|\mathcal{A} - \hat{\mathcal{A}}\|_F^2$, where $\hat{\mathcal{A}}$ is the 3rd order hypermatrix in the right-hand side of (8.6.3). This can be achieved with the *Alternating Least Squares*(ALS) algorithm [CC70, Har70, FBH03]. At each iteration, the ALS algorithm solves for one component matrix, keeping the others fixed. For example, to determine $\mathbf{Z}$, keeping $\mathbf{X}$ and $\mathbf{Y}$ fixed, one has to solve for [DKK11]:

$$\mathbf{Z}^* = \underset{\mathbf{Z}}{\arg\min} \|\mathbf{A}_{(3)} - \mathbf{Z}\,(\mathbf{Y} \odot \mathbf{X})^T\|_F^2, \tag{8.6.5}$$

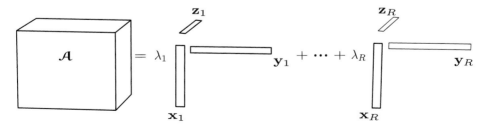

FIGURE 8.6.2: CP applied to the adjacency hypermatrix $\mathcal{A}$.

Algorithm 8.6.1: CP decomposition using the ALS algorithm.

Input: Hypermatrix $\mathcal{A} \in \mathbb{R}^{N \times N \times K}$, number of factors $R > 0$, maximum number of iterations T_{iter}, and stopping criterion $\epsilon > 0$.

Output: Vector $\boldsymbol{\lambda} \in \mathbb{R}^R$ and matrices $\mathbf{X} \in \mathbb{R}^{N \times R}$, $\mathbf{Y} \in \mathbb{R}^{N \times R}$, and $\mathbf{Z} \in \mathbb{R}^{K \times R}$.

1: Initialize: iteration $t = 0$,
   $\mathbf{X} = R$ principal eigenvectors of $\mathbf{A}_{(1)} \mathbf{A}_{(1)}^T$ and
   $\mathbf{Y} = R$ principal eigenvectors of $\mathbf{A}_{(2)} \mathbf{A}_{(2)}^T$
2: **while** not converged **do**
3:    Solve for $\mathbf{Z}$: $\mathbf{Z} = \mathbf{A}_{(3)} \, (\mathbf{Y} \odot \mathbf{X}) \, \left((\mathbf{Y}^T\mathbf{Y}) * (\mathbf{X}^T\mathbf{X}) \right)^{-1}$
4:    Normalize the columns of $\mathbf{Z}$ to unit ℓ_2 norm
5:    Solve for $\mathbf{Y}$: $\mathbf{Y} = \mathbf{A}_{(2)} \, (\mathbf{Z} \odot \mathbf{X}) \, \left((\mathbf{Z}^T\mathbf{Z}) * (\mathbf{X}^T\mathbf{X}) \right)^{-1}$
6:    Normalize the columns of $\mathbf{Y}$ to unit ℓ_2 norm
7:    Solve for $\mathbf{X}$: $\mathbf{X} = \mathbf{A}_{(1)} \, (\mathbf{Z} \odot \mathbf{Y}) \, \left((\mathbf{Z}^T\mathbf{Z}) * (\mathbf{Y}^T\mathbf{Y}) \right)^{-1}$
8:    Store the ℓ_2 norms of the column vectors of $\mathbf{X}$ into $\boldsymbol{\lambda}$
9:    Normalize the columns of $\mathbf{X}$ to unit ℓ_2 norm
10:   Compute the approximation $\hat{\mathcal{A}} = \sum_{r=1}^{R} \lambda_r \, \mathbf{x}_r \circ \mathbf{y}_r \circ \mathbf{z}_r$
11:   Check convergence conditions
       $t \leq T_{\text{iter}}$ and $\|\hat{\mathcal{A}} - \mathcal{A}\|_F^2 < \epsilon$
12:   $t \leftarrow t + 1$.
13: **end while**

---

where λ_r have been absorbed into the ℓ_2 norm of the column vectors of $\mathbf{X}$ and $\odot$ denotes the *Khatri-Rao product* defined as:

$$\mathbf{Y} \odot \mathbf{X} = [\mathbf{y}_1 \otimes \mathbf{x}_1 | \mathbf{y}_2 \otimes \mathbf{x}_2 | \dots | \mathbf{y}_R \otimes \mathbf{x}_R], \qquad (8.6.6)$$

with $\otimes$ referring to the Kronecker product between two vectors. The solution of the least squares problem (8.6.5) is:

$$\mathbf{Z} = \mathbf{A}_{(3)} \left[(\mathbf{Y} \odot \mathbf{X})^\dagger \right]^T = \mathbf{A}_{(3)} \, (\mathbf{Y} \odot \mathbf{X}) \, \left((\mathbf{Y}^T\mathbf{Y}) * (\mathbf{X}^T\mathbf{X}) \right)^{-1}, \qquad (8.6.7)$$

where $*$ denotes the Hadamard product between two matrices of the same size (i.e., element-wise product) and $\dagger$ denotes the Moore-Penrose pseudo-inverse of a matrix. Let $\boldsymbol{\Xi}$ be the square matrix of size $R \times R$ inside the big parentheses, i.e., $\boldsymbol{\Xi} = (\mathbf{Y}^T\mathbf{Y}) * (\mathbf{X}^T\mathbf{X})$. Frequently, $\boldsymbol{\Xi}^\dagger$ is used instead of $\boldsymbol{\Xi}^{-1}$ in (8.6.7). Assume the eigen-decomposition of $\boldsymbol{\Xi}$, $\boldsymbol{\Xi} = \mathbf{U}\,\boldsymbol{\Sigma}\,\mathbf{U}^T$, and $\Gamma \leq R$ be the rank of $\boldsymbol{\Xi}$. If $\Gamma < R$, $\boldsymbol{\Xi}$ is rank-deficient (i.e., a singular matrix). In this case, $\boldsymbol{\Xi}$ is replaced by its Γ-rank best approximation prior to matrix inversion. The ALS algorithm for obtaining the CP decomposition is summarized in Algorithm 8.6.1. Let us comment on the computational complexity of step 3: To obtain the Khatri-Rao product $\mathcal{O}(N^2 R)$ multiplications are needed, while matrix inversion inherits the cost of an SVD, i.e., $\mathcal{O}(R^3)$. Since $R << N$, the cost for any Khatri-Rao product is heavier than any matrix inversion. The cumulative number of multiplications needed in steps 3-11 is:

$$R(2K + 3)N^2 + 2R(2R + K + 3))N + 3(\eta_1 + \eta_2 + 2)R^3 + (2K + 3)R^2 + 2(K + 2)R, \quad (8.6.8)$$

where $\eta_1 = 4$ and $\eta_2 = 22$ [GL06]. The storage requirements become more prominent when large matrices are employed. In the latter case, more sophisticated approaches are needed [SPF14].

The rank-one factors reveal communities within the data. The largest entries in each factor $(\mathbf{x}_r, \mathbf{y}_r, \mathbf{z}_r)$ correspond to interlinked entries in the data [DKK11]. That is, top-ranking vertices in $\mathbf{x}_r$ are connected to top-ranking vertices in $\mathbf{y}_r$ with top-ranking link types in $\mathbf{z}_r$. If the top ranked scores in $\mathbf{z}_r$ refer to link types $k = 1, 2, 3$, capturing symmetric relations, then the documents associated to top ranked scores in $\mathbf{x}_{r'}$ and $\mathbf{y}_{r'}$ are expected to be nearly identical and to form a community [DKK11]. If the top ranked score in $\mathbf{z}_{r'}$ refers to the 5th link type (i.e., citations), which is an asymmetric relation, the following interpretation is given: the top ranked documents in $\mathbf{x}_r$ cite the top ranked documents in $\mathbf{y}_r$.

The matrices $\mathbf{X}$ and $\mathbf{Y}$ provide latent representations for each document vertex, which can be exploited to compute document similarities. Let $\mathbf{\Lambda} = \text{diag}(\boldsymbol{\lambda})$ be the diagonal matrix, having the elements of $\boldsymbol{\lambda}$ in its main diagonal. The matrix $\mathbf{\Upsilon} = \frac{1}{2}\mathbf{X}\mathbf{\Lambda}\mathbf{X}^T + \frac{1}{2}\mathbf{Y}\mathbf{\Lambda}\mathbf{Y}^T$ captures the similarity between the documents in a way similar to the latent semantic analysis (LSA). Moreover, having identified the documents containing a particular term or phrase in either the title, abstract, or keywords, the centroids $\mathbf{g}_X \in \mathbb{R}^R$ and $\mathbf{g}_Y \in \mathbb{R}^R$ are computed by averaging the associated row vectors of the matrices $\mathbf{X}$ and $\mathbf{Y}$. The vector $\boldsymbol{v} = \frac{1}{2}\mathbf{X}\mathbf{g}_X + \frac{1}{2}\mathbf{Y}\mathbf{g}_Y$ captures the similarities between all documents and the centroids. It is useful to reveal the documents related to the term under study. The method can be extended to finding the most similar papers to those written by a specific author. It is worth noting that the results heavily depend on the number of factors R used in the approximation (8.6.3) [DKK11].

8.6.2 Hypergraph matching for object recognition

Optimization-based approaches to graph matching for object recognition and scene analysis date back to 80s [BB82, Ch. 11]. Combinatorial or mixed continuous/combinatorial optimization techniques for feature matching were exploited in [BBM05, LH05, ZD06]. Feature matching is strongly related to content-based image retrieval, i.e., how one can establish matches between pairs or triples of scale-invariant feature transform (SIFT) [Low04] descriptors in two images. Formally, let N_1 and N_2 be the number of points (i.e., the output of a Hessian-affine keypoint detector) in image 1 and 2, respectively. Let $\mathbf{X} \in \{0, 1\}^{N_1 \times N_2}$ denote an assignment matrix, such that $x_{i_1 i_2}$ equals 1 when point P_{i_1} of image 1 is matched to point P_{i_2} of image 2 and 0 otherwise. Although a point from the first image is matched to exactly one point from the second image, two points from the second image could match an arbitrary number of points in the first image, if the constraint of a unit sum is imposed to each column, i.e., $\sum_{i_1=1}^{N_1} x_{i_1 i_2} = 1$. The matching between a pair of points (P_{i_1}, P_{j_1}) in image 1 and a pair of points (P_{i_2}, P_{j_2}) in image 2 is formulated as maximization of a score function:

$$\text{score}(\mathbf{X}) = \sum_{i_1=1}^{N_1} \sum_{i_2=1}^{N_2} \sum_{j_1=1}^{N_1} \sum_{j_2=1}^{N_2} a_{i_1 i_2 j_1 j_2} \, x_{i_1 i_2} \, x_{j_1 j_2}, \qquad (8.6.9)$$

where $a_{i_1 i_2 j_1 j_2}$ is a potential corresponding to the aforementioned pairs of points. Whenever there is a perfect match between the pair of points (P_{i_1}, P_{j_1}) and (P_{i_2}, P_{j_2}), $x_{i_1 i_2}$ and $x_{j_1 j_2}$ will be 1 and the score$(\mathbf{X})$ will be increased by a potential, admitting high values. The graph matching problem (8.6.9) is an integer quadratic programming problem with no known polynomial-time algorithm [DBKP11]. If the maximization is confined to the set of matrices such that $||\mathbf{X}||_F = \sqrt{N_2}$, i.e., when:

$$\underset{||\mathbf{X}||_F=\sqrt{N_2}}{\text{argmax}} \ \text{score}(\mathbf{X}), \qquad (8.6.10)$$

the classical Rayleigh quotient problem emerges, because score$(\mathbf{X})$ can be rewritten as $\tilde{\mathbf{x}}^T \tilde{\mathbf{A}} \tilde{\mathbf{x}}$, where $\tilde{\mathbf{x}} \in \mathbb{R}^{N_1 N_2}$ is obtained by lexicographically ordering $\mathbf{X}$ and $\tilde{\mathbf{A}}$ is a $N_1 N_2 \times$

$N_1 N_2$ symmetric matrix. Search for correspondences in a framework that accommodates both local geometric invariants and image descriptors was cast as a hypergraph matching problem using third-order constraints instead of unary or pairwise ones in [DBKP11]. For a triple of points $\{i_1, j_1, k_1\}$ from image 1 and a triple of points $\{i_2, j_2, k_2\}$ from image 2, (8.6.9) was extended to [DBKP11]:

$$
\begin{aligned}
\text{score}(\mathbf{X}) &= \sum_{i_1=1}^{N_1} \sum_{i_2=1}^{N_2} \sum_{j_1=1}^{N_1} \sum_{j_2=1}^{N_2} \sum_{k_1=1}^{N_1} \sum_{k_2=1}^{N_2} a_{i_1 i_2 j_1 j_2 k_1 k_2} \, x_{i_1 i_2} \, x_{j_1 j_2} \, x_{k_1 k_2} \\
&= \mathcal{A} \times_1 \tilde{\mathbf{x}} \times_2 \tilde{\mathbf{x}} \times_3 \tilde{\mathbf{x}},
\end{aligned}
\tag{8.6.11}
$$

where $\mathcal{A} \in \mathbb{R}^{N_1 N_2 \times N_1 N_2 \times N_1 N_2}$ is a 3rd-order $N_1 N_2$ dimensional super-symmetric hypermatrix representing the affinity between the triples of points. Since any triple of points defines a triangle, one might choose as affinity the cosine similarity between the vectors formed by the sines of the three angles in the triangle of image 1 and the triangle in image 2 or the normalized cross-correlation of intensity patterns inside properly normalized triangles in images 1 and 2. The hypergraph matching problem (8.6.11) was formulated as maximization of a multilinear objective function over all feature permutations. The maximization of (8.6.11) is equivalent to computing the rank-1 approximation of the hypermatrix $\mathcal{A}$. A power iteration can easily be devised [DBKP11]. In the same framework, one may impose the constraint for unit norm to the columns of $\mathbf{X}$, merge several hypermatrices $\mathcal{A}_t$ capturing potentials of different orders, or impose unit ℓ_1 norms to the columns of $\mathbf{X}$ [DBKP11].

In general, the hypergraph matching problem aims at finding a vertex-to-vertex mapping between two hypergraphs, such that the overall discrepancy between the corresponding matching hyperedges is minimized [ZS08]. The number of the vertices in the two hypergraphs may not be equal. Accordingly, the matching problem aims at finding an optimal matching sub-graph.

Formally, let $\mathcal{H} = (\mathbb{V}, \mathbb{E})$ and $\mathcal{H}' = (\mathbb{V}', \mathbb{E}')$ be two uniform hypergraphs, where the degree of each hyperedge is κ. A matching between $\mathcal{H}$ and $\mathcal{H}'$ is a vertex-to-vertex mapping $\varphi : \mathbb{V} \to \mathbb{V}'$. If $e = \{v_{i_1}, v_{i_2}, \ldots, v_{i_\kappa}\} \in \mathbb{E}$, then the vertex matching induces a hyperedge matching as well. That is, $\varphi(e) = \{\varphi(v_{i_1}), \varphi(v_{i_2}), \ldots, \varphi(v_{i_\kappa})\} \in \mathbb{E}'$.

The input to the hypergraph matching problem is a $|\mathbb{V}|^\kappa \times |\mathbb{V}'|^\kappa$ matrix $\mathbf{S}$ having as elements the probability of match between pairs of hyperedges, i.e., $S(e, e') = \text{Prob}(\varphi(e) = e' | \mathcal{H}, \mathcal{H}')$. The output of the problem under study is the $|\mathbb{V}| \times |\mathbb{V}'|$ matrix $\mathbf{X}$ having as elements the probability of match between pairs of vertices, i.e., $X(v, v') = \text{Prob}(\varphi(v) = v' | \mathcal{H}, \mathcal{H}')$. For a valid soft matching, $\mathbf{X}$ has to be doubly semi-stochastic, i.e., $\mathbf{X} \geq 0$, $\mathbf{X} \mathbf{1} \leq \mathbf{1}$, and $\mathbf{X}^T \mathbf{1} \leq \mathbf{1}$, where $\mathbf{1}$ denotes the vector of ones of appropriate size. The relationship between the input matrix $\mathbf{S}$ and the output matrix $\mathbf{X}$ can be expressed in a compact manner, if the matches are assumed pairwise conditionally independent. If this assumption holds, it can be proven that [ZS08]:

$$
\mathbf{S} = \underbrace{\mathbf{X} \otimes \mathbf{X} \otimes \cdots \otimes \mathbf{X}}_{\kappa \text{ times}} \stackrel{\triangle}{=} \otimes^\kappa \mathbf{X},
\tag{8.6.12}
$$

where $\otimes$ denotes the Kronecker product. Accordingly, to find $\mathbf{X}$ from $\mathbf{S}$ one needs to solve the optimization problem:

$$
\underset{\mathbf{X}}{\text{argmin}} \ \ \text{dist}(\mathbf{S}, \otimes^\kappa \mathbf{X})
$$

$$
\text{s.t. } \mathbf{X} \geq 0, \ \mathbf{X} \mathbf{1} \leq \mathbf{1}, \text{ and } \mathbf{X}^T \mathbf{1} \leq \mathbf{1},
\tag{8.6.13}
$$

for a proper distance function, e.g., the relative entropy error measure $D(\mathbf{S} \| \otimes^\kappa \mathbf{X})$. The complexity of the problem is greatly reduced, if $\mathbf{S}$ is marginalized, obtaining the matrix $\mathbf{Y}$

of size $|\mathbb{V}| \times |\mathbb{V}'|$ having elements:

$$Y(v, v') = \sum_{\substack{e|v \in e \\ e'|v' \in e'}} S_{ee'}. \tag{8.6.14}$$

The optimization problem (8.6.13) is equivalent to:

$$\begin{aligned} &\underset{\mathbf{X}}{\text{argmin}} \ \ \text{dist}(\mathbf{Y}, \mathbf{X}) \\ &\text{s.t. } \mathbf{X} \geq \mathbf{0}, \ \mathbf{X} \mathbf{1} \leq \mathbf{1}, \text{ and } \mathbf{X}^T \mathbf{1} \leq \mathbf{1}. \end{aligned} \tag{8.6.15}$$

Although (8.6.15) is convex, it can be further simplified by setting $\mathbf{1}^T \mathbf{X}^T \mathbf{1}$, which counts the number of matches, to a fixed value k and solving for:

$$\begin{aligned} \mathbf{X}^*(k) \ = \ &\underset{\mathbf{X} \geq \mathbf{0}}{\text{argmin}} \ \ \text{dist}(\mathbf{Y}, \mathbf{X}) \\ &\text{s.t. } \mathbf{X} \mathbf{1} \leq \mathbf{1}, \ \mathbf{X}^T \mathbf{1} \leq \mathbf{1}, \text{ and } \mathbf{1}^T \mathbf{X}^T \mathbf{1} = k, \end{aligned} \tag{8.6.16}$$

using a dual block update algorithm [ZS08]. Since $\mathbf{X}^*(k)$ is a convex function of k, this solution is used for minimizing over $0 \leq k \leq \min(|\mathbb{V}|, |\mathbb{V}'|)$.

8.6.3 Music recommendation and personalized music tagging

Music recommendation and *personalized music tagging* can be treated as ranking problems on arbitrary hypergraphs by enforcing structural constraints, as explained in Section 8.5.1.

A dataset was created by collecting real data from Last.fm in [TKP13]. In particular, to create the list of users, the 450 top artists were selected and their top 50 user fans where concatenated in the user set. This user set was later reduced based on the track and tag count of each user, yielding a final set of 1389 users. To create the track set, the 500 top played tracks for each user were concatenated in a list, from which 1765 unique tracks were selected based on their popularity among the users. Finally, tagging relations were collected for each user and the 1711 most frequent unique tags were retained. By using Porter's stemming algorithm [Por80] and calculating next the edit distance [RY98] between the tag pairs, all synonyms have been removed from the tags vocabulary (i.e., pairs, such as hardrock and hard rock or 90s and 1990s were merged). The final size of all sets is described in Table 8.6.1.

The vertex set of the hypergraph is defined as $\mathbb{V} = \mathbb{V}_u \cup \mathbb{V}_{ug} \cup \mathbb{V}_{ta} \cup \mathbb{V}_{tr}$. The incidence matrix of the hypergraph $\mathbf{H}$ is formed by concatenating the 5 hyperedge sets indicated as columns in Table 8.6.2. $\mathbf{H}$ has a size of 5867×146885 elements. In the following, the weights of the hyperedge sets $\mathbb{E}^{(1)}$, $\mathbb{E}^{(2)}$, and $\mathbb{E}^{(4)}$ are set equal to one.

TABLE 8.6.1: Music dataset objects, notations, and counts.

Objects	Notation	Count
Users	$\mathbb{V}_u$	1389
Groups	$\mathbb{V}_{ug}$	10
Tracks	$\mathbb{V}_{tr}$	1765
Tags	$\mathbb{V}_{ta}$	1711

$\mathbb{E}^{(1)}$ represents a pairwise friendship relation between users. The incidence matrix of the hypergraph with vertices $\mathbb{V}_u$ and hyperedges $\mathbb{E}^{(1)}$ has size 1389×13890.

$\mathbb{E}^{(2)}$ represents a group of users. It contains all the vertices of the corresponding users, as well as the vertices associated to the group object. The incidence matrix of the hypergraph with vertices $\mathbb{V}_u \cup \mathbb{V}_{ug}$ and hyperedges $\mathbb{E}^{(2)}$ has size 1399×13890.

$\mathbb{E}^{(3)}$ contains a user and a music track, representing a user-track listening relation. The hyperedge weight $w(e_{ij}^{(3)})$ is defined as the number of times the particular user u_i has listened to the track tr_j, normalized as follows to eliminate the bias:

$$w(e_{ij}^{(3)})' = \frac{w(e_{ij}^{(3)})}{\sqrt{\sum_{k=1}^{|\mathbb{V}_{tr}|} w(e_{ik}^{(3)})}\sqrt{\sum_{l=1}^{|\mathbb{V}_u|} w(e_{lj}^{(3)})}} \qquad (8.6.17)$$

and further scaled as $w(e_{ij}^{(3)})^* = \frac{w(e_{ij}^{(3)})'}{\text{ave}_j(w(e_{ij}^{(3)'}))}$, where $\text{ave}_j(w(e_{ij}^{(3)'}))$ is the average of normalized weights of the particular user u_i. The incidence matrix of the hypergraph with vertices $\mathbb{V}_u \cup \mathbb{V}_{tr}$ and hyperedges $\mathbb{E}^{(3)}$ has size 3154×68774.

$\mathbb{E}^{(4)}$ contains three vertices, a user, a tag, and a music track, capturing tagging relation. The incidence matrix of the hypergraph with vertices $\mathbb{V}_u \cup \mathbb{V}_{ta} \cup \mathbb{V}_{tr}$ and hyperedges $\mathbb{E}^{(4)}$ has size 4865×48566 elements.

$\mathbb{E}^{(5)}$ captures pairwise audio-track similarities. Such similarities are computed as follows. First, the 20 mel frequency cepstral coefficients (MFCCs) were employed to encode the timbral properties of the music signal. The MFCCs were calculated by using frames of duration 23 ms with a hop size of 11.5 ms and a 42-band filter bank. Next, a Gaussian mixture model (GMM) was created for each track with 30 components trained using the Expectation-Maximization (EM) algorithm, as in [Pam04]. The distance between two GMMs was computed by using the Earth Movers' Distance[RTG00], yielding the audio-track similarities. For a pair of tracks, tr_i and tr_j, the aforementioned distance was set as weight of the associated hyperedge $e_{ij}^{(5)}$, $w(e_{ij}^{(5)})$. The weights are then properly normalized to eliminate any bias: $w(e_{ij}^{(5)})' = \frac{w(e_{ij}^{(5)})}{max_{ij}(w(e_{ij}^{(5)}))}$. The incidence matrix of the hypergraph with vertices $\mathbb{V}_{tr}$ and hyperedges $\mathbb{E}^{(5)}$ has size 1765×1765. Frequently, only the K (e.g., $K=10$) most similar audio tracks to each track (i.e., its K nearest neighbors) are retained, yielding a sparse incidence matrix, where each row has only K nonzero entries.

Both music recommendation and personalized image tagging are cast as ranking on hypergraphs [TKP13, PK14a].

TABLE 8.6.2: The structure of the hypergraph incidence matrix $\mathbf{H}$ and its sub-matrices for music recommendation and tagging.

$\mathbb{E}^{(1)}$	$\mathbb{E}^{(2)}$	$\mathbb{E}^{(3)}$	$\mathbb{E}^{(4)}$	$\mathbb{E}^{(5)}$
$(\mathbb{V}_u, \mathbb{E}^{(1)})$	$(\mathbb{V}_u, \mathbb{E}^{(2)})$	$(\mathbb{V}_u, \mathbb{E}^{(3)})$	$(\mathbb{V}_u, \mathbb{E}^{(4)})$	$\mathbf{0}$
$\mathbf{0}$	$(\mathbb{V}_{ug}, \mathbb{E}^{(2)})$	$\mathbf{0}$	$\mathbf{0}$	$\mathbf{0}$
$\mathbf{0}$	$\mathbf{0}$	$\mathbf{0}$	$(\mathbb{V}_{ta}, \mathbb{E}^{(4)})$	$\mathbf{0}$
$\mathbf{0}$	$\mathbf{0}$	$(\mathbb{V}_{tr}, \mathbb{E}^{(3)})$	$(\mathbb{V}_{tr}, \mathbb{E}^{(4)})$	$(\mathbb{V}_{tr}, \mathbb{E}^{(5)})$

Let $\mathbf{y}$ be the query vector of size 4875×1. For recommendation, the query vector $\mathbf{y}$ is initialized by setting the entries corresponding to the target user and all objects (users, user groups, tags, and tracks), which are associated to this user to 1. Having set the query vector $\mathbf{y}$, the ranking vector $\mathbf{f}^*$ is given by either (8.5.7) or by solving (8.5.9). In the latter case, the regularization parameter ϑ, and the weights γ_r for the group objects should also be set. The ranking vector $\mathbf{f}^*$ has the same size and structure as $\mathbf{y}$. The values corresponding to music tracks are used only with the top ranked tracks being recommended for the test user.

For music tagging, the query vector can be initialized by setting the entry corresponding to the target user u and a certain track to (i.e., $y(v) = 1$, if $v = u$) and all other objects connected to the specific user (groups, tags, and tracks) to $\Theta(u, v)$ defined in (8.4.39), i.e., $y(v) = \Theta(u, v)$, if $v \neq u$. The values corresponding to tags are considered for personalized music tagging with the top ranked tags for a certain track, which was left out, being proposed to the user.

8.6.4   Simultaneous image tagging and geo-location prediction

In this section, simultaneous image tagging and geo-location prediction is treated within an arbitrary hypergraph framework as a ranking problem by enforcing structural constraints, as explained in Section 8.5.1.

An image dataset was collected from Flickr that contains both indoor and outdoor medium sized photos of popular Greek landmarks, various city scenes, and landscapes [PK14c]. Using FlickrApi (http://www.flickr.com/services/api), a large set of "geo-tagged" images was downloaded along with valuable information related to them (id, title, owner, latitude, longitude, tags, image views). The dataset was filtered with respect to image views (i.e., the times the specific image has been seen in Flickr) and owner's uploading statistics. It was assumed that images with many views normally depict landmarks worth seeing. Moreover, owners (i.e., users) with many uploaded images were considered more trustworthy as being more active, possessing more friends, and participating in more social groups. Then, social information related to social groups was crawled and only the groups that had at least 5 users from the data set as members were kept. The specific cardinalities are summarized in Table 8.6.3.

In order to form a proper set of tags, all the characters were converted to lower case, while unreadable symbols and redundant information were removed. Spelling mistakes were corrected and all morphological variations of the terms were merged to a single form, using the Edit Distance [RY98]. After having removed the terms used with low frequency, a vocabulary of unique words was created, along with their frequencies. The geo-tags were clustered into 125 different clusters using hierarchical clustering based on pairwise distances according to the "Haversine formula" (http://www.movable-type.co.uk/scripts/latlong.html).

The vertex set is defined as $\mathbb{V} = \mathbb{V}_{im} \cup \mathbb{V}_u \cup \mathbb{V}_{ug} \cup \mathbb{V}_{geo} \cup \mathbb{V}_{ta}$. The incidence matrix of the hypergraph $\mathbf{H}$ is formed by concatenating the 6 hyperedge sets indicated as columns

TABLE 8.6.3: Image dataset objects, notations, and counts.

Object	Notation	Count
Images	$\mathbb{V}_{im}$	1292
Users	$\mathbb{V}_u$	440
User Groups	$\mathbb{V}_{ug}$	1644
Geo-tags	$\mathbb{V}_{geo}$	125
Tags	$\mathbb{V}_{ta}$	2366

TABLE 8.6.4: The structure of the hypergraph incidence matrix $\mathbf{H}$ and its sub-matrices for image recommendation and geo-location prediction.

$\mathbb{E}^{(1)}$	$\mathbb{E}^{(2)}$	$\mathbb{E}^{(3)}$	$\mathbb{E}^{(4)}$	$\mathbb{E}^{(5)}$	$\mathbb{E}^{(6)}$
0	0	$(\mathbb{V}_{im}, \mathbb{E}^{(3)})$	$(\mathbb{V}_{im}, \mathbb{E}^{(4)})$	$(\mathbb{V}_{im}, \mathbb{E}^{(5)})$	$(\mathbb{V}_{im}, \mathbb{E}^{(6)})$
$(\mathbb{V}_{u}, \mathbb{E}^{(1)})$	$(\mathbb{V}_{u}, \mathbb{E}^{(2)})$	$(\mathbb{V}_{u}, \mathbb{E}^{(3)})$	$(\mathbb{V}_{u}, \mathbb{E}^{(4)})$	$(\mathbb{V}_{u}, \mathbb{E}^{(5)})$	0
0	$(\mathbb{V}_{ug}, \mathbb{E}^{(2)})$	0	0	0	0
0	0	0	$(\mathbb{V}_{geo}, \mathbb{E}^{(4)})$	0	0
0	0	0	0	$(\mathbb{V}_{ta}, \mathbb{E}^{(5)})$	0

in Table 8.6.4. $\mathbf{H}$ has a size of 5867×30924 elements. In the following, the weights of the hyperedge sets $\mathbb{E}^{(1)}$–$\mathbb{E}^{(5)}$ are set equal to one.

$\mathbb{E}^{(1)}$ represents a pairwise friendship relation between users. The incidence matrix of the hypergraph with vertices $\mathbb{V}_u$ and hyperedges $\mathbb{E}^{(1)}$ has size 440×2276.

$\mathbb{E}^{(2)}$ represents a user group. It contains all the vertices of the corresponding users as well as the ones corresponding to the user group. The incidence matrix of the hypergraph with vertices $\mathbb{V}_u \cup \mathbb{V}_{ug}$ and hyperedges $\mathbb{E}^{(2)}$ has size 2084×1644.

$\mathbb{E}^{(3)}$ contains a user and an uploaded image, representing a user-image possession relation. Each image has only one owner. The incidence matrix of the hypergraph with vertices $\mathbb{V}_u \cup \mathbb{V}_{im}$ and hyperedges $\mathbb{E}^{(3)}$ has size 1732×1292.

$\mathbb{E}^{(4)}$ captures a geo-location relation. This hyperedge set contains triplets of im, u and geo. The incidence matrix of the hypergraph with vertices $\mathbb{V}_{im} \cup \mathbb{V}_u \cup \mathbb{V}_{geo}$ and hyperedges $\mathbb{E}^{(4)}$ has size 1857×125 elements.

$\mathbb{E}^{(5)}$ contains triplets of images, users, and tags. Each hyperedge represents a tagging relation. The incidence matrix of the hypergraph with vertices $\mathbb{V}_{im} \cup \mathbb{V}_u \cup \mathbb{V}_{ta}$ and hyperedges $\mathbb{E}^{(5)}$ has size 4098×19127 elements.

$\mathbb{E}^{(6)}$ contains pairs of vertices, which represent two images. The weight $w(e_{ij}^{(6)})$ is set as the normalized similarity between images i and j, i.e., $w(e_{ij}^{(6)})' = \frac{w(e_{ij}^{(6)})}{\max_{ij}(w(e_{ij}^{(6)}))}$. Both global and local features were used. Firstly, the 100 nearest neighbors to each image were identified using the GIST descriptors [OT06] and they were reduced to the 5 most similar images to the reference image by using SIFT descriptors [Low04]. The incidence matrix of the hypergraph with vertices $\mathbb{V}_{im}$ and hyperedges $\mathbb{E}^{(6)}$ has size 1292×6460.

Both image tagging and geo-location prediction are cast as ranking on hypergraphs. That is, we seek for $\mathbf{f}$ that minimizes either (8.5.6) or (8.5.9). The query vector $\mathbf{y}$ is initialized by setting the entry corresponding to the test image im and its owner u to 1. The entries of the query vector corresponding to the tags ta are set equal to $\Theta(im, ta)$ defined in (8.4.39). The entries of the query vector corresponding to ug and geo are set equal to $\Theta(u, ug)$ and $\Theta(u, geo)$, respectively. The query vector $\mathbf{y}$ has a length of 5867 elements. Having set the query vector $\mathbf{y}$, the ranking vector $\mathbf{f}^*$ is given by either (8.5.7) or by solving (8.5.9). In the latter case, the regularization parameter ϑ, and the weights γ_r for the group objects should also be set. The ranking vector $\mathbf{f}^*$ has the same size and structure as $\mathbf{y}$. The values corresponding to tags are used for image tagging with the top ranked tags being recommended for the test image. The values corresponding to geo are used for geo-location prediction with only the 3 top ranked geo-locations (i.e., geo-clusters) being recommended for the test image.

For evaluation purposes, a test set containing the 25% of the tags and a training set containing the remaining 75% were defined [PK14c]. During testing, the tags contained in the test set were not included in the training procedure. That is, the associated elements to the test image are set equal to zero in Θ and $\mathbf{y}$. The relations between test images im and geo-locations (i.e., geo-clusters geo) are set also to 0.

The averaged Recall-Precision and F_1 measure were used as figures of merit. Precision is defined as the number of correctly recommended tags divided by the number of all recommended tags. Recall is defined as the number of correctly recommended tags divided by the number of all tags the user has actually set. The F_1 measure is the weighted harmonic mean of precision and recall, which measures the effectiveness of recommendation when treating precision and recall as equally important. The ranking obtained by (8.5.7) is referred to as Image Tagging on Hypergraph (ITH), while that obtained by (8.5.9) is coined as Query Group Sparse Optimization (QGSO). The geo-location prediction is referred to as GPR. The results of the (ITH) are demonstrated in Figure 8.6.3, in which the averaged Recall-Precision curves are plotted by averaging the Recall-Precision curves over 1186 images with at least 4 tags. To calculate the recall and precision, the 15 top ranked tags are recommended to any test image. By enforcing group sparsity in the ranking problem, the performance is improved significantly, as shown in Figure 8.6.3. The weights for the 5 different object groups (images, users, user groups, geo-tags, and tags) were set to 0.9, 0.9, 0.6, 0.2, and 0.2, respectively. This choice was made empirically. The typical values of μ^0, ϵ, and ϑ are 10^{-6}, 10^{-8}, and 2, respectively.

For the GPR, only the 3 top ranked elements are taken from the part of the $\mathbf{f}$ vector associated to the geo-locations. In Figure 8.6.4, the results for the GPR are presented. These results are further compared with those obtained by exploiting geographical information deduced from the tags. Greek geo-names were collected from the GeoNames geographical database (http://www.geo-names.org) along with their geo-coordinates. A geo-location prediction was made for each image having tags matching the geo-names. The distance between the image geo-coordinates, treated as ground truth, and the ones associated to

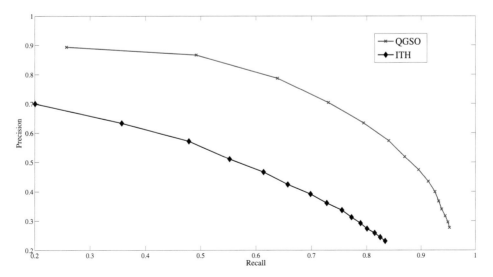

FIGURE 8.6.3: Averaged recall-precision curves for the ITH and the QGSO.

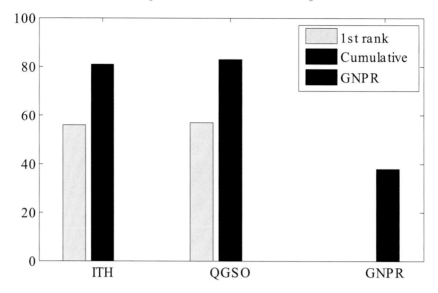

FIGURE 8.6.4: Geo-location prediction rates (in %) for the compared methods. **(See color insert.)**

the geo-name was computed by using the "Haversine formula." Predictions whose distances fall below a threshold of 500 m were considered as correct. Let us call the just described naive approach GNPR. Figure 8.6.4 depicts the geo-location prediction rates achieved by the ITH, the QGSO, and the GNPR.

8.6.5 Social image search exploiting joint visual-textual information

A method for social image search exploiting joint visual-textual information was proposed in [PK14d]. It is depicted in Figure 8.6.5.

The term frequency-inverse document frequency (TF-IDF) [SWY75] is a numerical statistic, quantifying the importance of a term in a document. It is the product of two statistics, namely the term frequency and the inverse document frequency. The term frequency weighs the terms more heavily which occur often in a specific document. On the other hand, the inverse document frequency down-weighs the terms, which tend to appear many times in several documents in the corpus. This way terms that are more common

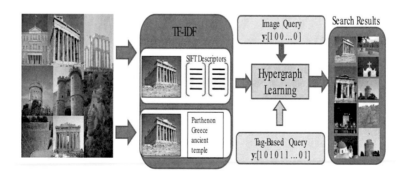

FIGURE 8.6.5: Description of the proposed social image search approach.

than others are handled effectively. By doing so, the terms that are truly representative of a document are given higher weights.

Let each social image $v_i \in \mathbb{V}_{im}$ denote a vertex in the hypergraph with $N = |\mathbb{V}_{im}|$. Each visual-textual term generates a hyperedge. Regarding the visual image content, SIFT is employed and SIFT descriptors are extracted from any image g. K-means is applied to the SIFT descriptors of any image v_i in order to quantize them to a predefined number (e.g., 200) of clusters represented by their mean vectors as codevectors. Accordingly, $v_i \in \mathbb{V}_{im}$ is represented by the concatenation of the codevectors instead of the concatenation of SIFT descriptors. K-means is applied to the set of the aforementioned quantized representations in order to create the visual word vocabulary. The indices of the resulting codevectors are treated as visual words $z_s \in \mathbb{Z}_s = \{z_{s_1}, z_{s_2}, \ldots, z_{s_R}\}$, where R is the size of visual word vocabulary. Let $\mathbf{B}_s \in \mathbb{R}^{N \times R}$ be the matrix having as elements the TF-IDF measurements, $B_s(i, j) = \varphi_{ij} \log_2 \frac{N}{N_j}$, where φ_{ij} is the frequency of visual term j in image i, N_j is the number of images that visual term j appears in, and N is the total number of images.

The textual information provided by the tags is captured by a bag-of-words representation as well. In order to form a proper text vocabulary, all characters are converted to lower case, unreadable symbols are removed and redundant information is eliminated. Next, a vocabulary of unique terms is generated along with their frequencies. Then, only the Q terms with the highest frequency are kept. The tags assigned to a social image are treated as a document annotating this image. Let $z_t \in \mathbb{Z}_t = \{z_{t_1}, z_{t_2}, \ldots, z_{t_Q}\}$ be the text vocabulary and $\mathbf{B}_t \in \mathbb{R}^{N \times Q}$ be the document-term matrix. Any social image $v_i \in \mathbb{V}_{im}$ is represented by a vector of size Q, having as elements the TF-IDF measurements obtained by taking into account the text information as well.

Here, the binary incidence matrix $\mathbf{H}$ used in [GWZ+13] is replaced by the fuzzy incidence matrix $\mathbf{H} = \left[\tilde{\mathbf{B}}_s \mid \tilde{\mathbf{B}}_t \right]$ of size $N \times M$, yielding a fuzzy hypergraph model, where $M = R + Q$. The matrices $\tilde{\mathbf{B}}_s$ and $\tilde{\mathbf{B}}_t$ are obtained by a min-max normalization of $\mathbf{B}_s$ and $\mathbf{B}_t$, respectively, so that their elements admit values in $[0, 1]$. As can be seen, the incidence matrix captures both the visual and the textual information, which is also inherited by the hypergraph Laplacian and the matrix $\boldsymbol{\Theta}$ (8.4.39), appearing in (8.5.7) as described in Section 8.5.1. To assess the impact of the fuzzy hypergraph incidence matrix the diagonal matrix $\mathbf{W}$ containing the hyperedge weights is set to $\mathbf{I}$ i.e., $w(e) = 1, \forall e \in \mathbb{E}$.

A query can be based on either tags or images. In the case of an image-based query, the query vector $\mathbf{y}$ is initialized by setting the entry corresponding to the query image to 1. In the case of a tag-based query, a simple tag-based search method is employed and the K top images are returned from all the images that include the query tag in their corresponding set of tags. Let $\Gamma = \{\gamma_1, \gamma_2, \ldots, \gamma_K\} \subset \mathbb{V}_{im}$ be the image set associated to this search. The query vector $\mathbf{y}$ is initialized as follows:

$$y(v) = \begin{cases} 1, & \text{if } v \in \Gamma \\ 0, & \text{otherwise.} \end{cases} \tag{8.6.18}$$

The ranking vector $\mathbf{f}^*$ is derived by solving (8.5.7), as detailed in Section 8.5.1. It has the same size and structure as $\mathbf{y}$.

The averaged Recall-Precision and the F_1 measure are used as figures of merit as in Section 8.6.4. Let us denote the proposed method as fuzzy hypergraph learning (FHL) and the one proposed in [GWZ+13] as HG-WE. The HG-WE was implemented by following precisely the details in [GWZ+13].

In order to evaluate the proposed method, a dataset of 3291 images, depicting 11 popular Greek sites (the old city of Rhodes, Santorini, the White Tower at Thessaloniki, Parthenon, Delphi, Meteora, the ancient Olympia, Sounio, Mycenae, the Greek Parliament, and

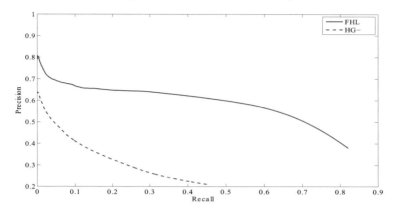

FIGURE 8.6.6: Averaged Recall-Precision curves for image-based queries.

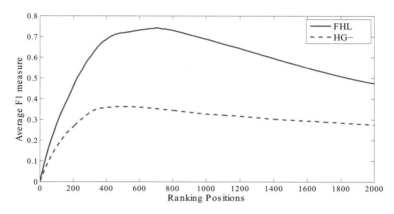

FIGURE 8.6.7: Averaged F_1 measure at several ranking positions for tag-based queries.

Epidaurus) was collected from Flickr. These Greek sites were used as query tags in the evaluation procedure. Both textual and visual vocabularies were derived and the typical values of R and Q were set to 3000 and 2000, respectively. Next, the dataset was further enriched by including 4986 unseen test images collected also from Flickr. Experiments were conducted for both image-based queries ($K = 1$) and tag-based queries ($K = 100$).

The FHL outperforms the HG-WE for both image-based and tag-based search, as shown in Figures 8.6.6 and 8.6.7, respectively. As is demonstrated in Figure 8.6.6, a precision rate of 77% is achieved for 1% recall for image-based queries. For tag-based queries, the maximum average F_1 measure equals 0.7425. It is obtained at the ranking position 704. The complete curve of the averaged F_1 measure per ranking position is displayed in Figure 8.6.7. It is worth mentioning that the FHL is much less computationally expensive than the HG-WE, as only one least squares minimization problem should be solved to obtain $\mathbf{f}^*$ in (8.5.7).

8.6.6   Annotation, classification, and tourism recommendation driven by probabilistic latent semantic analysis

In this section, we deal with image annotation, classification, and tourism recommendation [PK14b]. To make the discussion specific, we refer to a dataset of 50000 images related to Greek sites that was collected from Flickr. The geo-tags (GPS coordinates) of these images were clustered into 4660 clusters by means of hierarchical clustering applied to distances computed using the "Haversine formula." From these geo-clusters, only the 2000 most dense were considered as tourist places of interest (POIs), containing 45316 images. For each geo-cluster, a document was created by concatenating the text information (e.g., title, tags) of all the constituent images. Next, text information related to 150000 images was crawled in order to properly capture the context of the tourism application. All characters were converted to lower case. Unreadable symbols and redundant information were removed. Terms with frequency less than 100 were eliminated, yielding a vocabulary of 1901 terms.

Probabilistic latent semantic analysis (PLSA) performs a probabilistic mixture decomposition, which associates an unobserved class variable to co-occurrences of terms and documents. By applying the PLSA to a term-document matrix, the relations between the terms and the documents are captured by the probability distribution between the documents and the generated topics as well as between the topics and the terms. The PLSA models each term in a document as a sample from a mixture model [Hof99, BK14].

Let $t_a \in \mathbb{T}_a = \{t_{a_1}, t_{a_2}, \ldots, t_{a_k}\}$ be a vocabulary term and $d \in \mathbb{D} = \{d_1, d_2, \ldots, d_m\}$ denote a document. The joint probability model is defined by the mixture:

$$\begin{aligned}
P(t_a, d) &= P(d)P(t_a|d) \\[1em]
P(t_a|d) &= \sum_{z_a \in \mathbb{Z}_a} P(t_a|z_a)\, P(z_a|d),
\end{aligned}$$

(8.6.19)

where $z_a \in \mathbb{Z}_a = \{z_{a_1}, z_{a_2}, \ldots, z_{a_n}\}$ is an unobserved class variable representing the topics. As it is indicated in (8.6.19), the document specific term distribution $P(t_a|d)$ is a convex combination of the n topic dependent distributions $P(t_a|z_a)$. The annotation procedure is performed as follows:

1. PLSA is applied to a term-document matrix $\mathbf{B}_a \in \mathbb{R}^{k \times m}$. Here, the documents are formed by concatenating any terms in the tags or the title of the images that belong to a geo-cluster. Any document $d \in \mathbb{D}$ is represented by a vector of size k, having as elements the frequency of occurrence of each term in d.

2. For each document to be annotated, the most related topic is chosen, that with the highest probability, i.e., $z_a^* = \underset{z_a \in \mathbb{Z}_a}{\operatorname{argmax}} P(z_a|d)$.

3. The $k' << k$ most related terms to z_a^* are identified by sorting $P(t_a|z_a^*)$ in decreasing order of magnitude.

In particular, the term document matrix $\mathbf{B}_a$ is of size 1901×2000. Among the most descriptive terms of a document, those providing geographical information are identified using geogazetteers. Thus, a complete annotation model is built, which provides geographic information in addition to semantic information.

The semantic annotation is complemented by visual annotation based on scene classification. SIFT and GIST descriptors are extracted from any image. Different visual classes $c \in \mathbb{C} = \{c_1, c_2, \ldots, c_p\}$ have been defined, capturing the different themes pertaining the image dataset. The objective is to propagate the class label along with the associated tags to each image as visual annotation. To construct a proper visual word vocabulary, a small image subset $\mathbb{G} = \{g_1, g_2, \ldots, g_\nu\}$, made of images without occlusion or unwanted artifacts, is manually extracted and annotated using the p class labels. K means is applied to the

SIFT descriptors of any image in the controlled dataset $\mathbb{G}$ in order to quantize the descriptors to a predefined number (e.g., 200) of cluster mean vectors as codevectors. Any image $g \in \mathbb{G}$ is represented by the concatenation of the codevectors forming the vector $\tilde{\mathbf{g}}$. K-means is applied to the set of the aforementioned vectors $\tilde{\mathbf{g}}$ in order to define the visual word vocabulary. The indices of the resulting codevectors are treated as visual words $t_v \in \mathbb{T}_v = \{t_{v_1}, t_{v_2}, \ldots, t_{v_\iota}\}$, where ι is the size of visual word vocabulary. Let $\mathbf{B}_v \in \mathbb{R}^{\iota \times \nu}$ be the visual word-image matrix, having as columns the image representations built by measuring the frequency of visual words the reduced representations are quantized into. Similar to [BZM06], the PLSA is applied to $\mathbf{B}_v$ in order to calculate the conditional distributions $P(t_v|z_v)$ and $P(z_v|g)$, where $z_v \in \mathbb{Z}_v = \{z_{v_1}, z_{v_2}, \ldots, z_{v_l}\}$ are the visual latent topics. Having learned the aforementioned conditional distributions from $\mathbb{G}$, any test image g_{test} is represented by the conditional distribution $P(z_v|g_{test})$, obtained by running the M step of the EM algorithm for $P(z_v|g_{test})$ until convergence, keeping $P(t_v|z_v)$ fixed to those learned during the training. Next, the ι_G nearest neighbors of the GIST descriptor extracted from any test image g_{test} are identified, using the K-nearest neighbor classifier (KNN), which employs the Euclidean distances between g_{test} and the GIST descriptor of any image in the controlled subset $g \in \mathbb{G}$. Let $\mathbb{G}_{NN}(g_{test})$ be the set of nearest neighbors. $\mathbb{G}_{NN}(g_{test})$ is further narrowed to $\iota_{GR} << \iota_G$ training images by sorting the χ^2 distances between $P(z_v|g_{test})$ and $P(z_v|g)$, retaining the images associated to the ι_{GR} smallest distances. Let $\tilde{\mathbb{G}}_{NN}(g_{test})$ be the resulting narrow set. Finally, the test image is assigned to the visual class c being in majority within the narrow set $\tilde{\mathbb{G}}_{NN}(g_{test})$.

A hypergraph is created to capture the multi-link relations among the vocabulary terms t_a, the geo-clusters d, and the topics z_a. The incidence matrix of the hypergraph $\mathbf{H}$ has size 4251×6000 elements. It is formed by concatenating 2000 documents associated to the geo-clusters, 350 topics z_a, and 1901 vocabulary terms t_a. The vertex set is defined as $\mathbb{V} = \mathbb{D} \cup \mathbb{Z}_a \cup \mathbb{T}_a$.

1. For each document d_j associated to a geo-cluster, a hyperedge $e_1 \in \mathbb{E}_1$ is inserted, containing 1 in the j-th entry of $\mathbb{D}$, 1 for the most related topic $z_a^* \in \mathbb{Z}_a$ to d_j, and 30 1s for the 30 most descriptive terms $t_a \in \mathbb{T}_a$ for z_a^*. The weight for this hyperedge is $w(e_1) = P(z_a^*|d_j)$.

2. To capture the geographical proximity, hyperedges $e_2 \in \mathbb{E}_2$ are created. For each d_j corresponding to a specific geo-cluster, one hyperedge e_2 is inserted. It contains 1 to the j-th entry of $\mathbb{D}$ associated to d_j and 1 to the entries corresponding to geo-clusters being at a geographical distance less than 150 km. The weight for this hyperedge is set to 1.

3. In order to capture the visual similarity of the geo-clusters, the mean value of the GIST descriptors of all the images belonging in a geo-cluster is computed as a codevector. For each d_j, one hyperedge e_3 is inserted, having 1 to the j-th entry associated to d_j and 1 to the 10 nearest neighbor geo-clusters, identified by applying KNN to the aforementioned codevectors. The hyperedge weight is set to 1.

The structure of the hypergraph incidence matrix is summarized in Table 8.6.5.

Let Θ be defined as in (8.4.39) that can be found in Section 8.4. If d_j' is the geo-cluster where the test image g_{test} belongs to with respect to its geo-tag, the query vector $\mathbf{y} \in \mathbb{R}^{|V|}$ has elements:

$$y(v) = \begin{cases} 1, & \text{if } v = d_j' \\ \Theta(d_j', v), & \text{otherwise,} \end{cases} \qquad (8.6.20)$$

where $\Theta(d_j', v)$ is treated as a measure of relatedness between the vertices of the hypergraph.

TABLE 8.6.5: The hypergraph incidence matrix **H** for place of interest recommendation.

	$\mathbb{E}_1$	$\mathbb{E}_2$	$\mathbb{E}_3$
$\mathbb{D}$	$(\mathbb{D}, \mathbb{E}_1)$	$(\mathbb{D}, \mathbb{E}_2)$	$(\mathbb{D}, \mathbb{E}_3)$
$\mathbb{Z}_a$	$(\mathbb{Z}_a, \mathbb{E}_1)$	0	0
$\mathbb{T}_a$	$(\mathbb{T}_a, \mathbb{E}_1)$	0	0

The best ranking vector $\mathbf{f}^* \in \mathbb{R}^{|V|}$ is obtained by (8.5.7). The values corresponding to the first 2000 entries associated to geo-cluster documents are used as rankings for tourist destination recommendation. The top ranked geo-cluster documents are recommended as tourist POIs to the user, who has imported the test image g_{test}.

For evaluation purposes, a test set containing 205 images was randomly chosen and excluded from the training set along with any text associated to these images. The PLSA performance in semantic image annotation has been compared to that of the latent Dirichlet allocation (LDA) [BNJ03] and the term frequency-inverse document frequency (TF-IDF) [SWY75]. The average recall-precision curve is used as a figure of merit. As is shown in Figure 8.6.8, PLSA outperforms both the LDA and the TF-IDF. An average precision of 90% at 10% recall is reported, using the PLSA. It is worth noting, that the PLSA is much simpler than the LDA.

For visual image classification, the same test set was used. Each test image was assigned to one of 13 representative classes manually in order to create the ground truth. Visual classification accuracy is shown in Figure 8.6.9, when only the GIST descriptors were used and when both the SIFT and the GIST descriptors were employed. Better results were obtained by using both descriptors. Across the 205 test set images, the average accuracy of content-based image classification over 13 classes is 80%.

Two additional experiments were conducted to assess tourist POI recommendation. First, only hyperedges $e_1 \in \mathbb{E}_1$ were taken into account in hypergraph creation. Second, all the hyperedges were considered. In order to form the ground truth, relations were established manually among the geo-clusters, taking into account the distance, common geographical

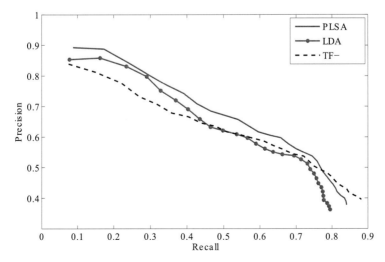

FIGURE 8.6.8: Recall-Precision curves for semantic image annotation by means of the PLSA, the LDA, and the TF-IDF.

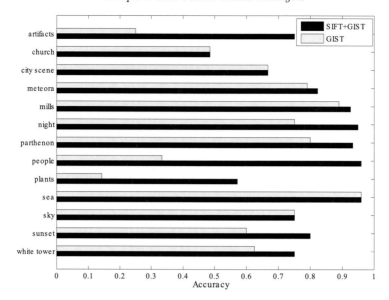

FIGURE 8.6.9: Accuracy results of the visual image classification.

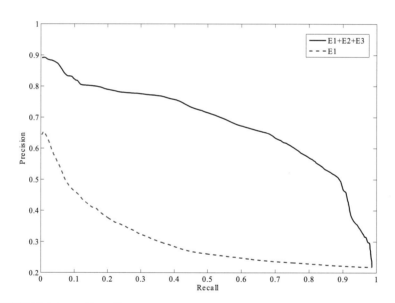

FIGURE 8.6.10: Recall-precision curves for tourist POI recommendation.

entities (e.g., mainland, island) and leisure activities. For this, various tourist related web sources were exploited, such as Trip Advisor (http://www.tripadvisor.com.gr/) and Travel Muse (http://www.travelmuse.com/). The associated recall-precision curves are plotted in Figure 8.6.10. As is clearly indicated, the results are increased when all the three types of hyperedges are considered, including the visual similarity between the geo-clusters. An average precision of 90% and 82% is reported at 1% and 10% recall, respectively.

8.7 Big data: Randomized methods for matrix/hypermatrix decompositions

From the applications described in Section 8.6, it has become evident that the normalized hypergraph Laplacian matrix or the adjacency hypermatrix of the hypergraph can admit stupendously big sizes. Classical linear algebra algorithms, such as the algorithms for eigen-analysis, SVD, or hypermatrix factorizations are not always adapted to deal with the aforementioned large-scale problems [HMT11]. Furthermore, one has to cope with missing or inaccurate data that frequently arise from increasing matrix sizes. The noise in the data and the propagation of rounding errors can become increasingly problematic. Nowadays, data transfer is a crucial factor. Indeed, communication becomes a bottleneck, because of the slow communication speed between different layers in memory hierarchy, latency in hard drives, inter-processor communications. Algorithms requiring few passes over the data have been proven much faster in practice despite the more floating point operations that are needed. Another practical necessity is the adaptation of algorithms to novel computer processor architectures, such as graphics processing units.

Randomized algorithms provide a powerful tool for constructing approximate matrix factorizations. They are simple, often faster, and perhaps surprisingly more robust than the standard deterministic algorithms. Frequently, they produce factorizations that are accurate to any specified tolerance above machine precision. Such algorithms are an example of the large-scale applicable global operations.

Standard deterministic matrix decompositions include the pivoted QR factorization, the eigenvalue decomposition, and the SVD [GL06]. The computational cost of the full QR factorization or the full SVD of an $N \times M$ matrix to double-precision accuracy is $\mathcal{O}(N\,M\,\min\{N,M\})$ flops. Truncated versions of these factorizations are often used to obtain a *low-rank approximation* of a given matrix $\mathbf{A} \in \mathbb{R}^{N \times M}$ of rank k, i.e.:

$$\mathbf{A} \approx \mathbf{B}\,\mathbf{C}, \tag{8.7.1}$$

where $\mathbf{B} \in \mathbb{R}^{N \times k}$ and $\mathbf{C} \in \mathbb{R}^{k \times M}$. When $k \ll N$, the factorization (8.7.1) allows the matrix to be stored inexpensively and to be multiplied rapidly with vectors or other matrices. Furthermore, the factorization (8.7.1) can be used for data interpretation or for solving least squares problems.

The most straightforward technique to obtain (8.7.1) is to compute the full SVD and to truncate it afterwards. However, one should know the rank k. There are two alternatives to estimate the rank: (i) to compute a partial QR decomposition; (ii) to apply a strong rank-revealing QR factorization. The classical algorithm to compute a partial QR decomposition is the Businger-Golub algorithm, which performs successive orthogonalization with pivoting on the columns of the matrix. The procedure stops after l steps, when the Frobenius norm of the remaining columns is less than a computational tolerance ϵ, i.e.:

$$\mathbf{A} = \mathbf{Q}\,\mathbf{R} + \mathbf{E}, \tag{8.7.2}$$

where $\mathbf{Q} \in \mathbb{R}^{N \times l}$ is a column orthonormal matrix, $\mathbf{R} \in \mathbb{R}^{l \times M}$ is a weakly upper-triangular matrix, and $\mathbf{E} \in \mathbb{R}^{N \times M}$ satisfies $||\mathbf{E}||_F \le \epsilon$. The computational cost is $\mathcal{O}(N\,M\,l)$. The number of steps l overestimates the rank k. The Gu-Eisenstat algorithm yields a strong-rank revealing QR decomposition (8.7.2) at a computational cost $\mathcal{O}(kNM)$, where ℓ_2 subordinate matrix norm $||E||_2 = \max_{\mathbf{x}\neq\mathbf{0}} \left(\frac{||\mathbf{E}\,\mathbf{x}||_2}{||\mathbf{x}||_2} \right)$ satisfies $||\mathbf{E}||_2 \le \sqrt{1 + 4\,k\,(M-k)}\,\sigma_{k+1}$, where σ_{k+1} is the $(k+1)$-th largest singular value of $\mathbf{E}$ that constitutes the minimal error possible in a rank-k approximation.

To compute the low-rank approximation of $\mathbf{A}$ (8.7.1), the first stage deals with the construction of a low-dimensional subspace that captures the action of $\mathbf{A}$. Formally, an approximate basis of the range of $\mathbf{A}$ is computed. That is, a column orthonormal matrix $\mathbf{Q} \in \mathbb{R}^{N \times k}$ (i.e., $\mathbf{Q}^T \mathbf{Q} = \mathbf{I}$) is sought such that:

$$\mathbf{A} \approx \mathbf{Q} \, \mathbf{Q}^T \, \mathbf{A}. \tag{8.7.3}$$

The basis matrix $\mathbf{Q}$ should contain as few columns as possible (i.e., $k \leq M$). But the primary objective is to yield an accurate approximation of $\mathbf{A}$. The second stage aims at restricting $\mathbf{A}$ to the subspace spanned by the columns of $\mathbf{Q}$ by constructing the small matrix of size $k \times M$ $\mathbf{B} = \mathbf{Q}^T \mathbf{A}$. Let $\mathbf{B} = \tilde{\mathbf{U}} \boldsymbol{\Sigma} \mathbf{V}^T$ denote the SVD of $\mathbf{B}$. To compute the column orthonormal matrices $\mathbf{U}$ and $\mathbf{V}$ as well as a nonnegative diagonal matrix $\boldsymbol{\Sigma}$, such that $\mathbf{A} \approx \mathbf{U} \boldsymbol{\Sigma} \mathbf{V}^T$, simply set $\mathbf{U} = \mathbf{Q} \tilde{\mathbf{U}}$, because $\mathbf{A} \approx \mathbf{Q} \mathbf{B}$. When $\mathbf{Q}$ has few columns, this procedure is efficient, because the small matrix $\mathbf{B}$ can easily be constructed and the SVD of $\mathbf{B}$ can be computed fast. In practice, the explicit construction of $\mathbf{B}$ can be avoided. It is not even necessary to revisit $\mathbf{A}$ by means of the so-called *single-pass* algorithms [HMT11]. Similar manipulations yield efficient algorithms for other standard factorizations, such as the pivoted QR and the eigen-decomposition.

Random sampling methods help to execute efficiently the first stage. That is, given $\mathbf{A} \in \mathbb{R}^{N \times M}$ of rank k known in advance and an oversampling parameter p, a column orthonormal matrix $\mathbf{Q} \in \mathbb{R}^{N \times l}$ is sought whose range approximates the range of $\mathbf{A}$, where $l = k + p$. This is achieved in rough terms by means of Algorithm 8.7.1.

Algorithm 8.7.1: Randomized range finder for the first stage.

Input: Matrix $\mathbf{A} \in \mathbb{R}^{N \times M}$, integer l.
Output: Column orthonormal matrix $\mathbf{Q} \in \mathbb{R}^{N \times l}$ whose range approximates the range of $\mathbf{A}$.

1: Draw an $M \times l$ random matrix $\boldsymbol{\Omega}$ whose elements are independent identically distributed Gaussian random variables with zero mean and unit variance.
2: Form the $N \times l$ matrix $\mathbf{Y} = \mathbf{A} \, \boldsymbol{\Omega}$.
3: Construct an $N \times l$ matrix $\mathbf{Q}$ whose columns form an orthonormal basis for the range of $\mathbf{Y}$, e.g., by means of the QR factorization of $\mathbf{Y} = \mathbf{Q} \, \mathbf{R}$.

The quality of the random number generator in Step 1 does not affect the quality of Algorithm 8.7.1. The structure of matrix $\mathbf{A}$ (i.e., whether it is sparse or structured) determines the speed of matrix-vector products in the Step 2 of Algorithm 8.7.1, which constitutes its computational bottleneck. The most important implementation issue is the basis calculation in Step 3. The columns of the sample matrix $\mathbf{Y}$ are almost linearly dependent. This calls for using stable methods to perform the orthonormalization. For example, the Gram-Schmidt procedure, augmented with the double orthogonalization [Bjö94] has been proven reliable in practice. The oversampling parameter p, which measures the discrepancy between l and the actual rank k depends on three factors:

The matrix dimensions. Very large matrices may require more oversampling.

The singular spectrum. The more rapid the decay of the singular values, the less oversampling is needed.

The random test matrix. Gaussian matrices with very little oversampling (e.g., $p=5$ or $p=10$) are found sufficient in practice. The structured random matrices that will be discussed next yield computational gains in certain settings at the expense of substantial oversampling.

The basis in Step 3 of Algorithm 8.7.1 can be generated incrementally, starting with an empty basis matrix $\mathbf{Q}^{(0)}$, as follows [HMT11]:

1: **for** $i = 1, 2, \ldots$ **do**
2:    Draw an $M \times 1$ Gaussian random vector $\boldsymbol{\omega}^{(i)}$ and set $\mathbf{y}^{(i)} = \mathbf{A}\,\boldsymbol{\omega}^{(i)}$.
3:    Compute $\tilde{\mathbf{q}}^{(i)} = \left(\mathbf{I} - \mathbf{Q}^{(i-1)}(\mathbf{Q}^{(i-1)})^T\right)\mathbf{y}^{(i)}$.
4:    Normalize $\mathbf{q}^{(i)} = \frac{\tilde{\mathbf{q}}^{(i)}}{||\tilde{\mathbf{q}}^{(i)}||_2}$, and form $\mathbf{Q}^{(i)} = \left[\mathbf{Q}^{(i-1)} | \mathbf{q}^{(i)}\right]$.
5: **end for**

For matrices whose singular values decay slowly, Algorithm 8.7.2 has been proven more efficient than Algorithm 8.7.1 [HMT11].

Algorithm 8.7.2: Randomized subspace iteration for the first stage.

Input: Matrix $\mathbf{A} \in \mathbb{R}^{N \times M}$, integers l and q.
Output: Column orthonormal matrix $\mathbf{Q} \in \mathbb{R}^{N \times l}$ whose range approximates the range of $\mathbf{A}$.

1: Draw an $M \times l$ random matrix $\boldsymbol{\Omega}$ whose elements are independent identically distributed Gaussian random variables with zero mean and unit variance.
2: Form $\mathbf{Y}_0 = \mathbf{A}\,\boldsymbol{\Omega}$ and compute its QR factorization $\mathbf{Y}_0 = \mathbf{Q}_0\,\mathbf{R}_0$.
3: **for** $j = 1, 2, \ldots, q$ **do**
4:    Form $\tilde{\mathbf{Y}}_j = \mathbf{A}^T\,\mathbf{Q}_{j-1}$ and compute its QR factorization $\tilde{\mathbf{Y}}_j = \tilde{\mathbf{Q}}_j\,\tilde{\mathbf{R}}_j$.
5:    Form $\mathbf{Y}_j = \mathbf{A}\,\tilde{\mathbf{Q}}_j$ and compute its QR factorization $\mathbf{Y}_j = \mathbf{Q}_j\,\mathbf{R}_j$.
6: **end for**
7: $\mathbf{Q} = \mathbf{Q}_q$.

For a dense matrix $\mathbf{A} \in \mathbb{R}^{N \times M}$, it is possible to obtain its approximate rank-l factorization can be computed in roughly $\mathcal{O}(N\,M\,\log l)$ flops, in contrast to the asymptotic cost $\mathcal{O}(N\,M\,l)$ required by the earlier methods. The key idea of the fast technique is to use a *structured random matrix* that allows us to compute the product $\mathbf{A}\,\boldsymbol{\Omega}$ in $\mathcal{O}(N\,M\,\log l)$ flops by employing a *subsampled random Fourier transform*(SRFT), yielding the $M \times l$ matrix:

$$\boldsymbol{\Omega} = \frac{M}{l}\,\mathbf{D}\,\mathbf{F}\,\mathbf{R}, \qquad (8.7.4)$$

where $\mathbf{D}$ is an $M \times M$ diagonal matrix, whose elements are independent random variables uniformly distributed on the complex unit circle, $\mathbf{F}$ is the $M \times M$ unitary discrete Fourier transform, whose elements are $f_{pq} = \frac{1}{\sqrt{M}} \exp(-\frac{j\,2\,\pi\,(p-1)\,(q-1)}{M})$, and $\mathbf{R}$ is an $M \times l$ matrix that samples l coordinates from M uniformly at random, i.e., its l columns are drawn randomly without replacement from the columns of the $M \times M$ identity matrix. If $\boldsymbol{\Omega}$ is given by (8.7.4), the product $\mathbf{Y} = \mathbf{A}\boldsymbol{\Omega}$ can be computed in $\mathcal{O}(N\,M\,l)$ flops via a subsampled Fast Fourier Transform [WLRT08].

In this chapter, we are interested in the eigenvalue decomposition of the normalized hypergraph Laplacian $\tilde{\mathbf{L}} \in \mathbb{R}^{N \times N}$, which is a symmetric matrix. Moreover, the normalized hypergraph Laplacian is a positive semidefinite. Accordingly, the discussion for the second stage will be focused to the case of a positive semidefinite matrix $\mathbf{A} = \tilde{\mathbf{L}}$. For positive semidefinite matrices, the *Nyström method* can be used to improve the quality of standard factorizations. The Nyström method builds a more sophisticated rank-k approximation, i.e.:

$$
\begin{aligned}
\mathbf{A} &\approx (\mathbf{A}\,\mathbf{Q})\left(\mathbf{Q}^T\,\mathbf{A}\,\mathbf{Q}\right)^{-1}(\mathbf{A}\,\mathbf{Q})^T \\
&= \left[(\mathbf{A}\,\mathbf{Q})\left(\mathbf{Q}^T\,\mathbf{A}\,\mathbf{Q}\right)^{-\frac{1}{2}}\right]\left[(\mathbf{A}\,\mathbf{Q})\left(\mathbf{Q}^T\,\mathbf{A}\,\mathbf{Q}\right)^{-\frac{1}{2}}\right]^T \\
&= \mathbf{G}\,\mathbf{G}^T, \qquad (8.7.5)
\end{aligned}
$$

where $\mathbf{G}$ is an approximate Cholesky factor of $\mathbf{A}$ with size $N \times k$. To compute the factor $\mathbf{G}$ numerically, first form the matrices $\mathbf{B}_1 = \mathbf{A}\,\mathbf{Q}$ and $\mathbf{B}_2 = \mathbf{Q}^T\,\mathbf{B}_1$. Then, the positive semidefinite matrix $\mathbf{B}_2$ is decomposed into its Cholesky factors, i.e., $\mathbf{B}_2 = \mathbf{C}^T\,\mathbf{C}$. It can easily be verified that:

$$
\begin{aligned}
\mathbf{B}_2 &= \mathbf{Q}^T\,\mathbf{A}\,\mathbf{Q} \\
\mathbf{C}^T\,\mathbf{C} &= \mathbf{Q}^T\,(\mathbf{G}\,\mathbf{G}^T)\,\mathbf{Q} \\
&= (\mathbf{Q}^T\,\mathbf{G})\,(\mathbf{G}^T\,\mathbf{Q}).
\end{aligned}
\tag{8.7.6}
$$

That is, $\mathbf{C} = \mathbf{G}^T\,\mathbf{Q}$. By pre-multiplying both sides of the previous equation by $\mathbf{G}$, we obtain:

$$
\mathbf{G}\,\mathbf{C} = \underbrace{\mathbf{G}\,\mathbf{G}^T}_{\mathbf{A}}\,\mathbf{Q} = \underbrace{\mathbf{A}\,\mathbf{Q}}_{\mathbf{B}_1}.
\tag{8.7.7}
$$

Accordingly, solve for $\mathbf{G}$ the linear system $\mathbf{G}\,\mathbf{C} = \mathbf{B}_1$, using a triangular solve, since $\mathbf{C}$ is a triangular matrix. The second stage is summarized in Algorithm 8.7.3.

Algorithm 8.7.3: Eigen-decomposition via the Nyström method for the second stage.

Input: Positive semidefinite matrix $\mathbf{A} \in \mathbb{R}^{N \times N}$ and a basis $\mathbf{Q} \in \mathbb{R}^{N \times l}$ satisfying (8.7.3).
Output: Approximate eigen-decomposition $\mathbf{A} \approx \mathbf{U}\,\mathbf{\Lambda}\,\mathbf{U}^T$, where $\mathbf{U}$ is an orthonormal matrix and $\mathbf{\Lambda}$ is a nonnegative diagonal matrix.

1: Form the matrices $\mathbf{B}_1 = \mathbf{A}\,\mathbf{Q}$ and $\mathbf{B}_2 = \mathbf{Q}^T\,\mathbf{B}_1$.
2: Perform a Cholesky factorization $\mathbf{B}_2 = \mathbf{C}^T\,\mathbf{C}$.
3: Solve for $\mathbf{G}$ the linear system $\mathbf{G}\,\mathbf{C} = \mathbf{B}_1$, using a triangular solve.
4: Compute an eigen-decomposition of $\mathbf{G} = \mathbf{U}\,\mathbf{\Lambda}\,\mathbf{U}^T$.

By merging Algorithms 8.7.1 and 8.7.3, the complete Algorithm 8.7.4 is obtained. For

---

Algorithm 8.7.4: Single pass algorithm.

Input: Positive semidefinite matrix $\mathbf{A} \in \mathbb{R}^{N \times N}$.
Output: Approximate eigen-decomposition $\mathbf{A} \approx \mathbf{U}\,\mathbf{\Lambda}\,\mathbf{U}^T$, where $\mathbf{U}$ is an $N \times l$ column orthonormal matrix and $\mathbf{\Lambda}$ is an $l \times l$ nonnegative diagonal matrix.

1: Draw an $N \times l$ random matrix $\mathbf{\Omega}$.
2: Form the $N \times l$ matrix $\mathbf{Y} = \mathbf{A}\,\mathbf{\Omega}$.
3: Find an $N \times l$ matrix $\mathbf{Q}$, such that $\mathbf{Y} \approx \mathbf{Q}\,\mathbf{Q}^T\,\mathbf{Y}$.
4: Solve for an $l \times l$ matrix $\mathbf{G}$ the linear system $\mathbf{G}\,(\mathbf{Q}^T\,\mathbf{\Omega}) = \mathbf{Q}^T\,\mathbf{Y}$.
5: Compute an eigen-decomposition of $\mathbf{G} = \tilde{\mathbf{U}}\,\mathbf{\Lambda}\,\tilde{\mathbf{U}}^T$.
6: Form $\mathbf{U} = \mathbf{Q}\,\tilde{\mathbf{U}}$.

---

a sparse or structured (e.g., Toeplitz) matrix $\mathbf{A}$, the matrix-vector multiplications needed in Step 2 of Algorithm 8.7.4 can be done much faster than $\mathcal{O}(N^2)$. It is not uncommon that $\mathcal{O}(2N)$ flops suffice. Let T_{mult} be the exact cost of matrix-vector multiplication. The computational complexity of Algorithm 8.7.4 is $2lT_{\mathrm{mult}} + \mathcal{O}(2k^2 N)$ [HMT11]. The interested reader may refer to [HMT11] for a detailed performance analysis that provides theoretical guarantees that the approximation error almost surely is less than a certain upper bound as well as a thorough derivation of the computational cost paid in various situations.

Many recent works have focused on the design and analysis of algorithms that efficiently create small *"sketches" of matrices and hypermatrices*. Such sketches have been used in eigen-decompositions [FKV98, AF07], semi definite programming solvers, and matrix completion. Sketches of hypermatrices have been exploited in decompositions [dKKV05, DM07, MD08]. Let $\mathbb{I}$ be the set of integers $\{1, 2, \ldots, N\}$ and $\times_d \mathbb{I} = \underbrace{\mathbb{I} \times \mathbb{I} \times \ldots \times \mathbb{I}}_{d \text{ times}}$.

Given an order-d hypermatrix $\mathcal{A} \in \mathbb{R}^{\times_d \mathbb{I}}$ and an error tolerance $\epsilon \geq 0$, a hypermatrix sketch $\tilde{\mathcal{A}} \in \mathbb{R}^{\times_d \mathbb{I}}$ of $\mathcal{A}$ is constructed such that:

$$||\mathcal{A} - \tilde{\mathcal{A}}||_2 \leq \epsilon \, ||\mathcal{A}||_2 \qquad (8.7.8)$$

and the number of non-zero entries in $\tilde{\mathcal{A}}$ is minimized, where $|| \, ||_2$ denotes the spectral norm of a hypermatrix. Algorithm 8.7.5 has been proposed that enjoys theoretical guarantees in [NDT15]. The algorithm zeroes out "small" elements of the hypermatrix $\mathcal{A}$, keeps "large" elements of $\mathcal{A}$, and randomly samples the remaining elements of $\mathcal{A}$ with probability that depends on their magnitude.

Algorithm 8.7.5: Hypermatrix sparsification.

Input: Order-d hypermatrix $\mathcal{A}$ and sampling parameter s.
Output: Order-d sketch $\tilde{\mathcal{A}}$ of $\mathcal{A}$.

1: **for** $i_1, i_2, \ldots i_d \in \mathbb{I} \times \mathbb{I} \times \ldots \times \mathbb{I}$ **do**
2:     **if** $a_{i_1, i_2, \ldots, i_d}^2 \leq \frac{\ln^2 N}{N^{\frac{d}{2}}} \frac{||\mathcal{A}||_F^2}{s}$ **then**
3:        $\tilde{a}_{i_1, i_2, \ldots, i_d} = 0$
4:     **else**
5:        **if** $a_{i_1, i_2, \ldots, i_d}^2 \geq \frac{||\mathcal{A}||_F^2}{s}$ **then**
6:           $\tilde{a}_{i_1, i_2, \ldots, i_d} = a_{i_1, i_2, \ldots, i_d}$
7:        **else**
8:

$$\tilde{a}_{i_1, i_2, \ldots, i_d} = \begin{cases} \frac{a_{i_1, i_2, \ldots, i_d}}{p_{i_1, i_2, \ldots, i_d}} & \text{with probability } p_{i_1, i_2, \ldots, i_d} = \frac{s \, a_{i_1, i_2, \ldots, i_d}^2}{||\mathcal{A}||_F^2} \\ 0 & \text{with probability } 1 - p_{i_1, i_2, \ldots, i_d} \end{cases}$$

9:        **end if**
10:     **end if**
11: **end for**

This section is concluded with a straightforward extension of the work in [AF07] to hypermatrices [Tso10] in conjunction with high-order SVD (HOSVD), summarized in Algorithm 8.7.6.

8.8 Conclusions

In this chapter, hypergraphs have been studied. It has been shown that the hypergraphs possess strong mathematical foundations and constitute a powerful concept, allowing us to cast a number of multimedia social search problems as clustering or ranking problems.

Starting with the κ-uniform hypergraphs, their spectra have been reviewed. Evolutionary stable strategies for clustering on κ-uniform hypergraphs have been shown to be closely related to a growth transformation for the maximization of a homogeneous polynomial. That is, such strategies constitute a special case of the Baum-Eagon theorem. Moreover,

---

Algorithm 8.7.6: MACH-HOSVD.

Input: Order-d hypermatrix $\mathcal{A}$, number of factors $R_1, R_2, \ldots, R_d$, and probability p.
Output: Core hypermatrix $\mathcal{G}$ and matrices $\mathbf{X}^{(j)}$, $j = 1, 2, \ldots, d$ made up of the R_j leading left singular vectors of the j-th mode matrices of the sketch hypermatrix $\tilde{\mathcal{A}}$.

1: **for** $i_1, i_2, \ldots i_d \in \mathbb{I} \times \mathbb{I} \times \ldots \times \mathbb{I}$ **do**
2:     Toss a coin with probability p of keeping it
3:     **if** success **then**
4:        $\tilde{a}_{i_1, i_2, \ldots, i_d} = \frac{a_{i_1, i_2, \ldots, i_d}}{p}$
5:     **else**
6:        $\tilde{a}_{i_1, i_2, \ldots, i_d} = a_{i_1, i_2, \ldots, i_d}$
7:     **end if**
8: **end for**
9: **for** $j = 1, 2, \ldots, d$ **do**
10:     Extract the leading R_j left singular vector of the j-th mode matrix $\tilde{\mathbf{A}}^{(j)}$ and store them in $\mathbf{X}^{(j)}$.
11: **end for**
12: Set $\mathcal{G} = \tilde{\mathcal{A}} \times_{j=1}^{d} \left(\mathbf{X}^{(j)} \right)^T$.

---

high-order decomposition techniques have been applied to the adjacency hypermatrix of κ-uniform hypergraphs, revealing their tight link to hypermatrix decompositions.

Next, spectral clustering for arbitrary hypergraphs has been studied. It has been shown that the underlined techniques stem from various transformations of hypergraphs into graphs, whose edge weights are expressed in terms of the hyperedge ones, notably the click expansion and the star expansion. Emphasis has been given to the hypergraph normalized cut criterion used more frequently in the literature. Its relationship with the star expansion has been established. All the aforementioned transformations describe the hypergraph in terms of the adjacency matrix of the induced graph and resort to the spectral analysis of the (normalized) Laplacian of the induced graph. In addition to clustering, ranking on hypergraphs has been reviewed, because recommendation and tagging can be addressed in that way. Proper optimization problems have been set in order to enforce structural constraints on the weights and their solution has been derived.

Several applications have been demonstrated. Emphasis has been given to music/image recommendation and tagging, geo-location prediction, and tourism recommendation.

Randomized algorithms for the factorization of big matrices and hypermatrices have been surveyed and efficient and fast algorithms have been outlined. The computational cost of the various operations has been explicitly stated.

To sum up, this chapter has unveiled the potential of hypergraphs to capture multimodal high-order relations in social media. Other important aspects of social media, such as temporal evolution, latent model adaptation, incremental spectral clustering, incremental hypermatrix factorizations, and scalability issues are studied in Chapter 11, complementing the topics presented in this chapter. There, parallel and distributed implementations of latent models, spectral analysis, and hypermatrix decompositions are also discussed.

8.9 Acknowledgments

This research has been co-financed by the European Union (European Regional Development Fund - ERDF) and Greek national funds through the Operation Program "Competitiveness-Cooperation 2011" - Research Funding Program: SYN-10-1730-ATLAS.

Bibliography

[ABB06]   S. Agarwal, K. Branson, and S. Belongie. Higher order learning with graphs. In *Proc. 23rd International Conference on Machine Learning*, pages 17–24, 2006.

[AF07]   D. Achlioptas and F. McSherry. Fast low-rank matrix approximations. *Journal of the ACM*, 54(2), 2007.

[AK95]   C. J. Alpert and A. B. Kahng. Recent directions in netlist partitioning: A survey. *Integration, the VLSI Journal*, 19(1-2):1–85, 1995.

[AKY99]   C. J. Alpert, A. B. Kahng, and S.-Z. Yao. Spectral partitioning with multiple eigenvectors. *Discrete Applied Mathematics*, 90:3–26, 1999.

[ALZM+05] S. Agarwal, J. Lim, L. Zelnik-Manor, P. Peronal, D. Kriegman, and S. Belongie. Beyond pairwise clustering. In *Proc. IEEE Conference on Computer Vision and Pattern Recognition*, volume 2, pages 838–845, 2005.

[BB82]   D. H. Ballard and C. M. Brown. *Computer Vision*. Prentice-Hall, 1982.

[BBM05]   A. C. Berg, T. L. Berg, and J. Malik. Shape matching and object recognition using low distortion correspondence. In *Proc. IEEE Conference on Computer Vision and Pattern Recognition*, pages 26–33, 2005.

[BE67]   L. E. Baum and J. A. Eagon. An inequality with applications to statistical estimation for probabilistic functions of Markov processes and to a model for ecology. *Bulletin American Mathematical Society*, 73:360–363, 1967.

[Ber89]   C. Berge. *Hypergraphs*. North Holland, 1989.

[Bjö94]   A. Björk. Numerics of Gram-Schmidt orthogonalization. *Linear Algebra Applications*, 197-198:297–316, 1994.

[BK14]   N. Bassiou and C. Kotropoulos. On-line PLSA: Batch updating techniques including out of vocabulary words. *IEEE Transactions Neural on Networks and Learning Systems*, 25(11):1953–1966, 2014.

[BM73]   C. Berge and E. Minieka. *Graphs and Hypergraphs*. North Holland, 1973.

[BNJ03]   D. M. Blei, A. Y. Ng, and M. I. Jordan. Latent Dirichlet Allocation. *Journal of Machine Learning Research*, 3:993–1022, 2003.

[Bol93]   M. Bolla. Spectral Euclidean representations for clustering of hypergraphs. *Discrete Mathematics*, 117(1-3):19–39, 1993.

[BP98]   S. Brin and L. Page. The anatomy of a large-scale hypertextual Web search engine. In *Proc. 7th International World Wide Web Conference*, pages 107–117, 1998.

[BP13]   S. R. Bulò and M. Pelillo. A game theoretic approach to hypergraph clustering. *IEEE Transactions on Pattern Analysis and Machine Intelligence*, 35(6):1312–1327, 2013.

[BTC+10]   J. Bu, S. Tan, C. Chen, C. Wang, H. Wu, Z. Lijun, and X. He. Music recommendation by unified hypergraph: Combining social media information and music content. In *Proc. ACM Multimedia Conference*, pages 391–400, 2010.

[BZM06]   A. Bosch, A. Zisserman, and X. Munoz. Scene classification via PLSA. In *Proc. European Conference on Computer Vision*, pages 517–530, 2006.

[CC70]   J. D. Carroll and J. J. Chang. Analysis of individual differences in multidimensional scaling with an N-way generalization of "Eckart-Young" decomposition. *Psychometrika*, 35:283–319, 1970.

[CD12]   J. Cooper and A. Dutle. Spectra of uniform hypergraphs. *Linear Algebra and Its Applications*, 436(9):3268–3292, 2012.

[Chu97]   F. R. K. Chung. *Spectral Graph Theory*, volume 92. Regional Conference Series in Mathematics American Mathematical Society, 1997.

[CQ14]   Z. Chen and L. Qi. Circulant tensors with applications to spectral hypergraph theory and stochastic process. *eprint arXiv:1312.2752v7 [Math.SP]*, pages 1–26, 2014.

[CZPA09]   A. Cichocki, R. Zdunek, A. H. Pan, and S.-I. Amari. *Nonnegative Matrix and Tensor Factorizations*. J. Wiley and Sons, Ltd., 2009.

[DBKP11]   O. Duchenne, F. Bach, I. Kweon, and J. Ponce. A tensor-based algorithm for high-order graph matching. *IEEE Transactions on Pattern Analysis and Machine Intelligence*, 33(12):2383–2395, 2011.

[DHZS02]   C. Ding, X. He, H. Zha, and H. Simon. PageRank, HITS, and a unified framework for link analysis. In *Proc. 25th ACM SIGIR Conference Research & Development on Information Retrieval*, pages 353–354, 2002.

[DKK11]   D. M. Dunlavy, T. G. Kolda, and W. P. Kegelmeyer. Multilinear algebra for analyzing data with multiple linkages. In J. Kepner and J. Gilbert, editors, *Graph Algorithms in the Language of Linear Algebra*, pages 85–114. SIAM, 2011.

[dKKV05]   W. F. de la Vega, R. Kannan, M. Karpinski, and S. Vempala. Tensor decomposition and approximation schemes for constraint satisfaction problems. In *Proc. 37th Annual ACM Symposium on Theory of Computing*, pages 747–754, 2005.

[DM07]   P. Drineas and M. W. Mahoney. A randomized algorithm for a tensor-based generalization of the singular value decomposition. *Linear Algebra and its Applications*, 420(2):553–571, 2007.

[ERV06]   E. Estrada and J. A. Rodriguez-Velazquez. Subgraph centrality and clustering in complex hyper-networks. *Physica A: Statistical Mechanics and its Applications*, 364:581–594, 2006.

[Fag83]   R. Fagin. Degrees of acyclicity for hypergraphs and relational database schemes. *Journal of the ACM*, 30(3):514–550, 1983.

[FBH03]   N. M. Faber, R. Bro, and P. K. Hopke. Recent developments in CANDE-COMP/PARAFAC algorithms: A critical review. *Chemometrics and Intelligent Laboratory Systems*, 65:119–137, 2003.

[FKV98]   A. Frieze, R. Kannan, and S. Vempala. Fast Monte Carlo algorithms for finding low-rank approximations. In *Proc. 39th Annual IEEE Symposium on Foundations of Computer Science*, pages 370–378, 1998.

[Fuk10]   T. Fukunaga. Computing minimum multiway cuts in hypergraphs from hypertree packing. In F. Eisenbrand and F. B. Shepherd, editors, *Integer Programming and Combinatorial Optimization*, volume LNCS 6080, pages 15–28. Springer, 2010.

[GL06]   G. H. Golub and C. F. Van Loan. *Matrix Computations*. The Johns Hopkins University Press, 3/e edition, 2006.

[Gov05]   V. M. Govindu. A tensor decomposition for geometric grouping and segmentation. In *Proc. IEEE International Conference on Computer Vision and Pattern Recognition*, pages 1150–1157, 2005.

[GWZ+13]   Y. Gao, M. Wang, Z. Zha, J. Shen, X. Li, and X. Wu. Visual-textual joint relevance learning for tag-based social image search. *IEEE Transactions on Image Processing*, 22(1):363–376, 2013.

[Har70]   R. A. Harshman. Foundations of the PARAFAC procedure: Models and conditions for an "explanatory" multimodal factor analysis. *UCLA Working Papers in Phonetics*, 16:1–84, 1970.

[HG07]   K. Heller and Z. Ghahramani. A nonparametric Bayesian approach to modeling overlapping clusters. In *Proc. International Conference on Artificial Intelligence and Statistics*, pages 187–194, 2007.

[HKKM98]   E.-H. Han, G. Karypis, V. Kumar, and B. Mobasher. Hypergraph based clustering in high-dimensional datasets: A summary of results. *IEEE Data Engineering Bulletin*, 21(1):15–22, 1998.

[HLL+11]   Y. Huang, Q. Liu, F. Lv, Y. Gong, and D. N. Metaxas. Unsupervised image categorization by hypergraph partition. *IEEE Transactions on Pattern Analysis and Machine Intelligence*, 33(6):1266–1273, 2011.

[HLZM10]   Y. Huang, Q. Liu, S. Zhang, and D. Metaxas. Image retrieval via probabilistic hypergraph ranking. In *Proc. IEEE International Conference on Computer Vision Pattern Recognition*, pages 3376–3383, 2010.

[HMT11]   N. Halko, P. G. Martinsson, and J. A. Tropp. Finding structure with randomness: Probabilistic algorithms for constructing approximate matrix decompositions. *SIAM Review*, 53(2):217–288, 2011.

[Hof99]   T. Hofmann. Probabilistic latent semantic indexing. In *Proc. 22nd International ACM SIGIR Conference Research and Development in Information Retrieval*, pages 50–57, 1999.

[HS98]   J. Hofbauer and K. Sigmund. *Evolutionary Games and Population Dynamics*. Cambridge University Press, 1998.

[HSJR13]   M. Hein, S. Setzer, L. Jost, and S. S. Rangapuram. The total variation on hypergraphs - learning on hypergraphs revisited. In C.J.C. Burges, L. Bottou, M. Welling, Z. Ghahramani, and K.Q. Weinberger, editors, *Advances in Neural Information Processing Systems*, volume 26, pages 2427–2435, 2013.

[KAKS97]   G. Karypis, R. Aggarwal, V. Kumar, and S. Shekhar. Multilevel hypergraph partitioning: application in VLSI domain. In *Proc. 34th ACM Annual Design Automation Conference*, pages 526–529, 1997.

[KB09]   T. G. Kolda and B. W. Bader. Tensor decompositions and applications. *SIAM Review*, 51:455–500, 2009.

[KBK05]   T. G. Kolda, B. W. Bader, and J. P. Kenny. Higher-order web link analysis using multilinear algebra. In *Proc. 5th IEEE International Conference on Data Mining*, pages 725–728, 2005.

[KHT06]   S. Klampt, U.-U. Haus, and F. Theis. Hypergraphs and cellular networks. *PLOS Computational Biology*, 5(5):e1000385, 2006.

[KK00]   G. Karypis and V. Kumar. Multilevel K-way hypergraph clustering. *VLSI Design*, 11(3):285–300, 2000.

[Kle99]   J. M. Kleinberg. Authoritative sources in hyperlinked environment. *Journal of the ACM*, 46:604–632, 1999.

[KSS13]   K. Kapoor, D. Sharma, and J. Srivastava. Weighted node degree centrality for hypergraphs. In *Proc. 2nd IEEE Workshop on Network Science*, pages 152–155, 2013.

[Kuh76]   H. W. Kuhn. Nonlinear programming: A historical view. *SIAM-AMS Proceedings*, IX:1–26, 1976.

[Lat97]   L. De Lathawuer. *Signal processing based on multilinear algebra*. PhD thesis, K. U. Leuven, Electrical Engineering Dept. (ESAT, 1997.

[Law73]   E. Lawler. Cutsets and partitions of hyperedges. *Networks*, 3(3):275–285, 1973.

[LH05]   M. Leordeanu and M. Hebert. A spectral technique for correspondence problems using pairwise constraints. In *Proc. International Conference on Computer Vision*, pages 1482–1489, 2005.

[Lim14]   L.-H. Lim. Tensors and hypermatrices. In L. Hogben, editor, *Handbook of Linear Algebra*, pages 15–1–15–30. Chapman & Hall/CRC, 2014.

[LLS11]   Z. Lin, R. Lui, and Z. Su. Linearized alternating direction method with adaptive penalty for low-rank representation. In *Advances in Neural Information Processing Systems*, pages 612–620, 2011.

[LLY10]   H. Liu, L. J. Latecki, and S. Yan. Robust clustering as ensembles of affinity relations. In *Advances in Neural Information Processing Systems*, pages 1414–1422, 2010.

[LM06]   A. N. Langville and C. D. Meyer. *Google's PageRank and Beyond*. Princeton University Press, 2006.

[Low04]   D. G. Lowe. Distinctive image features from scale-invariant keypoints. *International Journal Computer Vision*, 60(2):91–110, 2004.

[LPV13]   H. Lu, K. Plataniotis, and A. N. Venetsanopoulos. *Multilinear Subspace Learning: Dimensionality Reduction of Multidimensional Data*. Chapman & Hall/CRC Taylor and Francis Group, 2013.

[MD08]   M. W. Mahoney and P. Drineas. Tensor-CUR decompositions and data applications. *SIAM Journal on Matrix Analysis and Applications*, 30(2):957–987, 2008.

[MS01]   M. Meila and J. Shi. A random walk view of spectral segmentation. In *Proc. 8th International Workshop on Artificial Intelligence and Statistics*, 2001.

[NDT15]   N. Nguyen, P. Drineas, and T. Tran. Tensor sparsification via a bound on the spectral norm of random tensors. *Information and Inference: A Journal of the IMA*, 2015.

[NJW01]   A. Y. Ng, M. I. Jordan, and Y. Weiss. On spectral clustering: Analysis and an algorithm. In *Advances in Neural Information Processing Systems*, pages 849–856, 2001.

[OB12]   P. Ochs and T. Brox. Higher order motion models and spectral clustering. In *Proc. IEEE Conference on Computer Vision and Pattern Recognition*, pages 614–621, 2012.

[OT06]   A. Oliva and A. Torralba. Building the GIST of a scene: The role of global image features in recognition. *Progress in Brain Research*, 155:23–36, 2006.

[Pam04]   E. Pampalk. A MATLAB toolbox to compute music similarity from audio. In *Proc. 5th International Conference on Music Information Retrieval*, pages 254–257, 2004.

[PCAS14]   P. Purkait, T.-J. Chin, H. Ackermann, and D. Suter. Clustering with hypergraphs: The case for large hyperedges. In *Proc. European Conference on Computer Vision*, 2014.

[PF12]   L. Pu and B. Faltings. Hypergraph learning with hyperedge expansion. In P. A. Flach, T. De Bie, and N. Cristianini, editors, *European Conference on Machine Learning Principles and Practice of Knowledge Discovery in Databases*, volume LNCS 7523, Part I, pages 410–425. Springer, 2012.

[PK14a]   K. Pliakos and C. Kotropoulos. Personalized music tagging using ranking on hypergraphs. In *Proc. 2014 International Symposium on Communications, Control, and Signal Processing*, pages 681–684, 2014.

[PK14b]   K. Pliakos and C. Kotropoulos. PLSA driven image annotation, classification, and tourism recommendation. In *Proc. 2014 IEEE International Conference on Image Processing*, pages 3003–3007, 2014.

[PK14c]   K. Pliakos and C. Kotropoulos. Simultaneous image tagging and geo-location prediction within hypergraph ranking framework. In *Proc. 2014 IEEE International Conference on Audio, Speech, and Signal Processing*, pages 6944–6948, 2014.

[PK14d]   K. Pliakos and C. Kotropoulos. Social image search exploiting joint visual-textual information within a fuzzy hypergraph framework. In *Proc. IEEE 16th International Workshop on Multimedia Signal Processing*, 2014.

[Por80]   M. Porter. An algorithm for suffix stripping. *Program*, 14(3):130–137, 1980.

[PP07]   M. Pavan and M. Pelillo. Dominant sets and pairwise clustering. *IEEE Transactions on Pattern Analysis and Machine Intelligence*, 29(1):167–172, 2007.

[PZ13]   K. J. Pearson and T. Zhang. On spectral hypergraph theory of the adjacency tensor. *Graphs and Combinatorics*, pages 1–16, 2013.

[QS10]   Z. Qin and K. Scheinberg. Efficient block-coordinate descent algorithms for the group Lasso. *Industrial Engineering*, pages 1–21, 2010.

[Rod03]   J. Rodriguez. On the Laplacian spectrum and walk-regular hypergraphs. *Linear and Multilinear Algebra*, 51(3):285–297, 2003.

[RTG00]   Y. Rubner, C. Tomasi, and L. Guibas. The Earth Mover's distance as a metric for image retrieval. *International Journal on Computer Vision*, 40(2):99–121, 2000.

[RY98]   E. S. Ristad and P. Yianilos. Learning string-edit distance. *IEEE Transactions on Pattern Analysis and Machine Intelligence*, 20(5):522–532, 1998.

[SM00]   J. Shi and J. Malik. Normalized cuts and image segmentation. *IEEE Transactions on Pattern Analysis and Machine Intelligence*, 22(8):888–905, 2000.

[SPF14]   N. D. Sidiropoulos, E. E. Papalexakis, and C. Faloutsos. Parallel randomly compressed cubes: A scalable distributed architecture for big tensor decomposition. *IEEE Signal Processing Magazine*, 31(5):57–70, September 2014.

[SWY75]   G. Salton, A. Wong, and C. Yang. A vector space model for automatic indexing. *Communications of the ACM*, 18(11):613–620, 1975.

[SZH06]   A. Shashua, R. Zass, and T. Hazan. Multi-way clustering using super-symmetric non-negative tensor factorization. In *Proc. European Conference on Computer Vision*, volume 3954, pages 595–608, 2006.

[TCR10]   C. Tramasco, J.-P. Cointet, and C. Roth. Academic team formation as evolving hypergraphs. *Scientometrics*, 85(3):721–740, 2010.

[TKP13]   A. Theodorids, C. Kotropoulos, and Y. Panagakis. Music recommendation using hypergraphs and group sparsity. In *Proc. 2013 IEEE International Conference on Acoustics, Speech, and Signal Processing*, pages 56–60, 2013.

[TNW08]   H. K. Tan, C. W. Ngo, and X. Wu. Modeling video hyperlinks with hypergraph for web video reranking. In *Proc. 2008 ACM Multimedia Conference*, pages 659–662, 2008.

[Tso10]   C. E. Tsourakakis. MACH: Fast randomized tensor decompositions. In *Proc. SIAM Conference on Data Mining*, pages 689–700, 2010.

[WLRT08]   F. Woolfe, E. Liberty, V. Rokhlin, and M. Tygert. A fast randomized algorithm for the approximation of matrices. *Applied Computational Harmonic Analysis*, 25:335–366, 2008.

[YL06]   M. Yuan and Y. Lin. Model selection and estimation in regression with grouped variables. *Journal of the Royal Statistical Society: Series B (Statistical Methodology)*, 68(1):49–67, 2006.

[ZD06]   Y. Zheng and D. Doermann. Robust point matching for nonrigid shapes by preserving local neighborhood structures. *IEEE Transactions on Pattern Analysis and Machine Intelligence*, 28(4):643–649, 2006.

[ZHS06]   D. Zhou, J. Huang, and B. Schölkopf. Learning with hypergraphs: Clustering, classification, and embedding. In *Advances in Neural Information Processing Systems*, pages 1601–1608, 2006.

[ZS08]   R. Zass and A. Shashua. Probabilistic graph and hypergraph matching. In *Proc. IEEE Conference on Computer Vision and Pattern Recognition*, 2008.

[ZSC99]   J. Zien, M. Schlag, and P. Chan. Multilevel spectral hypergraph partitioning with arbitrary vertex sizes. *IEEE Transactions on Computer-Aided Design of Integrated Circuits and Systems*, 18(9):1389–1399, 1999.

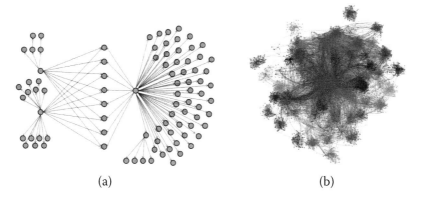

(a)   (b)

FIGURE 5.1.1: a) Recommendation network graph, b) YouTube video relation graph.

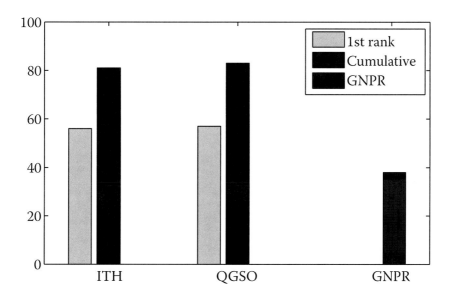

FIGURE 8.6.4: Geo-location prediction rates (in %) for the compared methods.

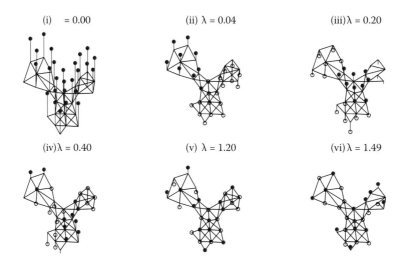

FIGURE 9.2.1: Stem plots of eigen vectors of Laplacian matrix $\mathcal{L}$ for an example graph. The filled (empty) nodes contain positive (negative) value. The zero-crossings (edges between a positive and a negative sample) of the eigenvectors usually increase with increasing eigenvalues λ

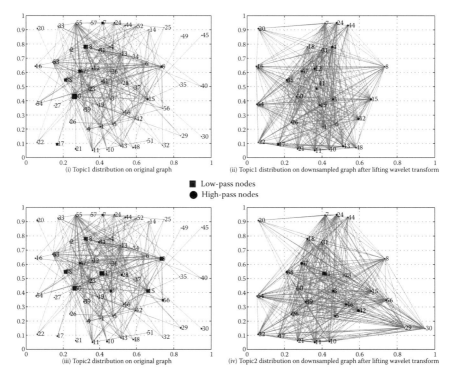

FIGURE 9.3.1: Graph representation of blogger network data of $N = 59$ blogger and $M = 3500$ blogs. (i) initial distribution of topic1 signal on the graph, (ii) transform coefficients after a two-level wavelet transform (blue-circles: low-pass nodes and red-circles: high-pass nodes) similarly for topic2 (iii) initial distribution (signal) and (iv) transform coefficients after a two-level wavelet transform.

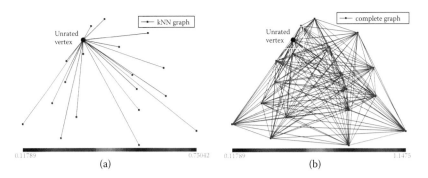

(a) (b)

FIGURE 9.3.3: An instance of predicting ratings of an unknown movie node (in red) using ratings of a known set of movie nodes (in blue), in MovieLens 100k dataset: (i) star graph and (ii) alternative graph that contains the star graphs and all the links between movies in the known set of movies.

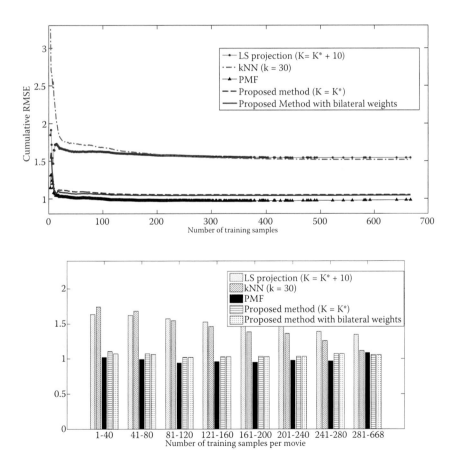

FIGURE 9.3.5: RMSE of different prediction algorithms with the number of training samples on MovieLens dataset.

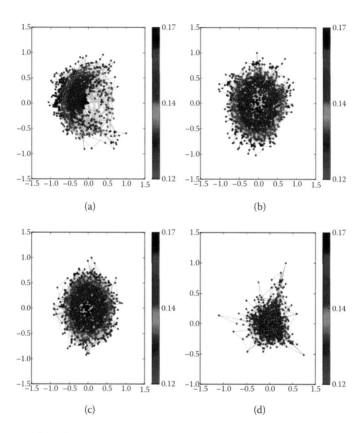

FIGURE 10.2.1: Graph embeddings for a Watts-Strogatz graph with $N = 2,000$: (a) Centrality-constrained embedding with $\mathbf{K}_1 = (-1/2)\mathbf{J}\boldsymbol{\Delta}^{(2)}\mathbf{J}$; (b) Centrality-constrained embedding with $\mathbf{K}_2 = \mathbf{A}\mathbf{A}^T$; (c) Centrality-constrained embedding with $\mathbf{K}_3 = \mathbf{L}^{\dagger}$; and (d) Centrality-agnostic embedding based on kernel matrix $\mathbf{K}_1$. The color bar maps node colors to varying centrality values.

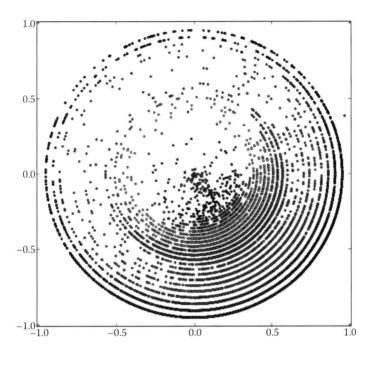

(a) Gnutella-04 (08/04/2012)

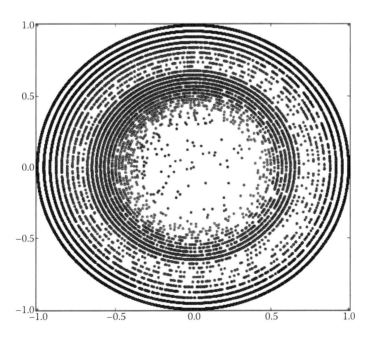

FIGURE 10.2.2: Visualization of two snapshots of the large-scale file-sharing network Gnutella [LKF07] based on degree centrality and $\mathbf{K} = \mathbf{A}\mathbf{A}^T$.

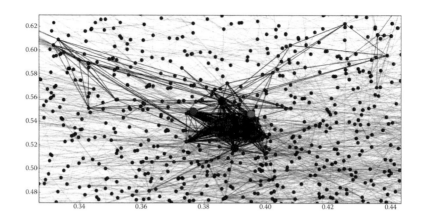

FIGURE 10.3.1: An anomalous (depicted in red) egonet for the arXiv General Relativity and Quantum Cosmology collaboration social graph [Les11]. The red egonet is flagged as anomalous by the proposed low-rank plus sparse matrix decomposition method.

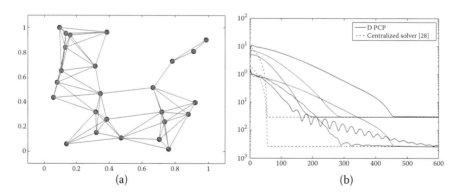

FIGURE 10.3.2: (a) A simulated small social network graph with $N = 25$ nodes. (b) Convergence of the D-PCP algorithm for different network sizes. D-PCP attains the same estimation error as the centralized solver.

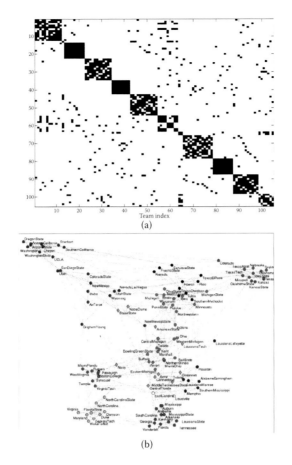

(a)

(b)

FIGURE 10.4.3: (a) Entries of **K** after removing the outliers, where rows and columns are permuted to reveal the clustering structure found by robust KPCA; and (b) Graph depiction of the clustered network [MG12b]. Teams belonging to the same estimated conference (cluster) are colored identically. The outliers are represented as diamond-shaped nodes.

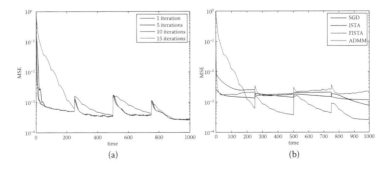

(a)   (b)

FIGURE 10.5.2: a) MSE (i.e., $\sum_{i,j}(\hat{a}_{ij}^t - a_{ij}^t)^2/N^2$) performance of Algorithm 10.5.1 versus time. For each t, (10.5.5) is solved "inexactly" for $k = 1$, 5, 10, and 15 inner iterations. It is apparent that $k = 5$ iterations suffice to attain convergence to the minimizer of (10.5.5) per t, especially after a short transient where the warm-restarts offer increasingly better initializations. b) MSE performance of real-time algorithms versus time. Real-time FISTA, Algorithm 10.5.2 (SGD), as well as inexact versions of Algorithm 10.5.1 (ISTA) and the ADMM solver in [BMG13] are compared.

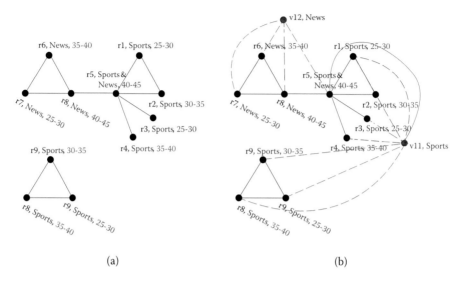

FIGURE 11.6.1: Graph examples for an article reader network with Two Attributes, "Topic" and "Age" [ZCX10]: a) Attributed graph and b) Attribute augmented graph: two attribute vertices v_{11} and v_{12} representing the topics "News" and "Sports" are added. Dashed lines connect authors with corresponding topics to the two vertices respectively. Attribute vertices and edges corresponding to the age attribute are omitted for the sake of clear presentation.

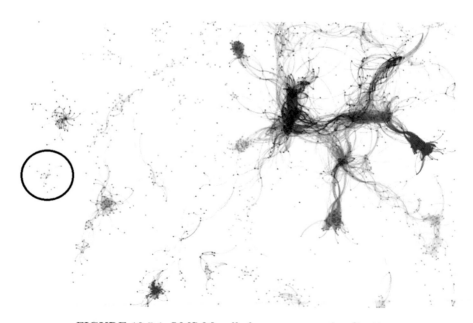

FIGURE 12.5.1: LMS Moodle log sequence visualization.

Chapter 9

Graph Signal Processing in Social Media

Sunil Narang

University of Southern California, U.S.A.

9.1 Motivation

Social media has become an integral part of modern society. There are general social networks with user bases larger than the population of most countries. There are niche sites for virtually every special interest out there, for example, to share photos, videos, status updates, sites for meeting new people, and sites to connect with old friends. The analysis of these networks has become an intriguing topic both due to monetary and management perspective as well theoretical challenges. In this chapter, we summarize some of the interesting problems in the analysis of social data, and the techniques used to solve them.

Mostly the social data have rich connections that can be naturally represented as graphs. Consider, for example, the Facebook friendship network where individual users (as nodes) are connected by friendship ties (as links). In some other applications social data can be represented as *point-clouds* of vectors and links are established between data sources based on the distance between their feature-vectors. For example, Tang et al. [HJCTH12] use graph representation of multimedia data for clustering and analysis. For brevity let's just call the datasets graphs. The links on the graph can be directed (as in Twitter follower/followee relations) or undirected (bi-directional). Both the nodes and links can have a number of numerical or categorical attributes. The problems of graph analysis can be roughly divided into two parts: a) those analyzing graph structure (e.g., segmentation, finding nearest neighbors, shortest distance between two nodes etc.) and b) those analyzing information available on the graph (filtering, denoising, missing value prediction, compression etc.). Graph analysis of social data dates back to pre-internet era.

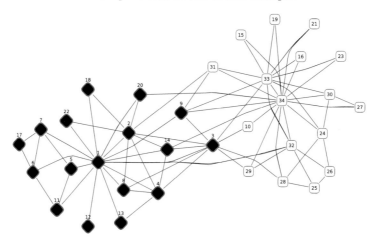

FIGURE 9.1.1: Zachary Karate Club network. The filled and empty nodes represent different factions formed the club.

A classic example of social graph analysis is the Zachary Karate Club data [Zac77], which was collected from the members of a university karate club by Wayne Zachary in 1977. Zachary used this data and an information flow model of network conflict resolution to explain the split-up of this group into two groups following disputes among the members. His analysis could correctly predict the alliance of 33 of the 34 members of the group. However, the modern day social data analysis problems are far more complex than that. Below we discuss a few key problems in graph analysis which have a lot of well-known applications.

Vector-Completion Problem: Vector-completion, or more generally matrix-completion, problems often arise in the context of recommendation systems. The users (in set U) preferences (on a scale of $0 - N$) for inventory items (in set I) are recorded in the form of a $U \times I$ matrix containing elements from finite set $\{0, 1, ..., N - 1\}$. Here, each row is a partially known preference vector of a user, which builds up as he or she views or accesses content. The problem then is to fill the remaining entries in the preference vector. The famous net-flix challenge [BLN07] is an example of the vector completion problem. The data can be represented as a weighted bipartite graph in which users' nodes U are connected to item nodes I. We will discuss some solutions to this problem in the application section.

Information Diffusion Pattern Analysis: Social media has become a major platform used by people to broadcast information. Given social media data, such as Twitter or blogs, we are interested in analyzing the diffusion patterns of different types of information (e.g., topics, emotions, sentiments etc.) within the same social network. One of the applications is topic sentiment analysis [WWL+11], where the goal is to find segmentation of graphs along topic/sentiment boundaries in Twitter or a citation network. Another application is emotion propagation [Kra12], which studies how emotional states can spread among people without their awareness (known as emotional contagion).

Multimedia data analysis: Another application of graph analysis is in representing digital multimedia content such as images and videos, where pixels can be connected with their neighbors to form graphs [SKN+10, MEO11]. Having a graph representation of images (or video) provides flexibility of adjusting link weights of the image graph and we provide a simple way to capture both directionality and intrinsic edge information of the image.

Graph Based Prediction: This problem is about predicting the value of a node based network configuration, and the value of other nearby nodes [GAO14]. This is useful for spam filtering, age or gender prediction, or identifying influential nodes.

In most of these cases the data on the nodes such as emotions, preferences, or age can be quantified and represented as discrete signals. Further, nodes connected to each other often have some similarity between them, which can be used for denoising, filtering, prediction etc. For example, the gender prediction problem, there is a strong correlation between the gender of a user and the gender of friends it is connected with. Thus, filtering or predicting the signal at a node w.r.t. the structure of the graph leads to efficient representation (for example sparse approximation) of the underlying signal. This interpretation has led to the design of a new set of tools and techniques [SM14] that analyzes graph signals using the extensions of tools and techniques in classical discrete signal processing (DSP) theory for time signals and images. This area is now called *graph signal processing* (GSP).

The GSP techniques help understand the structure behind the observations, hence come up with efficient methods to process the information by using tools defined on graphs. Major challenges are posed by the size of the datasets in these problems (number of nodes and links), making it difficult to visualize, process, analyze, and act on the information available. In distributed and cloud scenarios, the graph is stored on different machines and the data-exchanges between far-off nodes can be expensive (bandwidth, latency, energy constraints issues). Therefore, instead of operating on the original graph, it would be desirable to find and operate on smaller graphs with fewer nodes and data representing a smooth (more generally, it could be any sparse approximation of the original data) approximation of the original data. Moreover, such systems need to employ localized operations which could be computed at each node by using data from a small neighborhood of nodes around it. In this chapter, we describe GSP techniques which provide solutions to overcome both of the above challenges.

The remainder of this chapter is as follows: in Section 9.2 we describe important concepts of graph signal processing including spatial and spectral representation and vertex down-sampling schemes. In Section 9.3, we study two applications of graph analysis: a blogger network analysis to find topic based segmentation of the network, and the design of a movie recommendation system. Finally we conclude the chapter in Section 9.4 with a summary.

9.2 Graph signal processing (GSP)

Emerging data mining applications will have to operate on datasets defined on graphs. Examples of such datasets include online document networks, social networks, and knowledge bases etc. The data on these graphs can be visualized as a finite collection of samples, a *graph-signal* which can be defined as the information attached to each node (scalar or vector values mapped to the set of vertices/edges) of the graph. Two basic approaches have been used for signal processing on graphs. The first one uses the adjacency matrix of the underlying graph as its basic building block (for example, see [SM14, SM13] and subsequent extensions). The second approach adopts the graph Laplacian matrix as it's fundamental building block [NA12]. Both frameworks define several signal processing concepts similarly, but the difference in their foundation leads to different techniques for signal analysis and processing.

9.2.1 Basics of graph signal processing

A graph signal is a real-valued scalar function $f : \mathcal{V} \to \mathbb{R}$ defined on graph $G = (\mathcal{V}, E)$ such that $f(v)$ is the sample value of function at vertex $v \in \mathcal{V}$. The extension to complex or vector sample values $f(v)$ is straightforward but is not considered in this work. On a finite graph, the graph-signal can be viewed as a sequence or a vector $\mathbf{f} = [f(0), f(1), ..., f(N-1)]^T$, where the order of arrangement of the samples in the vector is arbitrary and neighborhood (or nearness) information is provided separately by the adjacency matrix $\mathbf{A}$. The matrix $\mathbf{A}$ is such that $\mathbf{A}(n, m)$ is the weight of the link(edge) between node n and m ($= 0$ if not connected). The remaining portion of this chapter deals with simple undirected graphs without self loops, which have a lot of interesting mathematical properties. Often, the directed graph is also processed as undirected graph by symmetrizing its links. Other work such as [SM14] implement analysis techniques directly on directed graphs. There is also some work on the analysis of multilayered graph in the context of social networks [OKH14] where different types of relationships (edges) between nodes are represented as different layers.

The $\mathbf{A}$ matrix is symmetric for undirected graphs, with 0 at the diagonal. We define $\mathbf{D}$ as the diagonal degree matrix where $D(n, n)$ is the degree (sum of the weight of the edges connected) of node n. The Laplacian matrix $\mathbf{L}$ is another fundamental matrix defined on graph, defined as $\mathbf{L} = \mathbf{D} - \mathbf{A}$. It is easy to see that $\mathbf{L}$ is a symmetric positive semi-definite matrix which means its eigenvalues are always non-negative. Define symmetric normalized adjacency matrix and symmetric normalized Laplacian matrix as $\mathcal{A} = \mathbf{D}^{-1/2}\mathbf{A}\mathbf{D}^{-1/2}$ and $\mathcal{L} = \mathbf{D}^{-1/2}\mathbf{L}\mathbf{D}^{-1/2}$, respectively. A j-hop neighborhood $\mathcal{N}_{j,n}$ around node n is the set of all nodes that are at most j edges apart from node n. It can be shown that:

$$\mathcal{N}_{j,n} = \{m : \mathbf{A}^j(n, m) \neq 0\}, \qquad (9.2.1)$$

which is an easy way to compute the j-hop neighborhood. Denote $< \mathbf{f}_1, \mathbf{f}_2 >$ as the inner product between signals $\mathbf{f}_1$ and $\mathbf{f}_2$. Graph-signals can, for example, be a set of measured values by sensor network nodes [WR06] or traffic measurement samples on the edges of an Internet graph [CK03][1] or information about the actors in a social network. Further, a graph based transform is defined as a linear transform $\mathbf{T} : \mathbb{R}^N \to \mathbb{R}^M$ in the N-node graph-signal space, such that the operation at each node n is a linear combination of the value of the sample at the node n and its j-hop neighborhood $\mathcal{N}_{j,n}$ for some j, i.e.,

$$y(m) = <\mathbf{t}_m \, \mathbf{f}> = T(m, n)f(n) + \sum_{o \in \mathcal{N}_{j,n}} T(m, o)f(o), \qquad (9.2.2)$$

where $\mathbf{t}_m$ is the mth row of the transform $\mathbf{T}$. The transform is called undersampled, critically sampled or oversampled depending upon the relation between input and output size ($M < N$, $M = N$, $M > N$, respectively). In analogy to the 1-D regular case, we would sometimes refer to graph-transforms as graph-filters and the elements $T(n, m)$ for $m = 1, 2, ...N$ as the filter coefficients at the nth node [2] A desirable feature of graph filters is *spatial localization*, which typically means that the energy of each filter (i.e., each row) of the graph filter is concentrated in a local region around a node. Let us define $\Delta_k^2(\mathbf{t}_n)$ given as:

$$\Delta_k^2(\mathbf{t}_n) = \frac{1}{||\mathbf{t}_n||^2} \sum_{l \in \mathcal{N}_{k,n}} T(n, l)^2 \qquad (9.2.3)$$

to be the fraction of energy of nth basis function (i.e., n^{th} row $\mathbf{t}_n$), in the k-hop neighborhood around node n. A graph transform is said to be strictly k-hop localized, or having a

---

[1]The graph in this case was chosen to be the line graph of the network graph. The line graph contains edges of the original graph as nodes and the network load as the graph signal.

[2]Not every linear transform is a graph-transform, since graph-transforms, by definition, are defined along the edges in the graph. For example, filter-coefficient $T(n, m)$ can be non-zero only if nodes n and m are connected, i.e., $d(n, m) < \infty$, and the magnitude of $T(n, m)$ usually decreases with increasing distance $d(n, m)$.

compact support in the spatial domain, if $\Delta_k^2(\mathbf{t}_n) = 1$ for all $n = 1, 2, ..N$. Note that spatial localization can also be applied in a weaker sense in which $\Delta_k^2(\mathbf{t}_n)$ is not exactly 1 but very close to it for all $n = 1, 2, ...N$. In this tutorial, we will study graph-filters with *compact support*, which are scalable with the size of the graph. Examples of compact support transforms are lifting transforms [JNS09, NO09] and the transforms proposed in [CK03] and [WR06]. Examples of non-compact transforms are diffusion wavelets [CM06] and spectral wavelets [HVG11].

9.2.2 Spectral representation of graph signals

Intuitively, a signal is smooth or *lowpass*, if the difference of sample values between a node and its neighbors is small. Thus, the smoothness of a graph signal depends on the underlying graph topology, and is usually described in terms of eigenvalues and eigenvectors of a fundamental symmetric graph matrix. The choice of this matrix differs among researchers, for example, Laplacian matrix is used in [HVG11, NA12], and adjacency matrix is used in [SM14]. We use Laplacian matrix in this chapter. Let us denote spectral decomposition of the graph G as the set of eigenvalues $\sigma(G)$, and the corresponding eigenvectors $\mathbf{u}_\lambda, \lambda \in \sigma(G)$ of the graph Laplacian matrix. Both the original Laplacian matrix $\mathbf{L}$ and the normalized form $\mathcal{L}$ are symmetric positive semidefinite matrices, and therefore can be used for this purpose. From the spectral projection theorem, there exists a real unitary matrix $\mathbf{U}$ which diagonalizes $\mathcal{L}$ (or $\mathbf{L}$), such that $\mathbf{U}^T \mathcal{L} \mathbf{U} = \Lambda = diag\{\lambda_i\}$ is a non-negative diagonal matrix. Let us use the symmetric normalized Laplacian matrix $\mathcal{L}$, which is more closely related to random walks in the graphs. This is because matrix $\mathcal{L}$ is *similar* to the random walk Laplacian matrix $\mathbf{Q} = \mathbf{I} - \mathbf{D}^{-1}\mathbf{A}$, and thus has identical set of eigenvalues (with identical algebraic multiplicities). In fact, the random The eigenvalues of $\mathcal{L}$ are in "normalized" form, i.e., if $\lambda \in \sigma(G)$ then $0 \le \lambda \le 2$, and are thus consistent with the eigenvalues in the stochastic processes. Refer to [Chu97] for details. This leads to an *eigenvalue decomposition* of matrix $\mathcal{L}$ given as:

$$\mathcal{L} = \mathbf{U}\Lambda\mathbf{U}^T = \sum_{i=1}^{N} \lambda_i \mathbf{u}_i \mathbf{u}_i^T, \tag{9.2.4}$$

where the eigenvectors $\mathbf{u}_1, \mathbf{u}_2, ..., \mathbf{u}_N$, which are columns of $\mathbf{U}$, form an orthogonal basis in $\mathbb{R}^N$ and $\{0 \le \lambda_1 \le \lambda_2... \le \lambda_N\}$ are corresponding eigenvalues. Thus, every graph-signal $\mathbf{f} \in \mathbb{R}^N$ can be decomposed into a linear combination of eigenvectors $\mathbf{u}_i$ given as $\mathbf{f} = \sum_{n=1}^{N} \bar{f}(n)\mathbf{u}_n$. It has been shown in [Chu97] that the eigenvectors of Laplacian matrix provide a harmonic analysis of graph signals which gives a Fourier-like interpretation. The eigenvectors act as the *natural vibration modes* of the graph, and the corresponding eigenvalues as the associated *graph-frequencies*. The mapping $\mathbf{u}_n \to \mathcal{V}$ associates the real numbers $u_n(i), i = \{1, 2, ..., N\}$, with the vertices $\mathcal{V}$ of G. The numbers $u_n(i)$ will be positive, negative or zero. The frequency interpretation of eigenvectors can thus be understood in terms of number of zero-crossings (pair of connected nodes with different signs) of eigenvector $\mathbf{u}_n$ on the graph G. For any finite graph the eigenvectors with large eigenvalues have more zero-crossings (hence high-frequency) than eigenvectors with small eigenvalues (see Figure 9.2.1 for example.). These results are related to "nodal domain theorems" and readers are directed to [DGLS01] for more details. The *spectrum* $\sigma(G)$ of a graph is defined as the set of eigen-values of its normalized Laplacian matrix, and it is always a subset of closed set $[0, 2]$ for any graph G. For the purpose of this chapter, an eigenvector $\mathbf{u}_\lambda$ is either considered to be a "lowpass" eigenvector if eigenvalue $\lambda \le 1$, or "highpass" eigenvector if $\lambda > 1$. The *graph Fourier transform* (GFT), denoted as $\bar{\mathbf{f}}$, is defined in [HVG11] as the projections of

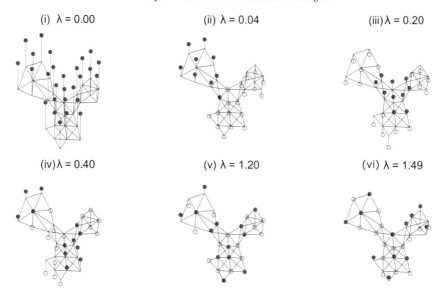

FIGURE 9.2.1: Stem plots of eigen vectors of Laplacian matrix $\mathcal{L}$ for an example graph. The filled (empty) nodes contain positive (negative) value. The zero-crossings (edges between a positive and a negative sample) of the eigenvectors usually increase with increasing eigenvalues λ. **(See color insert.)**

a signal $\mathbf{f}$ on the graph G onto the eigenvectors of G, i.e.,:

$$\bar{f}(\lambda) = <\mathbf{u}_\lambda \; , \; \mathbf{f}> = \sum_{i=1}^{N} f(i) u_\lambda(i). \tag{9.2.5}$$

Note that GFT is an energy preserving transform. A signal is considered "lowpass" (or "high-pass") if the energy $|\bar{f}(\lambda)|^2 \approx 0$ for all $\lambda > 1$ (or for all $\lambda \le 1$). In case of eigenvalues with multiplicity greater than 1 (say $\lambda_1 = \lambda_2 = \lambda$) the eigenvectors $\mathbf{u}_1, \mathbf{u}_2$ are unique up to a unitary transformation in the eigenspace $V_\lambda = V_{\lambda_1} = V_{\lambda_2}$. In this case, we can choose $\lambda_1 \mathbf{u}_1 \mathbf{u}_1^T + \lambda_2 \mathbf{u}_2 \mathbf{u}_2^T = \lambda \mathbf{P}_\lambda$ where $\mathbf{P}_\lambda$ is the projection matrix for eigenspace V_λ. Note that for all symmetric matrices, the dimension of eigenspace V_λ (geometric multiplicity) is equal to the multiplicity of eigenvalue λ (algebraic multiplicity) and the spectral decomposition in (9.2.4) can be written as:

$$\mathcal{L} = \sum_{\lambda \in \sigma(G)} \lambda \sum_{\lambda_i = \lambda} \mathbf{u}_i \mathbf{u}_i^T = \sum_{\lambda \in \sigma(G)} \lambda \mathbf{P}_\lambda. \tag{9.2.6}$$

The eigenspace projection matrices are idempotent and $\mathbf{P}_\lambda$ and $\mathbf{P}_\gamma$ are orthogonal if λ and γ are distinct eigenvalues of the Laplacian matrix, i.e.,:

$$\mathbf{P}_\lambda \mathbf{P}_\gamma = \delta(\lambda - \gamma) \mathbf{P}_\lambda, \tag{9.2.7}$$

where $\delta(\lambda)$ is the Kronecker delta function.

9.2.3 Downsampling in graphs

Downsampling is done to reduce the cost of storage and analysis of large graphs. A downsampling operation on the graph $G = (\mathcal{V}, E)$ can be defined as choosing a subset H of vertex set $\mathcal{V}$ such that all samples of the graph signal $\mathbf{f}$, corresponding to indices not

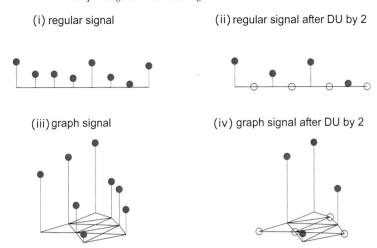

FIGURE 9.2.2: DU operation on a graph G. The empty circle nodes form the set L, whose values are discarded and replaced by 0 after DU operation.

in H, are discarded. A subsequent upsampling operation projects the downsampled signal back to original $\mathbb{R}^N$ space by inserting zeros in place of discarded samples in $H^c = L$. A comparison of downsampling a regular signal vs. a graph signal is shown in Figure 9.2.2.

Given such a set H we define a *downsampling function* $\boldsymbol{\beta}_H \in \{-1, +1\}$ given as:

$$\beta_H(n) = \begin{cases} 1 & \text{if } n \in H \\ -1 & \text{if } n \in L. \end{cases} \quad (9.2.8)$$

While several downsampling techniques have been proposed in literature, we only discuss the ones where the goal is to reconstruct the signal from the downsampled upsampled (DU) signal by interpolating values on L set using values in the H set. In general, downsampling always leads to signal distortion. However, its been proven in [Pes08] that for a special class of bandlimited signals in the Paley-Weiner space or PW_ω, there exist a set H such that the values at the discarded nodes L can be reconstructed exactly without any information loss. Furthermore, in the case of bipartite graphs, downsampling on one of the colored partitions leads to an effect analogous to frequency folding [NO11]. A bipartite graph $G = (L, H, E)$ is a graph whose vertices can be divided into two disjoint sets L and H, such that every link connects a vertex in L to one in H. Bipartite graphs are also known as *two-colorable graphs* since the vertices can be colored perfectly into two colors so that no two connected vertices are of the same color. This gives the cut-off frequency and also suggests a natural sampling strategy. The downsampled graph is constructed by reconnecting the downsampled vertices H with an edge whose weight is equal to the number of their common neighbors in the original graph. [3]. It can also be seen that undirected trees are bipartite graphs where nodes at alternative levels (nodes at even distance from root node and nodes at odd distance from root node) form the two color sets. Therefore, the downsampling technique proposed in [GNC10], and in [SO08] can be seen as a specialized case of downsampling on bipartite graphs. In [EFAR13], the results from downsampling on bipartite graphs [NO11] are generalized to circulant graphs. For one-step interpolation in arbitrary graphs, [Pes08] gives a sufficient condition that the sampling set needs to satisfy for unique recovery. Using

---

[3]Note that the reduced graph is not bipartite. Therefore, iterative downsampling techniques require additional steps to convert downsampled graph into one or more bipartite graphs. More explanation of the iterative downsampling techniques can be found in [NA12, SFV13]

this condition, a bound on the maximum bandwidth of all recoverable signals is given in [NGO13].

9.2.4   Graph wavelets and filterbanks

Wavelet transforms have been widely used as a signal processing tool for a sparse representation of signals. Wavelet based transforms split the sample space into an approximation and a detail subspace. The approximation subspace contains a smoother version of the original signal and the details of the signal are contained in the detail subspace. A discussion of wavelets and wavelet transforms on regular signals can be found in standard textbooks such as [VK95]. Here we only describe techniques used to extend classical wavelets to graphs. While wavelet-based techniques would seem well suited to provide efficient local analysis, a major obstacle to their application to graphs is that these, unlike images, are not regularly structured. Therefore properties of regular wavelets like locality and smoothness, do not have an obvious extension in the graph case. Moreover, classical wavelet transforms are critically sampled as they use local filtering operations followed by downsampling. In a graph, there is no obvious way to downsample nodes in a regular manner, since these neighborhoods vary in size and orientation.

Recently, a variety of designs of the wavelets and wavelet transforms on graphs have been proposed by researchers. These wavelet designs can be broadly classified into a) spatially localized designs and b) spectral designs. The spatial designs are designed in the spatial domain, i.e., directly on the nodes and their neighborhood, with spatial localization as their focus. These include lifting wavelets [JNS09, SO08, NO09], wavelets on hierarchical trees [GNC10], and wavelets for spatial traffic analysis [CK03]. The spectral designs, on the other hand, are designed in the spectral domain of the graph and then translated to spatial domains. Notable examples are: diffusion wavelets [CM06], spectral graph wavelets [HVG11], graph-QMF and graph-Bior filterbanks [NA12, NA13]. Usually, these wavelet functions are not the translates or dilates of a single mother wavelet function. However, we would like to preserve as many properties of classical wavelets as possible, for example orthogonality, sampling ratio, response to a constant (DC) signal, and reconstruction error. A comparison of some of these designs can be found in [NA12].

9.3   Applications

9.3.1   Information diffusion pattern analysis

Social media has become a major platform for people to broadcast information. Given social media data, such as Twitter or blogs, we are interested in analyzing whether different types of information (e.g., different topics) have similar diffusion patterns within the same social network. More specifically, we are interested in investigating combining topic models, such as the Latent Dirichlet Allocation (LDA) model, with our wavelet algorithms for graph analysis. The basic idea is as follows: for each topic generated by the LDA model, we run the wavelet analysis on the citation network or retweet graphs of users (in which the value of the node represents the relative frequency of the topics in the posts written by the user, and the edges represent the frequency of citations or retweet) so that we can capture the "high-pass" information of the graph associated with each topic, showing that there exist "boundaries" in how a topic propagates (these boundaries are discovered where high-pass information in the graph signal is significant). We expect to use this approach at several resolutions to compare information diffusion patterns and potentially provide insights to discoveries in social science.

Blogger Network Analysis Given large-scale linked unstructured data, such as a collection of blog posts or a research literature archive, there are two fundamental problems that have generated a lot of interest in the research community. One is to identify a set of high-level topics covered by the documents in the collection using unstructured data only; the other is to uncover and analyze the social network of the authors using the graph data only. So far, these problems have been viewed as separate problems and considered independently from each other. In this section, we argue that these two problems are in fact interdependent and should be addressed together, and we propose a wavelet graph analysis approach to solve the problem. The graph-based wavelet analysis is useful because it can reveal both spatial localization (spread) and spectral localization (uniformity of usages) of the topics simultaneously on the graph. As an example we collect data from a political blogger network. In these blogs we identify about 10000 keywords of interest and extract $T = 10$ topics by using the LDA model [BNJ03]. Each topic then corresponds to a distribution vector of word-frequencies (fractions). Further, the distribution of topics, i.e., projections of a word frequency vector used by a blogger onto the topic word frequency vectors, can be represented as a signal on this graph. In Figure 9.3.1, the distribution of a topic is interpreted as a signal mapped to nodes of the graph and represented as a circle on each node with the size of the circle being proportional to the magnitude of the signal at that node.

We demonstrate our analysis by choosing two example topics. The topics are represented as graph-signals on the graph in Figures 9.3.1(i) and (iii) and we observe no discernible difference in the input distributions of the two topics. We then apply a two-level lifting wavelet filterbank [NO09] to these signals. In this design, the vertex set is first partitioned into sets of even and odd nodes $\mathcal{V} = \mathcal{O} \cup \mathcal{E}$. The odd nodes compute their prediction coefficients using their own data and data from their even neighbors followed by even nodes computing their update coefficients using their own data and prediction coefficient of their neighboring odd nodes. The even nodes are reconnected for higher resolution lifting steps to create a new downsampled graph and the above lifting steps are repeated. We observe that in the spatial domain (graph) representation of the signal, the low-pass transform coefficients of topic2 have significantly higher magnitudes than the low-pass coefficients of topic1 especially around the node 41 (compare Figures 9.3.1 ii and 9.3.1 iv). Since the low-pass coefficients in a wavelet-transform provide average approximation of original signal over a certain (in this case a 2-hop) neighborhood, we surmise that *topic2 is more uniformly distributed in a 2-hop neighborhood around each blogger whereas the usage of topic1 fluctuates in the neighborhood of each node.* Further the node 41 and its 2-hop neighborhood is a major source of topic2 generation. In the spectral domain representation of the graph-signals of the two topics (Figure 9.3.2) we observe that the fraction of energy situated in lower (higher) spectral modes of the graph is higher (lower) for topic2 than topic1 (compare Figures 9.3.2 a and 9.3.2(d)). According to the spectral interpretation of graph-signals, this implies that *globally topic2 is more uniformly distributed on the graph than topic1.* Further, Figure 9.3.2 shows the energy distribution of the graph signals reconstructed from only low-pass coefficients in the spectral domain. The spectral coefficients α_n are obtained by projecting topic-signals onto n^{th} eigenvector of the graph. The y-axis in all the plots in Figure 9.3.2 shows energy in each coefficient. In Figures 9.3.2(b) and 9.3.2(e), we observe that the signal corresponding to topic2 has more energy concentrated at low frequencies (confirmed by the higher magnitude of low-frequency coefficients for topic2). This is consistent with the facts, since the top keywords corresponding to topic1 consist of specific names or nouns and hence likely to be used sparingly and by small focused groups of bloggers (therefore non-uniform distributed) and the top keywords for topic2 are relatively commonplace and likely to be used more or less uniformly by all bloggers (hence uniformly distributed). Refer to Table 9.3.1.

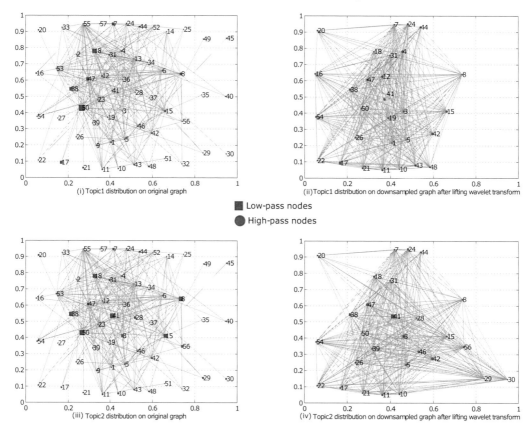

FIGURE 9.3.1: Graph representation of blogger network data of $N = 59$ blogger and $M = 3500$ blogs. (i) initial distribution of topic1 signal on the graph, (ii) transform coefficients after a two-level wavelet transform (blue-circles: low-pass nodes and red-circles: high-pass nodes) similarly for topic2 (iii) initial distribution (signal) and (iv) transform coefficients after a two-level wavelet transform. **(See color insert.)**

TABLE 9.3.1: Top keywords corresponding to the example topics chosen for analysis.

Topic	Top Keywords
topic1	*clinton, obam, huckab, mccain, delegat, poll, romney, prim, sen, win, msnbc, memorabl, caucus, race, hill*
topic2	*peopl, year, govern, use, law, work, said, americ, iraq, need, make, world, case, reason, right*

9.3.2 Interpolation in graphs

An important area of research is the interpolation problem in graph structured data. This arises in many guises, such as in semi-supervised learning of categorical data (see [KL02]) and missing value prediction such as matrix-completion problems [BLN07]. A common theme in all these applications is that the goal is to predict the property of some nodes (class, ranking or function), by interpolating the property values from a known set of nodes. The accuracy of all linear and non-linear interpolation methods on graphs rely on the implicit

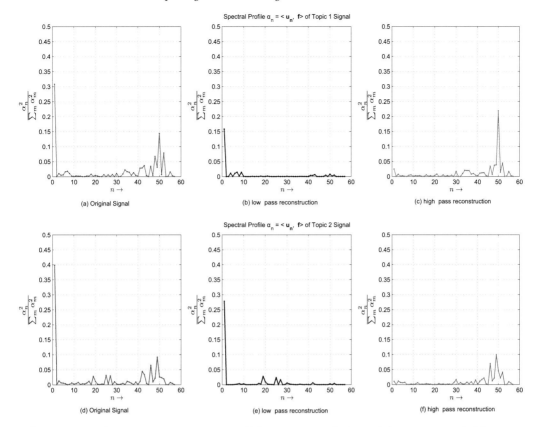

FIGURE 9.3.2: Spectral representation of blogger network data: (a) topic1 signal (b) reconstructed topic1 signal from only low-pass coefficients of the lifting transform and (c) from only high-pass coefficients of lifting transform, (d) topic2 signal (e) reconstructed topic2 signal from only low-pass coefficients and (f) from only high-pass coefficients of lifting transform.

assumption that nodes close to each other (in terms of the similarity captured by link-weights in the graph) usually have similar signal values. For example, in an item-item graph in a recommendation system, a typical user rates two similar items with similar ratings. In the same way, when predicting the functions of unannotated proteins based on a protein network, one relies on some notions of "closeness" or "distance" among the nodes. In other words, the graph functions of interest are slowly-varying or "smooth" on the graph. Smoothness is well studied in the classical signal processing domain, where it is measured in terms of the frequency contents of a signal. For example, there are well known sampling results for recovering smooth bandlimited signals from only a few of their samples via interpolation.

We focus on the development of similar interpolation techniques for graph signals. The main objective in designing an interpolation method is, of course, to achieve high accuracy. A major challenge in achieving this goal is managing to do so with reasonable computational complexity. Graph based interpolation approaches can be broadly divided into two categories: a) local methods and b) global methods. In local interpolation methods, such as k-nearest neighbors (kNN) methods, the predicted value at an unknown node is computed as a weighted combination of k-nearest known samples [CFS09]. Although the method is computationally simple and efficient, it does not capture global information, as well as the

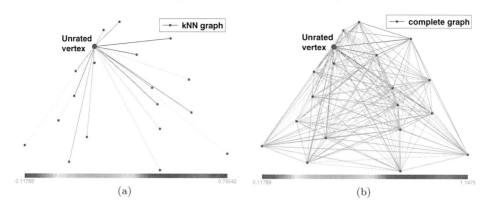

FIGURE 9.3.3: An instance of predicting ratings of an unknown movie node (in red) using ratings of a known set of movie nodes (in blue), in MovieLens $100k$ dataset: (i) star graph and (ii) alternative graph that contains the star graphs and all the links between movies in the known set of movies. **(See color insert.)**

dependencies that exist between known samples. On the other hand, global methods, such as [BN04, GS03], predict the value of all unknown nodes at once, by selecting as the solution a function that matches the values at known nodes while satisfying certain global "smoothness" conditions. Global methods are computationally more expensive but provide more accurate results. However, it is not clear how to optimize the choice of objective function. Although the graph signal processing method described below leads to a similar least-squares based interpolation as in [BN04], the signal processing perspective allows us to choose an optimal objective criterion, which will be shown to lead to better interpolation. Another problem is the choice of graph to interpolate the data upon. The example in Figure 9.3.3 demonstrates the choice of either operating on a bare-bones star graph commonly used in kNN [CFS09] prediction methods or an alternative graph that contains the star graphs and all the links between movies in the known set of movies.

Interpolation can be defined on both graphs, so it is unclear which of the two (or any other graph) should be chosen. The approach described below takes inspiration from signal processing techniques to formulate the partially known graph function as a downsampled-upsampled (DU) signal. In the regular signal domain, the original signal is recovered from its DU signal by applying a low-pass filter. Similarly, in the graph domain, we design low-pass filters to recover the original graph signal from its DU signal (see Figure 9.3.4).

9.3.2.1 Movie recommendation system

Let us design an interpolation method for collaborative filtering in recommendation systems. The input in this problem is a partially observed user-item rating matrix $\mathbf{R}$, such that $\mathbf{R}(u, m)$ is the rating given by user u to the movie m. Based on this information, the system predicts new user-item ratings. This problem is usually solved in two parts. During training, we create a model for how the items or users are correlated with each other, either by clustering them into latent dimensions as in [SM08] methods, or by explicitly finding the correlation between them as described below. In testing or application, we take the known ratings of a given users and then propagate to unknown items via interpolation. Specifically, let's choose the MovieLens $100k$ [MOV03] dataset containing $100k$ user-movie-rating triplets from $N = 943$ users and $M = 1682$ movies. The ratings are integer values between 1 and 5.

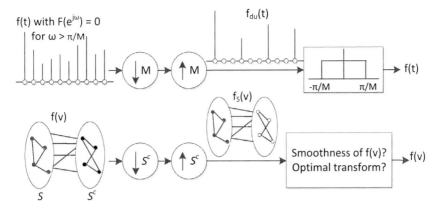

FIGURE 9.3.4: Graph interpolation problem represented as reconstruction of DU graph signal.

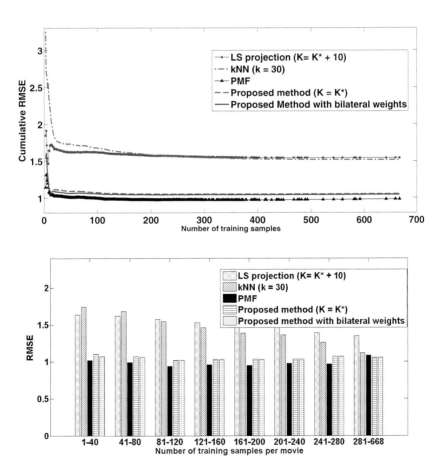

FIGURE 9.3.5: RMSE of different prediction algorithms with the number of training samples on MovieLens dataset. **(See color insert.)**

We use the 5-fold cross validation data available in [MOV03], which consists of 5 disjoint random sets of $20k$ triplets. At each iteration, one of these sets is used for testing and remaining sets for training.

We first compute a movie graph G_0 from the training samples, by computing cosine similarity [SKKR01] between every pair of movies. For each test user u, we define $\mathcal{S}$ to be the set of movies with known ratings, and $\mathcal{U}$ to be the set of test movies. We compute the subgraph $G_u = (\mathcal{S} \cup \mathcal{U}, E_u)$, corresponding to subset $\mathcal{S} \cup \mathcal{U}$ of nodes. We define DU signal $\mathbf{f}_u$ for u to be of size $|\mathcal{U} \cup \mathcal{S}|$, with $\mathbf{f}_u(\mathcal{U}) = 0$ and $\mathbf{f}_u(\mathcal{S})$ equal to known ratings. The goal now is to predict the value of $\mathbf{f}_u(\mathcal{U})$.

Subsequently, we compute interpolated signal $\hat{\mathbf{f}}_u$ by using a method proposed in [NGO13], with G_u and $\mathbf{f}_u$ as inputs. This method is very similar to the least squares (LS) interpolation of [BN04], except that here we operate on a normalized Laplacian matrix, and choose K^* specified by in [NGO13]. To show that this K is a good choice, we also implement the method in [BN04] with $K = K^* + 10$ (i.e., with 10 additional eigenvectors), respectively. Additionally, since the smoothness of a graph signal depends both on the signal values and the underlying graph, we can modify the graph to adapt to the given signal so that the signal is more band-limited on the simplified graph, thus leading to less interpolation error. The simplification makes sense in many cases such as in recommendation systems, where the underlying graph is the result of observing average correlation over a set of training users (multiple instances), and the signal corresponds to a single test user. This kind of signal adaptive filtering is achieved by replacing edge weights with *bilateral filters* [TM98].

The comparison of the performance of GSP method with two most popular algorithms for collaborative filtering: 1) kNN method [SKKR01] (with $k = 30$), and 2) probabilistic matrix factorization (PMF) [SM08] with 10 latent features is shown in Figure 9.3.5. We use root mean square error (RMSE) between predicted values and actual values to measure performance. The predicted values less than 0 (more than 5) are set to 0 (5) before computing RMSE. Figure 9.5(a) plots the RMSE of prediction as a function of number of training samples available. Observe that kNN method perform the worst of all method, and PMF method performs the best. The GSP methods, both with or without bilateral weighting, are very close to PMF method, with the interpolation with bilateral weights performing slightly better. We also observed that choosing $K = K^* + 10$ in LS method leads to poorer results. The Figure 9.5(b) shows another RMSE plot where users are grouped by the number of training samples, with x-axis showing those groups. We observe that both the LS method and kNN method perform significantly worse when the number of available training samples are small. The effect of applying bilateral weighting in our proposed methods is also most visible here.

The PMF method predicts the ratings of all movies for all users simultaneously by factorizing the whole $N \times M$ rating matrix. It is based on an iterative update rule and requires $\mathcal{O}(NMP)$ operations per iteration where P is the size of the latent space. Theoretically, any change to the rating matrix would require the PMF system to be retrained on all users. However, in our method, the process of computing the movie graph is decoupled from the process of predicting ratings for a given user. Accommodating a few new ratings in the systems is fast, as it only affects a local portion of the movie graph. Once the movie graph is fixed, the proposed method allows us to predict the ratings of movies for each user separately in $\mathcal{O}(K^* M^2)$ operations. Thus, assuming $K^* \approx P$, the proposed method is faster (*i.e.*, $M^2 < NM$) than PMF, when ratings of items change frequently and the recommendations need to be calculated only when a user requests them. The comparison of computational complexity is given in Table 9.3.2. Further, it may be possible to reduce complexity in our method by using simplified filtering operations.

TABLE 9.3.2: Computational Complexity of Interpolation Methods. $M = $ #(movies), $N = $ #(users), $R = $ #(latent factors), $K = $ #(eigenvalues less than cut-off)

	PMF	Graph Signal Interpolation
Training	#(iterations) $\times O(NMR)$	$O(NM^2)$
Update (new user/movie added)	#(iterations) $\times O(NMR)$	$O(1)$
Prediction (per user, per movie)	$O(R)$	$O(KM)$

9.4 Conclusions

In this chapter, we learned about graph signal processing (GSP) techniques for analyzing social media. We described important concepts of GSP, including filtering, vertex sampling, and interpolation. We described two applications of these techniques: one in analyzing blogger network data and second in designing a movie recommendation system. The preliminary research in GSP breaks ground for many new and interesting problems. Both a signal on a graph with N vertices and a classical discrete-time signal with N samples can be viewed as vectors in R^N. However, there is a large gap in the knowledge between DSP techniques, and the techniques designed for GSP. The future direction of GSP is to narrow this gap by developing a more systematic understanding of fundamental concepts, such as translation, modulation, sampling, and filtering etc. for graphs. Graph based data is found in diverse applications such as healthcare, social networks, intelligence, system biology, power grids, and large scale simulations, which are full of semantically rich relationships. In the future, GSP techniques will solve challenging problems in Big-Data using GSP ideas.

Bibliography

[BLN07]   J. Bennett, S. Lanning, and N. Netflix. The Netflix prize. In *Proc. Knowledge Discovery and Data Mining Cup and Workshop*, 2007.

[BN04]   M. Belkin and P. Niyogi. Semi-supervised learning on Riemannian manifolds. *Machine Learning*, 56(1-3):209–239, 2004.

[BNJ03]   D. M. Blei, A. Y. Ng, and M. I. Jordan. Latent Dirichlet allocation. *Journal of Machine Learning Research*, 3:993–1022, 2003.

[CFS09]   J. Chen, H. Fang, and Y. Saad. Fast approximate kNN graph construction for high dimensional data via recursive Lanczos bisection. *Journal of Machine Learning Research*, 10:1989–2012, 2009.

[Chu97]   F. R. K. Chung. *Spectral Graph Theory*. American Mathematical Society, 1997.

[CK03]   M. Crovella and E. Kolaczyk. Graph wavelets for spatial traffic analysis. In *Proc. IEEE INFOCOM*, volume 3, pages 1848–1857, 2003.

[CM06]   R. R. Coifman and M. Maggioni. Diffusion wavelets. *Applied and Computational Harmonic Analysis*, 21(1):53 – 94, 2006.

[DGLS01]   E. B. Davies, G. M. L. Gladwell, J. Leydold, and P. F. Stadler. Discrete nodal domain theorems. *Linear Algebra and its Applications*, 336(1-3):51–60, 2001.

[EFAR13]   V. Ekambaram, G. Fanti, B. Ayazifar, and K. Ramchandran. Multiresolution graph signal processing via circulant structures. In *Proc. 2013 IEEE Digital Signal Processing and Signal Processing Education Meeting (DSP/SPE)*, pages 112–117, 2013.

[GAO14]   A. Gadde, A. Anis, and A. Ortega. Active semi-supervised learning using sampling theory for graph signals. In *Proc. of the 20th ACM SIGKDD International Conference on Knowledge Discovery and Data Mining*, pages 492–501, 2014.

[GNC10]   M. Gavish, B. Nadler, and R. R. Coifman. Multiscale wavelets on trees, graphs and high dimensional data: Theory and applications to semi supervised learning. In *Proc. of the 27th International Conference on Machine Learning (ICML-10)*, pages 367–374, 2010.

[GS03]   L. Grady and E. L. Schwartz. Anisotropic interpolation on graphs: The combinatorial Dirichlet problem. Technical report, Boston University, 2003.

[HJCTH12]   M. Hasegawa-Johnson, S. M. Chu, H. Tang, and T. S. Huang. Partially supervised speaker clustering. *IEEE Transactions on Pattern Analysis and Machine Intelligence*, 34(5):959–971, 2012.

[HVG11]   D. K. Hammond, P. Vandergheynst, and R. Gribonval. Wavelets on graphs via spectral graph theory. *Applied and Computational Harmonic Analysis*, 30(2):129–150, 2011.

[JNS09]   M. Jansen, G. P. Nason, and B. W. Silverman. Multiscale methods for data on graphs and irregular multidimensional situations. *Journal of the Royal Statistical Society. Series B (Statistical Methodology)*, 71(1):97–125, 2009.

[KL02]   R. Kondor and J. Lafferty. Diffusion kernels on graphs and other discrete structures. In *Proc. International Conference on Machine Learning*, pages 315–322, 2002.

[Kra12]   A. D. Kramer. The spread of emotion via Facebook. In *Proc. of the SIGCHI Conference on Human Factors in Computing Systems*, pages 767–770, 2012.

[MEO11]   E. Martinez-Enriquez and A. Ortega. Lifting transforms on graphs for video coding. In *Proc. Data Compression Conference (DCC), 2011*, pages 73–82, 2011.

[MOV03]   MovieLens dataset, as of 2003. http://www.grouplens.org/.

[NA12]   S. Narang and O. A. Perfect reconstruction two-channel wavelet filter-banks for graph structured data. *IEEE Transactions on Signal Processing*, 60(6), 2012.

[NA13]   S. Narang and O. A. Compact support biorthogonal wavelet filterbanks for arbitrary undirected graphs. *IEEE Transactions on Signal Processing*, 61(19), 2013.

[NGO13]   S. K. Narang, A. Gadde, and A. Ortega. Signal processing techniques for interpolation in graph structured data. In *Proc. IEEE International Conference on Acoustics, Speech and Signal Processing (ICASSP)*, pages 5445–5449, 2013.

[NO09]   S. K. Narang and A. Ortega. Lifting based wavelet transforms on graphs. *Asia-Pacific Sig. and Information Proc. Association, 2009, (APSIPA ASC' 09)*, pages 441–444, 2009.

[NO11]   S. Narang and A. Ortega. Downsampling graphs using spectral theory. In *Proc. IEEE International Conference on Acoustics, Speech and Signal Processing (ICASSP)*, pages 4208–4211, 2011.

[OKH14]   B. Oselio, A. Kulesza, and A. Hero. Multi-layer graph analysis for dynamic social networks. *IEEE Journal of Selected Topics in Signal Processing*, 8(4):514–523, 2014.

[Pes08]   I. Pesenson. Sampling in Paley-Wiener spaces on combinatorial graphs. *Transactions American Mathematical Society*, 360(10):5603–5627, 2008.

[SFV13]   D. I. Shuman, M. J. Faraji, and P. Vandergheynst. A framework for multiscale transforms on graphs. *Computing Research Repository*, 2013.

[SKKR01]   B. Sarwar, G. Karypis, J. Konstan, and J. Riedl. Item-based collaborative filtering recommendation algorithms. In *Proc. of the 10th International Conference on World Wide Web*, pages 285–295, 2001.

[SKN$^+$10] G. Shen, W. Kim, S. Narang, A. Ortega, J. Lee, and H. Wey. Edge-adaptive transforms for efficient depth map coding. In *Proc. Picture Coding Symposium (PCS), 2010*, 2010.

[SM08]   R. Salakhutdinov and A. Mnih. Probabilistic matrix factorization. In *Proc. Advances in Neural Information Processing Systems*, volume 20, pages 1257–1264, 2008.

[SM13]   A. Sandryhaila and J. Moura. Discrete signal processing on graphs: Graph filters. In *Proc. IEEE International Conference on Acoustics, Speech and Signal Processing (ICASSP)*, pages 6163–6166, 2013.

[SM14]   A. Sandryhaila and J. Moura. Discrete signal processing on graphs: Frequency analysis. *IEEE Transactions on Signal Processing*, 62(12):3042–3054, 2014.

[SO08]   G. Shen and A. Ortega. Optimized distributed 2D transforms for irregularly sampled sensor network grids using wavelet lifting. In *Proc. IEEE International Conference on Acoustics, Speech and Signal Processing (ICASSP)*, pages 2513–2516, 2008.

[TM98]   C. Tomasi and R. Manduchi. Bilateral filtering for gray and color images. In *Proc. 6th International Conference on Computer Vision*, pages 839–846, 1998.

[VK95]   M. Vetterli and J. Kovačevic. *Wavelets and subband coding*. Prentice-Hall, Inc., 1995.

[WR06]   W. Wang and K. Ramchandran. Random multiresolution representations for arbitrary sensor network graphs. In *Proc. of the 5th International Conference on Information Processing in Sensor Networks*, pages 102–108, 2006.

[WWL⁺11] X. Wang, F. Wei, X. Liu, M. Zhou, and M. Zhang. Topic sentiment analysis in Twitter: A graph-based hashtag sentiment classification approach. In *Pro. of the 20th ACM International Conference on Information and Knowledge Management*, pages 1031–1040, 2011.

[Zac77] W. W. Zachary. An information flow model for conflict and fission in small groups. *Journal of Anthropological Research*, 1977.

Chapter 10

Big Data Analytics for Social Networks

Brian Baingana, Panagiotis Traganitis, Georgios Giannakis

University of Minnesota, U.S.A.

Gonzalo Mateos

University of Rochester, U.S.A.

10.1 Introduction

The last few years have witnessed a dramatic upswing in the volume, variety, and acquisition rate of data from disparate sources. Among other factors, this "data deluge" has been fueled by the ubiquity of the web, pervasive deployment of low-cost sensors, cheaper storage memory, and growing awareness that data is a key asset from which valuable insights can be unlocked. Social networks lie at the forefront of this revolution, presenting ample opportunities to address several challenges associated with big data. Examples include micro-blogging services (e.g., Twitter), web-based friendship networks (e.g., Facebook), and online product reviews (e.g., Yelp).

Exploiting the immense big data opportunities comes at the cost of overcoming significant challenges. First, traditional analytics are ill-equipped to cope with the sheer volume of data. Many distributed platforms (e.g., Hadoop/MapReduce and GraphLab) have emerged to tame the scale of data. Nevertheless, only a subset of algorithms are readily implementable on these platforms, and development of more versatile architectures is an active research area.

In addition, most social data are high-dimensional, with many features. In order to finesse the curse of dimensionality, parsimonious models must be devised for feature subset selection. Since most big data are acquired sequentially as streaming inputs, batch learning and optimization approaches are impractical. For example, securities traders interested in capturing market sentiment from real-time tweets are driven by the need for split-second buy/sell decisions. Furthermore, big data acquisition pipelines are plagued by measurement inaccuracies, misses, and incompleteness due to, e.g., privacy concerns.

10.1.1 Signal processing for big data

Signal processing (SP) provides a principled framework within which big data challenges can be readily addressed [SGM14]. Advances in SP have formalized parsimonious models that exploit key properties inherent to big data, e.g., sparsity, low-rank, and manifold structures. For example, several contemporary models jointly capture low rank and sparsity (see, e.g., [MMG13c]), subsuming several learning paradigms such as principal component analysis (PCA) [HTF09], dictionary learning (DL) [OF97], and compressive sampling (CS) [CW08].

Scaling to very large problem instances, while attaining real-time operation are noteworthy themes that have shaped the direction of contemporary research efforts. The *alternating direction method of multipliers (ADMM)* has enjoyed growing popularity in decentralized learning and optimization algorithms, especially in settings where data resides over a network of computing nodes [MBG10, BPC$^+$11, MMG13a].

For streaming data, *online learning* has emerged as a powerful framework for real-time analytics [MBPS10, KST11, MMG13b]. Popular online algorithms including online mirror descent, and online gradient descent have been studied and deployed in practical social network settings; see e.g., [SS11] for a comprehensive review. Figure 10.1.1 depicts various themes for which SP and learning offer a comprehensive framework.

In addition to decentralized computation, significant improvements in running time have been realized through random sampling and random projection algorithms [Mah11]. These methods operate on a randomized sketch of an input data matrix by either: i) sampling a small subset of rows of the matrix that are most informative (random sampling); or ii) linearly combining a small number of rows of the matrix (random projection). Under

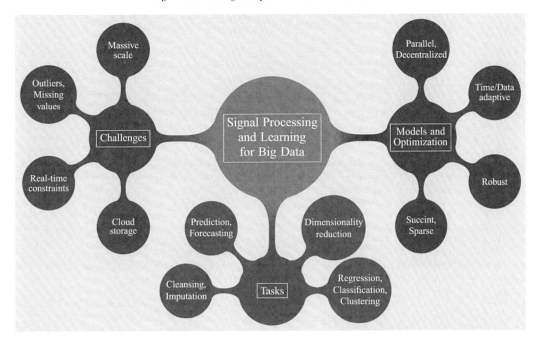

FIGURE 10.1.1: Signal processing in the context of Big Data analytics [SGM14].

reasonable conditions, these randomized approaches can in theory attain asymptotically faster worst-case running times than deterministic alternatives.

10.1.2 Social network analytics problems

Initial social network studies were exploratory in nature and unveiled surprising properties common to large networks e.g., the existence of power laws, small-world properties, and preferential attachment strategies for network growth [New10, EK10]. The exponential growth of the web coupled with ground-breaking advances in fields as diverse as computational biology and epidemiology, have led to a dramatic change in the scope and size of problems studied. A few examples include the ranking of web pages [BP98, CDK$^+$99], the discovery of causal interactions in gene regulatory networks [CBG13], and the prediction of the spread of infectious diseases [VR07, Jac10].

Looking at network science through the lens of statistical learning, several contemporary problems boil down to (non-)parametric regression, dimensionality reduction, clustering, or (semi-)supervised classification. "Work-horse" dimensionality reduction approaches such as multidimensional scaling (MDS) have been advocated for network visualization tasks [KK89, BP11, BG13]. Topology inference problems from network processes have been the focus of several recent works [RLS10, BMG14]. Another line of research has concentrated on virus (also information, buying patterns) propagation models over complex networks [Het00, EK10]. Other interesting topics include community discovery [GN02, For10], prediction of network processes from partial observations [Kol09, FRG14], and detection of anomalies in social networks [MG12b]. This chapter is representative of a subset of problems, for which the identified big data challenges are addressed by leveraging statistical learning advances.

Section 10.2 focuses on scalable visualization of social networks via nonlinear dimensionality reduction. Inference and imputation of corrupted signals over social graphs is the topic of Section 10.3. This is followed by Section 10.4, which presents algorithms for community discovery in big social networks. Finally, Section 10.5 presents recent algorithms for tracking topologies of dynamic social networks that facilitate diffusion of network processes. Admittedly, these approaches represent a small fraction of the gamut of social network analytics methods. Nevertheless, they are representative of a broader class of contemporary tools that are relevant to big data analytics in social networks.

10.2   Visualizing and reducing dimension in social nets

Network visualization is often accomplished through graph embedding, which entails mapping each node to a point in Euclidean space. Due to the data deluge spawned by modern network-based phenomena, the rising complexity and sheer volume of networks present new opportunities and challenges for graph embedding tools that capture global patterns and use visual metaphors to convey meaningful structural information e.g., hierarchy, node similarity, and natural communities [HL12].

Most traditional visualization algorithms trade off the clarity of structural characteristics of the underlying network for aesthetic requirements like minimal edge crossing and fixed edge lengths; e.g., [KK89, PT98, DBD13]. Although efficient for small graphs (hundreds of nodes), embeddings for larger graphs generated by classical approaches are seldom structurally informative. To this end, several approaches have been developed for embedding graphs while preserving specific structural properties. Pioneering methods (e.g., [KK89]) have resorted to *multidimensional scaling (MDS)*, which seeks a low-dimensional representation of high-dimensional data so that pairwise dissimilarities are preserved through Euclidean distances in the embedding [BG05, BG13]. In this case, the vertex dissimilarity structure is preserved through pairwise distance metrics in the embedding. Spectral embeddings whose coordinates consist of entries in the leading principal components of the network adjacency matrix are advocated in [LWH03]. The structure-preserving embedding algorithm solves a semidefinite program with linear topology constraints that emphasize reconstructability of the graph through neighborhood methods [SJ09]. Other visualization methods emphasize community structure [YLZ+13], while concentric layouts emphasize node hierarchy by placing the highest ranked nodes at the center of the embedding [AHDBV06, BP11, BG13].

Despite the rich history associated with graph drawing methods, development of visualization techniques that effectively capture hierarchical structure and other global patterns remains a challenging and active area of research. This section showcases a recently developed kernel-based visualization approach that leverages *local linear embedding (LLE)*, a popular manifold learning technique [BG15]. Similar to recent works on graph embedding, node importance is captured through *centrality* constraints [BP11, BG13]. In general, centrality measures provide a means to assign a level of importance to each node in a network [Sab66, Fre77]. For instance, betweenness centrality describes the extent to which information is routed through a specific node by measuring the fraction of all shortest paths traversing it; see e.g., [Kol09, p. 89]. Other measures include closeness and eigenvalue centrality.

10.2.1   Kernel-based graph embedding

Consider a network represented by an undirected graph $G = (\mathcal{V}, \mathcal{E})$, where $\mathcal{E}$ denotes the set of edges, and $\mathcal{V}$ the set of vertices with cardinality $|\mathcal{V}| = N$. Suppose the structure of

G is captured by its so-termed adjacency matrix $\mathbf{A}$ whose (i,j)-th entry (hereafter denoted by a_{ij}) is zero only if edge $(i,j) \notin \mathcal{E}$, otherwise it denotes the weight of (i,j). Given G and a prescribed embedding dimension p (typically $p \in \{2,3\}$), the graph embedding task is tantamount to searching for the set of $p \times 1$ vectors $\mathcal{X} := \{\mathbf{x}_i\}_{i=1}^N$ which "effectively" capture the underlying local graph structure.

Suppose $\{\mathbf{y}_i \in \mathbb{R}^q\}_{i=1}^N$ are data points sampled from a nonlinear manifold. LLE seeks the low-dimensional vectors $\{\mathbf{x}_i \in \mathbb{R}^p\}_{i=1}^N$ ($p \ll q$) that preserve the local neighborhood structure on the manifold. First, the neighborhoods $\{\mathcal{N}_i\}_{i=1}^N$ are constructed per datum by selecting the K-nearest neighbors of i, or setting $\mathcal{N}_i := \{\mathbf{y}_j \in \mathbb{R}^q : \|\mathbf{y}_i - \mathbf{y}_j\|_2 \leq \epsilon, \epsilon > 0, j = 1, \ldots, N\}$. Assuming $|\mathcal{N}_i| = K$, each point is then fit to a linear combination of its neighbors by solving the following constrained least-squares (LS) optimization problem

$$\underset{\substack{w_{i1},\ldots,w_{iK} \\ \{\sum_{j \in \mathcal{N}_i} w_{ij}=1\}}}{\arg\min} \left\| \mathbf{y}_i - \sum_{j \in \mathcal{N}_i} w_{ij}\mathbf{y}_j \right\|_2^2 \quad i = 1, \ldots, N, \qquad (10.2.1)$$

where $\{w_{ij}\}_{j=1}^K$ are the reconstruction weights for point i, while the constraint enforces shift invariance. Setting $w_{ij} = 0$ for $j \notin \mathcal{N}_i$, the final step determines $\{\mathbf{x}_i \in \mathbb{R}^p\}_{i=1}^N$ so that reconstruction weights are preserved by solving:

$$\underset{\substack{\mathbf{x}_1,\ldots,\mathbf{x}_N \\ \sum_{i=1}^N \mathbf{x}_i = \mathbf{0} \\ \frac{1}{N}\sum_{i=1}^N \mathbf{x}_i\mathbf{x}_i^T = \mathbf{I}}}{\arg\min} \sum_{i=1}^N \left\| \mathbf{x}_i - \sum_{j=1}^N w_{ij}\mathbf{x}_j \right\|_2^2. \qquad (10.2.2)$$

The equality constraints are included to eliminate the trivial all-zero solution, and also to eliminate shift and rotation ambiguities.

In order to tailor LLE for graph embedding where only weights $\{a_{ij}\}$ are available, one must contend with the general non-existence of high dimensional vectors $\{\mathbf{y}_i\}_{i=1}^N$ defined per node. Fortunately, the optimization problem (10.2.1) can be cast in terms of the inner products $\mathbf{y}_i^T\mathbf{y}_j$ for all $i,j \in \{1, \ldots, N\}$. This brings to bear the merits of kernel methods which entail computations on inner products of transformed feature vectors, $\phi(\mathbf{y})$, namely $k_{ij}(\mathbf{y}_i, \mathbf{y}_j) = \phi^T(\mathbf{y}_i)\phi(\mathbf{y}_j)$. However, this flexibility comes at the challenge of selecting the best kernel matrix $\mathbf{K} \in \mathbb{R}^{N \times N}$ where $[\mathbf{K}]_{ij} := k_{ij}(\mathbf{y}_i, \mathbf{y}_j)$. A few choices of $\mathbf{K}$ include:

i. The doubly-centered dissimilarity matrix $\mathbf{K} = -(1/2)\mathbf{J}\mathbf{\Delta}^{(2)}\mathbf{J}$, where $[\mathbf{\Delta}^{(2)}]_{ij}$ denotes the squared geodesic distance between nodes i and j, or any other dissimilarity metric on the graph, and $\mathbf{J} := \mathbf{I} - N^{-1}\mathbf{1}\mathbf{1}^T$ denotes the centering operator ($\mathbf{I}$ is the identity matrix and $\mathbf{1}$ is the all-one column vector) [BG05]. In this case, $\mathbf{K}$ is reminiscent of the kernel adopted by classical MDS.

ii. The Penrose-Moore pseudoinverse of the graph Laplacian; that is $\mathbf{K} = \mathbf{L}^\dagger$, where $\mathbf{L} := \mathbf{A} - \mathbf{D}$, $\mathbf{A} \in \{0,1\}^{N \times N}$ denotes the binary graph adjacency matrix, and $\mathbf{D} := \text{diag}(\mathbf{A}\mathbf{1})$. It turns out that $\mathbf{L}^\dagger$ admits an intuitive interpretation as a similarity matrix based on random walk distances on graphs [FPRS07].

iii. Matrix $\mathbf{K} = \mathbf{A}\mathbf{A}^T$, where $\mathbf{A} \in \{0,1\}^{N \times N}$, and $[\mathbf{A}\mathbf{A}^T]_{ij}$ counts the number of single-hop neighbors shared by nodes i and j.

Neighborhood selection in traditional LLE entails $\mathcal{O}(qN^2)$ complexity with $q \gg$. This bottleneck can be overcome by setting $\mathcal{N}_i$ to the single-hop neighbors per node. Let $\mathbf{Y}_i :=$ $[\phi(\mathbf{y}_1^i), \ldots, \phi(\mathbf{y}_{d_i}^i)]$ collect the "virtual" transformed vectors associated per $\mathcal{N}_i$, where d_i denotes the degree of node i. Letting $\mathbf{w}_i := [w_{i1}, \ldots, w_{id_i}]^T$, the constrained LS fit (10.2.1) can be written as:

$$\mathbf{w}_i = \underset{\{\mathbf{w}:\, \mathbf{1}^T \mathbf{w} = 1\}}{\arg\min} \ \mathbf{w}^T \mathbf{K}_i \mathbf{w} - 2\mathbf{w}^T \mathbf{k}_i, \tag{10.2.3}$$

where $\mathbf{K}_i := \mathbf{Y}_i^T \mathbf{Y}_i$ and $\mathbf{k}_i := \mathbf{Y}_i^T \phi(\mathbf{y}_i)$ are submatrices of $\mathbf{K}$ indexed by elements of $\mathcal{N}_i$. Resorting to Lagrange multiplier theory, one can readily solve for $\mathbf{w}_i$ in (10.2.3) in closed form. Moreover for large-scale graph embedding, (10.2.3) can be easily parallelized over clusters of computing nodes. Each subproblem entails $\mathcal{O}(d_i^3)$ complexity, which is manageable because typically $d_i \ll N$.

The low-dimensional graph embedding can be evaluated from the reconstruction weights via (10.2.2), by solving for:

$$\underset{\left\{ \begin{array}{c} \mathbf{X} \\ \sum_{i=1}^N \mathbf{x}_i = \mathbf{0} \\ \frac{1}{N} \sum_{i=1}^N \mathbf{x}_i \mathbf{x}_i^T = \mathbf{I} \end{array} \right\}}{\arg\min} \quad \mathrm{Tr}\left[\mathbf{X}^T (\mathbf{I} - \mathbf{W})^T (\mathbf{I} - \mathbf{W}) \mathbf{X} \right], \tag{10.2.4}$$

where $\mathbf{W}^T := [\tilde{\mathbf{w}}_1, \ldots, \tilde{\mathbf{w}}_N]$, $\tilde{w}_{ij} = w_{ij}$ if $j \in \mathcal{N}_i$ otherwise $\tilde{w}_{ij} = 0$, $\mathbf{X}^T := [\mathbf{x}_1, \ldots, \mathbf{x}_N]$, and $\mathrm{Tr}(.)$ denotes matrix trace. The solution comprises the 2nd to the $(p+1)$st least dominant eigenvectors of $(\mathbf{I} - \mathbf{W})^T (\mathbf{I} - \mathbf{W})$. For large graphs, $\mathbf{X}$ can be efficiently computed via orthogonal iterations which are amenable to decentralization, entailing $\mathcal{O}(pN^2)$ complexity; see e.g., [KM08].

Although this approach preserves the local graph topology defined by single-hop neighbors, the spectral decomposition is generally not scalable for very large networks. Moreover for large network visualization tasks, one is often more interested in conveying global properties such as node hierarchy. To this end, the next subsection modifies the embedding step to enforce centrality constraints.

10.2.2   Centrality-constraints

Large-scale settings call for emphasis on structural properties such as node hierarchy over aesthetics. Centrality measures impose a hierarchical ordering among nodes by quantifying the relative importance of nodes over their peers e.g., degree, closeness, and betweenness centralities [Kol09]. As a result, graph embedding under centrality constraints yields very informative network visualizations. Let $\mathcal{C}(\mathcal{G}) := \{c_i\}_{i=1}^N$ denote the set of centralities of $\mathcal{G}$, with c_i representing the centrality measure of node i. The goal is to determine the embedding $\{\mathbf{x}_i\}_{i=1}^N$ that effectively "preserves" the centrality ordering in $\mathcal{C}(\mathcal{G})$.

Modifying the final step in (10.2.2) to incorporate centrality constraints yields the following optimization problem:

$$\underset{\mathbf{x}_1, \ldots, \mathbf{x}_N}{\arg\min} \quad \sum_{i=1}^N \left\| \mathbf{x}_i - \sum_{j=1}^N w_{ij} \mathbf{x}_j \right\|_2^2$$

$$\text{s. t.} \quad \|\mathbf{x}_i\|_2^2 = f^2(c_i), \ i = 1, \ldots, N, \tag{10.2.5}$$

where $f(c_i)$ is a monotone decreasing function of c_i ensuring that more central nodes are placed closer to the center. The dropped $\mathbf{0}$-mean constraint can be compensated for by a post-processing centering operation upon determination of $\{\hat{\mathbf{x}}_i\}_{i=1}^N$.

Problem (10.2.5) is non-convex without global optimality guarantees. Fortunately, the problem decouples over vectors $\{\mathbf{x}_i\}_{i=1}^N$ motivating a block coordinate descent (BCD) approach [Ber99]. The optimization variables can now be partitioned into N blocks with $\mathbf{x}_i$ corresponding to block i. Consequently, during iteration r one cycles through all blocks by solving:

$$\mathbf{x}_i^r = \arg\min_{\mathbf{x}} \quad \left\| \mathbf{x} - \sum_{j<r} w_{ij}\mathbf{x}_j^r - \sum_{j>r} w_{ij}\mathbf{x}_j^{r-1} \right\|_2^2$$

$$\text{s. t.} \quad \|\mathbf{x}\|_2^2 = f^2(c_i). \tag{10.2.6}$$

Letting $\mathbf{v}_i^r := \sum_{j<r} w_{ij}\mathbf{x}_j^r + \sum_{j>r} w_{ij}\mathbf{x}_j^{r-1}$ and $\lambda \geq 0$ denote a Lagrange multiplier, the solution:

$$\mathbf{x}_i^r = \arg\min_{\mathbf{x}} \quad \|\mathbf{x} - \mathbf{v}_i^r\|_2^2 + \lambda(\|\mathbf{x}\|_2^2 - f^2(c_i)) \tag{10.2.7}$$

is given by:

$$\mathbf{x}_i^r = \frac{\mathbf{v}_i^r}{1 + \lambda}. \tag{10.2.8}$$

Upon substitution of (10.2.8) into the equality constraint in (10.2.6), one obtains the closed-form per-iteration update:

$$\mathbf{x}_i^r = \begin{cases} \frac{\mathbf{v}_i^r}{\|\mathbf{v}_i^r\|_2} f(c_i), & \text{if } \|\mathbf{v}_i^r\|_2 > 0 \\ \mathbf{x}_i^{r-1}, & \text{otherwise.} \end{cases} \tag{10.2.9}$$

If $\mathbf{X}^r$ denotes the embedding matrix after r BCD iterations, the operation $\mathbf{X} = (\mathbf{I} - N^{-1}\mathbf{1}\mathbf{1}^T)\mathbf{X}^r$ centers $\{\mathbf{x}_i^r\}_{i=1}^N$ to the origin in order to satisfy the shift invariance property of the embedding. Algorithm 10.2.1 summarizes the steps outlined for the centrality-constrained graph embedding scheme. The only inputs to the algorithm are the graph topology $\mathcal{G}$, the centrality measures, $\{c_i\}_{i=1}^N$, the graph embedding dimension p, and the kernel matrix $\mathbf{L}$. Note that Algorithm 10.2.1 scales well to big data settings since both steps can be computed in parallel. Moreover, (10.2.3) entails $\mathcal{O}(d_i^3)$ complexity, with $d_i \ll N$, and the dimensionality of (10.2.6) is $p \in \{2, 3\}$.

Algorithm 10.2.1: Centrality-constrained graph embedding

1: **Input:** $\mathcal{G}$, $\mathcal{C}(\mathcal{G})$, $\mathbf{K}$, ϵ, p
2: **for** $i = 1 \ldots N$ (in parallel) **do**
3:    Set $\mathcal{N}_i$ to single-hop neighbors of i
4:    Extract $\mathbf{K}_i$ and $\mathbf{k}_i$ from $\mathbf{K}$
5:    Solve $\mathbf{w}_i = \arg\min_{\mathbf{w}} \mathbf{w}^T\mathbf{K}_i\mathbf{w} - 2\mathbf{w}^T\mathbf{k}_i$ s. t. $\mathbf{1}^T\mathbf{w} = 1$
6:    Set $w_{ij} = 0$ for $j \notin \mathcal{N}_i$
7: **end for**
8: Initialize $\mathbf{X}^0$, $r = 0$
9: **repeat**
10:    $r = r + 1$
11:    **for** $i = 1 \ldots N$ (in parallel) **do**
12:      Compute $\mathbf{x}_i^r$ according to (10.2.9)
13:      $\mathbf{X}^r(i, :) = (\mathbf{x}_i^r)^T$
14:    **end for**
15: **until** $\|\mathbf{X}^r - \mathbf{X}^{r-1}\|_F \leq \epsilon$
16: $\mathbf{X} = (\mathbf{I} - \frac{1}{N}\mathbf{1}\mathbf{1}^T)\mathbf{X}^r$

10.2.3 Numerical tests

This section presents numerical tests conducted on a synthetic small-world network generated by the Watts-Strogatz model [WS98] and Gnutella, a real-world file-sharing network. Given the number of nodes N, average degree $\bar{d}$, and $\beta \in [0, 1]$, the Watts-Strogatz model constructs a $\bar{d}$-regular ring lattice and rewires each edge with probability β. The synthetic graph was generated with $N = 2 \times 10^3$, $\bar{d} = 4$, and $\beta = 0.3$. Several centrality measures are available with emphasis on different importance criteria. For instance, *closeness centrality* captures the extent to which a particular node is close to all other nodes in the network, and it is commonly defined as $c_i := 1/(\sum_{j \in \mathcal{V}} d_{ij})$, where d_{ij} denotes the geodesic distance between nodes i and j. For the experiments, closeness centralities were transformed as follows:

$$f(c_i) = \left(\frac{c_{\max} - c_i}{c_{\max} - c_{\min}} \right), \tag{10.2.10}$$

where $c_{\max} := \max_i \ c_i$, and $c_{\min} := \min_i \ c_i \ \forall i = 1, \ldots, N$.

Figure 10.2.1 depicts visualizations of the Watts-Strogatz network obtained by setting $N = 2,000$. In Figure 10.2.1 (a), (b), and (c), centrality-constrained embeddings are plotted with kernel matrices $\mathbf{K}_1 = (-1/2)\mathbf{J}\mathbf{\Delta}^{(2)}\mathbf{J}$, $\mathbf{K}_2 = \mathbf{A}\mathbf{A}^T$, and $\mathbf{K}_3 = \mathbf{L}^\dagger$, respectively. The appeal for centrality-constrained embeddings is clear when compared with Figure 10.2.1(d), which depicts a "centrality-agnostic" graph embedding using $\mathbf{K}_1$. Here, the final weight preservation step entailed spectral decomposition of $(\mathbf{I} - \mathbf{W})^T(\mathbf{I} - \mathbf{W})$, consistent with the original LLE algorithm. It is clear that even for a moderately sized synthetic graph, little meaningful information can be conveyed visually from the centrality-agnostic embedding. For instance, it is not obvious how an analyst would discern which nodes are most accessible to peers, or those whose removal would compromise the rate of information propagation over the network.

Figure 10.2.2 depicts visualizations of snapshots of the Gnutella peer-to-peer file-sharing network [LKF07]. Directed edges represent connections between hosts. The snapshots were captured on Aug. $4, 2012$ ($N = 10, 876$, $|\mathcal{E}| = 39, 994$) and Aug. $24, 2012$ ($N = 26, 518$, $|\mathcal{E}| = 65, 369$), respectively. The adjacency matrices were symmetrized to obtain undirected versions of the network. In this case $\mathbf{K} = \mathbf{A}\mathbf{A}^T$ and $\mathcal{C}$ was set to the node degrees. It is clear from the network drawings that despite the dramatic growth in the number of edges over a 20 day span, most new nodes had low degree, and are located far from the center.

10.2.4 Visualization of dynamic social networks

This section has so far focused on addressing issues concerning structurally-informative visualization of large static networks. However, these issues are exacerbated in settings involving dynamic networks whose topologies evolve over time [BW97, BW98, BC03, MMBd05, MMBd06, LS08, FT08]. Typically, one is given a time series of graphs $\{G_t\}_{t=1}^T$, representing static snapshots indexed by time intervals $t = 1, \ldots, T$. Although it is possible to compute a sequence of static embeddings, this approach does not scale to big data where graph snapshots may encode millions of nodes, and may be acquired in a streaming fashion.

Moreover, the sequence of embeddings must "respect" the viewer's *mental map*, which captures a sense of stability of the underlying structure of the network across time intervals. Sequential embeddings with drastic changes for a significant number of node positions violate a viewer's expectations, rendering mental reconciliation of temporal network changes difficult. Pioneering works have advocated a Bayesian framework that facilitates modeling dependencies between consecutive graph layouts [BW97, BW98]. A more recent visualization framework advocates *regularized graph layouts*, which involve optimizing a cost function augmented with a penalty that enforces stability across timeslots [XKI12].

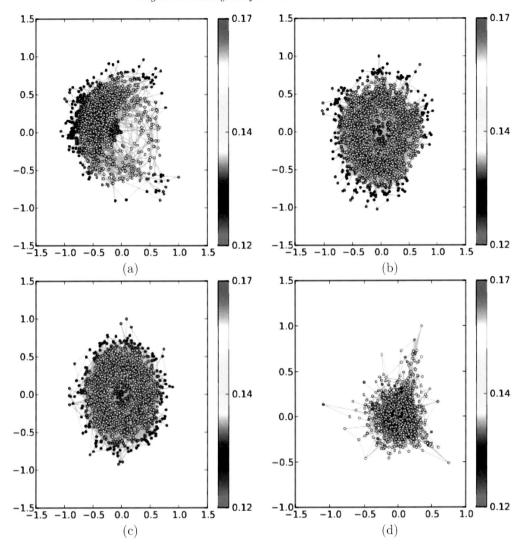

FIGURE 10.2.1: Graph embeddings for a Watts-Strogatz graph with $N = 2,000$: (a) Centrality-constrained embedding with $\mathbf{K}_1 = (-1/2)\mathbf{J}\mathbf{\Delta}^{(2)}\mathbf{J}$; (b) Centrality-constrained embedding with $\mathbf{K}_2 = \mathbf{A}\mathbf{A}^T$; (c) Centrality-constrained embedding with $\mathbf{K}_3 = \mathbf{L}^\dagger$; and (d) Centrality-agnostic embedding based on kernel matrix $\mathbf{K}_1$. The color bar maps node colors to varying centrality values. **(See color insert.)**

Instead of generating a sequence of stable embeddings, a recent approach promotes static displays that capture temporal variations using non-negative matrix factorization for dimensionality reduction [MM13]. It tacitly assumes that the number of nodes remains fixed but the edge connections vary with time. Static visual plots of rank-one matrix factors capture the temporal evolution of node importance. Despite numerous efforts, most contemporary approaches are not tailored for big networks with streaming (and possibly missing) inputs, a setup of growing research activity.

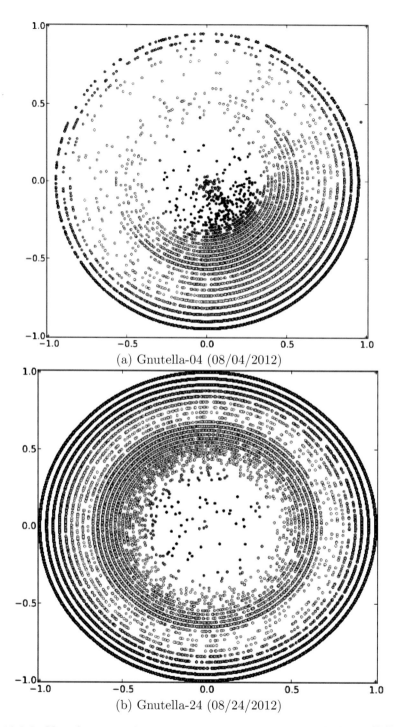

(a) Gnutella-04 (08/04/2012)

(b) Gnutella-24 (08/24/2012)

FIGURE 10.2.2: Visualization of two snapshots of the large-scale file-sharing network Gnutella [LKF07] based on degree centrality and $\mathbf{K} = \mathbf{A}\mathbf{A}^T$. (**See color insert.**)

10.3 Inference and imputation on social graphs

Inference over graphs is a rich subject, and includes the typical (non-) parametric regression, classification, and clustering tasks. Here, we will focus on anomaly detection, interpolation (a.k.a. imputation), and extrapolation (a.k.a. prediction).

10.3.1 Distributed anomaly detection for social graphs

This section deals with graph (network) anomaly identification. On the one hand, existing approaches have dealt with the pursuit of abnormal behavior exhibited by processes that evolve over graphs, such as Internet traffic flows [LCD04, ZGGR05, MMG13b, MMG13c, MR13], or spatiotemporal energy consumption profiles [CLL+10, MG12a, MG13], to name a couple. But also of great interest for spam, fraud and network intrusion detection is to consider the structure of the underlying graphs, and determine whether e.g., nodes, edges, or egonets are anomalous in the sense that they deviate from postulated network models; see e.g., [NC03, SQCF05, EH07, MT08, TL11, AMF12, MAB13] for noteworthy contributions. In the sequel, a novel approach to social graph anomaly detection is outlined, which is based on contemporary low-rank plus sparse matrix decompositions. A distributed algorithm is then developed leveraging the alternating-directions method of multipliers (ADMM); see e.g., [BT99, BPC+11].

10.3.1.1 Anomaly detection via sparse plus low-rank decomposition

The idea here is to consider the egonets in a social graph (i.e., each node's induced single-hop subgraph), and for each of them evaluate structural features or graph invariants such as, average degree, number of nodes, principal eigenvalue, and average clustering coefficient, to name a few. It is thus possible to collect all these structural quantities in a feature×egonet matrix $\mathbf{Y}$. Specifically, let the $D \times 1$ vector $\mathbf{y}_n := [y_{1,n}, \ldots, y_{D,n}]^T$ collect D different graph features, calculated for each egonet $n \in [1, N]$ in a graph of N vertices. Consider the $D \times N$ graph feature matrix $\mathbf{Y} := [\mathbf{y}_1, \ldots, \mathbf{y}_N]$. The d-th row $\mathbf{y}^T(d)$ of $\mathbf{Y}$ is the networkwide sequence corresponding to feature d, measured for all egonets in the graph. It may be useful to explicitly account for missing data, which could arise because not all features can be calculated for all egonets, or when a subsampling strategy is implemented to reduce the computational complexity in forming $\mathbf{Y}$. To this end, consider the set $\Omega \subseteq \{1, \ldots, D\} \times \{1, \ldots, N\}$ of index pairs (d, n) defining a sampling pattern (or mask) of the entries of $\mathbf{Y}$. Introducing the matrix sampling operator $\mathcal{P}_\Omega(\cdot)$, which sets the entries of its matrix argument not indexed by Ω to zero and leaves the rest unchanged, the (possibly) incomplete matrix of egonet features in the presence of outliers can be modeled as:

$$\mathcal{P}_\Omega(\mathbf{Y}) = \mathcal{P}_\Omega(\mathbf{X} + \mathbf{O} + \mathbf{E}), \tag{10.3.1}$$

where $\mathbf{X}$, $\mathbf{O}$, and $\mathbf{E}$ denote the nominal feature matrix, the outliers (i.e., anomalies), and small approximation errors, respectively. For nominal egonet features $y_{d,n} = x_{d,n} + e_{d,n}$, one has no anomaly; that is $o_{d,n} = 0$.

The model is inherently under-determined, since even for the (most favorable) case of full data, i.e., $\Omega \equiv \{1, \ldots, D\} \times \{1, \ldots, N\}$, there are twice as many unknowns in $\mathbf{X}$ and $\mathbf{O}$ as there is data in $\mathbf{Y}$. Estimating $\mathbf{X}$ and $\mathbf{O}$ becomes even more challenging when data are missing, since the number of unknowns remains the same, but the amount of data is reduced. In any case, estimation of $\{\mathbf{X}, \mathbf{O}\}$ from $\mathcal{P}_\Omega(\mathbf{Y})$ is an ill-posed problem unless one introduces extra structural assumptions on the model components to reduce the effective degrees of freedom. To this end, two cardinal properties of $\mathbf{X}$ and $\mathbf{O}$ will prove instrumental. First, a key observation is that for "nominal" complex networks most of these features obey power laws [FFF99, Kol09, EK10], and hence $\mathbf{Y}$ (or its entrywise logarithm) will be approximately *low rank*. Second, anomalies (or outliers) only occur sporadically across egonets and features, yielding a *sparse* matrix $\mathbf{O}$.

An estimator matching nicely the specifications of the graph anomaly detection problem stated, is the so-termed "robust" (stable) principal components pursuit (PCP, also known as RPCA for robust principal component analysis) [ZLW^{+}10, CLMW11, CSPW11, MG12b], that will be outlined here for completeness. PCP seeks estimates $\{\hat{\mathbf{X}}, \hat{\mathbf{O}}\}$ as the minimizers of:

$$\text{(P1)} \qquad \min_{\{\mathbf{X}, \mathbf{O}\}} \|\mathcal{P}_{\Omega}(\mathbf{Y} - \mathbf{X} - \mathbf{O})\|_F^2 + \lambda_* \|\mathbf{X}\|_* + \lambda_1 \|\mathbf{O}\|_1 ,$$

where the ℓ_1-norm $\|\mathbf{O}\|_1 := \sum_{d,n} |o_{d,n}|$ and the nuclear norm $\|\mathbf{X}\|_* := \sum_i \sigma_i(\mathbf{X})$ ($\sigma_i(\mathbf{X})$ denotes the i-th singular value of $\mathbf{X}$) are utilized to promote sparsity in the number of outliers (nonzero entries) in $\mathbf{O}$, and the low rank of $\mathbf{X}$, respectively. The nuclear and ℓ_1-norms are the closest convex surrogates to the rank and cardinality functions, which albeit the most natural criteria they are in general NP-hard to optimize [CG84, Nat95]. The tuning parameters $\lambda_1, \lambda_* \geq 0$ control the tradeoff between fitting error, rank, and sparsity level of the solution. When an estimate $\hat{\sigma}_v^2$ of the noise variance is available, guidelines for selecting λ_* and λ_1 have been proposed in [ZLW^{+}10].

The location of nonzero entries in $\hat{\mathbf{O}}$ reveals "anomalies" across both features and egonets, while their amplitudes quantify the magnitude of deviation. Clearly, it does not make sense to flag outliers in data that has not been observed, namely for $(d, n) \notin \Omega$. In those cases (P1) yields $\hat{o}_{d,n} = 0$ since both the Frobenius and ℓ_1-norms are separable across the entries of their matrix arguments.

A numerical test on the arXiv General Relativity and Quantum Cosmology collaboration social graph [Les11] is depicted in Figure 10.3.1. The graph has $5,242$ vertices (authors) and $14,496$ edges (indicating that two authors collaborated on at least one paper), and there is no missing data. The red egonet is flagged as anomalous, and it is apparent that edge density (number of collaborations) is markedly larger than e.g., the other highlighted peers (purple, green and magenta egonets). Beyond egonets, it is also possible to devise algorithms to unveil anomalous graphs at a macro level. To this end, the relevant graph invariants (rows of $\mathbf{Y}$) should be evaluated for the whole network using e.g., the scalable graph minining package Pegasus [KTF09], and across networks (columns of $\mathbf{Y}$) for all those graphs of interest in the analysis.

Being convex (P1) is computationally appealing, and it has been shown to attain good performance in theory and practice. For instance, in the absence of noise and when there is no missing data, identifiability and exact recovery conditions were reported in [CLMW11] and [CSPW11]. Even when data are missing, it is possible to recover the low-rank component under some technical assumptions [CLMW11]. Theoretical performance guarantees in the presence of noise are also available [ZLW^{+}10]. Regarding *batch centralized* algorithms, a PCP solver based on the accelerated proximal gradient method was put forth in [LGW^{+}11, MMG13c], while the ADMM was employed in [YY13, MMG13c]. For a single but *dynamic* network, detection of structural changes in time can be naturally accommodated if the feature vectors (now time-indexed columns of $\mathbf{Y}$) are recalculated per time slot, and processed on-the-fly using online graph algorithms for streaming data; see also [PPY13, MAB13, MMG13b, WTPP14].

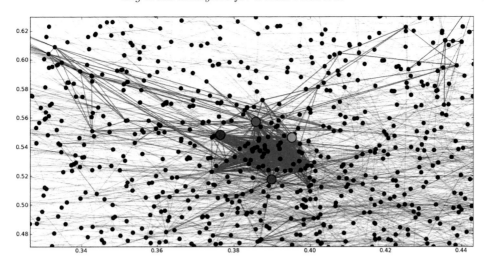

FIGURE 10.3.1: An anomalous (depicted in red) egonet for the arXiv General Relativity and Quantum Cosmology collaboration social graph [Les11]. The red egonet is flagged as anomalous by the proposed low-rank plus sparse matrix decomposition method. **(See color insert.)**

10.3.1.2   In-network processing algorithm

Increasingly-large graphs and computational challenges arising with big data motivate well devising *fully-distributed* iterative algorithms for unveiling anomalies in social graphs. In a nutshell, per iteration $k = 1, 2, \ldots$ nodes n (i.e., graph vertices) carry out simple computational tasks locally, relying on their own local feature vectors $\mathbf{y}_n$. Subsequently, local estimates are refined after exchanging messages only with directly connected neighbors in the vertex set $\mathcal{N}_n$, which facilitates percolation of local information to the whole network. The end goal is for each node to form local estimates $\mathbf{x}_n[k]$ and $\mathbf{o}_n[k]$ that coincide with the n-th columns of $\hat{\mathbf{X}}$ and $\hat{\mathbf{O}}$ as $k \to \infty$, where $\{\hat{\mathbf{X}}, \hat{\mathbf{O}}\}$ is the solution of (P1) obtained when all data $\mathcal{P}_\Omega(\mathbf{Y})$ are centrally available.

In its present form (P1) is not amenable for distributed implementation due to the non-separable nuclear norm present in the cost function. If an upper bound $\text{rank}(\hat{\mathbf{X}}) \leq \rho$ is a priori available, (P1)'s search space is effectively reduced and one can factorize the decision variable as $\mathbf{X} = \mathbf{P}\mathbf{Q}^T$, where $\mathbf{P}$ and $\mathbf{Q}$ are $D \times \rho$ and $N \times \rho$ matrices, respectively. Next, consider the following alternative characterization of the nuclear norm (see e.g. [SS05, SRJ04, RR13]):

$$\|\mathbf{X}\|_* := \min_{\{\mathbf{P}, \mathbf{Q}\}} \quad \frac{1}{2}\left(\|\mathbf{P}\|_F^2 + \|\mathbf{Q}\|_F^2\right)$$
$$\text{s. t.} \quad \mathbf{X} = \mathbf{P}\mathbf{Q}^T, \tag{10.3.2}$$

where the optimization is over all possible bilinear factorizations of $\mathbf{X}$, so that the number of columns ρ of $\mathbf{P}$ and $\mathbf{Q}$ is also a variable. Leveraging (10.3.2), the following equivalent reformulation of (P1) provides an important first step towards obtaining a distributed algorithm for graph anomaly detection

$$\min_{\{\mathbf{P}, \mathbf{Q}, \mathbf{A}\}} \sum_{n=1}^{N} \left[\|\mathcal{P}_{\Omega_n}(\mathbf{y}_n - \mathbf{P}\mathbf{q}_n - \mathbf{o}_n)\|^2 + \frac{\lambda_*}{2N}\left(\|\mathbf{P}\|_F^2 + N\|\mathbf{q}_n\|^2\right) + \lambda_1 \|\mathbf{o}_n\|_1 \right], \tag{10.3.3}$$

which is non-convex due to the bilinear terms $\mathbf{x}_n = \mathbf{P}\mathbf{q}_n$, and where $\mathbf{Q}^T := [\mathbf{q}_1, \ldots, \mathbf{q}_N]$.

Adopting the separable Frobenius-norm regularization in (10.3.3) comes with no loss of optimality relative to (P1), provided $\text{rank}(\hat{\mathbf{X}}) \leq \rho$. By finding the global minimum of (10.3.3) [which could have considerably less variables than (P1)], one can recover the optimal solution of (P1). But since (10.3.3) is non-convex, it may have stationary points which need not be globally optimum. Interestingly, as asserted in [MMG13a, Prop. 1] if a stationary point $\{\bar{\mathbf{P}}, \bar{\mathbf{Q}}, \bar{\mathbf{O}}\}$ of (10.3.3) satisfies $\|\mathcal{P}_\Omega(\mathbf{Y} - \bar{\mathbf{P}}\bar{\mathbf{Q}}^T - \bar{\mathbf{O}})\| < \lambda_*$, then $\{\hat{\mathbf{X}} := \bar{\mathbf{P}}\bar{\mathbf{Q}}^T, \hat{\mathbf{O}} := \bar{\mathbf{O}}\}$ is the globally optimal solution of (P1).

To decompose the cost in (10.3.3), in which summands inside the square brackets are coupled through the global variable $\mathbf{P}$, introduce auxiliary copies $\{\mathbf{P}_n\}_{n=1}^N$ representing local estimates of $\mathbf{P}$, one per node n. These local copies along with *consensus* constraints yield the distributed estimator:

$$\min_{\{\mathbf{P}_n, \mathbf{q}_n, \mathbf{o}_n\}} \sum_{n=1}^N \left[\|\mathcal{P}_{\Omega_n}(\mathbf{y}_n - \mathbf{P}_n \mathbf{q}_n - \mathbf{o}_n)\|^2 \right.$$
$$\left. + \frac{\lambda_*}{2N} \left(\|\mathbf{P}_n\|_F^2 + N\|\mathbf{q}_n\|^2 \right) + \lambda_1 \|\mathbf{o}_n\|_1 \right] \qquad (10.3.4)$$
$$\text{s. t.} \quad \mathbf{P}_n = \mathbf{P}_m, \ m \text{ linked with } n \in \mathcal{N},$$

which is equivalent to (10.3.3) provided the network topology graph is connected. Even though consensus is a fortiori imposed within neighborhoods, it extends to the whole (connected) network, and local estimates agree on the global solution of (10.3.3). Exploiting the separable structure of (10.3.4), a general framework for in-network sparsity-regularized rank minimization was put forth in [MMG13a], whereas a distributed algorithm for PCP (D-PCP) can be found in [MG13]. Specifically, distributed iterations were obtained after adopting the ADMM, an iterative Lagrangian method well-suited for parallel processing [BT99, BPC$^+$11]. In a nutshell, local tasks per iteration $k = 1, 2, \ldots$ entail solving small unconstrained quadratic programs to refine the projections $\mathbf{q}_n[k]$ on the nominal feature subspace $\mathbf{P}_n[k]$, in addition to soft-thresholding operations to update the egonet anomaly vectors $\mathbf{o}_n[k]$ per node; see [MG13] for further details. Per iteration, graph nodes exchange their subspace estimates $\mathbf{P}_n[k]$ only with directly connected neighbors. This way the communication overhead stays affordable, and independent of the network size N.

When employed to solve non-convex problems such as (10.3.4), so far ADMM offers no convergence guarantees. However, there is ample experimental evidence in the literature that supports empirical convergence of ADMM, especially when the non-convex problem at hand exhibits "favorable" structure. For instance, (10.3.4) is a linearly constrained bi-convex problem with potentially good convergence properties – extensive numerical tests in [MG13, MMG13a] demonstrate that this is indeed the case. While establishing convergence remains an open problem, one can still prove that upon convergence the distributed iterations attain consensus and global optimality, offering the desirable centralized performance guarantees [MMG13a].

10.3.1.3   Numerical tests

A test social network of $N = 25$ nodes is generated as a realization of the random geometric graph model, meaning nodes are randomly placed on the unit square and two nodes communicate with each other if their Euclidean distance is less than a prescribed communication range of 0.4; see Figure 10.3.2 (a). The number of egonet features is $D = 20$. Entries of $\mathbf{E}$ are independent and identically distributed (i.i.d.), zero-mean, Gaussian with variance $\sigma^2 = 10^{-3}$; i.e., $e_{d,n} \sim \mathcal{N}(0, \sigma^2)$. A simulated nominal egonet feature matrix with rank $r = 3$ is generated from the bilinear factorization model $\mathbf{X} = \mathbf{W}\mathbf{Z}^T$, where $\mathbf{W}$ and $\mathbf{Z}$ are $D \times r$ and $N \times r$ matrices with i.i.d. entries drawn from Gaussian distributions $\mathcal{N}(0, 100/D)$

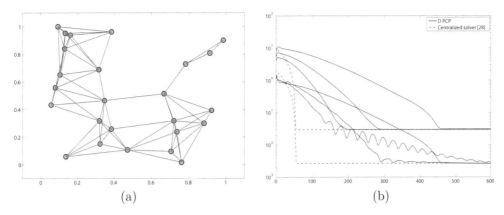

FIGURE 10.3.2: (a) A simulated small social network graph with $N = 25$ nodes. (b) Convergence of the D-PCP algorithm for different network sizes. D-PCP attains the same estimation error as the centralized solver. (**See color insert.**)

and $\mathcal{N}(0, 100/N)$, respectively. Every entry of $\mathbf{O}$ is randomly drawn from the set $\{-1, 0, 1\}$ with $\Pr(o_{d,n} = -1) = \Pr(o_{d,n} = 1) = 5 \times 10^{-2}$. To simulate missing data, a sampling matrix $\mathbf{\Omega} \in \{0, 1\}^{D \times N}$ is generated with i.i.d. Bernoulli distributed entries $o_{d,n} \sim \mathrm{Ber}(0.7)$ (30% missing data on average). Finally, measurements are generated as $\mathcal{P}_\Omega(\mathbf{Y}) = \mathbf{\Omega} \odot (\mathbf{X} + \mathbf{O} + \mathbf{E})$ [cf. (10.3.1)], and node n has available the n-th column $\mathbf{y}(n)$ of $\mathcal{P}_\Omega(\mathbf{Y})$.

To experimentally corroborate the convergence and optimality of the D-PCP algorithm for graph anomaly detection, the distributed iterations are run and compared with the centralized benchmark (P1), obtained using the solver in [YY13]. Parameters $\lambda_1 = 0.0141$ and $\lambda_* = 0.346$ are chosen as suggested in [ZLW+10]. For both schemes, Figure 10.3.2b shows the evolution of the global estimation errors $e_X[k] := \|\mathbf{X}[k] - \mathbf{X}\|_F / \|\mathbf{X}\|_F$ and $e_O[k] := \|\mathbf{O}[k] - \mathbf{O}\|_F / \|\mathbf{O}\|_F$. It is apparent that the D-PCP algorithm converges to the centralized estimator, and as expected convergence slows down due to the delay associated with the information flow throughout the network. The test is also repeated for network sizes of $N = 15$ and 35, to illustrate that the time till convergence scales gracefully as the network size increases.

10.3.2 Prediction from partially-observed network processes

Understanding the influence of topology on network processes has relied on measuring and monitoring the network itself. In practice however, gathering network-wide measurements scales poorly with the network size and may thus be impractical for various networks of interest. For instance, large social network surveys also pose a major logistic issue due to, for example, limited availability of individuals included in the survey. Tools such as `tcpdump` provide detailed packet-level information in Internet protocol (IP) networks, but collecting all these data can demand excessive power and bandwidth. Moreover, errors due to measurement and data-handling are more likely to emerge as the amount of data collected increases. A similar challenge arises when capturing spatial and temporal structures in big data. There, one may be forced to rely on partial (random) observations of the data so that inference algorithms remain operational while coping with the data deluge. With these motivating challenges in mind, this section surveys a recent joint topology- and data-driven algorithm to enable network-wide prediction of dynamical processes based on partial network observations, that is, measurements collected only at a subset of network nodes [FRG14]. The known (graph-induced) network structure and historical data are leveraged to design a dictionary for representing the network process. The novel approach draws from semi-supervised learning to enable learning the dictionary with only partial network observations. Once the dictionary is learned, network-wide prediction becomes possible via a regularized LS estimate which exploits the parsimony encapsulated in the design of the dictionary.

Consider an undirected weighted graph $G(\mathcal{V}, \mathcal{E})$, where $\mathcal{V}$ is the vertex set with cardinality $N = |\mathcal{V}|$ and $\mathcal{E}$ is the edge set. The connectivity and edge strenghts of G are characterized by the weighted adjacency matrix $\mathbf{A} \in \mathbb{R}^{N \times N}$, where the entry $a_{i,j} := [\mathbf{A}]_{i,j} > 0$ if nodes v_i and v_j are connected, and $a_{i,j} = 0$ otherwise. At time instant $t \in \mathbb{N}$, corresponding to each vertex $v_n \in \mathcal{V}$ there is a scalar variable $x_{n,t} \in \mathbb{R}$, which represents the network-wide dynamical process of interest. All node variables are collected in a single vector $\mathbf{x}_t := [x_{1,t} \ldots x_{N,t}]^T \in \mathbb{R}^N$. To account for missing data, it is assumed that $M < N$ vertices are measured at any given time. For simplicity in exposition, the number of observed vertices M is assumed fixed. However, expressions and algorithms derived in the subsequent sections can be readily modified to allow for time-varying M. Let $\mathbf{M}_t \in \mathbb{R}^{M \times N}$ denote a binary measurement matrix with $0 - 1$ entries selecting the measured components of $\mathbf{x}_t$. Each row of $\mathbf{M}_t$ corresponds to a vector of the canonical basis for $\mathbb{R}^N$, i.e., each row has only one nonzero entry, which takes the value of 1, while all other entries are set to 0. The $M \times 1$ measurement vector at time t is modeled as:

$$\mathbf{y}_t = \mathbf{M}_t \mathbf{x}_t + \boldsymbol{\epsilon}_t, \quad t = 1, 2, \ldots, \tag{10.3.5}$$

where $\boldsymbol{\epsilon}_t$ is a random error term capturing measurement imperfections.

Recently, a network process prediction algorithm was put forth in [FRG14], where missing entries of $\mathbf{x}_t$ are estimated from historical measurements in $\mathcal{T}_M := \{\mathbf{y}_t\}_{t=1}^T$ by leveraging the structural regularity of $\mathbf{x}_t$ (induced by the underlying graph) through a semi-supervised dictionary learning (DL) approach. Under the DL framework, *data-driven* dictionaries for *sparse* signal representation are adopted as a versatile means of capturing parsimonious signal structures; see e.g., [TF10] for a tutorial treatment. Propelled by the success of compressive sampling (CS) [Don06], sparse signal modeling has led to major advances in several machine learning, audio and image processing tasks [HTF09, TF10]. Motivated by these ideas, it is postulated in [FRG14] that graph signals can be represented as a linear combination $\mathbf{x}_t = \mathbf{B}\mathbf{s}_t$ of a few ($\ll Q$) columns of an over-complete dictionary (basis) matrix $\mathbf{B} := [\mathbf{b}_1, \ldots, \mathbf{b}_Q] \in \mathbb{R}^{N \times Q}$, where $\mathbf{s}_t \in \mathbb{R}^Q$ is a *sparse* vector of expansion coefficients. Many signals including speech and natural images admit sparse representations even under generic predefined dictionaries, such as those based on the Fourier and the wavelet bases, respectively [TF10]. Like audio and natural images, vertex variables can exhibit strong correlations induced from the structure of the underlying graph. For instance, Internet traffic volumes on two links are highly correlated if they both carry common end-to-end flows, as indicated by the corresponding routing matrix. DL schemes are attractive due to their flexibility, since they utilize training data to *learn* an appropriate over-complete basis customized for the data at hand. However, the use of DL for modeling network data is well motivated but so far relatively unexplored.

10.3.2.1 Semi-supervised prediction of network processes

Suppose for now that either a learnt, or, a suitable pre-specified dictionary $\mathbf{B}$ is available, and consider predicting the process on the unobserved vertices. Data-driven learning of dictionaries from historical data will be addressed in the ensuing subsection. To cope with the absence of some entries of $\mathbf{x}_t$ not present in $\mathbf{y}_t$, the idea here is to capitalize on the topology of G. To that end, suppose $w_{i,j}$ represents a similarity weight between the time-dependent variables associated with nodes v_i and v_j; e.g., the correlation between $x_{i,t}$ and

$x_{j,t}$. The topology of G, and thus the spatial correlation of the process, is captured by its Laplacian matrix $\mathbf{L} := \mathrm{diag}(\mathbf{A}\mathbf{1}_N) - \mathbf{A}$. Given $\mathbf{B}$, $\mathbf{L}$ and the measurements $\mathbf{y}_t$, contemporary tools developed in the area of CS and semi-supervised learning can be used to form $\hat{\mathbf{x}}_t$, which includes estimates for the missing $N - M$ vertex observations [Don06, BNS06, HTF09].

Given a snapshot of incomplete measurements $\mathbf{y}_t$ during the *operational phase* (where a suitable basis $\mathbf{B}$ is available), the sparse basis expansion coefficient vector $\mathbf{s}_t$ is estimated as:

$$\hat{\mathbf{s}}_t := \arg\min_{\mathbf{s}_t} \|\mathbf{y}_t - \mathbf{M}_t\mathbf{B}\mathbf{s}_t\|_2^2 + \lambda_s \|\mathbf{s}_t\|_1 + \lambda_w \mathbf{s}_t^T\mathbf{B}^T\mathbf{L}\mathbf{B}\mathbf{s}_t, \tag{10.3.6}$$

where λ_s and λ_w are tunable regularization parameters. The criterion in (10.3.6) consists of a LS error between the observed and postulated network measurements, along with two regularizers. The ℓ_1-norm $\|\mathbf{s}_t\|_1$ encourages sparsity in the coefficient vector $\hat{\mathbf{s}}_t$ [Don06, HTF09]. With $\mathbf{x}_t := [x_{1,t}, \ldots, x_{N,t}]^T$ given by $\mathbf{x}_t = \mathbf{B}\mathbf{s}_t$, the Laplacian regularization can be explicitly written as $\mathbf{s}_t^T\mathbf{B}^T\mathbf{L}\mathbf{B}\mathbf{s}_t = (1/2)\sum_{i=1}^{N}\sum_{j=1}^{N} a_{i,j}(x_{i,t} - x_{j,t})^2$. It is thus apparent that $\mathbf{s}_t^T\mathbf{B}^T\mathbf{L}\mathbf{B}\mathbf{s}_t$ encourages the vertex variables to be close if their corresponding weights are large. Typically adopted for semi-supervised learning, such a regularization term encourages $\mathbf{B}\mathbf{s}_t$ to lie on a smooth manifold approximated by G, which constrains how the measurements relate to $\mathbf{x}_t$ [BNS06, RBL$^+$07]. It is also common to use normalized variants of the Laplacian instead of $\mathbf{L}$ [Kol09, p. 46].

The cost in (10.3.6) is convex but non-smooth, and customized solvers developed for ℓ_1-norm regularized optimization can be employed here as well, e.g., [HTF09, p. 92]. Once $\hat{\mathbf{s}}_t$ is available, an estimate of the full vector of network samples is readily obtained as $\hat{\mathbf{x}}_t := \mathbf{B}\hat{\mathbf{s}}_t$. It is apparent that the quality of the imputation depends on the chosen $\mathbf{B}$, and DL from historical network data in $\mathcal{T}_M$ is described next.

10.3.2.2 Data-driven dictionary learning

In its canonical form, DL seeks a (typically fat) dictionary $\mathbf{B}$ so that training data $\mathcal{T}_N := \{\mathbf{x}_t\}_{t=1}^T$ are well approximated as $\mathbf{x}_t \approx \mathbf{B}\mathbf{s}_t$, $t = 1, \ldots, T$, for some sparse vectors $\mathbf{s}_t$ of expansion coefficients [TF10]. Standard DL algorithms cannot, however, be directly applied to learn $\mathbf{B}$ since they rely on the entire vector $\mathbf{x}_t$. To learn the dictionary in the *training phase* using incomplete measurements $\mathcal{T}_M$ instead of $\mathcal{T}_N$, the idea is to capitalize on the structure in $\mathbf{x}_t$, of which G is an abstraction [FRG14]. To this end, one can adopt a similar cost function as in the operational phase [cf. (10.3.6)], yielding the data-driven basis and the corresponding sparse representation:

$$\{\hat{\mathbf{S}}, \hat{\mathbf{B}}\} := \arg\min_{\mathbf{S},\mathbf{B}:\{\|\mathbf{b}_q\|_2 \leq 1\}_{q=1}^Q} \sum_{t=1}^{T} \left[\|\mathbf{y}_t - \mathbf{M}_t\mathbf{B}\mathbf{s}_t\|_2^2 + \lambda_s\|\mathbf{s}_t\|_1 + \lambda_w\mathbf{s}_t^T\mathbf{B}^T\mathbf{L}\mathbf{B}\mathbf{s}_t \right], \tag{10.3.7}$$

where $\hat{\mathbf{S}} := [\hat{\mathbf{S}}_1, \ldots, \hat{\mathbf{S}}_T] \in \mathbb{R}^{Q \times T}$. The constraints $\{\|\mathbf{b}_q\|_2 \leq 1\}_{q=1}^Q$ remove the scaling ambiguity in the products $\mathbf{B}\mathbf{s}_t$, and prevent the entries in $\mathbf{B}$ from growing unbounded. Again, the combined regularization terms in (10.3.7) promote both sparsity in $\mathbf{s}_t$ through the ℓ_1-norm, and smoothness across the entries of $\mathbf{B}\mathbf{s}_t$ via the Laplacian $\mathbf{L}$. The regularization parameters λ_s and λ_w are typically cross-validated [HTF09, Ch. 7]. Although (10.3.7) is non-convex, a BCD solver still guarantees convergence to a stationary point [BT99]. The BCD updates involve solving for $\mathbf{B}$ and $\mathbf{S}$ in an alternating fashion, both doable efficiently via convex programming [FRG14]. Alternatively, the online DL algorithm in [MBPS10] offers enhanced scalability by sequentially processing the data in $\mathcal{T}_S$. The training and operational (prediction) phases are summarized in Figure 10.3.3, where $C_t(\mathbf{B}, \mathbf{s})$ denotes the t-th summand from the cost in (10.3.7), and $k = 1, 2, \ldots$ indicate iterations of the BCD solver employed during the training phase.

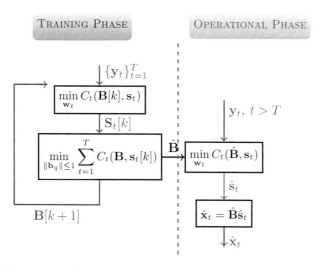

FIGURE 10.3.3: Training and operational phases of the semi-supervised DL approach for prediction of network processes that evolve over graphs [FRG14].

The explicit need for Laplacian regularization is apparent from (10.3.7). Indeed, if measurements from a certain vertex are not present in $\mathcal{T}_M$, the corresponding row of $\mathbf{B}$ may still be estimated with reasonable accuracy because of the third term in $C_t(\mathbf{B}, \mathbf{s})$. On top of that, it is because of Laplacian regularization that the prediction performance degrades gracefully as the number of missing entries in $\mathbf{y}_t$ increases; see also Figure 10.3.4. It is worth stressing that the time series $\{\mathbf{y}_t\}$ need not be stationary or even contiguous in time. The network-process prediction approach described so far can also be adapted to accommodate time-varying network topologies, using a time-dependent Laplacian $\mathbf{L}_t$. A word of caution is due however, since drastic changes in either $\mathbf{L}_t$ or in the statistical properties of the underlying process $\mathbf{x}_t$, will necessitate re-training $\mathbf{B}$ to attain satisfactory performance.

10.3.2.3   Numerical tests

Next, a numerical test on link count data from the Internet2 measurement archive [Int] is outlined. Consider an IP network comprising N nodes and L links, carrying the traffic of F origin-destination flows (network connections). Let $x_{l,t}$ denote the traffic volume (in bytes or packets) passing through link $l \in \{1, \dots, L\}$ over a fixed interval of time $(t, t + \Delta t)$. Link counts across the entire network are collected in the vector $\mathbf{x}_t \in \mathbb{R}^L$, e.g., using the ubiquitous SNMP protocol. Since measured link counts are both unreliable and incomplete due to hardware or software malfunctioning, jitter, and communication errors [ZRWQ09, Rou10], they are expressed as noisy versions of a subset of $S < L$ links

$$\mathbf{y}_t = \mathbf{M}_t \mathbf{x}_t + \boldsymbol{\epsilon}_t, \quad t = 1, 2, \dots,$$

where $\mathbf{M}_t$ is an $S \times L$ selection matrix with 0-1 entries whose rows correspond to rows of the identity matrix of size L, and $\boldsymbol{\epsilon}_t$ is an $S \times 1$ zero-mean noise term with constant variance accounting for measurement and synchronization errors. Given $\mathbf{y}_t$ the aim is to form an estimate $\hat{\mathbf{x}}_t$ of the full vector of link counts $\mathbf{x}_t$, which in this case defines the network state.

The data consists of link counts, sampled at 5 minute intervals, collected over several weeks. For the purposes of comparison, the training phase consisted of $2,000$ time slots, with a random subset of 50 links measured (out of $L = 54$ per time slot). Performance of the learned dictionary is then assessed over the next $T_0 = 2,000$ time slots. Each test

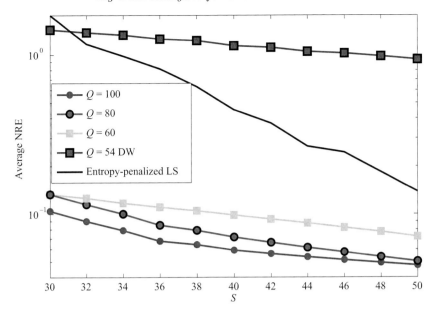

FIGURE 10.3.4: Link-traffic cartography of Internet2 data [Int]. Comparison of NRE for different values of S [FRG14].

vector $\mathbf{y}_t$ is constructed by randomly selecting S entries of the full link count vector $\mathbf{x}_t$. The tuning parameters are chosen via cross-validation ($\lambda_s = 0.1$ and $\lambda_w = 10^{-5}$). Figure 10.3.4 shows the normalized reconstruction error (NRE), evaluated as $(LT_0)^{-1} \sum_{t=1}^{T_0} \|\mathbf{y}_t - \hat{\mathbf{x}}_t\|^2$ for different values of Q and S. For comparison, the prediction performance with a fixed diffusion wavelet matrix [CPR07] (instead of the data-trained dictionary), as well as that of the entropy-penalized LS method [ZRLD05] is also shown. The latter approach solves a LS problem augmented with a specific entropy-based regularizer, that encourages the traffic volumes at the source/destination pairs to be stochastically independent. The DL-based method markedly outperforms the competing approaches, especially for low values of S. Furthermore, note how performance degrades gracefully as S decreases. Remarkably, the predictions are close to the actual traffic even when using only 30 link counts during the prediction phase.

10.4 Unveiling communities in social networks

Social networks generally exhibit community structure, which is characterized by the existence of groups within which the edge density is relatively high compared to the edge density between groups [GN02]. Communities are indicative of common functional roles or similar node behavior e.g., co-workers on Facebook, or followers of a cause on Twitter.

The community detection task has attracted significant attention from different disciplines, with several algorithms developed to tackle it. Traditional approaches resort to graph partitioning and data clustering algorithms such as k-means, hierarchical and spectral clustering [For10]. Graph partitioning algorithms such as Min-Cut and Ratio-Cut divide the graph into k parts of known size [For10]. However, these algorithms are not practical for community detection because they are limited to settings where the number and size of communities are known. On the other hand, hierarchical clustering, which is sensitive to selection of a similarity metric, is well-motivated because node similarity in social networks can be succinctly defined [HTF09].

Both k-means and spectral clustering are "workhorse" methods for (non)linearly separable data clustering. Nevertheless, spectral clustering is more appealing for community detection where the similarity graph is given. More recently, *modularity* methods have emerged, which entail optimization of a graph-clustering quality function [GN02]. Fortunato's community detection survey catalogs most of the contemporary community detection approaches [For10].

In general, most modern social networks are extremely large with millions of nodes and edges, and attempts to efficiently unveil their communities need to cope with typical big data challenges. To this end, a number of approaches based on parallelization, random sampling, and random projections have been advocated for clustering big data. Among these, parallelization is a relatively mature technology as tasks in k-means can be easily distributed over a computing cluster. However, randomized algorithms can reduce the computational load per node, and therefore require fewer nodes.

This section focuses on spectral clustering and its connections to the more general kernel k-means. In light of the aforementioned big data challenges, a discussion of recent random sampling extensions to such settings is given. It is also worth noting that the equivalence of spectral clustering with kernel k-means can prove useful to reduction of the computational load.

10.4.1   Big data spectral clustering

The emphasis of this subsection will be on big data spectral clustering. Although originally developed for data clustering, spectral clustering finds applications in community detection as it exhibits good performance in arbitrary cluster configurations. Spectral clustering exploits the properties of the similarity graph Laplacian $\mathbf{L}$ to group the vertices into a prescribed number of k clusters. Let $\mathbf{A} \in \mathbb{R}^{N \times N}$ denote the (weighted) adjacency matrix. The graph Laplacian is defined as $\mathbf{L} := \mathbf{D} - \mathbf{A}$, where $\mathbf{D}$ is a diagonal matrix with $[\mathbf{D}]_{ii} = \sum_{j=1}^{N}[\mathbf{A}]_{ij}$, and $[\mathbf{A}]_{ij}$ denotes (i, j) entry of $\mathbf{A}$. The key property of $\mathbf{L}$ is the equivalence of algebraic multiplicity of the zero eigenvalue to the number of connected components. The corresponding eigenvectors are the indicator vectors of each connected component. This can be verified by considering an eigenvalue $\lambda = 0$ of $\mathbf{L}$ and its corresponding eigenvector $\mathbf{v} \in \mathbb{R}^{N}$. Then:

$$\mathbf{L}\mathbf{v} = 0 \Rightarrow \mathbf{v}^{T}\mathbf{L}\mathbf{v} = 0 = \frac{1}{2}\sum_{i,j=1}^{N}[\mathbf{A}]_{ij}(v_i - v_j)^2, \qquad (10.4.1)$$

with v_i denoting the i-th entry of $\mathbf{v}$. As all terms of the sum must vanish and $[\mathbf{A}]_{ij} \geq 0$, the entries of $\mathbf{v}$, for which $[\mathbf{A}]_{ij} \neq 0$ must be equal. Thus $\mathbf{v}$ should have constant entries corresponding to vertices of the connected component. In a network with k completely separated clusters, the graph will have k connected components, and hence $\mathbf{L}$ will have k zero eigenvalues. The corresponding eigenvectors suffice to reveal the clusters in the network. This however is not the case in social networks where the graph can be connected with communities linked to each other by a few edges. Thus, $\mathbf{L}$ will have a single all-ones ($\mathbf{1} \in \mathbb{R}^{N}$) eigenvector, and k eigenvalues close to zero. While the eigenvectors corresponding to these k eigenvalues will not be indicator vectors, they can still be used to separate the clusters.

Spectral clustering algorithms find the k smallest, non-zero, eigenvalues $\{\lambda_i\}_{i=1}^{k}$ of $\mathbf{L}$, and their corresponding eigenvectors $\{\mathbf{v}_i \in \mathbb{R}^{N}\}_{i=1}^{k}$. With $\mathbf{V} := [\mathbf{v}_1, ..., \mathbf{v}_k]$, vertex i is mapped to row i of $\mathbf{V}$. This change of representation enhances the separability of clusters in the graph, which can be recovered using simple algorithms such as k-means. The eigenvalue computation can be performed using efficient methods such as the power iteration [GL12]. Algorithm 10.4.1 depicts the unnormalized spectral clustering algorithm. In addition to the basic definition of $\mathbf{L}$, certain spectral clustering approaches leverage normalized graph Laplacians, which are equivalent to Min-Cut and Ratio-Cut [SM00, NJW+02].

Although the number of clusters is not necessarily known, one can deduce k by comparing the magnitudes of the eigenvalues of $\mathbf{L}$. Since eigenvalues corresponding to clusters are close to zero, one can assess the value of k by finding the "jump" in the spectrum of the eigenvalues.

Spectral clustering also has strong connections to kernel PCA [HTF09] and kernel k-means (Algorithm 10.4.2). Kernel k-means [DGK04] extends the classic k-means algorithm, and is able to cluster even non-linearly separable data. This is accomplished by mapping each datum to a higher-dimensional space $\mathcal{F}$, using a function $\phi : \mathbb{R}^D \to \mathcal{F}$. The premise is that a mapping exists to render the dataset linearly separable, and hence amenable to simple and heuristic yet effective algorithms such as k-means. Even if $\mathcal{F}$ is infinite dimensional, the Representer theorem [Wah90] guarantees that inner products between data points on $\mathcal{F}$ suffice to perform clustering. Kernel k-means on a N-point dataset with k clusters aims to minimize the following objective function:

$$D = \sum_{j=1}^{k} \sum_{i=1}^{N} \| \phi(\mathbf{x}_i) - \boldsymbol{\mu}_{C_j} \|_2^2, \tag{10.4.2}$$

where $\phi(\mathbf{x}_i)$ is the feature space representation of data point $\mathbf{x}_i$, and the centroid $\boldsymbol{\mu}_{C_j} := \frac{1}{|C_j|} \sum_{j \in C_j} \phi(\mathbf{x}_j)$ is the sample mean of the points in cluster C_j. Using the Representer theorem the distance of each point in $\mathcal{F}$ from the centroid in (10.4.2) can be rewritten as:

$$\| \phi(\mathbf{x}_i) - \mu_C \|_2^2 = [\mathbf{K}]_{ii} - \frac{2}{|C|} \sum_{j \in C} [\mathbf{K}]_{ij} + \frac{1}{|C|^2} \sum_{j,l \in C} [\mathbf{K}]_{jl}, \tag{10.4.3}$$

where $\mathbf{K} \in \mathbb{R}^{N \times N}$ is the Gramian of the kernel used, and $[\mathbf{K}]_{ij}$ denotes the inner product between $\phi(\mathbf{x}_i)$ and $\phi(\mathbf{x}_j)$. Consequently, minimization of (10.4.2) can be written as:

$$\min_{\mathbf{U},\mathbf{C}} \ \mathrm{tr}(\mathbf{K}) - \mathrm{tr}(\mathbf{C}^{1/2} \mathbf{U}^T \mathbf{K} \mathbf{U} \mathbf{C}^{1/2}) \iff \max_{\hat{\mathbf{U}}} \ \mathrm{tr}(\hat{\mathbf{U}}^T \mathbf{K} \hat{\mathbf{U}}), \tag{10.4.4}$$

where $\mathbf{U} := [\mathbf{u}_i \ldots \mathbf{u}_k]$ is a cluster membership matrix with $\mathbf{u}_i \in \{0,1\}^N$, $[\mathbf{u}_i]_j = 1$, if point j belongs to cluster i; the diagonal matrix $\mathbf{C} \in \mathbb{R}^{k \times k}$ collects inverses of cluster cardinalities, $\mathbf{C} = \mathrm{diag}(\frac{1}{|C_1|}, ..., \frac{1}{|C_k|})$; and, $\hat{\mathbf{U}} := \mathbf{C}^{1/2} \mathbf{U}$. The optimization problem in (10.4.4) is non-convex as $\hat{\mathbf{U}}$ is binary. By relaxing the binary constraint, and requiring $\hat{\mathbf{U}}^T \hat{\mathbf{U}} = \mathbf{I}$, the problem can be recast as the following convex surrogate with a well-known solution [GL12]:

$$\max_{\hat{\mathbf{U}}^T \hat{\mathbf{U}} = \mathbf{I}} \ \mathrm{tr}(\hat{\mathbf{U}}^T \mathbf{K} \hat{\mathbf{U}}) = \sum_{i=1}^{k} \lambda_i, \tag{10.4.5}$$

where $\hat{\mathbf{U}}^* = \mathbf{V}\mathbf{Q}$, $\{\lambda_i\}_{i=1}^{k}$ are the largest eigenvalues of $\mathbf{K}$, $\mathbf{Q} \in \mathbb{R}^{k \times k}$ denotes an arbitrary orthonormal matrix, and the columns of $\mathbf{V} \in \mathbb{R}^{N \times k}$ are formed with the k eigenvectors corresponding to $\{\lambda_i\}_{i=1}^{k}$. This is equivalent to finding the k trailing eigenvectors of $\mathbf{I} - \mathbf{K}$. Due to the relaxation, the columns of $\hat{\mathbf{U}}^*$ most likely do not represent natural clusterings, and as such post-processing is required. Similarly, Ratio-Cut and Min-Cut can be converted to trace maximization problems [Lux07]. Furthermore, kernel PCA finds the k largest eigenvectors of the centered Gramian $\hat{\mathbf{K}}$.

Algorithm 10.4.1: Unnormalized spectral clustering

Require: k, $\mathbf{L} \in \mathbb{R}^{N \times N}$.
Ensure: Clustered vertices.
1: Compute the k smallest eigenvectors $\{\mathbf{v}_i\}_{i=1}^k$ of $\mathbf{L}$.
    Let $\mathbf{V} := [\mathbf{v}_1, \mathbf{v}_2, ..., \mathbf{v}_k] \in \mathbb{R}^{N \times k}$.
2: Let $\{\mathbf{x}_i \in \mathbb{R}^k\}_{i=1}^N$ be the rows of $\mathbf{V}$; $\mathbf{x}_i$ corresponds to the i-th vertex.
3: Group $\{\mathbf{x}_i\}_{i=1}^N$ into k clusters $\{C_i\}_{i=1}^k$.

Algorithm 10.4.2: Kernel k-means

Require: k, $\mathbf{K} \in \mathbb{R}^{N \times N}$, maximum number of iterations T
Ensure: Clustered points
1: Randomly assign points to clusters.
2: **repeat**
3:     **for** $i = 1$ **to** N **do**
4:         For point $\phi(\mathbf{x}_i)$ calculate closest centroid using equation 10.4.3
5:     **end for**
6:     Update point assignments; Assign each point to the cluster whose centroid is closest
7:     $t \leftarrow t + 1$; update iteration counter
8: **until** No changes in assignments or $t > T$

In large social networks, $\mathbf{L}$ is presumed to be sparse. In this case, methods such as Arnoldi/Lanczos iterations [GL12] can be used to efficiently compute the trailing eigenspace of $\mathbf{L}$, as usually only matrix-vector products are required. Readily available packages that tackle large-scale sparse eigenvalue problems exist [LSY98]. Distributed eigensolvers and parallel versions of k-means [ZMH09] can also be used. Care should be taken when the number of communities k is very large. While the trailing eigenvectors of $\mathbf{L}$ can be computed efficiently, the final clustering step would require clustering N k-dimensional vectors, which can prove challenging even for distributed versions of k-means. Multiple approaches that aim to tackle specifically large-scale spectral clustering tasks are available. These approaches come in three major flavors: Parallelization/distributed processing, random sampling and random projections.

A useful overview of performing spectral clustering for sparse Laplacian matrices in parallel is given in [CSB+11]. The parallelization can be performed using either MapReduce [DG08] or MPI [Sni98]. Pre-processing of the data using k-means and random projection trees is investigated in [YHJ09]. In this method the preprocessing step reduces the original N datapoints to $M < N$ representatives and performs spectral clustering on these M representatives, which results in a reduced graph which speeds up execution of the spectral clustering algorithm. However, as usually only similiraties between datapoints are given and not the datapoints themselves, algorithms such as kernel k-means or k-medoids, that can work using only similarities have to be employed.

Random sampling and random projections of the data are advocated in [SI09]. The random projection step involves projecting the data points to a lower dimensional space and computing the similarity matrix from these lower-dimensional representations of the data points. Again, as usually only the similarities are given and not the datapoints themselves, this part of the algorithm cannot provide the needed computational time reduction. Afterwards, entries of the similarity matrix are randomly sampled and spectral clustering is performed using this reduced similarity matrix. Nyström's method is proposed in [WLRB09] and [FBCM04], to form a low-rank similarity matrix, which is enabled by sampling the original similarity matrix and using the similarities between the sampled and non-sampled points. Finding the eigenspace of the new low-rank similarity matrix is much more efficient; however, one should perform the sampling carefully, as it can drastically influence the final result. When the similarity matrix is sparse, the Nyström eigenspace can be very similar to the original eigenspace, leading to highly accurate clustering.

Random sampling of the similarity matrix is explored in [ST11], whereby entries of the similarity matrix are sampled randomly, based on a budget constraint, and all other entries are set to zero. This sparsifies the similarity matrix and leads to faster computation of the eigenvectors. Random sketching is promoted in [GKB13] and [LC10]. Entries of the similarity matrix are randomly sketched using a random projection matrix to reduce the size of the similarity matrix. This reduction allows for faster computation of the eigenspace.

Large-scale kernel k-means methods can also significantly speed-up the clustering process. Results in [DGK07] have shown that using kernel k-means instead of spectral clustering can reduce the computation time required. Again care should be taken when using kernel k-means. While the Laplacian matrix of a social network might be sparse, the similarity matrix in general is not, possibly increasing the clustering time of kernel k-means (compared to spectral clustering methods), especially in cases where the number of clusters k is small. Methods to scale the kernel k-means algorithm are also available. Random sampling of the kernel matrix is investigated in [CJHJ11], where the centroids (cluster representatives) are forced to reside on the subspace spanned by those sampled points. Simulated tests demonstrate that the resultant algorithm can tackle large datasets effectively. Parallelization of the kernel k-means algorithm is proposed in [EFKK14]. Here, low-dimensional embeddings allow kernel k-means to be used in a distributed fashion using the MapReduce framework.

A more recent method, called Sketching and Validation (SkeVa [TSG15]) proposes taking multiple sketches (random samples of $M < N$ entries) of the similarity matrix, performing kernel k-means to find clusterings, and relies on different sets of random samples to validate these sketches by assigning a score to them. Afterwards, the sketch that yielded the highest score is used to cluster the remaining data. This structured trial-and-error approach has shown promising results with respect to clustering accuracy and reduced computational time. Furthermore, as each of the sketching and validation runs is independent, this approach admits easy parallelization, thereby combining random sketching approaches and distributed computing, making it attractive for the task at hand. Simulated and real data tests indicate that this algorithm can tackle large-scale datasets much faster than traditional kernel k-means.

10.4.1.1 Numerical tests

Three methods are compared in this section with respect to clustering accuracy and required time: Spectral clustering, kernel k-means, and the Sketching and Validation method for kernel k-means (Kernel SkeVa) introduced in [TSG15]. Regarding spectral clustering, the normalized version of $\mathbf{L}$ is used [NJW$^+$02], and Lanczos iterations are employed to evaluate the eigenspace of $\mathbf{L}$. The kernel used for kernel k-means and kernel SkeVa is the shortest path distance kernel $\mathbf{K} = (-1/2)\mathbf{JDJ}$, where $\mathbf{D}$ contains the shortest path distances between every node pair in the graph and $\mathbf{J} = \mathbf{I} - \frac{1}{N}\mathbf{1}\mathbf{1}^T$ is the double centering operator, while $\mathbf{1}$ denotes the all-ones vector. Figure 10.4.1 shows the community detection result for the different algorithms on the largest connected component of a Facebook egonet with $N = 744$ vertices, and $30,023$ edges containing $k = 5$ communities [Les12]. Vertices of the graph represent friends of a particular user, and edges between the vertices indicate whether two people are friends with each other. All methods are able to distinguish the clearly defined communities in this graph. Kernel SkeVa misclassifies some nodes when only 150 nodes are sampled, but this is to be expected as not all nodes are sampled. Since the size of this network is small all methods require similar amounts of time to perform the clustering (see

TABLE 10.4.1: Clustering times for Facebook egonet.

	Spectral Clustering	Kernel k-means	SkeVa (150 samples)	SkeVa (350 samples)
Time(s)	0.067	0.074	0.031	0.066

TABLE 10.4.2: Clustering times for arXiv General relativity collaboration network.

	Spectral Clustering	Kernel k-means	SkeVa (500 samples)	SkeVa (1,000 samples)
Time(s)	3.1	2.51	0.4	0.85

Table 10.4.1). Figure 10.4.2 shows the community detection result on the largest connected component of an arXiv collaboration network (General Relativity) with $N = 4,158$ vertices, and $13,422$ edges [Les11]. Vertices represent paper authors and edges indicate whether two people have co-authored a paper. It is assumed that $k = 36$ communities are present in this graph. Similar to the Facebook network, all algorithms are able to recognize the tight communities of this network, however the time required for kernel SkeVa is one order of magnitude lower.

10.4.2 Robust kernel PCA

Kernel (K)PCA is a generalization of (linear) PCA, seeking principal components in a *feature space* nonlinearly related to the *input space*, where the data in $\mathcal{T}_x$ live [SSM98]. KPCA has been shown effective in performing nonlinear feature extraction for pattern recognition [SSM98]. In addition, connections between KPCA and spectral clustering [HTF09, p. 548] motivate well the KPCA method outlined in this section, to robustly identify cohesive subgroups (communities) from social network data.

Consider a nonlinear function $\phi : \mathbb{R}^D \to \mathcal{H}$, that maps elements from the input space $\mathbb{R}^D$ to a feature space $\mathcal{H}$ of arbitrarily large — possibly infinite — dimensionality. Given transformed training data $\mathcal{T}_{\mathcal{H}} := \{\phi(\mathbf{y}_n)\}_{n=1}^N$, the proposed approach to robust KPCA fits the model [cf. the low-rank subspace model in (10.3.3)]:

$$\phi(\mathbf{y}_n) = \mathbf{m} + \mathbf{Pq}_n + \mathbf{o}_n + \mathbf{e}_n, \quad n = 1, \ldots, N, \tag{10.4.6}$$

where, again, $\mathbf{o}_n$ is an outlier vector, and $\mathbf{m}$ denotes the location (mean) vector in feature space $\mathcal{H}$. A natural criterion is ($\mathbf{\Phi} := [\phi(\mathbf{y}_1), \ldots, \phi(\mathbf{y}_N)]$ and $\mathbf{1}_N^T$ is the $N \times 1$ row vector of all ones):

$$\min_{\mathbf{m}, \mathbf{P}, \mathbf{Q}, \mathbf{O}} \|\mathbf{\Phi} - \mathbf{m}\mathbf{1}_N^T - \mathbf{PQ}^T - \mathbf{O}\|_F^2 + \frac{\lambda_*}{2}(\|\mathbf{P}\|_F^2 + \|\mathbf{Q}\|_F^2) + \lambda_2 \sum_{n=1}^N \|\mathbf{o}_n\|_2, \tag{10.4.7}$$

where $\sum_{n=1}^N \|\mathbf{o}_n\|_2$ is the so-termed group Lasso penalty [YL06]. It is a high-dimensional extension of the ℓ_1-norm, that encourages columnwise (vector) sparsity on the estimator of $\mathbf{O}$ [cf. entrywise sparsity with the ℓ_1-norm]. This way, one can declare whether the corresponding training vector $\phi(\mathbf{y}_1)$ is an outlier or not. Except for the principal components' matrix $\mathbf{Q} \in \mathbb{R}^{N \times \rho}$, both the data and the unknowns in (10.4.7) are now vectors/matrices of generally infinite dimension. In principle, this challenges the optimization task since it is impossible to store, or, perform updates of such quantities directly. For these reasons, assuming zero-mean data $\phi(\mathbf{y}_n)$, or, the possibility of mean compensation for that matter,

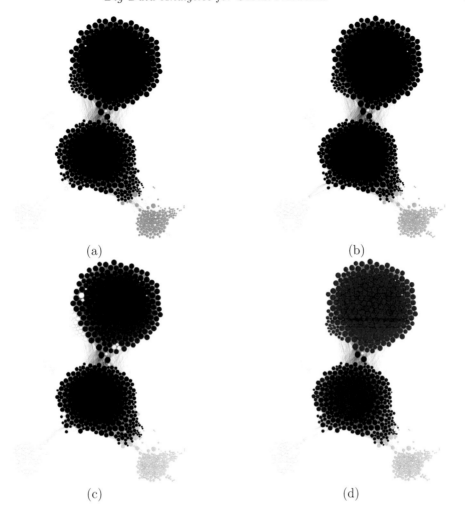

FIGURE 10.4.1: Community detection results for a Facebook egonet with $N = 744$ nodes and $k = 5$ communities using: (a) Normalized spectral clustering; (b) Kernel k-means; (c) Kernel SkeVa, where 150 nodes are sampled; and (d) Kernel SkeVa, where 350 nodes are sampled. Different shades of gray represent different communities.

cannot be taken for granted here. Thus, it is important to explicitly consider the estimation of $\mathbf{m}$ [which for instance, was not explicitly accounted for in (10.3.3)].

Interestingly, this hurdle can be overcome by endowing $\mathcal{H}$ with the structure of a reproducing kernel Hilbert space (RKHS), where inner products between any two members of $\mathcal{H}$ boil down to evaluations of the reproducing kernel $K_{\mathcal{H}} : \mathbb{R}^D \times \mathbb{R}^D \to \mathbb{R}$, i.e., $\langle \phi(\mathbf{y}_i), \phi(\mathbf{y}_j) \rangle_{\mathcal{H}} = K_{\mathcal{H}}(\mathbf{y}_i, \mathbf{y}_j)$. Specifically, it is possible to form the kernel matrix $\mathbf{K} := \mathbf{\Phi}^T \mathbf{\Phi} \in \mathbb{R}^{N \times N}$, without directly working with the vectors in $\mathcal{H}$. This so-termed *kernel trick* is the crux of most kernel methods in machine learning [HTF09], including kernel PCA [SSM98]. The problem of selecting $K_{\mathcal{H}}$ (and ϕ indirectly) will not be considered here.

Building on these ideas, it is shown in [MG12b] that natural alternating-minimization (AM) iterations one can devise to optimize (10.4.7) can be *kernelized*, to solve (10.4.7) at affordable computational complexity and memory storage requirements that do not depend on the dimensionality of $\mathcal{H}$. Specifically, for $k \geq 1$, [MG12b] shows that the sequence of AM iterates obtained to solve (10.4.7) can be written as $\mathbf{m}(k) = \mathbf{\Phi}\boldsymbol{\mu}(k)$, $\mathbf{P}(k) = \mathbf{\Phi}\mathbf{\Pi}(k)$,

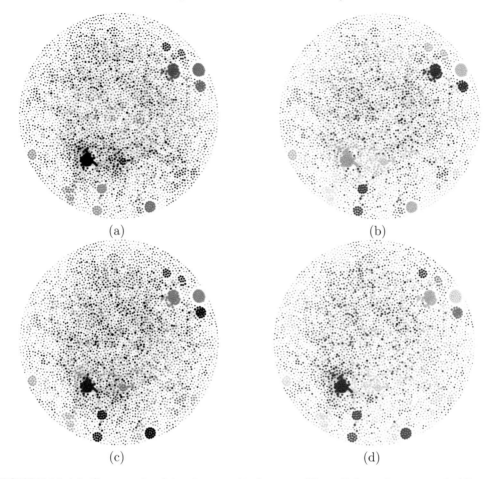

(a)

(b)

(c)

(d)

FIGURE 10.4.2: Community detection results for an arXiv collaboration network (General Relativity) with $N = 4,158$ nodes, and $k = 36$ communities using: (a) Normalized spectral clustering; (b) Kernel k-means; (c) Kernel SkeVa where 500 nodes are sampled; and (d) Kernel SkeVa where 1,000 nodes are sampled. Different shades of gray represent different communities.

and $\mathbf{O}(k) = \mathbf{\Phi}\mathbf{\Omega}(k)$. The quantities $\boldsymbol{\mu}(k) \in \mathbb{R}^N$, $\mathbf{\Pi}(k) \in \mathbb{R}^{N \times \rho}$, and $\mathbf{\Omega}(k) \in \mathbb{R}^{N \times N}$ are then recursively updated as in Algorithm 10.4.3, without the need of operating with vectors in $\mathcal{H}$.

In order to run the robust KPCA algorithm (tabulated as Algorithm 10.4.3), one does not have to store or process the quantities $\mathbf{m}(k)$, $\mathbf{U}(k)$, and $\mathbf{O}(k)$. As per [MG12b, Prop. 4], the iterations can be equivalently carried out by cycling through *finite-dimensional* 'sufficient statistics' $\boldsymbol{\mu}(k) \to \mathbf{\Pi}(k) \to \mathbf{Q}(k) \to \mathbf{\Omega}(k)$. In other words, the iterations of the robust kernel PCA algorithm are devoid of algebraic operations among vectors in $\mathcal{H}$. Recall that the size of matrix $\mathbf{Q}$ is independent of the dimensionality of $\mathcal{H}$.

Because $\mathbf{O}(k) = \mathbf{\Phi}\mathbf{\Omega}(k)$ and upon convergence of the algorithm, the outlier vector norms are computable in terms of $\mathbf{K}$, i.e., $[\|\mathbf{o}_1(\infty)\|_2^2, \ldots, \|\mathbf{o}_N(\infty)\|_2^2]^T = \text{diag}[\mathbf{\Omega}^T(\infty)\mathbf{K}\mathbf{\Omega}(\infty)]$. These are critical to identifying outlying vectors $\mathbf{y}_n$, since for those $\|\mathbf{o}_n(\infty)\|_2 > 0$. Moreover, the principal component corresponding to any given new data point $\mathbf{y}$ is obtained through the projection $\mathbf{1} = \mathbf{P}^T(\infty)[\boldsymbol{\phi}(\mathbf{y}) - \mathbf{m}(\infty)] = \mathbf{\Pi}^T(\infty)\mathbf{\Phi}^T\boldsymbol{\phi}(\mathbf{x}) - \mathbf{\Pi}^T(\infty)\mathbf{K}\boldsymbol{\mu}(\infty)$, which is again computable after N evaluations of the kernel function $K_{\mathcal{H}}$.

Algorithm 10.4.3: : Robust KPCA solver

Initialize $\mathbf{\Omega}(0) = \mathbf{0}_{N \times N}$, $\mathbf{Q}(0)$ randomly, and form $\mathbf{K} = \mathbf{\Phi}^T \mathbf{\Phi}$.

for $k = 1, 2, \ldots$ **do**

    Update $\boldsymbol{\mu}(k) = [\mathbf{I}_N - \mathbf{\Omega}(k-1)]\mathbf{1}_N/N$.

    Form $\mathbf{\Phi}_o(k) = \mathbf{I}_N - \boldsymbol{\mu}(k)\mathbf{1}_N^T - \mathbf{\Omega}(k-1)$.

    Update $\mathbf{\Pi}(k) = \mathbf{\Phi}_o(k)\mathbf{Q}(k-1)[\mathbf{Q}^T(k-1)\mathbf{Q}(k-1) + (\lambda_*/2)\mathbf{I}_\rho]^{-1}$.

    Update $\mathbf{Q}(k) = \mathbf{\Phi}_o^T(k)\mathbf{K}\mathbf{\Pi}(k)[\mathbf{\Pi}^T(k)\mathbf{K}\mathbf{\Pi}(k) + (\lambda_*/2)\mathbf{I}_\rho]^{-1}$.

    Form $\boldsymbol{\delta}_n(k) = \mathbf{e}_{N,n} - \boldsymbol{\mu}(k) - \mathbf{\Pi}(k)\mathbf{q}_n(k)$, $n = 1, \ldots, N$

    Form $\mathbf{\Delta}(k) = \mathrm{diag}\left(\frac{(\boldsymbol{\delta}_1^T(k)\mathbf{K}\boldsymbol{\delta}_1(k) - \frac{\lambda_2}{2})_+}{\boldsymbol{\delta}_1^T(k)\mathbf{K}\boldsymbol{\delta}_1(k)}, \ldots, \frac{(\boldsymbol{\delta}_N^T(k)\mathbf{K}\boldsymbol{\delta}_N(k) - \frac{\lambda_2}{2})_+}{\boldsymbol{\delta}_N^T(k)\mathbf{K}\boldsymbol{\delta}_N(k)} \right)$.

    Update $\mathbf{\Omega}(k) = [\mathbf{I}_N - \boldsymbol{\mu}(k)\mathbf{1}_N^T - \mathbf{\Pi}(k)\mathbf{Q}^T(k)]\mathbf{\Delta}(k)$.

end for

10.4.2.1 Numerical tests

Here robust KPCA is used to identify communities and outliers in a social network of $N = 115$ college football teams, by capitalizing on the connection between KPCA and spectral clustering [HTF09, p. 548]. Nodes in the network graph represent teams belonging to eleven conferences (plus five independent teams), whereas (unweighted) edges joining pairs of nodes indicate that both teams played against each other during the Fall 2000 Division I season [GN02]. The kernel matrix used to run robust KPCA is $\mathbf{K} = \zeta\mathbf{I}_N + \mathbf{D}^{-1/2}\mathbf{A}\mathbf{D}^{-1/2}$, where $\mathbf{A}$ and $\mathbf{D}$ denote the graph adjacency and degree matrices, respectively; while $\zeta > 0$ is chosen to render $\mathbf{K}$ positive semi-definite. The choice of the normalized graph Laplacian as kernel matrix is at the heart of the equivalence between KPCA and spectral clustering [HTF09]. The tuning parameters are chosen as $\lambda_2 = 1.297$ so that $\|\hat{\mathbf{O}}\|_0 = 10$, while $\lambda_* = 1$, and $\rho = 3$. Figure 10.4.3 (a) shows the entries of $\mathbf{K}$, where rows and columns are permuted to reveal the clustering structure found by robust KPCA (after removing the outliers); see also Figure 10.4.3 (b). The quality of the clustering is assessed through the adjusted rand index (ARI) after excluding outliers [FKG11], which yielded the value 0.8967. Four of the teams deemed as outliers are Connecticut, Central Florida, Navy, and Notre Dame, which are indeed independent teams not belonging to any major conference. The community structure of traditional powerhouse conferences such as Big Ten, Big 12, ACC, Big East, and SEC was identified exactly.

10.5 Topology tracking from information cascades

It has been observed in many settings that information often spreads in *cascades* by following implicit links between nodes in a network. Examples include the propagation of viral news events between blogs, adoption of emerging fashion trends within an age group, or acquisition of new buying habits by consumer groups. Consider the example of a terrorist attack reported within minutes on mainstream news websites. An information cascade may emerge because these websites' readership includes bloggers who subsequently write about the attack, influencing their own readers in turn to do the same. The underlying dynamics for propagation of such information are remarkably similar to those governing the rapid spread of infectious diseases within a population, leading to the so-termed *contagions* [Rog95, EK10, BMG14]. In general, one is only able to observe the nodes of such networks and the times when they got "infected" by a contagion, but not their link topology.

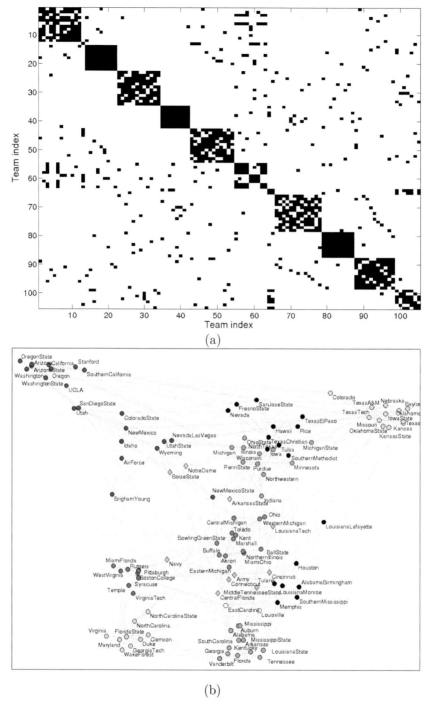

(a)

(b)

FIGURE 10.4.3: (a) Entries of **K** after removing the outliers, where rows and columns are permuted to reveal the clustering structure found by robust KPCA; and (b) Graph depiction of the clustered network [MG12b]. Teams belonging to the same estimated conference (cluster) are colored identically. The outliers are represented as diamond-shaped nodes. **(See color insert.)**

Knowledge of the network topology is crucial for several reasons. Viral web advertising can be more effectively achieved if a small set of influential initiators are identified through the link structure. Furthermore, knowledge of the structure of hidden needle-sharing networks among communities of injecting drug users can aid formulation of policies for curbing contagious diseases. Other examples include assessment of the reliability of heavily interconnected systems such as the power grid, or, estimating risk exposure among investment banks in a highly interdependent global economy. In general, unveiling the network topology can be used to predict the behavior of complex systems [Kol09, RLS10, BMG14].

Key to topology identification from a cascade is the ease of observation of its evolution over the unknown network. Indeed, this is tantamount to simply recording the times when each node got infected by a cascade. Well-studied diffusion models based on epidemiological studies have been put forth to identify an underlying network topology [VR07, EK10, Jac10]. Undoubtedly a challenging inference task, analysis of information cascades over modern social networks leads to the fundamental big data challenges. Cascades typically propagate over very large web-scale networks, and are acquired sequentially in infinite streams. More importantly, the underlying network topology is generally dynamic and varies over time.

Network inference from temporal traces of infection events has recently emerged as an active area of research. According to the taxonomy in [Kol09, Ch. 7], this can be viewed as a problem involving inference of *association networks*. Two other broad classes of network topology identification problems entail (individual) link prediction, or, *tomographic inference*. Several approaches postulate probabilistic models and rely on maximum likelihood estimation (MLE) to infer static edge weights as pairwise transmission rates between nodes [RBS11, ML13]. MLE-based stochastic gradient descent (SGD) iterations have been leveraged for inference of temporal diffusion networks [RLS10]. Most contemporary diffusion models attribute node infection events to the network topology alone (*endogenous* factors), and ignore *exogenous* factors such as non-topological information sources. Modeling causal endogenous and exogenous factors is the mainstay of *structural equation models (SEMs)*, and the rest of this section will focus on such a general approach that was recently advocated in [BMG14].

10.5.1 Dynamic SEMs for tracking cascades

Structural equation modeling refers to a family of statistical methods that model causal relationships between interacting variables in a complex system; see e.g., [Kap09]. Their appeal can be attributed to simplicity and the inherent ability to capture edge directionalities in graphs. They have been adopted in economics, psychometrics [Mut84], social sciences [Gol72], and genetics [LdH08, CBG13], among others.

Reasoning that infection times depend on both topological (endogenous) and external (exogenous) influences, a novel SEM-based scheme was proposed in [BMG14] for cascade modeling. Topological influences are modeled as linear combinations of infection times of other nodes in the network, whose weights correspond to entries in a time-varying asymmetric adjacency matrix. External influences such as those due to on-site reporting in news propagation contexts are useful for model identifiability, as they have been shown necessary to resolve directional ambiguities [BBG13]. It is assumed that the network varies slowly with time, facilitating adaptive parameter estimation by minimizing a sparsity-promoting exponentially-weighted LS criterion. Furthermore, inherent sparse connectivity of social networks is accounted for by ℓ_1-norm regularization [CGH09, ABG10, KST11, AG11].

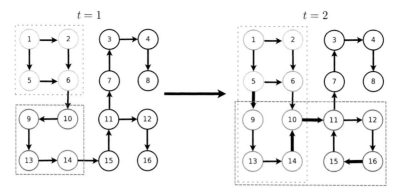

FIGURE 10.5.1: Two cascades propagating over a dynamic directed 16-node network during time intervals $t = 1$ and $t = 2$. Both cascades are initially observable at 4 nodes and they propagate to include 4 extra nodes during $t = 2$. Changes to the network topology are depicted by the new, thicker edges during $t = 2$.

10.5.1.1   Model and problem statement

Consider a dynamic N-node network observed over time intervals $t = 1, \ldots, T$, represented by a graph whose temporal topology is encoded through a time-series of unknown, time-indexed, and weighted adjacency matrices $\{\mathbf{A}^t \in \mathbb{R}^{N \times N}\}_{t=1}^T$. Per convention in network studies, entry (i, j) of $\mathbf{A}^t$ (henceforth denoted by a_{ij}^t) is nonzero only if a directed edge connects nodes i and j during the time interval t. The network topology is assumed to remain fixed per time interval t, but can change across intervals.

Over the course of the observation interval, many contagions propagate over the time-varying network as illustrated in the 16-node directed network in Figure 10.5.1. Suppose a fixed number of contagions C is sampled, and the difference between infection time of node i by contagion c and the earliest observation time is denoted by $y_{ic}^t \geq 0$. For uninfected nodes at interval t, y_{ic}^t is infinite and will be set to a large positive value for practical considerations. Assume that the susceptibility x_{ic} of node i to external (non-topological) infection by contagion c is known and time invariant over the observation interval. In the web context, x_{ic} can be set to the search engine rank of website i w.r.t. keywords associated with c.

The model in [BMG14] postulates that y_{ic}^t is linearly related to x_{ic} and the infection times of its single-hop neighbors. Events adhering to this model of network-facilitated propagation abound on the web where mention of e.g., a major baseball event by a blog will not only depend on the times when similar blogs first reported the event, but also the level of interest of the blogger in baseball as a sport. Similarly, in epidemiological studies an individual's infection time by an infectious disease depends on both the infection times of her immediate contacts as well as her immunity level to the disease. As a result y_{ic}^t is modeled according to the following linear *dynamic* structural equation model (SEM):

$$y_{ic}^t = \sum_{j \neq i} a_{ij}^t y_{jc}^t + b_{ii}^t x_{ic} + e_{ic}^t, \tag{10.5.1}$$

where b_{ii}^t captures the time-varying level of influence of external sources, and e_{ic}^t accounts for measurement errors and unmodeled dynamics. It follows from (10.5.1) that if $a_{ij}^t \neq 0$, then y_{ic}^t is affected by the value of y_{jc}^t. Rewriting (10.5.1) for the entire network leads to the vector model:

$$\mathbf{y}_c^t = \mathbf{A}^t \mathbf{y}_c^t + \mathbf{B}^t \mathbf{x}_c + \mathbf{e}_c^t, \tag{10.5.2}$$

where the $N \times 1$ vector $\mathbf{y}_c^t := [y_{1c}^t, \ldots, y_{Nc}^t]^T$ collects the node infection times by contagion c during interval t, $\mathbf{B}^t := \mathrm{diag}(b_{11}^t, \ldots, b_{NN}^t)$; and likewise $\mathbf{x}_c := [x_{1c}, \ldots, x_{Nc}]^T$ and $\mathbf{e}_c^t := [e_{1c}^t, \ldots, e_{Nc}^t]^T$. Collecting observations for all C contagions yields the dynamic matrix SEM:

$$\mathbf{Y}^t = \mathbf{A}^t \mathbf{Y}^t + \mathbf{B}^t \mathbf{X} + \mathbf{E}^t, \tag{10.5.3}$$

where $\mathbf{Y}^t := [\mathbf{y}_1^t, \ldots, \mathbf{y}_C^t]$, $\mathbf{X} := [\mathbf{x}_1, \ldots, \mathbf{x}_C]$, and $\mathbf{E}^t := [\mathbf{e}_1^t, \ldots, \mathbf{e}_C^t]$ are all $N \times C$ matrices. Given $\{\mathbf{Y}^t\}_{t=1}^T$ and $\mathbf{X}$, the goal is to track the underlying network topology $\{\mathbf{A}^t\}_{t=1}^T$, and the effect of external influences $\{\mathbf{B}^t\}_{t=1}^T$. In order to cope with constraints due to limited measurement budgets, it is desirable that $C \ll N$. Unfortunately without further constraints, this compromises the identifiability of (10.5.3). The approach outlined next overcomes this limitation by leveraging the edge sparsity inherent to social networks.

10.5.1.2 Exponentially-weighted least-squares estimator

For the sake of exposition, consider the *static* setting with all $\{\mathbf{Y}^t\}_{t=1}^T$ available. Leveraging the squared error cost leads to the batch problem:

$$\{\hat{\mathbf{A}}, \hat{\mathbf{B}}\} = \arg\min_{\mathbf{A}, \mathbf{B}} \quad \frac{1}{2} \sum_{t=1}^T \|\mathbf{Y}^t - \mathbf{A}\mathbf{Y}^t - \mathbf{B}\mathbf{X}\|_F^2 + \lambda \|\mathbf{A}\|_1$$
$$\text{s. t.} \quad a_{ii} = 0, \; b_{ij} = 0, \; \forall i \neq j, \tag{10.5.4}$$

where $\lambda > 0$ controls the sparsity level of $\hat{\mathbf{A}}$. Reasonably assuming the absence of a self-loop at node i leads to the constraint $a_{ii} = 0$, while having $b_{ij} = 0$, $\forall i \neq j$, ensures that $\hat{\mathbf{B}}$ is diagonal as in (10.5.2). Note that the estimator (10.5.4) tacitly assumes equal residual variances since the infection times per cascade result from the same contagion over the entire network.

In big data settings, measurements are more likely to be acquired sequentially over large social networks ($\geq 10^6$ nodes), motivating online estimation algorithms with minimal storage requirements. As a result, preference is given to recursive solvers facilitating sequential topology inference. Incorporating a "forgetting factor" that assigns more weight to more recent residuals then makes it possible to track slow temporal topological variations. Note that the batch estimator (10.5.4) yields the single estimates $\{\hat{\mathbf{A}}, \hat{\mathbf{B}}\}$ that best fit the data $\{\mathbf{Y}^t\}_{t=1}^T$ and $\mathbf{X}$ over the entire measurement horizon $t = 1, \ldots, T$, and as such (10.5.4) neglects potential network variations across time intervals.

For $t = 1, \ldots, T$, the *sparsity-regularized exponentially-weighted LS estimator (EWLSE)* is given by:

$$\{\hat{\mathbf{A}}^t, \hat{\mathbf{B}}^t\} = \arg\min_{\mathbf{A}, \mathbf{B}} \quad \frac{1}{2} \sum_{\tau=1}^t \beta^{t-\tau} \|\mathbf{Y}^\tau - \mathbf{A}\mathbf{Y}^\tau - \mathbf{B}\mathbf{X}\|_F^2 + \lambda_t \|\mathbf{A}\|_1$$
$$\text{s. t.} \quad a_{ii} = 0, \; b_{ij} = 0, \; \forall i \neq j, \tag{10.5.5}$$

where $\beta \in (0, 1]$ is the forgetting factor that forms estimates $\{\hat{\mathbf{A}}^t, \hat{\mathbf{B}}^t\}$ using all measurements acquired until time t. Whenever $\beta < 1$, past data are exponentially discarded thus enabling tracking of dynamic network topologies. The first summand in the cost corresponds to an exponentially-weighted moving average (EWMA) of the squared model residuals norms. The EWMA can be seen as an average modulated by a sliding window of equivalent length $1/(1 - \beta)$, which clearly grows as $\beta \to 1$. In the so-termed infinite-memory setting whereby $\beta = 1$, (10.5.5) boils down to the batch estimator (10.5.4). Notice that λ_t is allowed to vary with time in order to capture the generally changing edge sparsity level. In a linear regression context, a related EWLSE was put forth in [ABG10] for adaptive estimation of sparse signals; see also [KST11] for a projection-based adaptive algorithm.

10.5.2 Topology tracking algorithm

Proximal gradient (PG) algorithms have been popularized for ℓ_1-norm regularized linear regression problems, through the class of *iterative shrinkage-thresholding algorithms (ISTA)*; see e.g., [DDM04] and [PB13] for a comprehensive tutorial treatment. The main advantage of ISTA over off-the-shelf interior point methods is its computational simplicity. Iterations boil down to matrix-vector multiplications involving the regression matrix, followed by a soft-thresholding operation [HTF09, p. 93].

Introducing the optimization variable $\mathbf{V} := [\mathbf{A}\ \mathbf{B}]$, it follows that the gradient of $f(\mathbf{V}) := \frac{1}{2}\sum_{\tau=1}^{t}\beta^{t-\tau}\|\mathbf{Y}^\tau - \mathbf{A}\mathbf{Y}^\tau - \mathbf{B}\mathbf{X}\|_F^2$ is Lipschitz continuous, i.e., $\|\nabla f(\mathbf{V}_1) - \nabla f(\mathbf{V}_2)\| \leq L_f\|\mathbf{V}_1 - \mathbf{V}_2\|$, $\forall \mathbf{V}_1, \mathbf{V}_2$ in the domain of f. The Lipschitz constant L_f is time varying, but its dependence on t is kept implicit for notational convenience. Instead of directly optimizing the cost in (10.5.5), PG algorithms minimize a sequence of overestimators evaluated at judiciously chosen points.

Let $k = 1, 2, \ldots$ denote iterations and define $g(\mathbf{V}) := \lambda_t\|\mathbf{A}\|_1$, PG algorithms iteratively solve:

$$\mathbf{V}[k] := \arg\min_{\mathbf{V}}\left\{\frac{L_f}{2}\|\mathbf{V} - \mathbf{G}(\mathbf{V}[k-1])\|_F^2 + g(\mathbf{V})\right\}, \tag{10.5.6}$$

where $\mathbf{G}(\mathbf{V}[k-1]) := \mathbf{V}[k-1] - (1/L_f)\nabla f(\mathbf{V}[k-1])$ corresponds to a gradient-descent step taken from $\mathbf{V}[k-1]$, with step-size equal to $1/L_f$. The optimization problem (10.5.6) is known as the *proximal operator* of the function g/L_f evaluated at $\mathbf{G}(\mathbf{V}[k-1])$, and is denoted as $\mathrm{prox}_{g/L_f}(\mathbf{G}(\mathbf{V}[k-1]))$. Henceforth adopting the notation $\mathbf{G}[k-1] := \mathbf{G}(\mathbf{V}[k-1])$ for convenience, the PG iterations can be compactly rewritten as $\mathbf{V}[k] = \mathrm{prox}_{g/L_f}(\mathbf{G}[k-1])$.

A key element to the success of PG algorithms stems from the possibility of efficiently evaluating the proximal operator (cf. (10.5.6)). Specializing to (10.5.5), note that (10.5.6) decomposes into:

$$\mathbf{A}[k] := \arg\min_{\mathbf{A}}\left\{\frac{L_f}{2}\|\mathbf{A} - \mathbf{G}_A[k-1]\|_F^2 + \lambda_t\|\mathbf{A}\|_1\right\}$$

$$= \mathcal{S}_{\lambda_t/L_f}(\mathbf{G}_A[k-1]) \tag{10.5.7}$$

$$\mathbf{B}[k] := \arg\min_{\mathbf{B}}\left\{\|\mathbf{B} - \mathbf{G}_B[k-1]\|_F^2\right\} = \mathbf{G}_B[k-1], \tag{10.5.8}$$

subject to the constraints in (10.5.5), which so far have been left implicit, and $\mathbf{G} := [\mathbf{G}_A\mathbf{G}_B]$. Letting $\mathcal{S}_\mu(\mathbf{M})$ with (i,j)-th entry given by $\mathrm{sign}(m_{ij})\max(|m_{ij}| - \mu, 0)$ denote the soft-thresholding operator, it follows that $\mathrm{prox}_{\lambda_t\|\cdot\|_1/L_f}(\cdot) = \mathcal{S}_{\lambda_t/L_f}(\cdot)$, e.g., [DDM04, HTF09]. Because there is no regularization on the matrix $\mathbf{B}$, the corresponding update (10.5.8) boils-down to a simple gradient-descent step.

What remains now is to obtain expressions for the gradient of $f(\mathbf{V})$ with respect to $\mathbf{A}$ and $\mathbf{B}$, which are required to form the matrices $\mathbf{G}_A$ and $\mathbf{G}_B$. To this end, note that by incorporating the constraints $a_{ii} = 0$ and $b_{ij} = 0$, $\forall j \neq i$, $i = 1, \ldots N$, one can simplify the expression of $f(\mathbf{V})$ as:

$$f(\mathbf{V}) := \frac{1}{2}\sum_{\tau=1}^{t}\sum_{i=1}^{N}\beta^{t-\tau}\|(\mathbf{y}_i^\tau)^T - \mathbf{a}_{-i}^T\mathbf{Y}_{-i}^\tau - b_{ii}\mathbf{x}_i^T\|_F^2, \tag{10.5.9}$$

where $(\mathbf{y}_i^\tau)^T$ and $\mathbf{x}_i^T$ denote the i-th row of $\mathbf{Y}^\tau$ and $\mathbf{X}$, respectively; while $\mathbf{a}_{-i}^T$ denotes the $1 \times (N-1)$ vector obtained by removing entry i from the i-th row of $\mathbf{A}$, and likewise $\mathbf{Y}_{-i}^\tau$ is the $(N-1) \times C$ matrix obtained by removing row i from $\mathbf{Y}^\tau$. It is apparent from (10.5.9) that $f(\mathbf{V})$ is separable across the trimmed row vectors $\mathbf{a}_{-i}^\top$, and the diagonal entries b_{ii},

$i = 1, \ldots, N$. The sought gradients are:

$$\nabla_{\mathbf{a}_{-i}} f(\mathbf{V}) = \mathbf{\Sigma}_{-i}^t \mathbf{a}_{-i} + \bar{\mathbf{Y}}_{-i}^t \mathbf{x}_i b_{ii} - \boldsymbol{\sigma}_{-i}^t \qquad (10.5.10)$$

$$\nabla_{b_{ii}} f(\mathbf{V}) = \mathbf{a}_{-i}^T \bar{\mathbf{Y}}_{-i}^t \mathbf{x}_i + \frac{1 - \beta^t}{1 - \beta} b_{ii} \|\mathbf{x}_i\|_2^2 - (\bar{\mathbf{y}}_i^\tau)^T \mathbf{x}_i, \qquad (10.5.11)$$

where $(\bar{\mathbf{y}}_i^t)^T$ denotes the i-th row of $\bar{\mathbf{Y}}^t := \sum_{\tau=1}^t \beta^{t-\tau} \mathbf{Y}^\tau$, and $\bar{\mathbf{Y}}_{-i}^t := \sum_{\tau=1}^t \beta^{t-\tau} \mathbf{Y}_{-i}^\tau$. Similarly, $\boldsymbol{\sigma}_{-i}^t := \sum_{\tau=1}^t \beta^{t-\tau} \mathbf{Y}_{-i}^\tau \mathbf{y}_i^\tau$ and $\mathbf{\Sigma}_{-i}^t$ is obtained by removing the i-th row and i-th column from $\mathbf{\Sigma}^t := \sum_{\tau=1}^t \beta^{t-\tau} \mathbf{Y}^\tau (\mathbf{Y}^\tau)^T$. From (10.5.7)-(10.5.8) and (10.5.10)-(10.5.11), the parallel ISTA iterations:

$$\nabla_{\mathbf{a}_{-i}} f[k] = \mathbf{\Sigma}_{-i}^t \mathbf{a}_{-i}[k] + \bar{\mathbf{Y}}_{-i}^t \mathbf{x}_i b_{ii}[k] - \boldsymbol{\sigma}_{-i}^t \qquad (10.5.12)$$

$$\nabla_{b_{ii}} f[k] = \mathbf{a}_{-i}^T[k] \bar{\mathbf{Y}}_{-i}^t \mathbf{x}_i + \frac{(1 - \beta^t)}{1 - \beta} b_{ii}[k] \|\mathbf{x}_i\|_2^2 - (\bar{\mathbf{y}}_i^t)^T \mathbf{x}_i \qquad (10.5.13)$$

$$\mathbf{a}_{-i}[k+1] = \mathcal{S}_{\lambda_t / L_f} \left(\mathbf{a}_{-i}[k] - (1/L_f) \nabla_{\mathbf{a}_{-i}} f[k] \right) \qquad (10.5.14)$$

$$b_{ii}[k+1] = b_{ii}[k] - (1/L_f) \nabla_{b_{ii}} f[k] \qquad (10.5.15)$$

are provably convergent to the globally optimal solution $\{\hat{\mathbf{A}}^t, \hat{\mathbf{B}}^t\}$ of (10.5.5), as per the general convergence results available for PG methods and ISTA in particular [DDM04, PB13].

Computation of the gradients in (10.5.12)-(10.5.13) requires one matrix-vector mutiplication by $\mathbf{\Sigma}_{-i}^t$ and one by $\bar{\mathbf{Y}}_{-i}^t$, in addition to three vector inner-products, plus a few (negligibly complex) scalar and vector additions. Both the update of $b_{ii}[k+1]$ as well as the soft-thresholding operation in (10.5.14) entail negligible computational complexity. Per iteration, the actual rows of the adjacency matrix are obtained by zero-padding the updated $\mathbf{a}_{-i}[k]$, namely setting:

$$\mathbf{a}_i^T[k] = [a_{-i,1}[k] \ldots a_{-i,i-1}[k] \; 0 \; a_{-i,i}[k] \ldots a_{-i,N}[k]]. \qquad (10.5.16)$$

This way, the desired SEM parameter estimates at time t are given by $\hat{\mathbf{A}}^t = [\mathbf{a}_1^\top[k], \ldots, \mathbf{a}_N^\top[k]]^T$ and $\hat{\mathbf{B}}^t = \mathrm{diag}(b_{11}[k], \ldots, b_{NN}[k])$, for k large enough so that convergence has been attained.

Solving (10.5.5) **over the entire time horizon** $t = 1, \ldots, T$. To track the dynamically-evolving network topology, one can go ahead and solve (10.5.5) sequentially for each $t = 1, \ldots, T$ as data arrive, using (10.5.12)-(10.5.15). Because the network is assumed to vary slowly across time, it is convenient to warm-restart the ISTA iterations, that is, at time t initialize $\{\mathbf{A}[0], \mathbf{B}[0]\}$ with the solution $\{\hat{\mathbf{A}}^{t-1}, \hat{\mathbf{B}}^{t-1}\}$. This way, for smooth network variations one expects convergence to be attained after few iterations.

To obtain the new SEM parameter estimates via (10.5.12)-(10.5.15), it suffices to update (possibly) λ_t and the Lipschitz constant L_f, as well as the data-dependent EWMAs $\mathbf{\Sigma}^t$, and $\bar{\mathbf{Y}}^t$. Interestingly, the potential growing-memory problem in storing the entire history of data $\{\mathbf{Y}^t\}_{t=1}^T$ can be avoided by performing the recursive updates:

$$\mathbf{\Sigma}^t = \beta \mathbf{\Sigma}^{t-1} + \mathbf{Y}^t (\mathbf{Y}^t)^T, \quad \bar{\mathbf{Y}}^t = \beta \bar{\mathbf{Y}}^{t-1} + \mathbf{Y}^t. \qquad (10.5.17)$$

The complexity in evaluating the Gram matrix $\mathbf{Y}^t (\mathbf{Y}^t)^T$ dominates the per-iteration computational cost of the algorithm. To circumvent the need of recomputing the Lipschitz constant per time interval, the step-size $1/L_f$ in (10.5.14)-(10.5.15) can be selected by a line search [PB13]. One choice is the backtracking step-size rule [BT09], for which convergence to $\{\hat{\mathbf{A}}^t, \hat{\mathbf{B}}^t\}$ can be established as well.

Algorithm 10.5.1 summarizes the steps outlined in this section for tracking the dynamic

---

Algorithm 10.5.1: Pseudo real-time ISTA for topology tracking

---

Require: $\{\mathbf{Y}^t\}_{t=1}^T, \mathbf{X}, \beta$.

1: Initialize $\hat{\mathbf{A}}^0 = \mathbf{0}_{N \times N}, \hat{\mathbf{B}}^0 = \mathbf{\Sigma}^0 = \mathbf{I}_N, \bar{\mathbf{Y}}^0 = \mathbf{0}_{N \times C}, \lambda_0$.
2: **for** $t = 1, \ldots, T$ **do**
3:     Update λ_t, L_f and $\mathbf{\Sigma}^t$, $\bar{\mathbf{Y}}^t$ via (10.5.17).
4:     Initialize $\mathbf{A}[0] = \hat{\mathbf{A}}^{t-1}$, $\mathbf{B}[0] = \hat{\mathbf{B}}^{t-1}$, and set $k = 0$.
5:     **while** not converged **do**
6:         **for** $i = 1 \ldots N$ (in parallel) **do**
7:             Compute $\mathbf{\Sigma}_{-i}^t$ and $\bar{\mathbf{Y}}_{-i}^t$.
8:             Form gradients at $\mathbf{a}_{-i}[k]$ and $b_{ii}[k]$ via (10.5.12)-(10.5.13).
9:             Update $\mathbf{a}_{-i}[k+1]$ via (10.5.14).
10:            Update $b_{ii}[k+1]$ via (10.5.15).
11:            Update $\mathbf{a}_i[k+1]$ via (10.5.16).
12:         **end for**
13:         $k = k + 1$.
14:     **end while**
15:     **return** $\hat{\mathbf{A}}^t = \mathbf{A}[k]$, $\hat{\mathbf{B}}^t = \mathbf{B}[k]$.
16: **end for**

---

network topology, given temporal traces of infection events $\{\mathbf{Y}^t\}_{t=1}^T$ and susceptibilities $\mathbf{X}$. It is termed *pseudo real-time* ISTA, since in principle one needs to run multiple (inner) ISTA iterations till convergence per time interval $t = 1, \ldots, T$. This will in turn incur an associated delay, that may (or may not) be tolerable depending on the specific network inference problem at hand. Nevertheless, numerical tests indicate that in practice 5-10 inner iterations suffice for convergence; see [BMG14] for further details.

10.5.2.1   Accelerated convergence

For big data applications, first-order methods such as ISTA are often the only admissible option. Recently, several efforts have led to improvement of the sublinear global rate of convergence exhibited by PG algorithms while retaining their computational simplicity see e.g., [Nes83, Nes05, BT09] and references therein.

The so-termed *accelerated* (A)PG algorithm has been shown to remarkably attain convergence speedups in [Nes05]. APG algorithms generate the following sequence of iterates:

$$\mathbf{V}[k] = \arg \min_{\mathbf{V}} Q(\mathbf{V}, \mathbf{U}[k-1]) = \mathrm{prox}_{g/L_f}(\mathbf{G}(\mathbf{U}[k-1])),$$

where

$$\mathbf{U}[k] := \mathbf{V}[k-1] + \left(\frac{c[k-1]-1}{c[k]}\right)(\mathbf{V}[k-1] - \mathbf{V}[k-2]) \tag{10.5.18}$$

$$c[k] = \frac{1 + \sqrt{4c^2[k-1] + 1}}{2}. \tag{10.5.19}$$

The accelerated PG algorithm [a.k.a. fast (F)ISTA] utilizes a linear combination of the previous two iterates $\{\mathbf{V}[k-1], \mathbf{V}[k-2]\}$. The iteration-dependent combination weights are a function of the scalar sequence (10.5.19). FISTA affords a (worst-case) convergence rate guarantee of $\mathcal{O}(1/\sqrt{\epsilon})$ iterations to return an ϵ-optimal solution measured by its objective value (ISTA instead affords $\mathcal{O}(1/\epsilon)$) [BT09, Nes05]. With a few minor changes, (10.5.12)-(10.5.15) can be modified to attain accelerated convergence (see [BMG14] for details). A slight compromise to adopting FISTA is the increased memory cost for storing the two prior estimates of $\mathbf{A}$ and $\mathbf{B}$.

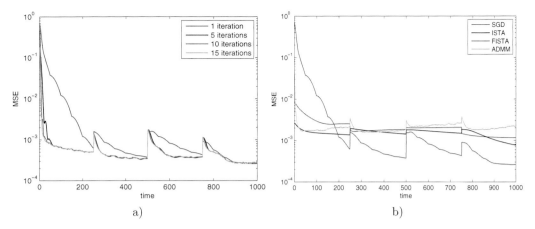

FIGURE 10.5.2: a) MSE (i.e., $\sum_{i,j}(\hat{a}_{ij}^t - a_{ij}^t)^2/N^2$) performance of Algorithm 10.5.1 versus time. For each t, (10.5.5) is solved "inexactly" for $k = 1, 5, 10,$ and 15 inner iterations. It is apparent that $k = 5$ iterations suffice to attain convergence to the minimizer of (10.5.5) per t, especially after a short transient where the warm-restarts offer increasingly better initializations. b) MSE performance of real-time algorithms versus time. Real-time FISTA, Algorithm 10.5.2 (SGD), as well as inexact versions of Algorithm 10.5.1 (ISTA) and the ADMM solver in [BMG13] are compared. **(See color insert.)**

10.5.3 Real-Time operation

Under streaming big data settings, it may be impractical to run multiple inner (F)ISTA iterations per time interval in the quest for convergence. In fact a high-quality answer obtained slowly may not be as valuable as a medium-quality answer that is obtained quickly. The remainder of this section focuses on strategies for online operation of the topology tracking algorithms, namely: i) termination of inner iterations prematurely; and ii) pursuing stochastic gradient iterations.

10.5.3.1 Premature termination

Consider a scenario where the underlying network processes are stationary (or piecewise stationary with sufficiently long coherence time). Premature termination is justified by the fact that the solution of (10.5.5) for each $t = 1, \ldots, T$ does not need to be very accurate since it is just an intermediate step in the outer loop matched to the time-instants of data acquisition. In fact, it may be reasonable to run a single inner-iteration (so that k coincides with the time index t).

For synthetically-generated data according to the setup described in [BMG14], Figure 10.5.2 shows the time evolution of the mean-square error (MSE) estimation performance upon running FISTA. For each time interval t, (10.5.5) is solved "inexactly" after running only $k = 1, 5, 10$ and 15 inner iterations. Note that realtime operation corresponds to $k = 1$.

10.5.3.2 Stochastic gradient descent iterations

Supposing $\beta = 0$ in (10.5.5), the resulting cost function can be expressed as $f_t(\mathbf{V}) + g(\mathbf{V})$, where $\mathbf{V} := [\mathbf{A}\ \mathbf{B}]$ and $f_t(\mathbf{V}) := (1/2)\|\mathbf{Y}^t - \mathbf{A}\mathbf{Y}^t - \mathbf{B}\mathbf{X}\|_F^2$ only accounts for data acquired during time interval t. Solving the simplified optimization problem based only on instantaneous data can be accomplished by following stochastic gradient descent (SGD) iterations whose simplicity and tracking capabilities are well documented. Thus, one obtains

---

Algorithm 10.5.2: SGD algorithm for topology tracking

Require: $\{\mathbf{Y}^t\}_{t=1}^T$, $\mathbf{X}$, η.
 1: Initialize $\mathbf{A}[1] = \mathbf{0}_{N \times N}$, $\mathbf{B}[1] = \mathbf{I}_N$, λ_1.
 2: **for** $t = 1, \ldots, T$ **do**
 3:　　Update λ_t.
 4:　　**for** $i = 1 \ldots N$ (in parallel) **do**
 5:　　　　Form gradients at $\mathbf{a}_{-i}[t]$ and $b_{ii}[t]$ via (10.5.20)-(10.5.21).
 6:　　　　Update $\mathbf{a}_{-i}[t+1]$ via (10.5.22).
 7:　　　　Update $b_{ii}[t+1]$ via (10.5.23).
 8:　　　　Update $\mathbf{a}_i[t+1]$ via (10.5.16).
 9:　　**end for**
10:　　**return** $\hat{\mathbf{A}}^t = \mathbf{A}[t+1], \hat{\mathbf{B}}^t = \mathbf{B}[t+1]$.
11: **end for**

---

the following updates:

$$\nabla_{\mathbf{a}_{-i}} f_t[t] = \mathbf{Y}_{-i}^t \left((\mathbf{Y}_{-i}^t)^T \mathbf{a}_{-i}[t] + \mathbf{x}_i b_{ii}[t] - \mathbf{y}_i^t \right) \tag{10.5.20}$$

$$\nabla_{b_{ii}} f_t[t] = \mathbf{a}_{-i}^T[t] \mathbf{Y}_{-i}^t \mathbf{x}_i + b_{ii}[t] \|\mathbf{x}_i\|^2 - (\mathbf{y}_i^t)^T \mathbf{x}_i \tag{10.5.21}$$

$$\mathbf{a}_{-i}[t+1] = \mathcal{S}_{\lambda_t/\eta} \left(\mathbf{a}_{-i}[t] - \eta \nabla_{\mathbf{a}_{-i}} f_t[t] \right) \tag{10.5.22}$$

$$b_{ii}[t+1] = b_{ii}[t] - \eta \nabla_{b_{ii}} f_t[t]. \tag{10.5.23}$$

Compared to the parallel ISTA iterations in Algorithm 10.5.1 [cf. (10.5.12)-(10.5.14)], three main differences are noteworthy: (i) iterations k are merged with the time intervals t of data acquisition; (ii) the stochastic gradients $\nabla_{\mathbf{a}_{-i}} f_t[t]$ and $\nabla_{b_{ii}} f_t[t]$ involve the (noisy) data $\{\mathbf{Y}^t(\mathbf{Y}^t)^T, \mathbf{Y}^t\}$ instead of their time-averaged counterparts $\{\mathbf{\Sigma}^t, \bar{\mathbf{Y}}^t\}$; and (iii) a generic constant step-size η is utilized for the gradient descent steps.

The overall SGD algorithm is tabulated under Algorithm 10.5.2. Accelerated versions could be developed as well, at the expense of a marginal increase in computational complexity and doubling of memory requirements.

10.5.4　Experiments on real data

The tracking algorithms were tested on real cascade data obtained by monitoring blog posts and news articles on the web between March 2011 and February 2012 (45 weeks) [LK14]. Popular textual phrases (a.k.a. *memes*) due to globally-popular topics during this period were identified and the times when they were mentioned on the websites were recorded as Unix timestamps (i.e., number of hours since midnight on January 1, 1970). In order to test the tracking algorithms, cascade traces related to two keywords were extracted: i) "Kim Jong-un" the current leader of North Korea whose popularity rose after the death of his father (and predecessor); and ii) "Reid Hoffman" the founder of *LinkedIn*. Only significant cascades that propagated to at least 7 websites were retained. This resulted in $N = 360$ websites, $C = 466$ cascades, and $T = 45$ weeks for "Kim Jong-un" memes. Similarly, $N = 125$, $C = 85$, and $T = 41$ weeks for "Reid Hoffman".

In cases where website i made no mention of cascade c during interval t, y_{ic}^t was set to $100t_{\max}$ (i.e., a large number), where $t_{\max}$ denotes the largest timestamp in the dataset. The entries of marix $\mathbf{X}$ typically capture prior knowledge about susceptibility of nodes to contagions. In the web context, $\mathbf{x}_{ic}$ could be aptly set to the average search engine ranking of website i on keywords pertaining to c. In the absence of such real data for the observation interval, the entries of $\mathbf{X}$ were uniformly sampled over the interval $[0, 0.01]$.

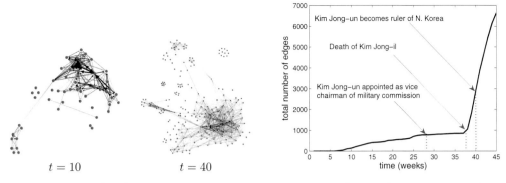

FIGURE 10.5.3: (Top) Visualization of estimated networks from information cascades related to the topic "Kim Jong-un" at $t = 10$ and $t = 40$ weeks. (Bottom) Evolution of total number of inferred edges [BMG14].

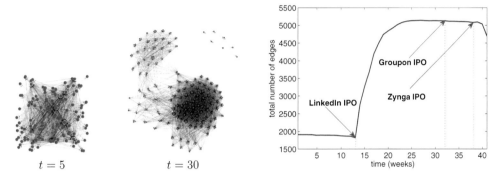

FIGURE 10.5.4: (Top) Visualization of estimated networks obtained by tracking "Reid Hoffman" cascades at $t = 5$ and $t = 30$ weeks. (Bottom) Evolution of total number of inferred edges [BMG14].

Experimental results. Algorithm 10.5.1 was run on both datasets with $\beta = 0.9$ and $\lambda_t = 100$. Figure 10.5.3 (top) depicts drawings of the inferred network for Kim Jong-un at $t = 10$ and $t = 40$ weeks. Speculation about the possible successor of the dying North Korean ruler, Kim Jong-il, rose until his death on December 17, 2011 (week 38). He was succeeded by Kim Jong-un on December 30, 2011 (week 40). The visualizations show an increasing number of edges over the 45 weeks, illustrating the growing interest of international news websites and blogs in the new ruler, about whom little was known in the first 10 weeks. Unfortunately, the observation horizon does not go beyond $T = 45$ weeks. A longer span of data would have been useful to investigate the rate at which global news coverage on the topic eventually subsided.

Figure 10.5.3 (bottom) depicts the time evolution of the total number of edges in the inferred dynamic network. Of particular interest are the weeks during which: i) Kim Jong-un was appointed as the vice chairman of the North Korean military commission; ii) Kim Jong-il died; and iii) Kim Jong-un became the ruler of North Korea. These events were the topics of many online news articles and political blogs, an observation that is reinforced by the experimental results shown in the plot.

The results of running Algorithm 10.5.1 on the second dataset are shown in Figure 10.5.4. Although Reid Hoffman was already popular in technology media coverage, his visibility in popular news and blogs increased tremendously following the highly successful initial public offering (IPO) of LinkedIn on May 19, 2011. Toward the end of 2011, a number of other successful technology companies like Groupon and Zynga went public, possibly stabilizing the amount of media coverage on Reid Hoffman. In fact, the drop in number of edges towards week 41 could be attributed to the captivation of media attention by the IPOs that occurred later in the year.

10.6 Conclusion

Big data analytics has risen to prominence within the last few years, and it remains a very active research area for the foreseeable future. In parallel, computational analysis of social networks has recently emerged as a versatile, cross-disciplinary field. Interestingly, a number of problems encountered in mining the web, learning and prediction of consumer behavior, and the dynamics of the spread of infectious diseases, all lie at the intersection of social networks, big data, and efficient (online) optimization.

Towards addressing these big data problems, signal processing and machine learning offer a robust framework for advanced data analytics. A fair and totally balanced survey of all pertinent issues and approaches is impossible within the scope of one chapter. Nevertheless, this chapter presented several interesting problems of recent interest within the social media community, namely: visualization of large social graphs, inference and imputation over social networks, community discovery, and tracking dynamic network topologies from information cascades. Efficient algorithms scaling well under big data settings along with experimental tests have been presented, and wherever possible, references to contemporary and prior approaches have been highlighted.

10.7 Acknowledgments

The authors wish to thank the following friends, colleagues, and co-authors who contributed to their joint publications from which the material of this chapter was extracted: Drs. J. A. Bazerque, P. Forero, S.-J. Kim, M. Mardani, K. Rajawat, and K. Slavakis. The work was supported in part by NSF grants 1343248, 1423316, 1442686, 1514056, and EAGER 1500713; as well as the MURI grant no. AFOSR FA9550-10-1-0567; and the NIH grant no. 1R01GM10GM4975-01. This work has also been supported by the European Social Fund and Greek national funds through the Operational Program "Education and Lifelong Learning" of the National Strategic Framework-Research Funding Program: Thalis-UoA-Secure Wireless Nonlinear Communications at the Physica.

Bibliography

[ABG10]   D. Angelosante, J. A. Bazerque, and G. B. Giannakis. Online adaptive estimation of sparse signals: where RLS meets the ℓ_1-norm. *IEEE Transactions on Signal Processing*, 58:3436–3447, 2010.

[AG11]   D. Angelosante and G. B. Giannakis. Sparse graphical modeling of piecewise-stationary time series. In *Proc. of International Conference on Acoustics, Speech and Signal Processing*, pages 1960–1963, 2011.

[AHDBV06] J. Alvarez-Hamelin, L. Dall'Asta, A. Barrat, and A. Vespignani. Large scale networks fingerprinting and visualization using the k-core decomposition. *Advances in Neural Information Processing Systems*, 18:41–50, 2006.

[AMF12]   L. Akoglu, M. McGlohon, and C. Faloutsos. OddBall: Spotting anomalies in weighted graphs. In *Proc. Pacific Asia Knowledge Discovery and Data Mining*, pages 410–421, 2012.

[BBG13]   J. A. Bazerque, B. Baingana, and G. B. Giannakis. Identifiability of sparse structural equation models for directed and cyclic networks. In *Proc. of Global Conference on Signal and Info. Processing*, pages 839–842, 2013.

[BC03]   U. Brandes and S. R. Corman. Visual unrolling of network evolution and the analysis of dynamic discourse. *Information Visualization*, 2(1):40–50, 2003.

[Ber99]   D. P. Bertsekas. *Nonlinear Programming*. Athena Scientific, 1999.

[BG05]   I. Borg and P. J. Groenen. *Modern Multidimensional Scaling: Theory and Applications*. Springer, 2005.

[BG13]   B. Baingana and G. B. Giannakis. Centrality-constrained graph embedding. In *Proc. of the 38th International Conference on Acoustics, Speech and Signal Processing*, pages 3113–3117, 2013.

[BG15]   B. Baingana and G. B. Giannakis. Kernel-based embeddings for large graphs with centrality constraints. In *Proc. of the 40th International Conference on Acoustics, Speech and Signal Processing*, 2015.

[BMG13]   B. Baingana, G. Mateos, and G. B. Giannakis. Dynamic structural equation models for tracking topologies of social networks. In *Proc. of 5th International Workshop on Computational Advances in Multi-Sensor Adaptive Processing*, pages 292–295, 2013.

[BMG14]   B. Baingana, G. Mateos, and G. B. Giannakis. Proximal-gradient algorithms for tracking cascades over social networks. *IEEE Journal of Selected Topics in Signal Processing*, 8(4):563–575, 2014.

[BNS06]   M. Belkin, P. Niyogi, and V. Sindhwani. Manifold regularization: A geometric framework for learning from labeled and unlabeled examples. *Journal of Machine Learning Research*, 7:2399–2434, 2006.

[BP98]   S. Brin and L. Page. The anatomy of a large scale hypertextual web search engine. *Computer Networks and ISDN Systems*, 30(1):107–117, 1998.

[BP11]   U. Brandes and C. Pich. More flexible radial layout. *Journal of Graph Algorithms and Applications*, 15:157–173, 2011.

[BPC+11]   S. Boyd, N. Parikh, E. Chu, B. Peleato, and J. Eckstein. Distributed optimization and statistical learning via the alternating direction method of multipliers. *Foundations and Trends in Machine Learning*, 3:1–122, 2011.

[BT99]   D. P. Bertsekas and J. N. Tsitsiklis. *Parallel and Distributed Computation: Numerical Methods*. Athena-Scientific, second edition, 1999.

[BT09]   A. Beck and M. Teboulle. A fast iterative shrinkage-thresholding algorithm for linear inverse problems. *SIAM Journal on Imaging Sciences*, 2:183–202, 2009.

[BW97]   U. Brandes and D. Wagner. A Bayesian paradigm for dynamic graph layout. In *Proc. of the 5th International Symposium on Graph Drawing*, pages 236–247, 1997.

[BW98]   U. Brandes and D. Wagner. Dynamic grid embedding with few bends and changes. In *Proc. of the 9th Annual International Symposium on Algorithms and Computation*, pages 89–98, 1998.

[CBG13]   X. Cai, J. A. Bazerque, and G. B. Giannakis. Inference of gene regulatory networks with sparse structural equation models exploiting genetic perturbations. *PLoS Computational Biology*, 9(5):1–13, 2013.

[CDK+99]   S. Chakrabarti, B. E. Dom, S. R. Kumar, P. Raghavan, S. Rajagopalan, A. Tomkins, D. Gibson, and J. Kleinberg. Mining the web's link structure. *IEEE Computer*, 32(8):60–67, 1999.

[CG84]   A. Chistov and D. Grigorev. Complexity of quantifier elimination in the theory of algebraically closed fields. In *Math. Found. of Computer Science*, volume 176 of *Lecture Notes in Computer Science*, pages 17–31. Springer, 1984.

[CGH09]   Y. Chen, Y. Gu, and A. O. Hero III. Sparse LMS for system identification. In *Proc. of International Conference on Acoustics, Speech and Signal Processing*, pages 3125–3128, 2009.

[CJHJ11]   R. Chitta, R. Jin, T. C. Havens, and A. K. Jain. Approximate kernel k-means: Solution to large scale kernel clustering. In *Proc. of the 17th ACM International Conference on Knowledge discovery and data mining*, pages 895–903, 2011.

[CLL+10]   J. Chen, W. Li, A. Lau, J. Cao, and K. Eang. Automated load curve data cleansing in power systems. *IEEE Transactions on Smart Grid*, 1(1):213–221, 2010.

[CLMW11]   E. J. Candès, X. Li, Y. Ma, and J. Wright. Robust principal component analysis? *Journal of the ACM*, 58(1):1–37, 2011.

[CPR07]   M. Coates, Y. Pointurier, and M. Rabbat. Compressed network monitoring for IP and all-optical networks. In *Proc. ACM Internet Measurement Conference*, pages 241–252, 2007.

[CSB+11]   W. Y. Chen, Y. Song, H. Bai, C. J. Lin, and E. Y. Chang. Parallel spectral clustering in distributed systems. *IEEE Transactions on Pattern Analysis and Machine Intelligence*, 33(3):568–586, 2011.

[CSPW11]   V. Chandrasekaran, S. Sanghavi, P. R. Parrilo, and A. S. Willsky. Rank-sparsity incoherence for matrix decomposition. *SIAM Journal on Optimization*, 21(2):572–596, 2011.

[CW08]   E. Candès and M. B. Wakin. An introduction to compressive sampling. *IEEE Signal Processing Magazine*, 25(2):21–30, 2008.

[DBD13]   U. Dogrusoz, M. E. Belviranli, and A. Dilek. Cise: A circular spring embedder layout algorithm. *IEEE Transactions on Visualization and Computer Graphics*, 19:953–966, 2013.

[DDM04]   I. Daubechies, M. Defrise, and C. D. Mol. An iterative thresholding algorithm for linear inverse problems with a sparsity constraint. *Communications on Pure and Applied Mathematics*, 57:1413–1457, 2004.

[DG08]   J. Dean and S. Ghemawat. MapReduce: Simplified data processing on large clusters. *Communications of the ACM*, 51(1):107–113, 2008.

[DGK04]   I. S. Dhillon, Y. Guan, and B. Kulis. Kernel k-means: Spectral clustering and normalized cuts. In *Proc. of the 10th ACM International Conference on Knowledge Discovery and Data Mining*, pages 551–556, 2004.

[DGK07]   I. S. Dhillon, Y. Guan, and B. Kulis. Weighted graph cuts without eigenvectors: A multilevel approach. *IEEE Transactions on Pattern Analysis and Machine Intelligence*, 29(11):1944–1957, 2007.

[Don06]   D. L. Donoho. Compressed sensing. *IEEE Transactions on Information Theory*, 52(4):1289 –1306, 2006.

[EFKK14]   A. Elgohary, A. K. Farahat, M. S. Kamel, and F. Karray. Embed and conquer: Scalable embeddings for kernel k-means on MapReduce. In *SIAM International Conference on Data Mining*, 2014.

[EH07]   W. Eberle and L. B. Holder. Discovering structural anomalies in graph-based data. In *Proc. of International Conference on Data Mining*, pages 393–398, 2007.

[EK10]   D. Easley and J. Kleinberg. *Networks, Crowds, and Markets: Reasoning About a Highly Connected World*. Cambridge University Press, 2010.

[FBCM04]   C. Fowlkes, S. Belongie, F. Chung, and J. Malik. Spectral grouping using the Nystrom method. *IEEE Transactions on Pattern Analysis and Machine Intelligence*, 26(2):214–225, 2004.

[FFF99]   M. Faloutsos, P. Faloutsos, and C. Faloutsos. On power-law relationships of the Internet topology. In *Proc. of Special Interest Group on Data Communications*, pages 251–262, 1999.

[FKG11]   P. Forero, V. Kekatos, and G. B. Giannakis. Outlier-aware robust clustering. In *Proc. of International Conference on Acoustics, Speech and Signal Processing*, pages 2244–2247, 2011.

[For10]   S. Fortunato. Community detection in graphs. *Physics Reports*, 486(3):75–174, 2010.

[FPRS07]   F. Fouss, A. Pirotte, J. M. Renders, and M. Saerens. Random-walk computation of similarities between nodes of a graph with application to collaborative recommendation. *IEEE Trans. on Knowledge and Data Engineering*, 19:355–369, 2007.

[Fre77]   L. C. Freeman. A set of measures of centrality based on betweenness. *Sociometry*, 40:35–41, 1977.

[FRG14]   P. Forero, K. Rajawat, and G. B. Giannakis. Prediction of partially observed dynamical processes over networks via dictionary learning. *IEEE Transactions on Signal Processing*, 62(13):3305–3320, 2014.

[FT08]   Y. Frishman and A. Tal. Online dynamic graph drawing. *IEEE Transactions on Visualization and Computer Graphics*, 14(4):727–740, 2008.

[GKB13]   A. Gittens, P. Kambadur, and C. Boutsidis. Approximate spectral clustering via randomized sketching. *Computing Research Repository*, 2013.

[GL12]   G. H. Golub and C. F. V. Loan. *Matrix Computations*, volume 3. JHU Press, 2012.

[GN02]   M. Girvan and M. E. J. Newman. Community structure in social and biological networks. *Proceedings of the National Academy of Sciences*, 99(12):7821–7826, 2002.

[Gol72]   A. S. Goldberger. Structural equation methods in the social sciences. *Econometrica*, 40:979–1001, 1972.

[Het00]   H. W. Hethcote. The mathematics of infectious diseases. *SIAM Review*, 42(4):599–653, 2000.

[HL12]   L. Harrison and A. Lu. The future of security visualization: Lessons from network visualization. *IEEE Network*, 26:6–11, 2012.

[HTF09]   T. Hastie, R. Tibshirani, and J. Friedman. *The Elements of Statistical Learning*. Springer, second edition, 2009.

[Int]   Internet2. Internet2 network flow data. http://www.internet2.edu. Accessed: 2014.

[Jac10]   M. O. Jackson. *Social and Economic Networks*. Princeton University Press, 2010.

[Kap09]   D. Kaplan. *Structural Equation Modeling: Foundations and Extensions*. Sage Publications, second edition, 2009.

[KK89]   T. Kamada and S. Kawai. An algorithm for drawing general undirected graphs. *Information Processing Letters*, 31:7–15, 1989.

[KM08]   D. Kempe and S. McSherry. A decentralized algorithm for spectral analysis. *Journal of Computer and Systems Sciences*, 74:70–83, 2008.

[Kol09]   E. D. Kolaczyk. *Statistical Analysis of Network Data: Methods and Models*. Springer, 2009.

[KST11]   Y. Kopsinis, K. Slavakis, and S. Theodoridis. Online sparse system identification and signal reconstruction using projections onto weighted ℓ_1 balls. *IEEE Transactions on Signal Processing*, 59:936–952, 2011.

[KTF09]   U. Kang, C. E. Tsourakakis, and C. Faloutsos. PEGASUS: A peta-scale graph mining system – implementation and observations. In *Proc. of International Conference on Data Mining*, pages 229–238, 2009.

[LC10]   F. Lin and W. W. Cohen. Power iteration clustering. In *Proc. of the 27th International Conference on Machine Learning*, pages 655–662, 2010.

[LCD04]   A. Lakhina, M. Crovella, and C. Diot. Diagnosing network-wide traffic anoma-
            lies. In *Proc. of the 2004 Conference on Applications, Technologies, Architec-
            tures, and Protocols for Computer Communications*, pages 219–230, 2004.

[LdH08]   B. Liu, A. de la Fuente, and I. Hoeschele. Gene network inference via structural
            equation modeling in genetical genomics experiments. *Genetics*, 178:1763–
            1776, 2008.

[Les11]   J. Leskovec. General relativity and quantum cosmology collaboration network.
            Stanford Network Analysis Project, 2011.

[Les12]   J. Leskovec. Social circles: Facebook. *Stanford Network Analysis Project*, 2012.

[LGW+11]   Z. Lin, A. Ganesh, J. Wright, L. Wu, M. Chen, and Y. Ma. Fast convex
            optimization algorithms for exact recovery of a corrupted low-rank matrix.
            UIUC Tech. Report UILU-ENG-09-2214, 2011.

[LK14]   J. Leskovec and A. Krevl. SNAP Datasets: Stanford large network dataset
            collection. http://snap.stanford.edu/data, June 2014.

[LKF07]   J. Leskovec, J. Kleinberg, and C. Faloutsos. Graph evolution: Densification
            and shrinking diameters. *ACM Transactions on Knowledge Discovery from
            Data*, 1(1), 2007.

[LS08]   L. Leydesdorff and T. Schank. Dynamic animations of journal maps: Indica-
            tors of structural changes and interdisciplinary developments. *Journal of the
            American Society for Information Science and Technology*, 59(11):1810–1818,
            2008.

[LSY98]   R. B. Lehoucq, D. C. Sorensen, and C. Yang. *ARPACK Users' Guide: Solution
            of Large-scale Eigenvalue Problems with Implicitly Restarted Arnoldi Methods*,
            volume 6. SIAM, 1998.

[Lux07]   U. V. Luxburg. A tutorial on spectral clustering. *Statistics and Computing*,
            17(4):395–416, 2007.

[LWH03]   B. Luo, R. C. Wilson, and E. R. Hancock. Spectral embedding of graphs.
            Pattern Recognition, 36:2213–2230, 2003.

[MAB13]   B. A. Miller, N. Arcolano, and N. T. Bliss. Efficient anomaly detection in
            dynamic, attributed graphs: Emerging phenomena and big data. In *Proc.
            International Conference Intelligence and Security Informatics*, pages 179–
            184, 2013.

[Mah11]   M. W. Mahoney. Randomized algorithms for matrices and data. *Foundations
            and Trends in Machine Learning*, 3(2):123–224, 2011.

[MBG10]   G. Mateos, J. A. Bazerque, and G. B. Giannakis. Distributed sparse linear
            regression. *IEEE Transactions on Signal Processing*, 58(10):5262–5276, 2010.

[MBPS10]   J. Mairal, F. Bach, J. Ponce, and G. Sapiro. Online learning for matrix factor-
            ization and sparse coding. *Journal of Machine Learning Research*, 11:19–60,
            2010.

[MG12a]   G. Mateos and G. B. Giannakis. Robust nonparametric regression via sparsity
            control with application to load curve data cleansing. *IEEE Transactions on
            Signal Processing*, 60(4):1571–1584, 2012.

[MG12b]   G. Mateos and G. B. Giannakis. Robust PCA as bilinear decomposition with outlier-sparsity regularization. *IEEE Transactions on Signal Processing*, 60(10):5176–5190, 2012.

[MG13]   G. Mateos and G. B. Giannakis. Load curve data cleansing and imputation via sparsity and low rank. *IEEE Transactions on Smart Grid*, 4(4):2347–2355, 2013.

[ML13]   S. Meyers and J. Leskovec. On the convexity of latent social network inference. In *Proc. of Neural Information Processing Systems*, pages 1741–1749, 2013.

[MM13]   S. Mankad and G. Michailidis. Structural and functional discovery in dynamic networks with non-negative matrix factorization. *Physical Review E*, 88(4):042812, 2013.

[MMBd05]   J. Moody, D. McFarland, and S. Bender-deMoll. Dynamic network visualization. *American Journal of Sociology*, 110(4):1206–1241, 2005.

[MMBd06]   J. Moody, D. McFarland, and S. Bender-deMoll. The art and science of dynamic network visualization. *Journal of Social Structure*, 7(2):1–38, 2006.

[MMG13a]   M. Mardani, G. Mateos, and G. B. Giannakis. Decentralized sparsity-regularized rank minimization: Algorithms and applications. *IEEE Transactions on Signal Processing*, 61:5374–5388, 2013.

[MMG13b]   M. Mardani, G. Mateos, and G. B. Giannakis. Dynamic anomalography: Tracking network anomalies via sparsity and low rank. *IEEE Journal of Selected Topics in Signal Processing*, 7:50–66, 2013.

[MMG13c]   M. Mardani, G. Mateos, and G. B. Giannakis. Recovery of low-rank plus compressed sparse matrices with application to unveiling traffic anomalies. *IEEE Transactions on Information Theory*, 59:5186–5205, 2013.

[MR13]   G. Mateos and K. Rajawat. Dynamic network cartography. *IEEE Signal Processing Magazine*, 30(3):129–143, 2013.

[MT08]   H. D. K. Moonesinghe and P.-N. Tan. OutRank: A graph-based outlier detection framework using random walks. *International Journal on Artificial Intelligence Tools*, 17(1):1–18, 2008.

[Mut84]   B. Muthén. A general structural equation model with dichotomous, ordered categorical, and continuous latent variable indicators. *Pyschometrika*, 49:115–132, 1984.

[Nat95]   B. K. Natarajan. Sparse approximate solutions to linear systems. *SIAM Journal on Computing*, 24:227–234, 1995.

[NC03]   C. C. Noble and D. J. Cook. Graph-based anomaly detection. In *Proc. of Special Interest Group on Knowledge Discovery and Data Mining*, pages 631–636, 2003.

[Nes83]   Y. Nesterov. A method of solving a convex programming problem with convergence rate $O(1/k^2)$. *Soviet Mathematics Doklady*, 27:372–376, 1983.

[Nes05]   Y. Nesterov. Smooth minimization of nonsmooth functions. *Mathematical Programming*, 103:127–152, 2005.

[New10]   M. E. J. Newman. *Networks: An Introduction.* Oxford University Press, 2010.

[NJW⁺02]   A. Y. Ng, M. I. Jordan, Y. Weiss, et al. On spectral clustering: Analysis and an algorithm. In *Advances in Neural Information Processing Systems*, volume 2, pages 849–856, 2002.

[OF97]   B. A. Olshausen and D. J. Field. Sparse coding with an overcomplete basis set: A strategy employed by v1? *Vision Research*, 37(23):3311–3325, 1997.

[PB13]   N. Parikh and S. Boyd. Proximal algorithms. *Foundations and Trends in Optimization*, 1:123–231, 2013.

[PPY13]   Y. Park, C. E. Priebe, and A. Youssef. Anomaly detection in times series of graphs using fusion of graph invariants. *IEEE Journal of Selected Topics in Signal Processing*, 7(1):67–75, 2013.

[PT98]   A. Papakostas and I. Tollis. Algorithms for area-efficient orthogonal drawings. *Computational Geometry: Theory and Applications*, 9:83–110, 1998.

[RBL⁺07]   R. Raina, A. Battle, H. Lee, B. Packer, and A. Y. Ng. Self-taught learning: Transfer learning from unlabeled data. In *Proc. of the 24th International Conference on Machine learning*, pages 759–766, 2007.

[RBS11]   M. G. Rodriguez, D. Balduzzi, and B. Scholkopf. Uncovering the temporal dynamics of diffusion networks. In *Proc. of 28th International Conference on Machine Learning*, 2011.

[RLS10]   M. G. Rodriguez, J. Leskovec, and B. Scholkopf. Structure and dynamics of information pathways in online media. In *Proc. of 6th ACM International Conference on Web Search and Data Mining*, 2010.

[Rog95]   E. M. Rogers. *Diffusion of Innovations.* Free Press, fourth edition, 1995.

[Rou10]   M. Roughan. A case study of the accuracy of SNMP measurements. *Journal of Electrical and Computer Engineering*, 2010.

[RR13]   B. Recht and C. Re. Parallel stochastic gradient algorithms for large-scale matrix completion. *Mathematical Programming Computation*, 5(2):201–226, 2013.

[Sab66]   G. Sabidussi. The centrality index of a graph. *Psychometrika*, 31:581–683, 1966.

[SGM14]   K. Slavakis, G. B. Giannakis, and G. Mateos. Modeling and optimization for big data analytics. *IEEE Signal Processing Magazine*, 31:18–31, 2014.

[SI09]   T. Sakai and A. Imiya. Fast spectral clustering with random projection and sampling. In *Proc. of the 6th International Conference on Machine Learning and Data Mining in Pattern Recognition*, pages 372–384, 2009.

[SJ09]   B. Shaw and T. Jebara. Structure preserving embedding. In *Proc. of International. Conference on Machine Learning*, pages 937–944, 2009.

[SM00]   J. Shi and J. Malik. Normalized cuts and image segmentation. *IEEE Transactions on Pattern Analysis and Machine Intelligence*, 22(8):888–905, 2000.

[Sni98]   M. Snir. *MPI–the Complete Reference: The MPI Core*, volume 1. MIT Press, 1998.

[SQCF05]   J. Sun, H. Qu, D. Chakrabarti, and C. Faloutsos. Neighborhood formation and anomaly detection in bipartite graphs. In *Proc. of International Conference on Data Mining*, 2005.

[SRJ04]   N. Srebro, J. Rennie, and T. S. Jaakkola. Maximum-margin matrix factorization. In *Proc. Advances in Neural Information Processing Systems*, pages 1329–1336, 2004.

[SS05]   N. Srebro and A. Shraibman. Rank, trace-norm and max-norm. In *Proc. of Learning Theory*, pages 545–560. Springer, 2005.

[SS11]   S. Shalev-Schwartz. Online learning and online convex optimization. *Foundations and Trends in Machine Learning*, 4(2):107–194, 2011.

[SSM98]   B. Scholkopf, A. J. Smola, and K. R. Muller. Nonlinear component analysis as a kernel eigenvalue problem. *Neural Computation*, 10:1299–1319, 1998.

[ST11]   O. Shamir and N. Tishby. Spectral clustering on a budget. In *Intl. Conf. on Artificial Intelligence and Statistics*, pages 661–669, 2011.

[TF10]   I. Tošić and P. Frossard. Dictionary learning. *IEEE Signal Processing Magazine*, 28:27–38, 2010.

[TL11]   H. Tong and C.-Y. Lin. Non-negative residual matrix factorization with application to graph anomaly detection. In *Proc. of SIAM Conference on Data Mining*, pages 143–153, 2011.

[TSG15]   P. Traganitis, K. Slavakis, and G. B. Giannakis. Big data spectral clustering via sketching and validation. *IEEE Journal of Selected Topics in Signal Processing*, 2015.

[VR07]   F. Vega-Redondo. *Complex Social Networks*. Cambridge University Press, 2007.

[Wah90]   G. Wahba. *Spline Models for Observational Data*, volume 59. SIAM, 1990.

[WLRB09]   L. Wang, C. Leckie, K. Ramamohanarao, and J. Bezdek. Approximate spectral clustering. In *Proc. of the 13th Pacific-Asia Conference on Knowledge Discovery and Data Mining*, pages 134–146, 2009.

[WS98]   D. J. Watts and S. H. Strogatz. Collective dynamics of small world networks. *Nature*, 393(6684):440–442, 1998.

[WTPP14]   H. Wang, M. Tang, Y. Park, and C. E. Priebe. Locality statistics for anomaly detection in time series of graphs. *IEEE Transactions on Signal Processing*, 62(3):703–717, 2014.

[XKI12]   K. S. Xu, M. Kliger, and A. O. H. III. A regularized graph layout framework for dynamic network visualization. *Data Mining and Knowledge Discovery*, 27(1):84–116, 2012.

[YHJ09]   D. Yan, L. Huang, and M. I. Jordan. Fast approximate spectral clustering. In *Proc. of the 15th ACM International Conference on Knowledge Discovery and Data Mining*, pages 907–916, 2009.

[YL06]   M. Yuan and Y. Lin. Model selection and estimation in regression with grouped variables. *Journal of the Royal Statistical Society: Series B*, 68:49–67, 2006.

[YLZ+13]   J. Yang, Y. Liu, X. Zhang, X. Yuan, Y. Zhao, S. Barlowe, and S. Liu. Piwi: Visually exploring graphs based on their community structure. *IEEE Transactions on Visualization and Computer Graphics*, 19:1034–1047, 2013.

[YY13]   X. M. Yuan and J. Yang. Sparse and low-rank matrix decomposition via alternating direction method. *Pacific Journal of Optimization*, 9(1):167–180, 2013.

[ZGGR05]   Y. Zhang, Z. Ge, A. Greenberg, and M. Roughan. Network anomography. In *Proc. of ACM Internet Measurement Conference*, pages 30–30, 2005.

[ZLW+10]   Z. Zhou, X. Li, J. Wright, E. Candès, and Y. Ma. Stable principal component pursuit. In *Proc. of International Symposium on Information Theory*, pages 1518–1522, 2010.

[ZMH09]   W. Zhao, H. Ma, and Q. He. Parallel k-means clustering based on MapReduce. In *Proc. of the 1st International Conference on Cloud Computing*, pages 674–679, 2009.

[ZRLD05]   Y. Zhang, M. Roughan, C. Lund, and D. L. Donoho. Estimating point-to-point and point-to-multipoint traffic matrices: An information-theoretic approach. *IEEE/ACM Transactions on Networking*, 13(5):947 – 960, Oct. 2005.

[ZRWQ09]   Y. Zhang, M. Roughan, W. Willinger, and L. Qiu. Spatio-temporal compressive sensing and Internet traffic matrices. In *Proc. of ACM SIGCOM Conference on Data Commun.*, pages 267–278, 2009.

Chapter 11

Semantic Model Adaptation for Evolving Big Social Data

Nikoletta Bassiou and Constantine Kotropoulos

Aristotle University of Thessaloniki, Greece

11.1 Introduction to Social Data Evolution

Social data mainly comprise open source social network information from newswire and social media. There are cases, however, where social data can stem from other sources of network data, such as smart phones, proximity sensors, simulated data, surveys, communication networks, private company data, social science research, and databases [CDW13]. Social data exhibit three characteristics that play a key role in social network analysis. First, social data are voluminous. Indeed, there are numerous on-line social networks that allow their users to easily create and share content. For example, there are more than 1.39 billion monthly active Facebook users [FbS15] and around 288 million monthly active Twitter users that generate 5800 tweets per second [Twi15], while 300 hours of video are uploaded to YouTube every minute [You15]. Second, social data are multi-faceted. They range from textual, image, audio, and video content to user metadata. They can be important in many fields of study besides computer science, such as sociology, politics, or marketing for example. Third, the data are dynamic. Structural changes can occur at multiple time scales or can be localized to a subset of users [LCSX11]. Consequently, social media data analysis needs to handle the data volume as well as the number and the diversity of the data facets in a dynamically evolving framework.

Graphs provide the most obvious and straightforward representation of a social network with the set of nodes/vertices corresponding to social entities (e.g., users) and the set of edges representing the associations among entities. The associations can be either explicit (e.g., a friendship or relationship between users) or implicit (e.g., a tag inserted to a news story). The connectivity between graph elements is usually encoded by an adjacency or similarity matrix. Thus, social network analysis exploits matrix algebra operations, such as *Singular Value Decomposition (SVD)* used in *spectral clustering*. Moreover, the *Latent Semantic Analysis (LSA)* that resorts to the truncated SVD and its probabilistic counterpart the *Probabilistic Latent Semantic Analysis (PLSA)*, are employed to extract the most important data aspects or latent data relationships. Additionally, different types of edges or entities in a social network are often represented by *tensors* (strictly speaking, the *hypermatrices* of Chapter 8). Tensors consist of the multi-dimensional extension of matrices, and have arisen as the most natural way to encode n-way relationships between entities simultaneously, including the temporal network dynamics.

A fundamental tool in the analysis of large complex networks is *community detection*. In the graph representation setting, communities are groups of nodes that exhibit high connectivity within a group and low connectivity across the groups [PKVS12]. In this sense, community detection is usually treated as a graph clustering problem. Depending on the definition of community and the methodological principle used, community detection methods are further classified into vertex/spectral clustering methods [PL05, Lux07], cohesive subgraph discovery methods [PDFV05, XYFS07], community quality optimization methods [SM00, New07, CZG09, KPSC10], divisive methods [FLG00, GN02, For10], and model-based methods [RAK07, LHLC09]. In more detail, model-based methods consider either an underlying statistical model to divide the network into communities, or dynamic processes. *Label propagation* is such a dynamic process that forms consensus on labels from the densely connected groups of nodes and reveals communities built from nodes sharing the same labels [PCW09].

Besides revealing the underlying structure in social networks, community detection is largely exploited for higher-level inference and knowledge-discovery tasks, such as to build recommender systems. Traditional *recommender systems* are mainly based on collaborative filtering methods that can be either memory-based or model-based [DGR07]. Memory-based methods exploit the underlying similarity between users [JZM04] or items [SKKR01, LSY03, DK04] for recommendations. That is, in user-based approaches, the ratings in groups of similar users help to predict the ratings of active users, while in item-based approaches, predictions are based on the information entailed in items similar to those chosen by the user [MZL+11]. Thus, memory-based methods largely employ clustering in order to build the groups of similar users and items, treating the recommendation problem as a classification problem. That is, an active user is classified in the group with the most similar users to exploit the group users' ratings in order to generate recommendations for him/her. Common similarity measures include the Pearson correlation coefficient [RIS+94] and the norms between the rating vectors in a proper vector space [BHK98].

In contrast to memory-based methods, the model-based approaches build a compact model from the observed user-item data in order to infer recommendations. The most representative algorithms used for this goal include the LSA [DDF+90, SKKR00], the PLSA [Hof04], the *Latent Dirichlet Allocation (LDA)* [BNJ03], the Bayesian clustering [BHK98], the multiple multiplicative factor model [MZ04], the Markov Decision process [SBH02], the Bayesian hierarchical model [ZK07], and the ranking model [LY08]. To enhance their performance, recommender systems also use tags, which constitute an additional source of information. These systems are usually referred to as tag-based recommender systems [IN14] and are largely built on tensor models for generating tags and item recommendations [RST10, SNM10, LDv12, RD13].

Considering the dynamic nature of social networks and the large amount of information they entail, all the aforementioned approaches need to be scalable and adaptive to the evolutionary nature of on-line social networks. Thus, recent research efforts mainly focus on the ability of the social network analysis tools to efficiently capture the network dynamics, while efforts more targeted to the efficient handling of large volumes of data (frequently met under the term "big data") also arise. This chapter summarizes the dynamic alternatives of the corresponding state-of-the-art approaches. In Section 11.2, latent semantic model-based methods, such as LSA, PLSA, and LDA, are discussed. In Section 11.3, incremental spectral clustering is analyzed, while in Section 11.4 tensor model adaptation methods are presented. In Section 11.5, methods that further optimize and adapt these state-of-the-art approaches for handling big data are reviewed and parallel distributed implementations are summarized. Representative applications, namely the incremental label propagation and the incremental graph clustering, to evolving social data analysis are described in Section 11.6. Conclusions are drawn in Section 11.7.

11.2 Latent Model Adaptation

Latent variable models (LSA, PLSA and LDA) are widely used in semantic analysis applications, since they successfully capture the underlying semantic relationships among data objects. In social networks, latent variable models are widely used for revealing the semantic relations between the social media textual data (i.e., word/document clustering) and for modeling the underlying latent topics [WAB12]. They are also used in applications, such as news recommendations and collaborative filtering [DGR07, WYL$^+$09], web-usage mining [JZM04, XZMZ05, XYW$^+$09] or co-citation and trends analysis [CH00, WYL$^+$09]. They are useful for data filtering by means of topic modeling [GLMY11] as well. In Sections 11.2.1-11.2.3, LSA, PLSA and LDA and their incremental counterparts are described within a text document framework. However, they can also be used to capture relations between other types of data, such as users, topics, tags, and ratings.

11.2.1 Incremental Latent Semantic Analysis

Latent Semantic Analysis (LSA) extends the vector space model where the dataset is represented as a term-document matrix, by employing a reduced dimension representation of the term-document relationship. This representation is built by means of the *Partial Singular Value Decomposition (PSVD)* of the term-document matrix [ZS99]. Let $\mathbf{A} \in \mathbb{R}^{M \times N}$ be the term-document matrix holding the co-occurrence of M terms in N documents. The PSVD of $\mathbf{A}$ is given by the best rank-k approximation $\mathbf{A}_k = \mathbf{U}_k \mathbf{\Sigma}_k \mathbf{V}_k^T$, where $\mathbf{U}_k$ and $\mathbf{V}_k$ are built by the first k columns of $\mathbf{U} \in \mathbb{R}^{M \times M}$ and $\mathbf{V} \in \mathbb{R}^{N \times N}$, holding the left and right singular vectors of $\mathbf{A}$, respectively. $\mathbf{\Sigma}_k$ is the k-th leading diagonal submatrix of $\mathbf{\Sigma} \in \mathbb{R}^{M \times N}$, having in its main diagonal the singular values of $\mathbf{A}$.

In the case of a frequently updated term-document collection with the addition of new documents and terms, LSA has to be also updated, reflecting these changes in the PSVD of the term-document matrix. The straightforward method of *recomputing the SVD* that performs the PSVD of the augmented term-document matrix $\tilde{\mathbf{A}} \in \mathbb{R}^{(M+M') \times (N+N')}$ suffers from time and memory constraints, especially for large data [BDO94, O'B94]. Instead of recomputing the PSVD of the augmented term-document matrix from scratch, the LSA updating methods resort to the existing PSVD in order to update the existing LSA model.

In more detail, *folding-in*, which is the simpler LSA updating method, handles new documents and terms by projecting the corresponding new document vectors forming $\mathbf{D} \in \mathbb{R}^{M \times N'}$ and the new term vectors forming $\mathbf{T} \in \mathbb{R}^{M' \times N}$ onto the existing k-dimensional subspace. That is, folding-in new documents and terms is given by $\mathbf{D}_k = \mathbf{D}^T \mathbf{U}_k \mathbf{\Sigma}^{-1}$ and $\mathbf{T}_k = \mathbf{T} \mathbf{V}_k \mathbf{\Sigma}^{-1}$, respectively. The projection $\mathbf{D}_k \in \mathbb{R}^{N' \times k}$ is folded-in to the existing PSVD by appending it to the bottom of $\mathbf{V}_k$, yielding the updated matrix $\tilde{\mathbf{V}}_k \in \mathbb{R}^{(N+N') \times k}$. Similarly, the projection $\mathbf{T}_k \in \mathbb{R}^{M' \times k}$ is folded-in to the existing PSVD by appending it to the bottom of $\mathbf{U}_k$, yielding the updated matrix $\tilde{\mathbf{U}}_k \in \mathbb{R}^{(M+M') \times k}$. Folding-in new documents has a computational complexity of $\mathcal{O}(2kMN')$, while folding-in new terms has $\mathcal{O}(2kM'N)$ complexity [BDO94].

SVD updating is a more complicated, but more robust, updating method than folding-in. It estimates the PSVD of the augmented term-document matrix $\tilde{\mathbf{A}}$ by updating the existing PSVD of the original term-document matrix $\mathbf{A}$ [BDO94, O'B94]. The method exploits the *QR decomposition* in the following three-step procedure. In all steps, $\tilde{\mathbf{A}}$ denotes the augmented term document matrix and $\hat{\mathbf{A}}$ is an intermediate matrix whose PSVD of order k is denoted by $\hat{\mathbf{A}}_k$.

- *Updating documents.* Let us examine what happens when $\tilde{\mathbf{A}}$ is the fat matrix obtained by appending $\mathbf{D} \in \mathbb{R}^{M \times N'}$ after $\mathbf{A} \in \mathbb{R}^{M \times N}$, i.e. $\tilde{\mathbf{A}} = [\mathbf{A}|\mathbf{D}]$. Then:

$$\tilde{\mathbf{A}} = [\mathbf{A}|\mathbf{D}] \simeq [\mathbf{A}_k|\mathbf{D}] = [\mathbf{U}_k|\mathbf{Q}_D] \underbrace{\begin{bmatrix} \mathbf{\Sigma}_k & \mathbf{U}_k^T \mathbf{D} \\ \mathbf{0} & \mathbf{R}_D \end{bmatrix}}_{\hat{\mathbf{A}}} \begin{bmatrix} \mathbf{V}_k^T & \mathbf{0} \\ \mathbf{0} & \mathbf{I}_{N'} \end{bmatrix}. \tag{11.2.1}$$

One can easily see that:

$$[\mathbf{U}_k|\mathbf{Q}_D] \, \hat{\mathbf{A}} \begin{bmatrix} \mathbf{V}_k^T & \mathbf{0} \\ \mathbf{0} & \mathbf{I}_{N'} \end{bmatrix} = [\mathbf{U}_k \mathbf{\Sigma}_k \mathbf{V}_k^T \, | \, \underbrace{\mathbf{U}_k \mathbf{U}_k^T \mathbf{D}}_{\mathbf{D} - \hat{\mathbf{D}}} + \underbrace{\mathbf{Q}_D \mathbf{R}_D}_{\hat{\mathbf{D}}}], \tag{11.2.2}$$

where $\hat{\mathbf{D}} = \mathbf{Q}_D \mathbf{R}_D = (\mathbf{I}_M - \mathbf{U}_k \mathbf{U}_k^T) \mathbf{D} \in \mathbb{R}^{M \times N'}$ is the projection of the columns of $\mathbf{D}$ onto the left nullspace of $\mathbf{A}$ (i.e., the orthogonal complement of the column space of $\mathbf{A}$).

Let also the SVD of $\hat{\mathbf{A}}$ be:

$$\hat{\mathbf{A}} = [\hat{\mathbf{U}}_k|\hat{\mathbf{U}}_{N'}] \begin{bmatrix} \hat{\mathbf{\Sigma}}_k & \mathbf{0} \\ \mathbf{0} & \hat{\mathbf{\Sigma}}_{N'} \end{bmatrix} [\hat{\mathbf{V}}_k|\hat{\mathbf{V}}_{N'}]^T, \tag{11.2.3}$$

where $\hat{\mathbf{U}}_k, \hat{\mathbf{V}}_k \in \mathbb{R}^{(k+N') \times k}$ and $\hat{\mathbf{\Sigma}}_k \in \mathbb{R}^{k \times k}$. The PSVD of order k of $\tilde{\mathbf{A}}$ is $\hat{\mathbf{A}}_k = \hat{\mathbf{U}}_k \hat{\mathbf{\Sigma}}_k \hat{\mathbf{V}}_k^T$. Then the PSVD of $\tilde{\mathbf{A}}$ in k dimensions is given by [TS07, ZS99]:

$$\tilde{\mathbf{A}}_k = \left([\mathbf{U}_k|\mathbf{Q}_D] \hat{\mathbf{U}}_k \right) \hat{\mathbf{\Sigma}}_k \left(\begin{bmatrix} \mathbf{V}_k & \mathbf{0} \\ \mathbf{0} & \mathbf{I}_{N'} \end{bmatrix} \hat{\mathbf{V}}_k \right)^T. \tag{11.2.4}$$

This updating procedure has a complexity of $\mathcal{O}(k^3 + (N+M)k^2 + (N+M)kN' + N'^3)$ [TS07].

- *Updating terms.* When $\tilde{\mathbf{A}}$ is the tall matrix obtained by appending $\mathbf{T} \in \mathbb{R}^{M' \times N}$ below $\mathbf{A} \in \mathbb{R}^{M \times N}$, i.e.:

$$\tilde{\mathbf{A}} = \begin{bmatrix} \mathbf{A} \\ \mathbf{T} \end{bmatrix} \simeq \begin{bmatrix} \mathbf{A}_k \\ \mathbf{T} \end{bmatrix} = \begin{bmatrix} \mathbf{U}_k & \mathbf{0} \\ \mathbf{0} & \mathbf{I}_{M'} \end{bmatrix} \underbrace{\begin{bmatrix} \boldsymbol{\Sigma}_k & \mathbf{0} \\ \mathbf{TV}_k & \mathbf{R}_T^T \end{bmatrix}}_{\hat{\mathbf{A}}} [\mathbf{V}_k | \mathbf{Q}_T]^T . \tag{11.2.5}$$

One can easily identify that:

$$\begin{bmatrix} \mathbf{U}_k & \mathbf{0} \\ \mathbf{0} & \mathbf{I}_{M'} \end{bmatrix} \hat{\mathbf{A}} \, [\mathbf{V}_k | \mathbf{Q}_T]^T = \begin{bmatrix} \mathbf{U}_k \boldsymbol{\Sigma}_k \mathbf{V}_k^T \\ \hline \underbrace{\mathbf{I}_{M'} \mathbf{TV}_k \mathbf{V}_k^T}_{\mathbf{T} - \hat{\mathbf{T}}^T} + \underbrace{\mathbf{I}_{M'} \mathbf{R}_T^T \mathbf{Q}_T^T}_{\hat{\mathbf{T}}^T} \end{bmatrix}, \tag{11.2.6}$$

where $\hat{\mathbf{T}} = (\mathbf{I}_N - \mathbf{V}_k \, \mathbf{V}_k^T) \, \mathbf{T}^T = \mathbf{Q}_T \, \mathbf{R}_T \in \mathbb{R}^{M' \times N}$ is the projection of the row vectors $\mathbf{T}$ onto the orthogonal complement of the row space of $\mathbf{A}$ (i.e., the nullspace of $\mathbf{A}$). Let also the SVD of $\hat{\mathbf{A}}$ be:

$$\hat{\mathbf{A}} = [\hat{\mathbf{U}}_k | \hat{\mathbf{U}}_{M'}] \begin{bmatrix} \hat{\boldsymbol{\Sigma}}_k & \mathbf{0} \\ \mathbf{0} & \hat{\boldsymbol{\Sigma}}_{M'} \end{bmatrix} [\hat{\mathbf{V}}_k | \hat{\mathbf{V}}_{M'}]^T, \tag{11.2.7}$$

where $\hat{\mathbf{U}}_k, \hat{\mathbf{V}}_k \in \mathbb{R}^{(k+M') \times k}$ and $\hat{\boldsymbol{\Sigma}}_k \in \mathbb{R}^{k \times k}$. Then, the PSVD of $\tilde{\mathbf{A}}$ in k dimensions is given by [TS07, ZS99]:

$$\tilde{\mathbf{A}}_k = \left(\begin{bmatrix} \mathbf{U}_k & \mathbf{0} \\ \mathbf{0} & \mathbf{I}_{M'} \end{bmatrix} \hat{\mathbf{U}}_k \right) \hat{\boldsymbol{\Sigma}}_k \left([\mathbf{V}_k | \mathbf{Q}_T] \hat{\mathbf{V}}_k \right)^T . \tag{11.2.8}$$

The computational complexity of this procedure is $\mathcal{O}(k^3 + (N+M)k^2 + (N+M)kM' + M'^3)$ [TS07].

- *Updating weights.* Let also $\tilde{\mathbf{A}} = \mathbf{A} + \mathbf{S}\,\mathbf{G}^T$ where $\mathbf{A} \in \mathbb{R}^{M \times N}$, $\mathbf{S} \in \mathbb{R}^{M \times p}$ and $\mathbf{G} \in \mathbb{R}^{N \times p}$. $\mathbf{S}$ can be treated as a selection matrix holding the p weights to be adjusted. $\mathbf{G}$ is a matrix having as columns the difference between the old term weights and the new term weights. If $\tilde{\mathbf{S}} = \mathbf{Q}_S \, \mathbf{R}_S = (\mathbf{I}_M - \mathbf{U}_k \, \mathbf{U}_k^T) \, \mathbf{S} \in \mathbb{R}^{M \times p}$ and $\hat{\mathbf{G}} = \mathbf{Q}_G \, \mathbf{R}_G = (\mathbf{I}_N - \mathbf{V}_k \, \mathbf{V}_k^T) \, \mathbf{G} \in \mathbb{R}^{N \times p}$ we obtain:

$$\tilde{\mathbf{A}} \simeq \mathbf{A}_k + \mathbf{SG}^T = [\mathbf{U}_k | \mathbf{Q}_S] \underbrace{\left(\begin{bmatrix} \boldsymbol{\Sigma}_k & \mathbf{0} \\ \mathbf{0} & \mathbf{0} \end{bmatrix} + \begin{bmatrix} \mathbf{U}_k^T \mathbf{S} \\ \mathbf{R}_S \end{bmatrix} \begin{bmatrix} \mathbf{V}_k^T \mathbf{G} \\ \mathbf{R}_G \end{bmatrix}^T \right)}_{\hat{\mathbf{A}}} [\mathbf{V}_k | \mathbf{Q}_G]^T . \tag{11.2.9}$$

Let also the SVD of $\hat{\mathbf{A}}$ be:

$$\hat{\mathbf{A}} = [\hat{\mathbf{U}}_k | \hat{\mathbf{U}}_p] \left(\begin{bmatrix} \hat{\boldsymbol{\Sigma}}_k & \mathbf{0} \\ \mathbf{0} & \hat{\boldsymbol{\Sigma}}_p \end{bmatrix} \right) [\hat{\mathbf{V}}_k | \hat{\mathbf{V}}_p]^T, \tag{11.2.10}$$

where $\hat{\mathbf{U}}_k, \hat{\mathbf{V}}_k \in \mathbb{R}^{(k+p) \times k}$ and $\hat{\boldsymbol{\Sigma}}_k \in \mathbb{R}^{k \times k}$. Then, the PSVD of $\tilde{\mathbf{A}}$ in k dimensions is given by [TS07, ZS99]:

$$\tilde{\mathbf{A}}_k = \left([\mathbf{U}_k | \mathbf{Q}_S] \hat{\mathbf{U}}_k \right) \hat{\boldsymbol{\Sigma}}_k \left([\mathbf{V}_k | \mathbf{Q}_G] \hat{\mathbf{V}}_k \right)^T . \tag{11.2.11}$$

This updating procedure has a complexity of $\mathcal{O}\left(k^3 + (N+M)k^2 + (N+M)kp + p^3\right)$ [TS07].

The folding-in method and the SVD updating method are combined in an alternate hybrid LSA updating method, called *folding-up*. The method alternates repeatedly between the folding-in and the SVD Updating in order update the existing LSA model. The complexity of the process is of the same order as SVD Updating, but reduced by a factor that is dependent on the number of iterations in which SVD updating is replaced by folding-in [TS07].

A more recent approach, called *Incremental LSI (ILSI)*, updates the existing LSA model $(\mathbf{A}_k = \mathbf{U}_k \boldsymbol{\Sigma}_k \mathbf{V}_k^T)$ for a new term-document matrix $\mathbf{A}'$ as follows [JNCJ08]:

1. Zero rows are appended to matrices $\mathbf{U}_k$ and $\mathbf{V}_k$, resulting in $\mathbf{U}' = \begin{bmatrix} \mathbf{U}_k \\ \mathbf{0}_{M' \times k} \end{bmatrix}$ and

   $\mathbf{V}' = \begin{bmatrix} \mathbf{V}_k \\ \mathbf{0}_{N' \times k} \end{bmatrix}$ matrices, respectively.

2. The new central matrix $\boldsymbol{\Sigma}' = \mathbf{U}'^T \mathbf{A}' \mathbf{V}'$ is computed.

3. The SVD decomposition of $\boldsymbol{\Sigma}' = \hat{\mathbf{U}}_k \hat{\boldsymbol{\Sigma}}_k \hat{\mathbf{V}}_k^T$ is estimated.

4. The updated SVD matrices $\mathbf{U}'_k = \mathbf{U}' \hat{\mathbf{U}}_k$, $\mathbf{V}'_k = \mathbf{V}' \hat{\mathbf{V}}_k$ and $\boldsymbol{\Sigma}'_k = \hat{\boldsymbol{\Sigma}}_k$ are estimated.

The total overhead complexity of incremental LSI over the traditional LSI lies in the complexity of Step 2 and Step 4 of the algorithm, which is $\mathcal{O}(k(M+M')(N+N'+k))$ and $\mathcal{O}(k^2(M+M'+N+N'))$ respectively, while Step 3 costs $\mathcal{O}(k^3)$. [JNCJ08].

11.2.2 Incremental Probabilistic Latent Semantic Analysis

Probabilistic Latent Semantic Analysis (PLSA) is a statistical latent variable model (or aspect model) [HP98], which associates an unobserved class variable to co-occurrence data. For example, in a text processing application, the co-occurrence data may refer to a training corpus X consisting of document-word pairs (d, w) collected from N documents $d \in D = \{d_1, d_2, \ldots, d_N\}$ with a vocabulary of M words/terms $w \in W = \{w_1, w_2, \ldots, w_M\}$ that were generated by topics $z \in Z = \{z_1, z_2, \ldots, z_K\}$. Following the aspect model independence assumptions [MB88], all the pairs (d, w) are assumed to be independent and identically distributed (the *"bag of words"* approach), and conditionally independent given the respective latent class z [Hof01]. It also holds that $|Z| \ll \min(|D|, |W|)$, where $|\cdot|$ stands for the cardinality of the corresponding set.

The data generation process can be better described by the following scheme [Hof01]: 1) select a document d with probability $P(d)$; 2) pick a latent topic z for the document with probability $P(z|d)$; and 3) generate a term w with probability $P(w|z)$. Accordingly, the joint distribution of a word w in a document d generated by a latent topic z is given by $P(d, w, z) = P(d)P(z|d)P(w|z)$ or after applying the Bayes rule by $P(d, w, z) = P(z)P(d|z)P(w|z)$ that resorts to an equivalent model which is perfectly symmetric in both entities documents and terms.

Asymmetric Formulation. The joint distribution of d and w is obtained by summing over all possible realizations of z:

$$P(d, w) = \sum_{z \in Z} P(d, w, z) = P(d) \underbrace{\sum_{z \in Z} P(z|d)P(w|z)}_{P(w|d)}. \tag{11.2.12}$$

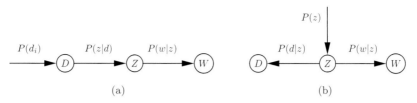

FIGURE 11.2.1: Graphical model representation of the aspect model in the (a) asymmetric and (b) symmetric formulation.

As can be seen from (11.2.12), the document-specific term distribution $P(w|d)$ is obtained by a convex combination of the $|Z|$ aspects/factors $P(w|z)$. In order to determine $P(z|d)$ and $P(w|z)$, the log-likelihood function of the training corpus $X = \{d, w\}$ using $\theta = \{P(w|z), P(z|d)\}$:

$$\mathcal{L} = \log P(X|\theta) = \sum_{d \in D} \sum_{w \in W} n(d, w) \log P(d, w) \tag{11.2.13}$$

has to be maximized with respect to all the aforementioned probabilities. That is, $\theta_{ML} = \arg\max_\theta \log P(X|\theta)$. In (11.2.13), $n(d, w)$ denotes the term-document frequency. The estimation of $P(d)$ can be carried out independently resulting in $P(d) = \frac{n(d)}{\sum_{d' \in D} n(d')}$. The conditional probabilities $P(z|d)$ and $P(w|z)$ are estimated by means of the *EM algorithm* [DLR77, Hof01], which alternates between the *Expectation (E)-step:*

$$\hat{P}_{ML}(z|d, w) = \frac{P(w|z)P(z|d)}{\sum_{z' \in Z} P(w|z')P(z'|d)} \tag{11.2.14}$$

and the *Maximization (M)-step:*

$$P(w|z) = \frac{\sum_{d \in D} n(d, w)\hat{P}_{ML}(z|d, w)}{\sum_{d \in D} \sum_{w' \in W} n(d, w')\hat{P}_{ML}(z|d, w')} \tag{11.2.15}$$

$$P(z|d) = \frac{\sum_{w \in W} n(d, w)\hat{P}_{ML}(z|d, w)}{n(d)}. \tag{11.2.16}$$

By alternating (11.2.14) with (11.2.15)-(11.2.16), a convergent procedure is obtained to a local maximum of the log-likelihood.

Symmetric Formulation. Following similar lines to the asymmetric model, the joint distribution of d and w is given by:

$$P(d, w) = \sum_{z \in Z} P(d, w, z) = \sum_{z \in Z} P(d|z)P(w|z)P(z). \tag{11.2.17}$$

and the corresponding log-likelihood function $\mathcal{L}$ of the training corpus $X = \{d, w\}$ using $\theta = \{P(w|z), P(d|z), P(z)\}$ has to be maximized. Let $R = \sum_{d \in D} \sum_{w \in W} n(d, w)$. The corresponding E-step and M-step are formulated as follows:

E-step:

$$\hat{P}_{ML}(z|d, w) = \frac{P(z)P(d|z)P(w|z)}{\sum_{z' \in Z} P(z')P(d|z')P(w|z')} \tag{11.2.18}$$

M-step:

$$P(w|z) \;=\; \frac{\sum_{d \in D} n(d,w)\hat{P}_{ML}(z|d,w)}{\sum_{d \in D} \sum_{w' \in W} n(d,w')\hat{P}_{ML}(z|d,w')} \tag{11.2.19}$$

$$P(d|z) \;=\; \frac{\sum_{w \in W} n(d,w)\hat{P}_{ML}(z|d,w)}{\sum_{d' \in D} \sum_{w \in W} n(d',w)\hat{P}_{ML}(z|d',w)} \tag{11.2.20}$$

$$P(z) \;=\; \frac{\sum_{d \in D} \sum_{w \in W} n(d,w)\hat{P}_{ML}(z|d,w)}{R}. \tag{11.2.21}$$

The symmetric PLSA formulation can be rewritten in matrix notation as $\mathbf{P} = \mathbf{U}_K \mathbf{S}_K \mathbf{V}_K^T$, where $\mathbf{U}_K$ is the $|W| \times |Z|$ matrix with jk element $P(w|z)$, $\mathbf{V}_K$ is the $|D| \times |Z|$ matrix with ik element $P(d|z)$, $\mathbf{S}_K$ is the $|Z| \times |Z|$ diagonal matrix having as elements on its main diagonal $P(z)$, $z \in Z$, and $\mathbf{P}$ is the $|W| \times |D|$ matrix with elements the probabilities $P(w,d)$. Such a decomposition looks like the partial singular value decomposition employed within the LSA as described in Section 11.2.1. Despite the resemblance, it should be stressed that the LSA and the PLSA solve different optimization problems. Indeed, the LSA minimizes the *Frobenius norm* between the original-term document matrix and its best K-rank approximation, while the PLSA maximizes the likelihood function of multinomial sampling. In other words, the PLSA minimizes the *cross entropy* (or *Kullback-Leibler divergence*) between the model and the empirical distribution. The graphical representation of the asymmetric and the symmetric formulations of the PLSA model is depicted in Figure 11.2.1. The time complexity of the standard PLSA is $\mathcal{O}(|Z|C_{W,D})$, where $|Z|$ is the number of latent topics and C is the number of total word-document co-occurrence, which is usually less than $|W||D|$.

When new data are added to the initial data collection, the aforementioned PLSA model has to be updated in order to reflect/assimilate these changes. Such a need emerges when documents and/or terms are added or deleted in document clustering or topic-detection. Various methods for updating the PLSA model exist that are frequently found in the literature with terms such as *on-line*, *incremental*, or *folding-in*.

The *PLSA folding-in* is the first and simplest method for updating the PLSA model when new documents ($d_{new} \in D_{new}$) are added in the initial document collection [Bra05, BTHC06, GH99]. It is based on an incremental variant of the EM algorithm [NH98] that recalculates only the probabilities of the topics given the new documents $P(z|d_{new})$ in the M-step leaving the probabilities of the words given the topics $P(w|z)$ unchanged. That is:

(E)-step:

$$\hat{P}_{ML}(z|d_{new},w) = \frac{P(w|z)P(z|d_{new})}{\sum_{z' \in Z} P(w|z')P(z'|d_{new})}. \tag{11.2.22}$$

(M)-step:

$$P(z|d_{new}) = \frac{\sum_{w' \in d_{new}} n(d_{new},w')\hat{P}_{ML}(z|d_{new},w')}{\sum_{z' \in Z} \sum_{w \in d_{new}} n(d_{new},w')\hat{P}_{ML}(z'|d_{new},w')}. \tag{11.2.23}$$

Usually, a very small number of iterations is needed for the EM to converge. The time complexity of the PLSA folding-in per EM iteration is $\mathcal{O}(|Z|C_{W,D_{new}})$, where $|Z|$ is the number of latent topics and C is the number of total word-document co-occurrence, which is usually less than $|W||D_{new}|$.

In another incremental approach [WZWC08], the PLSA model is updated by means of a modified EM scheme that is based on the *Generalized Expectation Maximization* [NH98]:

(E)-step:

$$\hat{P}_{ML}(z|d,w)_l = \frac{P(w|z)_l P(z|d)_l}{\sum_{z' \in Z} P(w|z')_l P(z'|d)_l}. \tag{11.2.24}$$

(M)-step:

$$P(z|d)_{l+1} = \frac{\sum_{w \in W} n(d,w) \hat{P}_{ML}(z|d,w)_{l+1}}{\sum_{z' \in Z} \sum_{w' \in d_{new}} n(d,w') \hat{P}_{ML}(z'|d,w')_{l+1}} \tag{11.2.25}$$

$$P(w|z)_{l+1} = \frac{\sum_{d \in D} n(d,w) \hat{P}_{ML}(z|d,w)_{l+1} + \alpha\, P(w|z)_l}{\sum_{d \in D} \sum_{w' \in d_{new}} n(d,w') \hat{P}_{ML}(z|d,w')_{l+1} + \alpha \sum_{w' \in W} P(w'|z)_l}. \tag{11.2.26}$$

The subscripts $l+1$ and l denote new and old model parameters, respectively. The algorithm exhibits a complexity of $\mathcal{O}(|Z|C_{W,D_{new}})$ per EM iteration.

In another approach, an incremental version of EM updates the PLSA model parameters using subsets of training data that are selected sequentially and cyclically at each iteration [XYW$^+$09], while an on-line EM algorithm works on the weighted mean values of the conditional probability $P(z|d_{new}) = \frac{P(z)P(d_{new}|z)}{P(d_{new})}$ in order to update the PLSA model parameters [XYW$^+$11] .

In a more sophisticated PLSA updating method, called *incremental PLSA*, a batch of new incoming documents is added and a batch of old documents is discarded in a document scope moving window framework [CC08]. The PLSA folding-in is used to fold in new terms and documents in four steps:

1. **Discard old documents and terms.** At each advance of the window, out-of-date documents d_{out} and the corresponding terms w_{out} are discarded. The corresponding PLSA parameters $P(w_{out}|z)$, $P(d_{out}|z)$, $P(z|w_{out})$ and $P(z|d_{out})$ are also removed and the conditional probabilities for the remaining documents and terms are estimated by renormalization.

2. **Fold-in new documents.** The new documents d_{new} are folded-in by means of the PLSA folding-in for the asymmetric formulation using (11.2.22) and (11.2.23)).

3. **Fold-in new terms.** The new terms w_{new} present in the new documents d_{new} are folded-in by means of the PLSA folding-in exploiting the symmetric formulation. That is, $P(d_{new}|z)$ is estimated as follows:

$$P(z|d_{new},w) = \frac{P(w|z)\, P(z|d_{new})}{\sum_{z' \in Z} P(w|z')\, P(z'|d_{new})} \tag{11.2.27}$$

$$P(d_{new}|z) = \frac{\sum_{w \in d_{new}} n(d_{new},w)\, P(z|d_{new},w)}{\sum_{d \in D_{new}} \sum_{w \in d} n(d,w')\, P(z'|d,w')}. \tag{11.2.28}$$

The probability $P(z|w_{new})$ for new terms is estimated by the EM folding-in process:

(E)-step:

$$P(z|d_{new},w_{new}) = \frac{P(d_{new}|z)\, P(z|w_{new})}{\sum_{z' \in Z} P(d_{new}|z')\, P(z'|w_{new})}. \tag{11.2.29}$$

(M)-step:

$$P(z|w_{new}) = \frac{\sum_{d \in D_{new}} n(d,w_{new})\, P(z|d_{new},w_{new})}{\sum_{d' \in D_{new}} n(d',w_{new})}. \tag{11.2.30}$$

The initial values of $P(z|w_{new})$ are set randomly and normalized.

4. **Update the PLSA parameters.** The values $P(w_{new}|z)$ that did not exist in the previous window step are calculated by using (11.2.29) to estimate $P(z|d,w)$. In a similar manner, the values $P(w_{old}|z)$ are adjusted using (11.2.14) to estimate $P(z|d,w)$. The probability values are then normalized so that their sum is equal to 1:

$$P(w|z) = \frac{\sum_{d\in D\cup D_{new}} n(d,w)\, P(z|d,w)}{\sum_{d'\in D\cup D_{new}} \sum_{w'\in d'} n(d',w')\, P(z|d',w')}. \tag{11.2.31}$$

Finally, all the PLSA parameters are revised by applying the original PLSA algorithm (11.2.14)-(11.2.16).

The time complexity of the incremental PLSA algorithm per EM iteration is $\mathcal{O}(|Z|C_{(|D_{new}|+|D|),(|W_{new}|+|W|)})$, where $C_{(D_{new}+D),(W_{new}+W)}$ is the number of total word-document co-occurrence, which is usually less than $(|W_{new}|+|W|)(|D_{new}|+|D|)$, and $|Z|$ is the number of latent topics. The computational complexity is equal to the computational complexity of the original PLSA algorithm executed in the augmented dataset. However, the incremental PLSA needs less EM iterations to converge than the original PLSA.

Two adaptation paradigms for PLSA, namely the *Quasi-Bayes (QB) PLSA* for incremental learning and the *MAP PLSA* for corrective training, are based on a Bayesian framework [CW08]. The aforementioned framework uses a Dirichlet density kernel as a prior. The MAP PLSA maximizes the posterior probability:

$$\theta_{MAP} = \arg\max_{\theta} P(\theta|X) = \arg\max_{\theta} \log P(X|\theta) + \log g(\theta), \tag{11.2.32}$$

where X are the adaptation data (new data) and $g(\theta)$ is the prior density of the model parameters. The prior density of the model parameters, under the assumption that $P(w|z)$ and $P(z|d)$ are independent, is expressed by:

$$g(\theta) \propto \prod_{z\in Z}\left[\prod_{w\in W} P(w|z)^{\alpha_{wz}-1} \prod_{d\in D} P(z|d)^{\beta_{zd}-1}\right], \tag{11.2.33}$$

where $\phi = \{\alpha_{wz}, \beta_{zd}\}$ are the hyperparameters of Dirichlet densities. Thus, the MAP-PLSA model parameters are estimated by [CW08]:

$$P_{MAP}(w|z) = \frac{\sum_{d\in D} n(d,w)\, P(z|d,w)+(\alpha_{wz}-1)}{\sum_{w'\in W}[\sum_{d\in D} n(d,w')\, P(z|d,w')+(\alpha_{w'z}-1)]} \tag{11.2.34}$$

$$P_{MAP}(z|d) = \frac{\sum_{w\in W} n(d,w)\, P(z|d,w)+(\beta_{zd}-1)}{n(d)+\sum_{z'\in Z}(\beta_{z'd}-1)}. \tag{11.2.35}$$

In MAP PLSA, no incremental learning mechanism is designed for continuously updating the model parameters with new words and topics, while fading away out-of-date words or documents. Incremental learning is achieved by QB PLSA that estimates the model parameters by maximizing the posterior probability using the sequence of adaptation documents for each epoch. Let $\chi^n = \{X_1,\ldots,X_n\}=\{(d^{(1)},w^{(1)}),\ldots,(d^{(n)},w^{(n)})\}$ be the adaptation data for the n-th epoch. The QB PLSA estimate is determined by maximizing the posterior probability:

$$
\begin{aligned}
\theta_{QB}^{(n)} &= \arg\max_{\theta} P(\theta|\chi^n) = \arg\max_{\theta} \log\left[P(X_n|\theta)\, P(\theta|\chi^{n-1})\right]\\
&\cong \arg\max_{\theta} \log\left[P(X_n|\theta)\, g(\theta|\phi^{(n-1)})\right],
\end{aligned} \tag{11.2.36}
$$

where $\phi^{(n-1)}$ are the hyperparameters estimated on previously seen documents. At each epoch n, the previous block of documents X_{n-1} is released and only the current block of documents X_n and the accumulated statistics $\phi^{(n-1)}$ are used for PLSA updating. Thus, the QB PLSA parameters at the n-th epoch are estimated by:

$$P_{QB}^{(n)}(w^{(n)}|z) = \frac{\sum_{d\in D} n(d^{(n)}, w^{(n)})\, P^{(n)}(z|d^{(n)}, w^{(n)}) + (\alpha_{wz}^{(n-1)} - 1)}{\sum_{w'\in W}[\sum_{d\in D} n(d, w'^{(n)})\, P^{(n)}(z|d^{(n)}, w'^{(n)}) + (\alpha_{w'z}^{(n-1)} - 1)]} \tag{11.2.37}$$

$$P_{QB}^{(n)}(z|d^{(n)}) = \frac{\sum_{w\in W} n(d^{(n)}, w^{(n)})\, P^{(n)}(z|d^{(n)}, w^{(n)}) + (\beta_{zd}^{(n-1)} - 1)}{n(d^{(n)}) + \sum_{z'\in Z}(\beta_{z'd}^{(n-1)} - 1)}, \tag{11.2.38}$$

where the hyperparameters' values at the n-th learning epoch are given by:

$$\alpha_{wz}^{(n)} = \sum_{d\in X_n} n(d^{(n)}, w^{(n)})\, P^{(n)}(z|d^{(n)}, w^{(n)}) + \alpha_{wz}^{(n-1)} \tag{11.2.39}$$

$$\beta_{zd}^{(n)} = \sum_{w\in X_n} n(d^{(n)}, w^{(n)})\, P^{(n)}(z|d^{(n)}, w^{(n)}) + \beta_{zd}^{(n-1)} \tag{11.2.40}$$

and the initial values are estimated on the initial data collection as follows:

$$\alpha_{wz}^{(0)} = 1 + \sum_{d\in D} n(d, w)\, P(z|d, w) \tag{11.2.41}$$

$$\beta_{zd}^{(0)} = 1 + \sum_{w\in W} n(d, w)\, P^{(n)}(z|d, w). \tag{11.2.42}$$

Given the $|Z|$ latent variables, the number N_n of new documents at each learning epoch n, the total number of all adaptation documents $N' = \sum_n N_n$, and $C_{W,N'}$ and C_{W,N_n} the numbers of non-zero entries in the corresponding word-by-document probability matrices, which are far fewer than $|W|N$ and $|W|N_n$, respectively, the computational complexity of MAP PLSA and QB PLSA per EM iteration is respectively $\mathcal{O}(|Z|C_{W,N'})$ and $\mathcal{O}(|Z|C_{W,N_n})$. That is, the time complexity of MAP PLSA is proportionally increased by the number of the observed events in the batch collection, while that of QB PLSA is only affected by the number of events added at the current epoch [CW08].

In the *on-line Probabilistic Latent Semantic Analysis (oPLSA)*, the PLSA parameters for both the asymmetric and the symmetric formulations are updated within the context of a varying document stream [BK14]. In particular, a fixed-size moving window is employed over a document stream in order to incorporate new documents and at the same time to discard old ones (i.e., documents that fall outside the scope of the window). In addition, the oPLSA assimilates new words that had not been previously seen (out-of-vocabulary words) and discards the words that exclusively appear in the documents to be thrown away. A schematic representation of the moving window over the word-document matrix is depicted in Figure 11.2.2.

The oPLSA method is formulated by taking into consideration the basic operations between two successive EM iterations. That is, for $w \in W$, $d \in D$ and $z \in Z$, the E-step at iteration $l + 1$ proceeding iteration l, when documents or words are neither removed nor added, is given by:

$$\hat{P}(z|d, w)_{l+1} = \frac{P(w|z)_l P(z|d)_l}{\sum_{z'\in Z} P(w|z')_l P(z'|d)_l}. \tag{11.2.43}$$

Let

$$P_1(w|z)_{l+1} = P(w|z)_l \sum_{d\in D} \frac{n(d, w)P(z|d)_l}{\sum_{z'\in Z} P(w|z')_l P(z'|d)_l}. \tag{11.2.44}$$

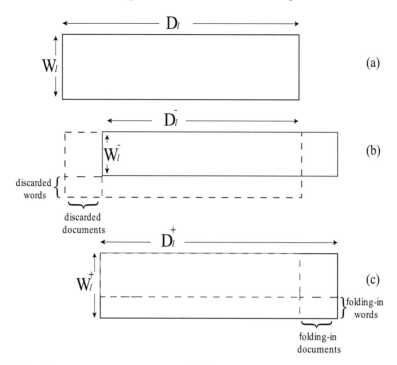

FIGURE 11.2.2: Schematic representation of (a) the initial term-document matrix, (b) the deletion of old documents and the words that appear exclusively in them, and (c) the insertion of new documents and their associated words in the word-document matrix as the window advances.

After the substitution of equation (11.2.43) into equations (11.2.15) and (11.2.16), the M-step equations are rewritten as:

$$P(w|z)_{l+1} = \frac{P_1(w|z)_{l+1}}{\sum_{w' \in W} P_1(w'|z)_{l+1}} \ , \tag{11.2.45}$$

$$P(z|d)_{l+1} = \frac{P_2(z|d)_{l+1}}{n(d)}, \tag{11.2.46}$$

where

$$P_2(z|d)_{l+1} = P(z|d)_l \sum_{w \in W} \frac{n(d,w)P(w|z)_l}{\sum_{z' \in Z} P(w|z')_l P(z'|d)_l}. \tag{11.2.47}$$

The oPLSA method is formulated by means of a fundamental operation, namely the addition/deletion of a **pivotal** document to the document collection. A pivotal document contains a single word appearing a number of times. Any document can be decomposed as a union of pivotal documents. The oPLSA method updates the existing PLSA model parameters, which are produced from the original PLSA algorithm executed on the initial word and document collections, for every advance of a window. That is, the PLSA model parameters at the $(l + 1)$-th position of the window are derived by the PLSA model parameters at the l-th position of the window, obtained after EM convergence. The algorithm performs three steps to update the existing PLSA model parameters between the l-th and $(l + 1)$-th position of the window. These steps are analyzed next for the asymmetric formulation of the PLSA model, while the corresponding steps for the symmetric formulation are derived similarly [BK14].

1. **Discard old documents and their exclusive words.**

   A single document d_{out} is discarded from the existing document collection D_l resulting in $D_l^- = D_l - \{d_{out}\}$. If the discarded document d_{out} contains only a single exclusive word w_{out}, (i.e., a word that does not appear in any document in D_l^-), this word is also discarded from W_l yielding the vocabulary of already seen words $W_l^- = W_l - \{w_{out}\}$. The corresponding PLSA model probabilities for d_{out}, $P_l(z|d_{out})$, and w_{out}, $P_l(w_{out}|z)$, are eliminated and the PLSA model parameters for the remaining documents $d \in D_l^-$ and words $w \in W_l^-$ are renormalized as follows:

   $$P^-(z|d)_l = \frac{P(z|d)_l}{\sum_{z' \in Z} P(z'|d)_l}, z \in Z, d \in D_l^- \tag{11.2.48}$$

   $$P^-(w|z)_l = \frac{P(w|z)_l}{\sum_{w' \in W_l^-} P(w'|z)_l}, z \in Z, w \in W_l^-. \tag{11.2.49}$$

   Clearly, (11.2.48) and (11.2.49) are still valid when more than one documents and words are to be discarded.

2. **Add a new word and document.**

   a. Let a new pivotal document d_{in}, which contains only a single word w_{in} appearing α times, be inserted in the already seen document collection D_l^-, yielding $D_l^+ = D_l^- + \{d_{in}\}$. The word, w_{in}, can be either a word from the vocabulary of already seen words W_l^- or a new word (*OOV word* at this point) denoted as w_{OOV}. In the latter case, the new word is inserted in W_l^-, expanding it into $W_l^+ = W_l^- + \{w_{OOV}\}$, with the entries of the augmented word-document matrix satisfying:

   $$n(d,w)_{l+1} = \begin{cases} \alpha, & \text{if } w = w_{in} \text{ and } d = d_{in} \\ n(d,w)_l, & \text{if } w \in W_l^- \text{ and } d \in D_l^- \\ 0, & \text{otherwise,} \end{cases} \tag{11.2.50}$$

   where $n(d,w)_l$ is the document-word matrix at the l-th position of the window.

   b. For already seen documents $d \in D_l^-$, the latent-variable probability is initialized by $P^+(z|d)_l = P^-(z|d)_l$. The corresponding probability $P^+(z|d_{in})_l$ for the new document d_{in} is initialized by $P^+(z|d_{in})_l = P^+(w_{in}|z)_l, \forall z \in Z$, based on the assumption that the assignment of the new document to the latent topics is driven by the single word w_{in} it contains.

   c. For already seen words $w \in W_l^-$, $P^+(w|z)_l = P^-(w|z)_l$. For an OOV word w_{OOV}, the corresponding conditional probability $P^+(w_{OOV}|z)_l$ is initialized by the *Good-Turing estimate* of the probability of unseen words [NH98]:

   $$p_{GT_l}(w_{OOV}) = \frac{n_1}{R_{l+1}} = \frac{\sum_{w \in W_l^-:n(w)=1} 1}{\sum_{d \in D_l^+} \sum_{w \in W_l^+} n(d,w)_{l+1}} = \frac{\sum_{w \in W_l^-:n(w)=1} 1}{R_l + a}, \tag{11.2.51}$$

   where $n(w) = \sum_{d \in D_l^-} n(d,w)_l$. The probability given by (11.2.51) is uniformly distributed among the topics:

   $$P^+(w_{OOV}|z)_l = \frac{1}{|Z|} p_{GT_l}(w_{OOV}) \tag{11.2.52}$$

   and the conditional probabilities of already seen words given the topics are renormalized using:

   $$P^+(w|z)_l = \left(1 - P^+(w_{OOV}|z)_l\right) \frac{P^-(w|z)_l}{\sum_{w \in W_l^-} P^-(w|z)_l}, \tag{11.2.53}$$

   where $w \in W_l^-$, so that $\sum_{w \in W_l^+} P^+(w|z)_l = 1$.

3. **Fold in the new word for pivotal documents.** The PLSA model probabilities at the window position $l + 1$ are estimated by updating the PLSA model probabilities at the window position l. To achieve this, the computations between two successive EM iterations, as described in (11.2.43)-(11.2.47), are taken into consideration [BK14]. Thus, the conditional probability $P_1^+(w|z)_{l+1}$ of the word $w \in W_l^+$ given the latent topic $z \in Z$ is given by:

$$P_1^+(w|z)_{l+1} = P_1(w|z)_{l+1} + \frac{n(d_{in}, w)_{l+1} P^+(w|z)_l P^+(z|d_{in})_l}{\sum_{z' \in Z} P^+(w|z')_l P^+(z'|d_{in})_l} \quad (11.2.54)$$

$$= \begin{cases} P_1(w|z)_{l+1}, \text{ if } w \neq w_{in} \\[2mm] P_1(w|z)_{l+1} + \frac{\alpha P^+(w|z)_l P^+(z|d_{in})_l}{\sum_{z' \in Z} P^+(w|z')_l P^+(z'|d_{in})_l}, \\[2mm] \qquad\qquad\qquad \text{if } w = w_{in} \text{ and } w_{in} \in W_l^- \\[4mm] \frac{\alpha P^+(w|z_k)_l P^+(z|d_{in})_l}{\sum_{z' \in Z} P^+(w|z')_l P^+(z'|d_{in})_l}, \text{if } w = w_{in} \text{ and } w_{in} = w_{OOV}, \end{cases}$$

where $P_1(w|z)_{l+1}$ is defined in (11.2.44).

By normalizing $P_1^+(w|z)_{l+1}$ as in (11.2.45), $P^+(w|z)_{l+1}$ is obtained:

$$P^+(w|z)_{l+1} = \frac{P_1^+(w|z)_{l+1}}{\sum_{w' \in W_l^+} P_1^+(w'|z)_{l+1}} = \begin{cases} \frac{P_1(w|z)_{l+1}}{A_{l+1}(w_{in})}, & \text{if } w \neq w_{in} \\[2mm] \frac{P_1^+(w|z)_{l+1}}{A_{l+1}(w_{in})}, & \text{if } w = w_{in}. \end{cases} \quad (11.2.55)$$

The denominator in eq.(11.2.55) is given by:

$$A_{l+1}(w_{in}) = \quad\quad\quad\quad\quad\quad\quad\quad\quad\quad\quad\quad\quad\quad\quad\quad (11.2.56)$$
$$\begin{cases} \sum_{w \in W_l^-} P_1(w|z)_{l+1} + P_1^+(w_{in}|z)_{l+1} - P_1(w_{in}|z)_{l+1}, \text{ if } w_{in} \in W_l^- \\[3mm] \sum_{w \in W_l^-} P_1(w|z)_{l+1} + P_1^+(w_{in}|z)_{l+1}, \text{if } w_{in} = w_{OOV}. \end{cases}$$

Similarly, the conditional probability $P_2(z|d)_{l+1}$ of the latent topic $z \in Z$ given the document $d \in D_l^+$ is given by:

$$P_2^+(z|d)_{l+1} = P^+(z|d)_l \sum_{w \in W_l^+} \frac{n(d, w)_{l+1} P^+(w|z)_l}{\sum_{z' \in Z} P^+(w|z')_l P^+(z'|d)_l} =$$

$$= \begin{cases} P_2(z|d)_{l+1}, & \text{if } d \in D_l^- \\[2mm] \frac{\alpha P^+(w_{in}|z)_l P^+(z|d)_l}{\sum_{z' \in Z} P^+(w_{in}|z')_l P^+(z'|d)_l}, & \text{if } d = d_{in}, \end{cases} \quad (11.2.57)$$

where $P_2(z|d)_{l+1}$ is defined in (11.2.47). When $d = d_{in}$ it holds that:

$$P_2^+(z|d_{in})_{l+1} = \begin{cases} P_1^+(w_{in}|z)_{l+1} - P_1(w_{in}|z)_{l+1}, & \text{if } w_{in} \in W_l^- \\ P_1^+(w_{in}|z)_{l+1}, & \text{if } w_{in} = w_{OOV}. \end{cases} \quad (11.2.58)$$

Finally, $P^+(z|d)_{l+1}$ is updated as follows:

$$P^+(z|d)_{l+1} = \frac{P_2^+(z|d)_{l+1}}{n(d)} = \begin{cases} \frac{P_2(z|d)_{l+1}}{n(d)} = P(z|d)_{l+1}, & \text{if } d \in D_l^- \\[2mm] \frac{P_2^+(z|d)_{l+1}}{\alpha}, & \text{if } d = d_{in}. \end{cases} \quad (11.2.59)$$

In the case of an actual (no pivotal) document, the updating procedure has to be repeated as many times as the number of the words in the document. Additionally, the insertion of more than one document during the transition phase from the window at the l-th position to the $(l+1)$-th position can be assimilated by document-wise repeating the updating equations of the PLSA parameters. Finally, after having absorbed all the documents in the transition phase, the standard PLSA algorithm is applied for refining the conditional probabilities estimated by the updating equations during the aforementioned transition phase.

11.2.3 Incremental Latent Dirichlet Allocation

Latent Dirichlet Allocation (LDA) is a generative model of documents [BNJ03]. Indeed, the documents in LDA are represented as random mixtures over the latent topics. To simplify the discussion, we represent each document $\mathbf{d}$ as a vector of word indices:

$$\mathbf{d} = \left(w_1^{(d)}, w_2^{(d)}, \dots, w_{|W^{(d)}|}^{(d)} \right)^T \quad w_i^{(d)} \in \{1, 2, \dots, M\}, \tag{11.2.60}$$

where $M = |W|$ is the size of the overall vocabulary and $|W^{(d)}|$ is the number of words in $\mathbf{d}$. Let us assume a collection of $|Z| = K$ topics. For each document $\mathbf{d}$, we have a distribution of topics $z^{(d)}$ with topic proportions $\theta_k^{(d)}$ such that $\sum_{k=1}^K \theta_k^{(d)} = 1$, which gives a latent description of the document in terms of its membership. To control complexity, one may use a *Dirichlet prior* to limit the number of topics active in any particular document, i.e.:

$$p(\boldsymbol{\theta}^{(d)}|\boldsymbol{\alpha}) = Dirichlet(\boldsymbol{\theta}^{(d)}|\boldsymbol{\alpha}) = \frac{\Gamma(\sum_{k=1}^K \alpha_k)}{\prod_{k=1}^K \Gamma(\alpha_k)} \theta_1^{\alpha_1 - 1} \theta_2^{\alpha_2 - 1} \dots \theta_K^{\alpha_K - 1} \tag{11.2.61}$$

where $\boldsymbol{\alpha} = (\alpha_1, \alpha_2, \dots, \alpha_k)^T$ has elements related to the number of topics and $\Gamma(\cdot)$ is the *Gamma function*. The Dirichlet prior is the conjugate prior of the multinomial distribution.

A generative model for sampling a document $\mathbf{d}$ with $|W^{(d)}|$ word positions is as follows. For each word position $w_i^{(d)}$, $i = 1, 2, \dots, |W^{(d)}|$:

a) choose a topic:

$$z_{w_i}^{(d)} \quad \sim \quad p(z_{w_i}^{(d)}|\boldsymbol{\theta^{(d)}}) = Multinomial(\boldsymbol{\theta}^{(d)})$$

$$= \frac{\sum_{k=1}^K \alpha_k!}{\alpha_1! \, \alpha_2! \dots \alpha_K!} \left(\theta_1^{(d)} \right)^{\alpha_1} \left(\theta_2^{(d)} \right)^{\alpha_2} \dots \left(\theta_K^{(d)} \right)^{\alpha_K}, \tag{11.2.62}$$

b) choose a word:

$$w_i^{(d)} \sim p\left(w_i^{(d)}|\vartheta_{\cdot|z_{w_i}^{(d)}} \right), \tag{11.2.63}$$

where $\vartheta_{\cdot|z_{w_i}^{(d)}}$ describes the distribution of words within the topic $z_{w_i^{(d)}}$.

The word probabilities are parameterized by a $K \times M$ matrix $\mathbf{B}$ with elements $\beta_{ij} = p(w_j = 1|z_i = 1)$. Given the parameters $\boldsymbol{\alpha}, \mathbf{B}$, the joint distribution of a topic mixture $\boldsymbol{\theta}$, a set of K topics $z^{(d)}$, and a set of $|W^{(d)}|$ words is given by:

$$p(\boldsymbol{\theta}, \mathbf{z}^{(d)}, \mathbf{w}^{(d)}|\boldsymbol{\alpha}, \mathbf{B}) = p(\boldsymbol{\theta}^{(d)}|\boldsymbol{\alpha}) \prod_{i=1}^{|W^{(d)}|} p(z_{w_i}^{(d)}|\boldsymbol{\theta}^{(d)}) \, p(w_i^{(d)}|z_{w_i}^{(d)}, \beta_{z_{w_i^{(d)}}}), \tag{11.2.64}$$

where the last conditional argument refers to the $z_{w_i^{(d)}}$ row of $\mathbf{B}$. Thus, the probability of observing the document d, denoted as a sequence of M words forming $\mathbf{w}$ is expressed as [BNJ03]:

$$P(\mathbf{w}|\boldsymbol{\alpha}, \mathbf{B}) = \int_{\boldsymbol{\Theta}} p(\boldsymbol{\theta}|\boldsymbol{\alpha}) \left(\prod_{m=1}^{M} \sum_{z_m=1}^{z_K} P(z_m, \boldsymbol{\theta})\, P(w_m|z_m, \mathbf{B}) \right) d\boldsymbol{\theta}. \qquad (11.2.65)$$

Under this generative process, a corpus of documents can be analyzed with the LDA by examining the posterior distribution of the topics $\mathbf{B}$, the topic proportions $\boldsymbol{\theta}$, and the topic assignments Z conditioned on the documents. This posterior cannot be computed directly [BNJ03, HBB10]. It is usually approximated by using *Markov Chain Monte Carlo (MCMC)* methods or *variational inference*.

Markov Chain Monte Carlo (MCMC) methods are sampling-based algorithms that attempt to collect samples from the posterior to approximate it with an empirical distribution. Let us assume symmetric Dirichlet priors for $\boldsymbol{\alpha}$ (i.e., $\alpha_1 = \alpha_2 = \ldots = \alpha_k = \alpha$) and each row of $\mathbf{B}$ (i.e., all elements of the row equal η). In the *collapsed Gibbs sampling (CGS)* the variable $\boldsymbol{\theta}$ is analytically integrated out of the model, and sampling of the topic assignments $\mathbf{z}_M$ is performed sequentially for each word w_j as follows [GS04]:

$$P(z_j|\mathbf{z}_{M \setminus j}, \mathbf{w}_M) \propto \frac{n_{z_j, M \setminus j}^{(w_j)} + \eta}{n_{z_j, M \setminus j}^{(\cdot)} + M\eta}\; \frac{n_{z_j, M \setminus j}^{(d_j)} + \alpha}{n_{\cdot, M \setminus j}^{(d_j)} + K\alpha}, \qquad (11.2.66)$$

where $\mathbf{z}_{M \setminus j} = \{z_1, \ldots, z_{j-1}, z_{j+1}, \ldots z_M\}$, M is the vocabulary size, K is the number of topics, $n_{z_j, M \setminus j}^{(w_j)}$ is the number of times word w_j is assigned to topic z_j, excluding the current word, $n_{z_j, M \setminus j}^{(\cdot)}$ is the total number of words assigned to topic z_j, excluding the current word, $n_{z_j, M \setminus j}^{(d_j)}$ is the number of times a word in document d_j is assigned to topic z_j, excluding the current word, and $n_{\cdot, M \setminus j}^{(d_j)}$ is the total number of words in document d_j, excluding the current word.

Variational methods are a deterministic alternative to sampling-based algorithms. Rather than approximating the posterior with samples, variational methods postulate a parameterized family of distributions over the hidden structure and then find the member of this family that is closest to the posterior. Thus, the inference problem is transformed to an optimization problem. That is, in batch *Variational Bayesian Inference (VB)*, the true posterior is approximated by a simpler distribution $q(\mathbf{z}, \boldsymbol{\theta}, \mathbf{B})$, which is indexed by a set of free parameters [Att00, JGJS99]. These parameters are optimized to maximize the *Evidence Lower BOund (ELBO)*, which is equivalent to minimizing the Kullbak-Leibler Divergence (KL) between $q(\mathbf{z}, \boldsymbol{\theta}, \mathbf{B})$ and the posterior $p(\mathbf{z}, \boldsymbol{\theta}, \mathbf{B}|\mathbf{w}, \alpha, \eta)$ [HBB10]. For symmetric Dirichlet priors, ELBO is defined as:

$$\log p(\mathbf{w}|\alpha, \eta) \geq \mathcal{L}(\mathbf{w}, \boldsymbol{\phi}, \boldsymbol{\gamma}, \boldsymbol{\lambda}) \triangleq \mathbb{E}_q[\log p(\mathbf{w}, \mathbf{z}, \boldsymbol{\theta}, \mathbf{B}|\alpha, \eta)] - \mathbb{E}_q[\log q(\mathbf{z}, \boldsymbol{\theta}, \mathbf{B})], \qquad (11.2.67)$$

where $\mathbb{E}_q$ is the expectation under q and choosing a fully factorized distribution of the form [BNJ03, HBB10]:

$$q(z^{(d)}) = \phi_w^{(d)} \quad q(\theta^{(d)}) = Dirichlet(\theta^{(d)}|\gamma^{(d)}) \quad q(\beta_k) = Dirichlet(\beta_z|\lambda_z), \qquad (11.2.68)$$

where the posterior over the per-word topic assignments z is parameterized by ϕ, the posterior over the per-document topic weights θ is parameterized by γ, and the posterior over the topics β is parameterized by λ. $\mathcal{L}$ can be optimized using coordinate ascent over the variational parameters $\boldsymbol{\phi}, \boldsymbol{\gamma}, \boldsymbol{\lambda}$ [BNJ03, HBB10]:

$$\phi_{wz}^{(d)} \propto \exp\{\mathbb{E}_q[\log \theta_z^{(d)}]\} + \exp\{\mathbb{E}_q[\log \beta_{zw}^{(d)}]\} \qquad (11.2.69)$$

$$\gamma_z^{(d)} = \alpha + \sum_{w \in W} n(d, w) \, \phi_{wz}^{(d)} \tag{11.2.70}$$

$$\lambda_{zw} = \eta + \sum_{d \in D} n(d, w) \, \phi_{wz}^{(d)}. \tag{11.2.71}$$

The expectations in (11.2.69) are estimated by means of the digamma function $\Psi(\cdot)$, i.e.:

$$\mathbb{E}_q[\log \theta_z^{(d)}] = \Psi(\gamma_z^{(d)}) - \Psi\left(\sum_{z' \in Z} \gamma_{z'}^{(d)}\right) \tag{11.2.72}$$

$$\mathbb{E}_q[\log \beta_{zw}^{(d)}] = \Psi(\lambda_{zw}) - \Psi\left(\sum_{w' \in W} \lambda_{zw'}\right). \tag{11.2.73}$$

The updates in (11.2.71)-(11.2.73) are guaranteed to converge to a stationary point of the ELBO. They can be partitioned into an (E)-step and (M)-step by analogy to the EM algorithm [DLR77]. In the (E)-step, γ and ϕ are iteratively updated until convergence, holding λ fixed, and in the (M)-step, λ is updated given ϕ [HBB10].

Both classes of approximation methods (MCMC and VB) are effective, but they present significant computational challenges in the face of massive datasets. Thus, developing scalable approximate inference methods for topic models is an active area of research [AWST09, NASW09, YMM09, YXQ09, HBB10]. In an online variational inference algorithm, called *Online VB for LDA* or *online LDA* [HBB10], the parameters of the variational posterior over the topic distributions $\mathbf{B}$, are approximated by maximizing the ELBO $\mathcal{L}$ setting for $\boldsymbol{\lambda}$. That is, having estimated the values $\gamma(n(d), \lambda)$ and $\phi(n(d), \lambda)$ of the per-document variational parameters $\gamma^{(d)}$ and $\phi^{(d)}$, respectively, in the (E)-step of the classic VB algorithm the ELBO maximizes:

$$\mathcal{L}(\mathbf{w}, \boldsymbol{\phi}, \boldsymbol{\gamma}, \boldsymbol{\lambda}) = \sum_{d \in D} \ell(n(d), \gamma(n(d), \boldsymbol{\lambda}), \phi(n(d), \boldsymbol{\lambda}), \boldsymbol{\lambda}), \tag{11.2.74}$$

where $\ell(n(d), \gamma(n(d), \boldsymbol{\lambda}), \phi(n(d), \boldsymbol{\lambda}), \boldsymbol{\lambda})$ denotes the contribution of the document d to the ELBO and $n(d)$ is the word count vector for document d [HBB10]. Thus, upon the observation of a document d indexed as d_i in the corpus having word counts $n^{(d)}$, an (E)-step is performed to find locally optimal values of $\gamma^{(d)}$ and $\phi^{(d)}$, holding $\boldsymbol{\lambda}$ fixed. Then, the optimal value $\tilde{\boldsymbol{\lambda}}$ of $\boldsymbol{\lambda}$ given $\phi^{(d)}$ is estimated by assuming that the entire corpus consists of a single document d repeated N times. That is:

$$\lambda_{zw} = \eta + N \, n(d_i, w) \, \phi_{wz}^{(d_i)}. \tag{11.2.75}$$

In the online case (i.e., $N \to \infty$), the empirical Bayes estimation of $\boldsymbol{\beta}$ emerges. Next, $\boldsymbol{\lambda}$ is updated as a weighted average of its previous value and its optimal value $\tilde{\boldsymbol{\lambda}}$, where a control function $\rho_i \triangleq (\tau_0 + i)^{-\kappa}$ with $\kappa \in (0.5, 1]$ and $\tau_0 \geq 0$, is used for weight assignment. The parameter κ controls the rate at which the old values of $\tilde{\boldsymbol{\lambda}}$ fade away and τ_0 slows down the early iterations of the algorithm. The just described online LDA algorithm corresponds to a stochastic natural gradient algorithm on the variational objective $\mathcal{L}$ [HBB10].

In order to reduce noise in the stochastic gradient estimation [BB08, LK09], the *online VB-LDA* method is additionally formulated in a mini-batch framework, where multiple observations per update are exploited. That is, in *mini-batch online VB-LDA* the document corpus D is chunked in mini-batches $B = \{B_1, \ldots, B_S\}$ of equal size $R = |B_S|$ and $\tilde{\boldsymbol{\lambda}}$ is estimated upon the mini-batches as follows:

$$\tilde{\boldsymbol{\lambda}} = \tilde{\lambda}_{zw} = \eta + \frac{N}{R} \sum_{d \in B_s} n(d, w) \, \phi_{wz}^{(d)}, \tag{11.2.76}$$

where $B_s \in B$. In the special case $R = N$ and $\kappa = 0$, the online VB-LDA degenerates to batch VB-LDA algorithm. Another online adaptation of LDA resorts to batch collapsed Gibbs sampling. The online algorithm, called *o-LDA* [BB07], is built on the incremental LDA model [SLTS05]. In the incremental LDA algorithm, batch LDA (i.e., a batch Gibbs sampler) initially runs on a small window of the incoming data stream. Then the topic of each new word w_r is sampled by conditioning on the words observed so far, instead of all other words in the corpus.

The performance of o-LDA depends critically on the accuracy of the topics inferred during the batch phase, since after the batch initialization, o-LDA applies (11.2.66) incrementally for each new word w_r, never resampling old topic variables. Thus, if the documents used to initialize o-LDA are not representative of the full dataset, a poor inference may result. Also, because each topic variable is sampled by conditioning only on previous words and topics, samples drawn with o-LDA are not distributed according to the true posterior distribution $P(\mathbf{z}_N|\mathbf{w}_N)$ [CSG09].

An alternative online MCMC approach revises the decisions about previous topic assignments, attempting to alleviate the shortcomings of the o-LDA. This approach, called *incremental LDA*, periodically resamples the topic assignments for the previously analyzed words by introducing the incremental Gibbs sampler [CSG09], an algorithm that rejuvenates old topic assignments in light of new data. Unlike o-LDA, the incremental Gibbs sampler does not have a batch initialization phase, but it samples topic variables of new words by means of (11.2.66). That is, after each step r, the incremental Gibbs sampler resamples the topics of some of the previous words, and the topic assignment z_j of each index j in the "rejuvenation sequence" $R(r)$ is drawn from its conditional distribution as follows:

$$P(z_j|\mathbf{z}_{r\setminus j}, \mathbf{w}_r) \propto \frac{n_{z_j,r\setminus j}^{(w_j)} + \eta}{n_{z_j,r\setminus j}^{(\cdot)} + M\eta} \frac{n_{z_j,r\setminus j}^{(d_j)} + \alpha}{n_{\cdot,r\setminus j}^{(d_j)} + K\alpha}, \tag{11.2.77}$$

where $\mathbf{z}_{r\setminus j} = \{z_1, \ldots, z_{j-1}, z_{j+1}, \ldots z_r\}$, $n_{z_j,r\setminus j}^{(w_j)}$ is the number of times word w_j is assigned to topic z_j, $n_{z_j,r\setminus j}^{(\cdot)}$ is the total number of words assigned to topic z_j, $n_{z_j,r\setminus j}^{(d_j)}$ is the number of times a word in document d_j is assigned to topic z_j, and $n_{\cdot,r\setminus j}^{(d_j)}$ is the total number of words in document d_j. All the counts are taken over the Gibbs sampler steps 1 through r, excluding the word at position j itself (hence the subscripts $r \setminus j$).

Instead of frequently resampling the previous topic assignments, another solution is to concurrently maintain multiple samples of z_r, rejuvenating them less frequently. In this way, the algorithm simultaneously explores several regions of the state space. In addition, it can be used in a multi-processor environment, since it is simpler to parallelize multiple samples dedicating each sample to a single machine than to parallelize operations on one sample. An ensemble of independent samples from the incremental Gibbs sampler could be used to approximate the posterior distribution $P(\mathbf{z}_N|\mathbf{w}_N)$, but if the samples are not rejuvenated often enough, they will not have the desired distribution. With this motivation, particle filters can be used to perform importance weighting on a set of sequentially-generated samples instead of the CGS [CSG09].

Besides the online methods, two approximate parallel CGS schemes for LDA with similar predictive performance on held-out documents to the batch CGS have been developed for large corpora [AWST09].

11.3 Incremental Spectral Clustering

Spectral clustering has attracted much attention thanks to its solid theoretical foundations in graph theory [Chu97] and its good performance. It has been applied in many areas, such as word and document clustering [ZHD01, JX06, BK11], Web/blog clustering [Din04, NXC$^+$07], computer vision [Wei99, SM00, PZK04, CY05, MTP10], speech recognition [JB03], and classification of biological data [PM05, SFSZ05]. Spectral clustering methods outperform commonly used methods, such as the k-means and the EM mixture models, in handling complex situations with unknown cluster shapes.

Spectral clustering exploits the information inherent in the spectrum of the data affinity matrix in order to detect the structure of the data distributions [KTS11]. The affinity matrix, which holds all the similarity information between the data points, is analyzed in terms of its eigenstructure expressed by the eigenvectors that correspond to the smallest eigenvalues of the graph Laplacian. There is no need for the data to be represented as coordinates in the Euclidean n-th dimensional space. Similarity can be expressed in terms of any measure between the data points.

Spectral clustering methods vary depending on the use of the *unnormalized graph Laplacian* [Lux07], the *normalized Laplacian* expressed as a random walk normalized Laplacian [SM00], or the *symmetric normalized Laplacian* [NJW01]. However, all these methods resort to traditional clustering algorithms, such as the k-means, for clustering the eigenvectors. By examining spectral clustering in a graph partitioning framework, the methods vary depending on the disassociation measure used between groups, namely the normalized cut criterion [SM00], the min-max cut criterion [Din04], or the ratio cut [WC89, HK92]. A comparison on spectral clustering methods can be found in [VM03, Lux07].

In more detail, let $G = G(\mathcal{E}, \mathcal{V}, \mathbf{W})$ be a weighted graph with node set $\mathcal{V}$, edge set $\mathcal{E}$, and a similarity matrix $\mathbf{W} \in \mathbb{R}^{n \times n}$ with elements w_{ij} that indicate the similarity between nodes v_i and v_j. Commonly used similarity metrics between two data points include the inner product of the feature vectors, $w_{ij} = \mathbf{x}_i^T \mathbf{x}_j$, the diagonally scaled Gaussian similarity, $w_{ij} = \exp\left(-\|\mathbf{x}_i - \mathbf{x}_j\|^2/\sigma^2\right)$, or the affinity matrix of the graph. The vector norm $\|\cdot\|$ is usually the l_2 norm, i.e., $\|\mathbf{x}\| = \sqrt{\mathbf{x}^T\mathbf{x}}$, and σ is a scaling parameter that determines how similarity depends on distance. That is, compared to the true scale of the problem, high σ values make most points to appear similar, while low σ values reduce the similarity between even close points. From the perspective of spectral clustering, graph G can be represented through its Laplacian matrix [Chu97], which can be either [Lux07]:

- *Unnormalized*: $\mathbf{L} = \mathbf{D} - \mathbf{W}$, or

- *Normalized*: $\mathbf{L}_{sym} = \mathbf{D}^{-1/2}\,\mathbf{L}\,\mathbf{D}^{-1/2} = \mathbf{I} - \mathbf{D}^{-1/2}\,\mathbf{W}\,\mathbf{D}^{-1/2}$ (symmetric matrix) or $\mathbf{L}_{rw} = \mathbf{D}^{-1}\,\mathbf{L} = \mathbf{I} - \mathbf{D}^{-1}\mathbf{W}$ (closely related to random walk),

where $\mathbf{D} = diag\{d_1, \ldots, d_n\}$ is the degree matrix with elements $d_i = \sum_{j=1}^{n} w_{ij}$ in the main diagonal.

In case of large datasets or dynamic data, the aforementioned spectral clustering methods are inefficient, since they cannot be applied in an incremental fashion. As a result, various solutions have emerged that either simulate the change of eigensystem in order to avoid re-computations in the presence of new data [NXC$^+$10, DGC14] or extract representative points to compress the dataset [VDL07, KTS11].

For methods simulating the change of eigensystem, an *incremental spectral clustering* method [NXC$^+$07, NXC$^+$10] updates the eigenvectors of the generalized eigenproblem [SM00] by finding the derivatives on the eigenvalues/vectors with respect to perturbations in all the quantities involved. An iterative refinement algorithm is given for the eigenvalues and eigenvectors given a change in the edges or vertices of a graph. The resulting k smallest eigenvectors are then clustered using the k-means clustering, while the eigenvectors are recomputed after every R-th graph in the sequence [NXC+10].

In detail, the data dynamics are fed in the eignvalue system by means of the *incidence vector/matrix*. An *incidence matrix* $\mathbf{R}$ is a matrix with elements the *incident vectors* $\mathbf{r}_{ij}(w)$. An *incident vector* is defined as $\mathbf{r}_{ij}(w) = \sqrt{w}\mathbf{u}_{ij}$, where $\mathbf{u}_{ij}$ is a column vector with only two non-zero elements: i-th element equal to 1 and j-th element -1. In other words, an incidence vector $\mathbf{r}_{ij}(w)$ assigns the similarity metric w between the two data points i and j or the weight of edge (i, j). Accordingly the incidence matrix is another representation of the similarity matrix [NXC+10]. Assuming $\mathbf{L} = \mathbf{R}\,\mathbf{R}^T$, whenever a change occurs in the graph due to the perturbation of the edge weight between the data points i and j by Δw_{ij} represented by the incidence vector $\mathbf{r}_{ij}(\Delta w_{ij})$, the new graph Laplacian is given by $\tilde{\mathbf{L}} = \tilde{\mathbf{R}}\tilde{\mathbf{R}}^T$, where $\tilde{\mathbf{R}} = [\mathbf{R}|\mathbf{r}_{ij}(\Delta w_{ij})]$. Therefore, the increment of the Laplacian matrix due to this modification is given by [NXC+10]:

$$\Delta\mathbf{L} = \tilde{\mathbf{L}} - \mathbf{L} = \Delta w_{ij}\,\mathbf{u}_{ij}\,\mathbf{u}_{ij}^T \tag{11.3.1}$$

$$\Delta\mathbf{D} = \Delta w_{ij}\,diag\{\mathbf{v}_{ij}\}, \tag{11.3.2}$$

where $\mathbf{v}_{ij}$ is a column vector with zero elements, except i-th and j-th elements that equal to 1.

Next, let us approximate the increments of the eigenvalues and the eigenvectors in the spectral clustering due to the perturbation $\mathbf{r}_{ij}(\Delta w_{ij})$. The generalized eigenvalue system of the normalized cut is $\mathbf{L}\,\mathbf{q} = \lambda\mathbf{D}\mathbf{q}$, where $(\lambda, \mathbf{q})$ is an eigenpair (i.e., the pair of a generalized eigenvalue and its associated eigenvector). By taking into account the second and third order error and using the existing eigenvector, the eigenvalue increment is approximated by [NXC+10]:

$$\Delta\lambda \quad = \quad \Delta w_{ij}\frac{a+b}{1+c+d}, \tag{11.3.3}$$

where

$$a \quad = \quad (q_i - q_j)^2 - \lambda\,(q_i^2 + q_j^2)$$

$$b \quad = \quad (q_i - q_j)\,(\Delta q_i - \Delta q_j) - \lambda(q_i\,\Delta q_i + q_j\,\Delta q_j)$$

$$c \quad = \quad \Delta w_{ij}\,(q_i^2 + q_j^2)$$

$$d \quad = \quad \sum_{k\in\mathcal{N}_{ij}} q_k\,d_k\,\Delta q_k,$$

with q_i and Δq_i denoting the i-th element of the generalized eigenvector $\mathbf{q}$ and its perturbation $\Delta\mathbf{q}$. d is an approximation of $\mathbf{q}^T\,\mathbf{D}\,\Delta\mathbf{q}$ under the assumption that the impact of similarity perturbation Δw_{ij} lies within the spatial neighbourhood $\mathcal{N}_{ij}$. The latter is defined as $\mathcal{N}_{ij} = \{k|w_{ik} > \tau \text{ or } w_{jk} > \tau\}$, where τ is a predefined threshold, which can become 0 for a sparse dataset. Let $\mathbf{K} = \mathbf{L} - \lambda\,\mathbf{D}$ and $\mathbf{h} = (\Delta\lambda\,\mathbf{D} + \lambda\,\Delta\mathbf{D} - \Delta\mathbf{L})\,\mathbf{q}$, where $\Delta\mathbf{D}$ and $\Delta\mathbf{L}$ are given by (11.3.1) and (11.3.2), respectively. Assuming that $\Delta q_k = 0$ if $k \notin \mathcal{N}_{ij}$ and eliminating these entries from $\Delta\mathbf{q}$ as well as the corresponding columns in $\mathbf{K}$, let $\Delta\mathbf{q}_{ij}$ and $\mathbf{K}_{\mathcal{N}_{ij}}$ denote the elements survived in both $\Delta\mathbf{q}$ and $\mathbf{K}$. It has been proved that [NXC+10]:

$$\Delta\mathbf{q}_{ij} = (\mathbf{K}_{\mathcal{N}_{ij}}^T\,\mathbf{K}_{\mathcal{N}_{ij}})^{-1}\,\mathbf{K}_{\mathcal{N}_{ij}}^T\,\mathbf{h}. \tag{11.3.4}$$

The incremental spectral clustering method is summarized in Algorithm 11.3.1.

The algorithm needs a constant running time to compute $\Delta\lambda$ and $\mathcal{O}(\bar{N}^2\, n) + \mathcal{O}(\bar{N}^3) + \mathcal{O}(\bar{N}\, n) + \mathcal{O}(\bar{N}^2)$ to compute $\Delta\mathbf{q}$, where $\bar{N}$ is the average size of the spatial neighborhood of a node, $\mathcal{O}(\bar{N}^2\, n)$ is needed to compute $\mathbf{K}_{\mathcal{N}_{ij}}^T\, \mathbf{K}_{\mathcal{N}_{ij}}$ in (11.3.4), $\mathcal{O}(\bar{N}^3)$ for the inversion

---

Algorithm 11.3.1: Incremental Spectral Clustering [NXC+07, NXC+10]

Input: Similarity matrix $\mathbf{W}$, degree matrix $\mathbf{D}$, Laplacian matrix $\mathbf{L}$, k number of clusters, and $n_{simChanges}$ similarity changes.

1. Suppose at time t, the dataset grows large enough. Solve $\mathbf{L}\,\mathbf{q} = \lambda\,\mathbf{D}\mathbf{q}$ using Algorithm 11.3.2 [SM00] for the eigenvectors with the smallest eigenvalues. This solution and the matrices $\mathbf{D}$ and $\mathbf{L}$ serve as initialization for the following steps.

2. Every time a similarity change occurs

   - Update the eigenvalues and eigenvectors using Algorithm 11.3.3.
   - Update matrices $\mathbf{L}$ and $\mathbf{D}$ as described in (11.3.1) and (11.3.2), respectively.

3. If a data point is added or deleted, it is decomposed into a sequence of similarity changes and Step 2 is repeatedly conducted.

4. After $n_{simChanges}$ similarity changes occur, re-initialize the spectral clustering by repeating Step 1 in order to eliminate any accumulated errors.

Algorithm 11.3.2: Normalized Spectral Clustering [SM00]

Input: Similarity matrix $\mathbf{W}$, degree matrix $\mathbf{D}$, Laplacian matrix $\mathbf{L}$, and k number of clusters.
Output: Clusters $A_1, \ldots, A_k$ with $A_i = \{j | v_j \in C_i\}$ being the set of indices of vertices v_j assigned to i-th cluster.

1. Compute the first k eigenvectors $\mathbf{q}_1, \ldots, \mathbf{q}_k$ of the generalized eigenproblem $\mathbf{L}\,\mathbf{q} = \lambda\,\mathbf{D}\,\mathbf{q}$.

2. Let $\mathbf{U} \in \mathbb{R}^{n \times k}$ be the matrix containing the vectors $\mathbf{q}_1, \ldots, \mathbf{q}_k$ as columns.

3. For $i = 1, \ldots, n$, let $\mathbf{u}_i \in \mathbb{R}^k$ be the vector corresponding to the i-th row of $\mathbf{U}$.

4. Cluster the points $(\mathbf{u}_i)_{i=1,\ldots,n}$ $i = 1, 2, \ldots, n$ with the k-means algorithm into clusters $C_1, \ldots, C_k$.

Algorithm 11.3.3: Iterative refinement of $\Delta\lambda$ and $\Delta\mathbf{q}$ [NXC+10]

Input: Δw_{ij}, number of iterations n_{iter}
Output: Refined $\Delta\lambda$ and $\Delta\mathbf{q}$.

1. Set $\Delta\mathbf{q} = 0$.

2. Estimate $\Delta\lambda$ by (11.3.3), using the existing $\Delta\mathbf{q}$.

3. Estimate $\Delta\mathbf{q}$ by (11.3.4), using the existing $\Delta\lambda$.

4. Repeat Steps 2 and 3 until there is no significant change in $\Delta\lambda$ and $\Delta\mathbf{q}$, or for at most n_{iter} iterations.

of $\mathcal{O}(\bar{N} n)$ for $\mathbf{K}^T_{\mathcal{N}_{ij}} \mathbf{h}$, and $\mathcal{O}(\bar{N}^2)$ for $\Delta \mathbf{q}_{ij}$ [NXC+10]. In Web applications, $\bar{N}$ is usually constant and small, so the running time of the incremental approach is $O(n)$. The time complexity can be tuned to some extent at the expense of accuracy adjusting $\bar{N}$ by tuning the threshold τ.

In the case of methods that extract representative points to compress the dataset, a self-adaptive incremental spectral clustering method [VDL07] handles the addition of new data by employing the spectral clustering algorithm [NJW01] on a few representatives for each cluster. In more detail, the incremental spectral clustering method starts with an empty dataset X and an empty affinity matrix $\mathbf{A}$. For each data point x_i added to the dataset X, the algorithm iteratively estimates a representative for each cluster, as the data point that is most similar to all other points. If the cluster representative is too far away from any point in the cluster with respect to a similarity threshold, the number of the clusters is increased and a new clustering is performed. In this case, the entries of the affinity matrix that have been assigned to such a cluster are replaced by a single cluster representative, while the original cluster contents are stored for future use in the computation of a new cluster representative, if necessary. The affinity matrix is thus shrunk to a smaller size. The process continues iteratively for all the data points to be added. Usually, the clusters determined are more than the actual ones, since the iterative method always increases the number of clusters k. To alleviate this problem, the number of clusters k should be regularly or randomly decreased by 1.

A more recent incremental spectral clustering approach, called *Incremental Approximate Spectral Clustering*, which is viewed as an incremental eigenvalue solution for the spectral clustering described in [NJW01], finds the approximate eigenvectors of a symmetric matrix given a change [DGC14]. The method is based on the positive semi-definite shifted Laplacian $\hat{\mathbf{L}} = 2\mathbf{I} - \mathbf{L}_{sym} = \mathbf{I} + \mathbf{D}^{-1/2} \mathbf{W} \mathbf{D}^{-1/2}$. In the initialization step, the shifted Laplacian matrix $\hat{\mathbf{L}}_1$ for the first graph G_1 is estimated, the largest k eigenvectors found are normalized, and k-means clustering is applied to the rows of the matrix having as column vectors the eigenvectors. Then, for every successive graph G_{t+1}, a rank-k eigendecomposition of $\hat{\mathbf{L}}_{t+1}$ is approximated with eigenupdate methods on the already known eigendecomposition of $\hat{\mathbf{L}}_t$ of graph G_t. The eigenupdate methods are based on the SVD updating, which is used for latent semantic indexing [ZS99] (Section 11.2.1). These updates are efficient as long as the change between $\hat{\mathbf{L}}_{t+1}$ and $\hat{\mathbf{L}}_t$ is small. However, as the number of iterations increases, they suffer from cumulative errors. To alleviate this, a recomputation of the eigenvectors of $\hat{\mathbf{L}}_t$ is performed after a predefined number of iterations.

11.4   Tensor Model Adaptation

11.4.1   Basic Tensor Concepts

Tensors (or more correctly *hypermatrices*) are multi-dimensional (multi-way) arrays that are widely used to represent multi-dimensional data [LPV13]. Data in multiple tensors can be modeled as *tensor sequences* or *tensor streams*. A sequence of N-th order tensors $\mathcal{X}_1, \ldots, \mathcal{X}_n$ where $\mathcal{X}_i \in \mathbb{R}^{I_1 \times \cdots \times I_N}$, $(1 \leq i \leq n)$ is defined as *tensor sequence*, when n is fixed and as a *tensor stream*, when n increases over time [STF06]. Tensor sequences can be efficiently handled by means of offline tensor analysis, while for tensor streams incremental tensor analysis is employed [STPF08].

11.4.2 Incremental Tensor Analysis

Offline Tensor Analysis (OTA).

OTA for tensor sequences is a generalization of *Principal Component Analysis (PCA)* for higher-order tensors. That is, it can be applied to general N-th order tensors instead of simple vectors (first-order tensors). Given a tensor sequence $\mathcal{X}_1, \ldots, \mathcal{X}_n$, where $\mathcal{X}_t|_{t=1}^n \in \mathbb{R}^{I_1 \times \cdots \times I_N}$, OTA finds the projection matrices $\mathbf{U}_k|_{k=1}^N \in \mathbb{R}^{I_k \times r_k}$ and a sequence of *core tensors*, $\mathcal{Y}_t|_{t=1}^n \in \mathbb{R}^{r_1 \times \cdots \times r_N}$ such that the reconstruction error $e = \sum_{t=1}^n \|\mathcal{X}_t - \mathcal{Y}_t \prod_{i=1}^N \times_i \mathbf{U}_i\|$ is minimized where $\|\cdot\|$ is the squared Frobenius norm. The core tensor can be viewed as a low-dimensional summary of an input tensor, while the projection matrices represent the transformation between the input tensor and the core tensor. Several solutions for approximating the projection matrices exist [LV00, KBK05, STPF08]. An iterative algorithm, which projects and matricizes along each mode for every tensor sequence and then performs PCA for finding the projection matrix for that mode, is depicted in Algorithm 11.4.1. An example of OTA over n second-order tensors is presented in Figure 11.4.1.

Algorithm 11.4.1: Offline Tensor Analysis (OTA)

Input: tensor sequence $\mathcal{X}_t|_{t=1}^n \in \mathbb{R}^{I_1 \times \cdots \times I_N}$, dimensionality $r_d|_{d=1}^N$ of the core tensors $\mathcal{Y}_t|_{t=1}^n$.
Output: Core tensors $\mathcal{Y}_t|_{t=1}^n \in \mathbb{R}^{r_1 \times \cdots \times r_N}$ and projection matrices $\mathbf{U}_d|_{d=1}^N \in \mathbb{R}^{I_d \times r_d}$.

1. Initialize projection matrix $\mathbf{U}_d|_{d=1}^N$ to be an $I_d \times r_d$ truncated identity matrix.

2. For each $d \in [1, N]$

    a. Initializae the covariance matrix $\mathbf{C}_d \in \mathbb{R}^{I_d \times I_d}$ with zero values.

    b. For every tensor $\mathcal{X}_t$ in sequence repeat Steps 2(b)i - 2(b)ii:

        i. Construct $\mathcal{Z}$ by projecting into all but the d-th projection matrices, $\mathcal{Z} = \mathcal{X}_t \prod_{j \neq d} \times_j \mathbf{U}_j^T$.

        ii. Update the covariance matrix for each mode of $\mathcal{Z}$, $\mathbf{C}_d = \mathbf{Z}_{(d)} \mathbf{Z}_{(d)}^T$

    c. Set $\mathbf{U}_d$ as the matrix having columns the top r_d eigenvectors of $\mathbf{C}_d$.

3. For every tensor $\mathcal{X}_t$ in sequence estimate the core tensors $\mathcal{Y}_t = \mathcal{X}_t \prod_{j=1}^N \times_j \mathbf{U}_j^T$.

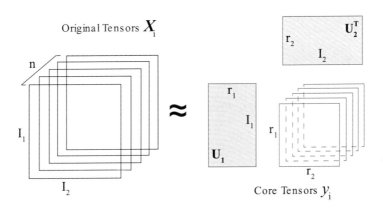

FIGURE 11.4.1: OTA process for n second-order tensors (all matrices $\mathbf{U}_d$ are tall, i.e., $I_d > r_d$) [STPF08].

The computational cost of OTA is:

$$\mathcal{O}\left(\sum_{d=1}^{N} n \sum_{\substack{i=1 \\ i \neq d}}^{N} \prod_{j=1}^{i} r_j \prod_{j=i}^{N} I_j + n I_d \prod_{j=1}^{N} I_j + I_d^3 \right) \simeq \mathcal{O}\left(n \prod_{j=1}^{N} I_j \sum_{d=1}^{N} I_d + \sum_{i=1}^{N} I_d^3 \right), \quad (11.4.1)$$

where $\sum_{\substack{i=1 \\ i \neq d}}^{N} \prod_{j=1}^{i} r_j \prod_{j=i}^{N} I_j$ is the cost of Step 2(b)i, $I_d \prod_{j=1}^{N} I_j$ is the computational cost of Step 2(b)ii and $\mathcal{O}(I_d^3)$ the cost of Step 2c in Algorithm 11.4.1 [STPF08].

Although Algorithm 11.4.1 is efficient for tensor sequences, it cannot be applied to dynamic environments (i.e., tensor streams), where new tensors are added to an initial tensor sequence. The straightforward application of OTA to every newly incoming tensor is computationally expensive. Therefore, *Incremental Tensor Analysis (ITA)* has been proposed to work efficiently on tensor streams. Given a new tensor $\mathcal{X}_t \in \mathbb{R}^{I_1 \times \cdots \times I_N}$ and the old projection matrices, ITA finds the new projection matrices $\mathbf{U}_k|_{k=1}^{N} \in \mathbb{R}^{I_k \times r_k}$ and the core tensors $\mathcal{Y}_t$, such that the reconstruction error:

$$e_t = \sum_{t=1}^{n} \|\mathcal{X}_t - \mathcal{Y}_t \prod_{i=1}^{N} \times_i \mathbf{U}_i \| \qquad (11.4.2)$$

is minimized.

There are three variants of incremental tensor analysis: *Dynamic Tensor Analysis (DTA)*, *Streaming Tensor Analysis (STA)*, and *Window-Based Tensor Analysis (WTA)*. In DTA and WTA, an incremental update of the covariance matrices is used, while in STA an incremental update of projection matrices is employed. Additionally, in order to assimilate changes over time DTA and STA work with a forgetting factor that fades away the covariance matrices of earlier steps, while WTA uses a sliding window. The ITA methods are presented in more detail next [STPF08]:

- **Dynamic Tensor Analysis (DTA).** Each mode of the tensor is processed at-a-time and the covariance of d-th mode is updated by $\mathbf{C}_d \leftarrow \lambda \mathbf{C}_d + \mathbf{X}_{(d)}\mathbf{X}_{(d)}^T$, where $\mathbf{X}_{(d)} \in I_d \times \mathbb{R}^{(\prod_{i \neq d} I_i)}$ is the mode-d matricizing of the tensor $\mathcal{X}$. The updated projection matrices are computed by diagonalization: $\mathbf{C}_d = \mathbf{U}_d\mathbf{S}_d\mathbf{U}_d^T$, where $\mathbf{U}_d$ is an orthogonal matrix and $\mathbf{S}_d$ is a diagonal matrix. The DTA process is illustrated in Figure 11.4.2, while the corresponding algorithm is outlined in Algorithm 11.4.2. The computational cost for updating the covariance matrix (Algorithm 11.4.2 - Step 2) is $\mathcal{O}\left(\sum_{i=1}^{N} I_i \prod_{j=1}^{N} I_j\right)$, while step 3 in Algorithm 11.4.2 requires N eigen-decompositions for matrices of size $I_d \times I_d$, $d = 1, 2, \ldots, N$ which costs $\mathcal{O}(I_d^3)$ each [STPF08]. Thus, the computational cost of DTA is $\mathcal{O}\left(\sum_{i=1}^{N}(I_i + I_i^2)\prod_{j=1}^{N} I_j\right)$.

- **Streaming Tensor Analysis (STA).** In the presence of a new tensor, STA aims to smoothly adjust the projection matrices. The STA procedure is presented in Figure 11.4.3. For every mode d of the new tensor $\mathcal{X}$, the matrix $\mathbf{X}_{(d)}$ is derived by unfolding, and the projection matrix $\mathbf{U}_d$ is adjusted by applying the *Streaming Pattern Discovery in Multiple Time-series (SPIRIT)* algorithm over the columns of $\mathbf{X}_{(d)}$[PSF05]. The STA algorithm is summarized in Algorithms 11.4.3 and 11.4.4. Taking into account that the computational cost of Steps 2 and 3 of Algorithm 11.4.4 is $\mathcal{O}(r(4I_i + 1))$ and $\mathcal{O}(2r_i^2 I_i)$, respectively, the computational complexity of STA is $\simeq \mathcal{O}\left(\sum_{i}^{N}(2r_i^2 I_i + 4r_i I_i)\prod_{\substack{j=1 \\ j \neq i}}^{N} I_j\right) = \mathcal{O}\left(\sum_{i}^{N}(2r_i^2 + 4r_i)\prod_{j=1}^{N} I_j\right)$. This complexity is smaller than that of DTA when $r_i \ll I_i$.

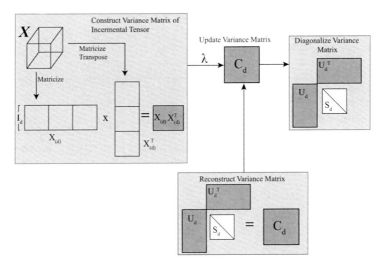

FIGURE 11.4.2: DTA process [STPF08].

Algorithm 11.4.2: Dynamic Tensor Analysis (DTA)

Input: new tensor $\mathcal{X} \in \mathbb{R}^{I_1 \times \ldots \times I_N}$, old projection matrices $\mathbf{U}_d|_{d=1}^{N} \in \mathbb{R}^{I_d \times r_d}$, old energy matrices $\mathbf{S}_d|_{d=1}^{N} \in \mathbb{R}^{r_d \times r_d}$, output ranks $r_i|_{i=1}^{N}$, and forgetting factor λ.

Output: new projection matrices $\mathbf{U}_d|_{d=1}^{N} \in \mathbb{R}^{I_d \times r_d}$ and core tensor $\mathcal{Y} \in \mathbb{R}^{r_1 \times \ldots \times r_N}$.

- For each tensor mode $d \in [1, N]$ repeat:

    1. Reconstruct the old covariance matrix $\mathbf{C}_d \leftarrow \mathbf{U}_d \mathbf{S}_d \mathbf{U}_d^T$.
    2. Update the old covariance matrix $\mathbf{C}_d \leftarrow \lambda \mathbf{C}_d + \mathbf{X}_{(d)} \mathbf{X}_{(d)}^T$.
    3. Set $\mathbf{U}_d$ as the matrix having the top r_d eigenvectors of $\mathbf{C}_d$ as columns.

- Compute the core tensor $\mathcal{Y} = \mathcal{X} \prod_{j=1}^{N} \times_j \mathbf{U}_j^T$.

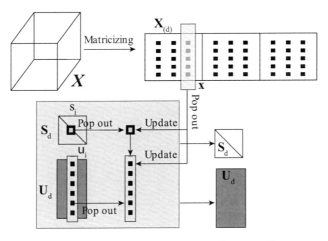

FIGURE 11.4.3: STA process [STPF08].

Algorithm 11.4.3: Streaming Tensor Analysis (STA)

Input: new tensor $\mathcal{X} \in \mathbb{R}^{I_1 \times \dots \times I_N}$, old projection matrices $\mathbf{U}_d|_{d=1}^N \in \mathbb{R}^{I_d \times r_d}$, old energy matrices $\mathbf{S}_d|_{d=1}^N \in \mathbb{R}^{r_d \times r_d}$, output ranks $r_i|_{i=1}^N$, and forgetting factor λ.
Output: new projection matrices $\mathbf{U}_d|_{d=1}^N \in \mathbb{R}^{I_d \times r_d}$ and core tensor $\mathcal{Y} \in \mathbb{R}^{r_1 \times \dots \times r_N}$.

- For each tensor mode $d \in [1, N]$ repeat:

    1. Reconstruct matrix $\mathbf{X}_{(d)}$ as mode-d unfolding of the new tensor $\mathcal{X}$.
    2. For each column vector $\mathbf{x}$ in $\mathbf{X}_{(d)}$ apply the SPIRIT algorithm (Algorithm 11.4.4): $(\mathbf{U}_d, \mathbf{S}_d) \leftarrow \text{TrackU}(\mathbf{U}_d, \mathbf{x}, \mathbf{S}_d, \lambda)$.

- Compute the core tensor $\mathcal{Y} = \mathcal{X} \prod_{j=1}^N \times_j \mathbf{U}_j^T$.

Algorithm 11.4.4: SPIRIT Algorithm - TrackU

Input: projection matrix $\mathbf{U} \in \mathbb{R}^{n \times r}$, new vector $\mathbf{x} \in \mathbb{R}^n$, energy matrix $\mathbf{S} \in \mathbb{R}^{r \times r}$, and forgetting factor λ.
Output: projection matrix $\mathbf{U}$ and energy matrix $\mathbf{S}$.

1. Initialize $\mathbf{x}' = \mathbf{x}$ and $\mathbf{s} = diag(\mathbf{S})$.

2. For each $i \in [1, r]$ repeat:

    - Estimate the projection $y_i = \mathbf{u}_i^T \mathbf{x}'_i$, where $\mathbf{u}_i$ is the i-th column of $\mathbf{U}$.
    - Estimate the energy as the i-th eigenvalue: $s_i \leftarrow \lambda s_i + y_i^2$.
    - Estimate the error $\mathbf{e}_i = \mathbf{x}'_i - y_i \mathbf{u}_i$.
    - Update the estimate of the principal component $\mathbf{u}_i \leftarrow \mathbf{u}_i + \frac{1}{s_i} y_i \mathbf{e}_i$.
    - Repeat with remainder of $\mathbf{x}'_{i+1} = \mathbf{x}'_i - y_i \mathbf{u}_i$.

3. Orthogonalize $\mathbf{u}_i$ by fixing $\mathbf{u}_1$.

4. Set $\mathbf{S}$ to be the diagonal matrix having $\mathbf{s}$ in its main diagonal.

- **Window-Based Tensor Analysis (WTA).** Instead of using a forgetting factor, as in DTA and STA, WTA handles time changes by means of a sliding window that assigns the same weight to all time-stamps in the window. A *tensor window* $\mathcal{D}(n, W)$ is defined as a subset of a tensor stream ending at time n with size W. That is, $\mathcal{D}(n, W) \in \mathbb{R}^{W \times I_1 \times \dots \times I_N} = \{\mathcal{X}_{n-W+1}, \dots, \mathcal{X}_n\}$, where each $\mathcal{X}_i \in \mathbb{R}^{I_1 \times \dots \times I_N}$. In order to incrementally extract patterns from tensor streams, two variant WTA methods exist. Namely, the *Independent-Window Tensor Analysis (IWTA)* and the *Moving-Window Tensor Analysis (MWTA)* [STPF08].

    Given a tensor window $\mathcal{D} \in \mathbb{R}^{W \times I_1 \times \dots \times I_N}$, IWTA finds the projection matrices $\mathbf{U}_0 \in \mathbb{R}^{W \times r_0}$ and $\mathbf{U}_i|_{i=1}^N \in \mathbb{R}^{I_i \times r_i}$, such that the error $e = \|\mathcal{D} - \mathcal{D} \prod_{i=1}^N \times_i (\mathbf{U}_i^T \mathbf{U}_i)\|$ is minimized [STPF08]. In that sense, each tensor window is treated independently of each other and is summarized into a core tensor $\mathcal{Y}$ associated with projection matrices $\mathbf{U}_i$ by $\mathcal{Y} = \mathcal{D} \prod_{i=0}^N \times_i \mathbf{U}_i^T$. At every time-stamp t, a tensor window $\mathcal{D}(n, W)$ is formed that includes the current tensor $\mathcal{D}_n$ and the $W - 1$ previous versions. The projection matrices $\mathbf{U}_i|_{i=0}^N$ are estimated by applying the alternating least squares method (i.e., the PARAFAC algorithm) [Tuc66]. Each projection matrix $\mathbf{U}_i|_{i=0}^N$ is initialized to be a $I_i \times r_i$ truncated identity matrix. An intuitive outline of IWTA is presented in Figure 11.4.4.

    The MWTA relaxes the time independence assumption of ITWA and uses the time-dependence structure on the tensor windows. In detail, given a tensor window

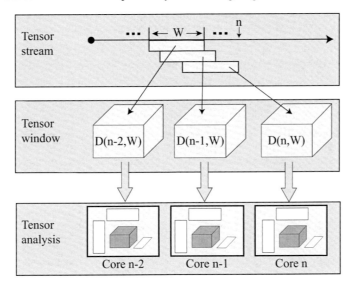

FIGURE 11.4.4: IWTA process: The core tensors and projection matrices are estimated for every tensor window separately [STPF08].

$\mathfrak{D}(n, W) \in \mathbb{R}^{W \times I_1 \times \ldots \times I_N}$ and the previous result $\mathfrak{D}(n-1, W) \in \mathbb{R}^{W \times I_1 \times \ldots \times I_N}$, WTA finds the new projection matrices $\mathbf{U}_0 \in \mathbb{R}^{W \times r_0}$ and $\mathbf{U}_i|_{i=1}^N \in \mathbb{R}^{I_i \times r_i}$, such that the error $e = \|\mathfrak{D}(n, W) - \mathfrak{D}(n, W) \prod_{i=1}^N \times_i (\mathbf{U}_i^T \mathbf{U}_i)\|$ is minimized. More specifically, MWTA exploits the overlapping information of two consecutive tensor windows to update the covariance matrices $\mathbf{C}_i|_{i=0}^N \in \mathbb{R}^{I_i \times I_i}$ by the iterative procedure that follows [STPF08]:

- Covariance matrices $\mathbf{C}_i$ updates for modes 1 to N:
  $\mathbf{C}_d^T \leftarrow \mathbf{C}_d - \mathbf{D}_{n-W,(d)} \mathbf{D}_{n-W,(d)}^T + \mathbf{D}_{n,(d)} \mathbf{D}_{n,(d)}$, where $\mathbf{D}_{n-W,(d)}$ and $\mathbf{D}_{n,(d)}$ is the mode-d unfolding matrix of tensors $\mathfrak{D}_{n-W}$ and $\mathfrak{D}_n$, respectively.

- Covariance matrix $\mathbf{C}_0$ update that corresponds to the time mode: The tensor window $\mathfrak{D}_{n-W}, \ldots, \mathfrak{D}_n$ is unfolded along the time mode giving the matrix $[\mathbf{x}|\mathbf{D}|\mathbf{y}]$, where $\mathbf{x}$ and $\mathbf{y}$ are the vectorizations of the older tensor $\mathfrak{D}_{n-W}$ and the new tensor $\mathfrak{D}_n$, respectively (i.e., the transposed time mode of the associated tensors) and $\mathbf{D}$ is the overlapping part between two tensor windows. Given that the old covariance matrix for tensor window $D(n-1, W)$ is:

$$\mathbf{C}_0^{old} = [\mathbf{x}|\mathbf{D}]^T [\mathbf{x}|\mathbf{D}] = \begin{bmatrix} \mathbf{x}^T\mathbf{x} & \mathbf{x}^T\mathbf{D} \\ \mathbf{D}^T\mathbf{x} & \mathbf{D}^T\mathbf{D} \end{bmatrix}, \qquad (11.4.3)$$

the new covariance matrix $\mathbf{C}_0^{new}$ for tensor window $D(n, W)$ can be incrementally computed as:

$$\mathbf{C}_0^{new} = \begin{bmatrix} \mathbf{D}^T\mathbf{D} & \mathbf{D}^T\mathbf{y} \\ \mathbf{y}^T\mathbf{D} & \mathbf{y}^T\mathbf{y} \end{bmatrix}. \qquad (11.4.4)$$

The upper left submatrix $\mathbf{D}^T\mathbf{D}$ in (11.4.4) is obtained from the lower right submatrix $\mathbf{C}_0^{old}$ in (11.4.3) without any computation and $\mathbf{y}^T\mathbf{y}$ is the inner product of the new data. $\mathbf{D}^T \mathbf{y}$ involves both the new tensor and all other tensors in the window. The update of the covariance matrix is not local to the added and deleted tensor, but has a global effect over the entire window.

The algorithm of MWTA is summarized in Algorithm 11.4.5.

Algorithm 11.4.5: Moving Window Tensor Analysis (MWTA)

Input: new tensor $\boldsymbol{\mathcal{D}}_n \in \mathbb{R}^{I_1 \times \dots \times I_N}$, old tensor $\boldsymbol{\mathcal{D}}_{n-W} \in \mathbb{R}^{I_1 \times \dots \times I_N}$, old covariance matrices $\mathbf{C}_d|_{d=1}^{N} \in \mathbb{R}^{I_d \times I_d}$.

Output: new covariance matrices $\mathbf{C}_d|_{d=1}^{N} \in \mathbb{R}^{I_d \times I_d}$, projection matrices $\mathbf{U}_d|_{d=1}^{N} \in \mathbb{R}^{I_d \times r_d}$ and core tensor $\boldsymbol{\mathcal{Y}} \in \mathbb{R}^{r_1 \times \dots \times r_N}$.

1. Initialize each mode except mode 0 that corresponds to time. That is, for each $d \in [1, N]$:

    - Estimate $\mathbf{D}_{n-W,(d)}(\mathbf{D}_{n,(d)})$ by mode-d unfolding of $\boldsymbol{\mathcal{D}}_{n-W}(\boldsymbol{\mathcal{D}}_n)$.
    - Update the old covariance matrix $\mathbf{C}_d \leftarrow \mathbf{C}_d - \mathbf{D}_{n-W,(d)} \mathbf{D}_{n-W,(d)}^T + \mathbf{D}_{n,(d)} \mathbf{D}_{n,(d)}^T$.
    - Perform diagonalization $\mathbf{C}_d = \mathbf{U}_d \Lambda_d \mathbf{U}_d^T$.
    - Truncate $\mathbf{U}_d$ to the first r_d columns.

2. Apply the iterative algorithm with the new initialization starting from the time mode.

Ignoring the cost of diagonalization, the computational complexity of IWTA and MWTA is $\mathcal{O}(W \sum_{i=1}^{N} I_i)$ and $\mathcal{O}(W \sum_{i=1}^{N} r_i + \sum_{i=1}^{N} I_i \; r_i + W \; r_0)$, respectively. MWTA is faster in practice, because it requires a smaller number of iterations to converge than IWTA [STPF08].

11.5   Parallel and Distributed Approaches for Big Data Analysis

The incremental methods described in Sections 11.2-11.4 can efficiently handle the dynamic nature of the social data and the need for real time processing (i.e., online algorithms). Additionally, they can deal with the big data volume, by incrementally processing the data in subsets rather than as a whole at once. However, for big data, where processing speed, memory, storage, and scalability are important issues, parallel and distributed approaches can be even more effective. In Sections 11.5.1-11.5.4, parallel approaches on latent variable models, spectral clustering, and tensor models are presented.

11.5.1   Parallel Probabilistic Latent Semantic Analysis

The computationally intensive EM algorithm entailed in PLSA makes the use of PLSA for large data collections rather impractical. Therefore, parallel implementations that distribute the algorithm across multiple processors with either shared or distributed memory have been proposed to overcome this PLSA limitation.

The underlying idea of such a parallel PLSA method is to partition the co-occurrence matrix into blocks to be queued and processed one-by-one by each available processor in a round-robin way minimizing at the same time the overall processors' idle time [HCZS08]. The block dividing algorithm splits the co-occurrence matrix along the dimension of the observed variables (i.e., words and documents) [HCZS08], or along the dimension of the latent variable (topics) [WAM09]. In the first case, an additional pre-processing step is needed to distribute the non-zero co-occurrence counts among blocks in order to achieve load balance during parallelization [HCZS08], while in the second case, when the latent variables are evenly divisible to the number of processors, a load balance among processors is achieved without any pre-processing step [WAM09]. Moreover, a more effective formulation of the EM algorithm that combines the expectation and the minimization step into a single step is used [WAM09]. Among the various parallel implementations, multiple processors with shared memory [HCZS08] and with distributed memory [WAM09] are considered, using the *OpenMP* [CMD⁺00] and the *Message Passing Interface (MPI)* [TRG05] programming model, respectively. In the case of distributed memory, the communication between processors is explicitly taken care of, while all the processors may have equal roles with all their peers, or one processor is assigned the role of a master and all the others that of a slave. In the first case, all the processors communicate directly with their peers, while in the second case the processors communicate only with the master processor. When the parallel method is implemented with a master processor and k slave processors, the master processor is responsible for initializing and normalizing the entailed conditional probabilities of the PLSA model, while the slave processors estimate the corresponding new conditional probabilities. The slave processors remain idle during the execution of the initialization step in the master processor, while the master processor is prevented from remaining idle when the slave processors are active by being assigned a work similar to that of a slave processor to perform. Transmissions among processors are blocked until the processors have the probabilities to be sent/received estimated [WAM09].

Two more recent parallel implementations of PLSA, P^2LSA and P^2LSA+, are based on the *MapReduce* model [DG08]. In P^2LSA, the E-step is performed by the Map function and the M-step is performed by the Reduce function. The intermediate results computed in the E-step have to be sent to the M-step, but exchanging large amounts of data between the Map and the Reduce function increases the network load and the overall running time [JGS⁺11]. The P^2LSA variant, P^2LSA+, alleviates this problem by performing the E-step and the M-step simultaneously in the Map function, thus reducing the data transfers between the EM steps [JGS⁺11].

11.5.2   Parallel Latent Dirichlet Allocation

The majority of the parallel implementations for LDA are built on the MCMC-based approximation rather on variational inference. *Dirichlet Compound Multinomial LDA (DCM-LDA)* is one of the first parallel LDA methods that performs Gibbs sampling on data subsets distributed in different processors. Each processor works independently without any communication with the others, while a global clustering of the topics is performed at the end of parallel processing in order to obtain a global solution [MM07].

In another approach, called *Approximate Distributed LDA (AD-LDA)*, a local Gibbs sampling iteration is performed in each processor followed by a global update using a reduce-scatter operation [NASW07]. In more detail, the $|D|$ documents are distributed among p processors, with $|D|/p$ documents in each processor. Similarly, the words $\mathbf{w}$ and the corresponding topic assignments $\mathbf{z}$ are respectively partitioned into $\mathbf{w} = \{\mathbf{w}_1, \mathbf{w}_2, \ldots, \mathbf{w}_p\}$ and $\mathbf{z} = \{\mathbf{z}_1, \mathbf{z}_2, \ldots, \mathbf{z}_p\}$ so that $\mathbf{w}_p$ and $\mathbf{z}_p$ exist only in processor p. The topic-document counts $n_{z_k}^{(d_j)}$ are likewise distributed, but each processor maintains its own copy of word-topic counts $n_{z_k|p}^{(w_i)}$ and topic counts $n_{z_k|p}^{(\cdot)}$. Each iteration of the algorithm is composed of a Gibbs sampling step and a synchronization step. In the sampling step, each processor p samples $\mathbf{z}_{M|p} = \{z_{1|p}, \ldots, z_{M|p}\}$, where M is the vocabulary size, using the global topics of the previous iteration. In the synchronization step, the local word-topic counts $n_{z_k|p}^{(w_i)}$ in each processor are aggregated to produce a global set of word-topic counts $n_{z_k}^{(w_i)}$. This process is repeated until convergence or for a fixed number of iterations [NASW07, WBSC09], providing substantial memory and time savings.

The synchronous algorithm AD-LDA requires global synchronization at each iteration. This is an important drawback of the method, since a global synchronization step may not be always feasible due to processors unavailability, or different processor speeds. In an asynchronous distributed version of LDA (*Async-LDA*) that follows a similar two-step process as AD-LDA, this drawback is tackled by means of a step where each processor communicates with another random processor after the local Gibbs sampling step [ASW08]. That is, during each iteration of Async-LDA, the processors perform a full sweep of collapsed Gibbs sampling over their local topic assignment variables $\mathbf{z}_{M|p}$ for each word w_j in a manner analogous to (11.2.66):

$$P(z_{j|p}|\mathbf{z}_{M\setminus j|p}, \mathbf{w}_{M|p}) \propto \frac{n^{(w_j)}_{(z_j, M\setminus j)|p} + n^{(w_j)}_{(z_j, M\setminus j)|\setminus p} + \eta}{n^{(\cdot)}_{(z_j, M\setminus j)|p} + n^{(\cdot)}_{(z_j, M\setminus j)|\setminus p} + M\eta} \left(n^{(d_j)}_{(z_j, M\setminus j)|p} + \alpha \right) \qquad (11.5.1)$$

where the notation $n_{(\ )|p}$ refers to the local counts in each processor p and $n_{(\ )|\setminus p}$ denotes the processor's p belief for the corresponding counts of all the other processors with which it has already communicated, excluding the processor's p local counts. Thus, the sampling in (11.5.1) is based on the processors "noisy view" of the global set of topics [ASW08]. In the communication step between two processors p and p', two cases are handled: (a) p and p' communicate for the first time and they simply exchange their local word-topics counts $n^{(\cdot)}_{z_k|p}$ and $n^{(\cdot)}_{z_k|p'}$ to be added to the corresponding global word-topic counts $n^{(\cdot)}_{z_k|\setminus p}$ and $n^{(\cdot)}_{z_k|\setminus p'}$ each processor monitors through communication with other processors; (b) p and p' have communicated in a previous step, so besides exchanging and adding their local counts they also remove from the global word-topic counts $n^{(\cdot)}_{z_k|\setminus p}$ and $n^{(\cdot)}_{z_k|\setminus p'}$ the influence of the other processor from their previous encounter. In the latter case, any over-influence among processors that happen to communicate more frequently is prevented. When there are no strict memory and bandwidth limitations, allowing each processor to cache the previous counts of other processors and forward its individual cached counts to the other processors it communicates with, the efficiency of Async-LDA can be significantly improved [ASW08].

Two parallel implementations of the synchronous AD-LDA [NASW07], use either the Message Passing Interface (MPI) [TRG05] or the MapReduce model [DG08], have been described [WBSC09]. In the *MPI-PLDA* implementation, a worker (i.e., a thread or a process that executes part of the parallel computing job), performs the necessary initializations, estimates the processor specific document and word-topic counts accumulated from local document topic assignments in each processor and performs the Gibbs sampling, as described in [NASW07]. At the end of each Gibbs sampling iteration, the procedure MPI_AllReduce is invoked in order to estimate the global word-topic counts and distribute them to all the other workers. The worker sleeps until the MPI implementation finishes AllReduce and the corresponding updated counts are stored in the worker's buffer. The implementation also has checkpoints testing for a machine failure.

The *PLDA* is implemented in the MapReduce framework by means of three procedures (PLDA-MapperStart, PLDA-Map and PLDA-MapperFlush) in the mapping phase and three procedures (PLDA-ReducerStart, PLDA-Reduce, and PLDA-ReducerFlush) in the reducing phase. Gibbs sampling is performed in the mapping phase, while model and topic assignment updates are performed in the reducing phase. Each map worker is assigned a fraction of $|D|/p$ documents in an input shard[1] and loads a local copy of the word-topic counts by calling PLDA-MapperStart. Then, it invokes a PLDA-Map for each local document in order to update the corresponding topic assignments $z_{j|p}$. After Gibbs sampling

---

[1]Shard is a local subset of input and output to a MapReduce process in the form of key-value pairs.

on all documents in a shard is finished, the PLDA-MapperFlush is invoked to output the updated local word-topic counts. In the reducing phase, two standard reducers are used: an IdentityReducer that copies each $z_{j|p}$ produced by Gibbs sampling to the disk and a VectorSummingReducer that outputs the aggregated word-topic counts by taking into account the local counts. All map and all reduce workers run in parallel and they communicate only in the shuffling phase [WBSC09].

PLDA+ consists an improved distributed algorithm of LDA based on PLDA. The PLDA+ reduces inter-computer communication time by employing four interdependent strategies (i.e., data placement, pipeline processing, word bundling, and priority-based scheduling) [LZCS11]. Other LDA parallelization methods include parallel algorithms of Gibbs sampling and variational inference for *GPUs (Graphics Processing Units)* [YXQ09], and parallel implementations of the variational EM algorithm in a multiprocessor architecture as well as in a distributed setting [NCL07].

11.5.3 Parallel Spectral Clustering

Spectral clustering described in Section 11.3 is effective for finding clusters, but its performance and its applicability to large scale datasets suffers from high computational complexity. More precisely, spectral clustering exhibits a quadratic time and space complexity, when constructing and storing the affinity matrix that holds the pair-wise similarities among the data instances. Time complexity becomes cubic when calculating the eigendecomposition of the Laplacian matrix, making the application of the spectral clustering algorithm to big data difficult. Several alternative methods have been proposed to address these computational and memory difficulties.

One of the most commonly used approaches to handle the memory bottleneck is to sparsify the affinity matrix and to use a sparse eigensolver for the decomposition of the resulting sparse Laplacian matrix. *Sparsification* is usually achieved by means of the t-nearest neighbor approach that considers only the significant relationships between the data instances, or by the ϵ-neighborhood approach that zeroes out those elements of the affinity matrix that are below a pre-defined threshold ϵ [CSBL11]. Both sparsification schemes keep the computational complexity high, since all the elements of the affinity matrix still need to be estimated. To alleviate this, some approaches consider zeroing out random entries in the affinity matrix, saving computational time at the expense of clustering performance [AMS01]. More elaborate methods focus on reducing the computational cost of the eigendecomposition of the graph Laplacian by using the classical *Nyström approximation* method on a dense submatrix of the original affinity matrix [FBCM04]. That is, data are randomly sampled in order to obtain small-size eigenvectors that are then used to estimate an approximation of the eigenvectors of the original matrix.

Similar to the idea of reducing the original dataset, k-means clustering is first applied on the dataset resulting in a large cluster number and the data points that are close to the centers (according to a pre-defined distance threshold) are removed [SS08]. *KASP* and *RASP* algorithms also attempt to perform fast approximate spectral clustering by collapsing all data points into centroids obtained through k-means or random projection trees respectively, and by applying eigen-decomposition only to the centroids [YXQ09].

Random projection is also used in order to reduce data dimensionality [SI09]. Early stopping strategies could be applied to speed up eigen-decompositions [CGL+06, LYZ+07] based on the observation that well-separated data points converge to the final solution more quickly. In *Landmark-based spectral clustering*, all data points are encoded by means of a codebook constructed by selecting landmark points among the data points [CC11].

In a slightly different method, the so-called *Efficient Spectral Clustering on Graphs (ESCG)* for large-scale graph data, the original graph is coarsened by generating supernodes linked to the nodes in the original graph and a bipartite structure is obtained which preserves the links between original graph nodes and the new supernodes. The super nodes are expected to behave as cluster indicators that guide the clustering of nodes in the original graph. Thus, the supernode clustering and the regular node clustering mutually induce each other. In this way, the clustering of the original graph can be solved by clustering the bipartite graph [LWDH13].

The most common approach to sparsifying the original affinity matrix with the t-nearest neighbor approach together with the sparse eigen-solver decomposition of the Laplacian matrix, has been recently formulated for parallel distributed processing under the MPI and the MapReduce programming models [CSBL11]. In short:

- The sparse similarity matrix construction using t-nearest neighbors entails three steps: a) distance computation among all the data points, b) the modification of the sparse matrix to a symmetric matrix, and c) the similarity computations using distances. All three steps are implemented using MapReduce as follows:

  a) Having n data points and a distributed environment with p processors, n/p rows of the distance matrix are constructed at each node. In the map phase, intermediate keys/values are created so that every n/p data points have the same key. In the reduce phase, these n/p data points, referred to as local points, are loaded to the memory of a node and the distances between the points of the whole dataset and the local points are estimated. n/p max heaps are also used to store a local data points t-nearest neighbors.

  b) To make the sparse distance matrix symmetric, the symmetric elements of the matrix are assigned the same value. In the map phase, two key/value pairs are generated for each non-zero element in the sparse distance matrix. The first key is the row ID of the element and the corresponding value is the column ID and the distance, while the second key is the column ID and the corresponding value is the row ID and the distance. In the reduction phase, elements having the same key (that correspond to values in the same row of the distance matrix) are collected. After symmetrization, each row contains at most $2t$ non-zero elements. Since $t \ll n$, the resulting symmetric matrix is still sparse.

  c) The Gaussian similarity between distances is estimated in a separate MapReduce step that self-tunes the scaling parameter σ, which controls how rapidly the similarity between two data points reduces with the distance between the data points. The estimate of the scaling parameter for each data point with t-nearest neighbors is usually defined as the average of t distances or as the median value of each row of the sparse similarity matrix. This estimate is performed in the map phase, while each reduce function obtains a row and all parameters.

- Among the various parallel eigen-decompositions of the sparse similarity matrix, a variant of the *Lanczos/Arnoldi factorization ARPACK*, called *PARPACK*, is considered [MS96, CSBL11]. The parallel implementation is based on the MPI programming model, where each MPI node stores n/p rows of the graph Laplacian matrix. Similarly, the eigen-vector matrix, estimated in the m-step of the Arnoldi factorization by calling ARPACK, is split into p partitions with n/p rows. The major communication overhead between the nodes resides to the estimate of the parallel sparse matrix-vector product of the Arnoldi factorization, since each node should dispatch its matrix rows to the other nodes. This task is performed by a gathering operation in MPI (MPI_AllGather) which is based on the recursive doubling algorithm [TRG05].

- The k eigen-vectors of the Laplacian matrix estimated by the parallel eigen-solver are distributedly stored, thus the corresponding normalized matrix can be computed in parallel and stored on p local machines. Each row of the normalized matrix is treated as a data point in the parallel k-means algorithm, which is implemented using MPI. In more detail, for the parallel k-means algorithm the master node randomly chooses a point (row of the normalized matrix) as the first cluster center and it broadcasts it to all the nodes. Then, each node identifies the most orthogonal point to this center by estimating the cosine similarity between its local points and the center. The p minimal cosine similarities are collected and the most orthogonal point to the first center is selected as the second center. This procedure is iteratively executed in order to produce k centers. After the definition of the k initial centers, each node assigns its local data to clusters and estimates local squared error function for each cluster. The master node receives the sum of all points in each cluster, calculates the new centers, and broadcasts them to all the nodes implemented by a MPI_AllReduce operation.

11.5.4 Distributed Tensor Decomposition

Assuming a three-way tensor $\mathbf{X} \in \mathbb{R}^{I_1 \times I_2 \times I_3}$, the *PARAFAC decomposition* of the tensor into R components is expressed by means of a triplet of matrices $\mathbf{A} \in \mathbb{R}^{I_1 \times R}$, $\mathbf{B} \in \mathbb{R}^{I_2 \times R}$, and $\mathbf{C} \in \mathbb{R}^{I_3 \times R}$ together with normalization factor vector $\boldsymbol{\lambda} \in \mathbb{R}^{R \times 1}$. Alternating Least Squares is the most popular algorithm for PARAFAC decomposition and entails three main steps. Each one of these steps is a conditional update of one of the three factor matrices given the other two followed by a normalization step of the columns of the updated matrix. These conditional updates are expressed by the equations [KPHF12]:

$$\mathbf{A} = \mathbf{X}_{(1)} \left(\mathbf{C} \odot \mathbf{B} \right) \left(\mathbf{C}^T \mathbf{C} * \mathbf{B}^T \mathbf{B} \right)^\dagger \tag{11.5.2}$$

$$\mathbf{B} = \mathbf{X}_{(2)} \left(\mathbf{C} \odot \mathbf{A} \right) \left(\mathbf{C}^T \mathbf{C} * \mathbf{A}^T \mathbf{A} \right)^\dagger \tag{11.5.3}$$

$$\mathbf{C} = \mathbf{X}_{(3)} \left(\mathbf{B} \odot \mathbf{A} \right) \left(\mathbf{B}^T \mathbf{B} * \mathbf{A}^T \mathbf{A} \right)^\dagger \tag{11.5.4}$$

where $\odot$ is the Khatri-Rao product, $\dagger$ is the pseudo-inverse of the corresponding matrix, $*$ is the Hadamard product, and $\mathbf{X}_{(1)}$, $\mathbf{X}_{(2)}$ and $\mathbf{X}_{(3)}$ are the mode-1, mode-2 and mode-3 unfolding matrices of tensor $\mathbf{X}$, respectively.

Large scale tensor decomposition, namely the PARAFAC decomposition, has been recently formulated in a parallel distributed framework using the MapReduce programming model. The new algorithm, called *GIGATENSOR*, exploits the sparseness of the real word tensors and redesigns the tensor decomposition algorithm in a way that avoids the intermediate data explosion problem [KPHF12]. GIGATENSOR focuses on making the estimates of the update rules in (11.5.2)-(11.5.4) more computationally efficient. To achieve this, the optimal ordering of computations is taken into consideration in order to minimize the number of floating point operations (flops) needed. This ordering is defined for the updating rule of $\mathbf{A}$ as follows:

$$\mathbf{M}_1 = \mathbf{X}_{(1)} \left(\mathbf{C} \odot \mathbf{B} \right) \tag{11.5.5}$$

$$\mathbf{M}_2 = \left(\mathbf{C}^T \mathbf{C} * \mathbf{B}^T \mathbf{B} \right)^\dagger \tag{11.5.6}$$

$$\mathbf{A} = \mathbf{M}_1 \mathbf{M}_2. \tag{11.5.7}$$

Similar equations hold for the orderings of $\mathbf{B}$ and $\mathbf{C}$.

In order to deal with the intermediate data explosion problem entailed in the Khatri-Rao product in (11.5.5), an equivalent algebraic computation is performed that decouples the two terms in the product [KPHF12]:

$$\mathbf{N}_1 = \mathbf{X}_{(1)} * (\mathbf{1}_{I_1} \circ (\mathbf{C}(:,r)^T) \otimes \mathbf{1}_{I_2}^T)) \tag{11.5.8}$$

$$\mathbf{N}_2 = (bin(\mathbf{X}_{(1)}) * (\mathbf{1}_{I_1} \circ (\mathbf{1}_{I_3}^T \otimes \mathbf{C}(:,r)^T)) \tag{11.5.9}$$

$$\mathbf{M}_1 = (\mathbf{N}_1 * \mathbf{N}_2) \, \mathbf{1}_{I_2 I_3}, \tag{11.5.10}$$

where $\mathbf{1}$ is a vector of ones with the size denoted in the subscript, $\circ$ is the outer product, $\otimes$ denotes the Kronecker product, and $bin(\ )$ is a function that converts any non-zero value into 1, preserving sparsity. Moreover, to make the computations in (11.5.6) more efficient, $\mathbf{C}^T\mathbf{C}$ and $\mathbf{B}^T\mathbf{B}$ are computed separately as the sum of outer products of the rows, i.e.:

$$\mathbf{C}^T\mathbf{C} = \sum_{k=1}^{I_3} \mathbf{C}(k,:)^T \circ \mathbf{C}(k,:)$$

$$\mathbf{B}^T\mathbf{B} = \sum_{k=1}^{I_2} \mathbf{B}(k,:)^T \circ \mathbf{B}(k,:), \tag{11.5.11}$$

where $\mathbf{C}(k,:)$ and $\mathbf{B}(k,:)$ are the kth row of the matrix $\mathbf{C}$ and $\mathbf{B}$, respectively. The final result is the obtained by performing the Hadamard product of the two resulting matrices of (11.5.11).

The three steps described in (11.5.5-11.5.7) are implemented in the MapReduce framework. To begin with, for (11.5.5) three different MapReduce algorithms are formulated that implement (11.5.8-11.5.10). In the mapping phase of the first MapReduce algorithm, each element of the mode-1 matrix $\mathbf{X}_{(1)}(i,j)$ is mapped on $\lceil j/I_2 \rceil$ and each element of the factor $\mathbf{C}(j,r)$ is mapped on j, such that tuples sharing the same key are shuffled to the same reducer. In the reducing face, the reducer takes the tuples of $\mathbf{C}(j,r)$ and of the positive values of $\mathbf{X}_{(1)}(i,j)$, joins them for Hadamard product and emits $\mathbf{X}_{(1)}(i,j)\,\mathbf{C}(j,r)$ for every i for which the corresponding $\mathbf{X}_{(1)}(i,j)$ is greater than zero. A similar task is performed in the second MapReduce algorithm for (11.5.9). That is, in the mapping phase each element of $\mathbf{X}_{(1)}(i,j)$ is mapped on $\lceil j/I_2 \rceil$ and each element of the factor $\mathbf{B}(j,r)$ is mapped on j, while in the reducing phase the reducer takes the tuples of $\mathbf{B}(j,r)$ and emits the corresponding values of $\mathbf{B}(j,r)$ for every i for which the corresponding $\mathbf{X}_{(1)}(i,j)$ is greater than zero. The third algorithm combines the outputs of the first and the second algorithms using the Hadamard product, and sums up each row to get the final result. In the mapping phase, $\mathbf{X}_{(1)}(i,j)\,\mathbf{C}(j,r)$ and $\mathbf{B}(j,r)$ are mapped on i, such that tuples with the same i are shuffled to the same reducer. In the reducing phase, the reducer takes $\mathbf{X}_{(1)}(i,j)\,\mathbf{C}(j,r)\,\mathbf{B}(j,r)$ for every j for which the corresponding $\mathbf{X}_{(1)}(i,j)$ is greater than zero and emits the column-sum $\sum_{j=1}^{I_2 I_3} \mathbf{X}_{(1)}(i,j)\,\mathbf{C}(j,r)\,\mathbf{B}(j,r)$.

The implementation of (11.5.6) in the MapReduce framework is performed by exploiting (11.5.11). Each factor matrix $\mathbf{C}$ ($\mathbf{B}$) is partitioned row-wise and is mapped on 0 so that all the output is shuffled to the only reducer, which together with a combiner takes the outer product $\sum_{k=1}^{I_3} \mathbf{C}(k,:)^T \circ \mathbf{C}(k,:)$ ($\sum_{k=1}^{I_3} \mathbf{B}(k,:)^T \circ \mathbf{B}(k,:)$) and emits the corresponding sum of these products. For the computation of the last step (11.5.7), only one MapReduce job is required, since the distributed cache multiplication [KMF11] is used in order to broadcast the second matrix $(\mathbf{C}^T\mathbf{C} * \mathbf{B}^T\mathbf{B})^\dagger$ to all the mappers that process the first matrix $\mathbf{X}_{(1)} (\mathbf{C} \odot \mathbf{B})$, performing then join in the first matrix [KPHF12].

11.6 Applications to Evolving Social Data Analysis

11.6.1 Incremental Label Propagation

Label propagation is widely used to reveal communities, i.e., local structure modules, in large real-world networks. In the basic *Label Propagation Algorithm (LPA)*, each node is initialized with a unique label. At every iteration, each node is assigned the label shared by most of its neighbors with ties broken uniformly at random [RAK07]. As the labels propagate through the network, densely connected groups of nodes form a consensus on their labels. At the end, the nodes having the same labels are grouped together as communities. Given a simple undirected graph $G = G(\mathcal{E}, \mathcal{V})$ with $\mathcal{V} = \{v_1, \ldots, v_{|V|}\}$ being the set of nodes, $\mathcal{E}$ denoting the set of edges, and $c_v(t)$ representing the community label of each node $v \in \mathcal{V}$ at iteration t, the LPA is summarized as follows [RAK07]:

1. Initialize the labels at all nodes in the network. For a given node v, $c_v(0) = v$. Set $t = 1$.

2. Arrange the nodes in the network in a random order and set it to $\mathcal{V}$.

3. For each $v \in \mathcal{V}$ chosen in that specific order, let v_i, $i = 1, \ldots, k$ be its neighbors, and $c_v(t) = f(v_1(t), v_2(t), \ldots, v_j(t), v_{j+1}(t-1), \ldots, v_k(t-1))$. Function $f(\cdot)$ returns the label occurring with the highest frequency among the neighbors of v. Any ties are broken uniformly. It is seen that the labels of the first j neighbors of v have already been updated, while the labels of the remaining $k - j$ neighbors have not been updated yet.

4. If every node has a label the same as the majority of its neighbors, then the algorithm stops. Else, set $t = t + 1$ and go to 3.

Step 3 of the just described algorithm is essentially an asynchronous label updating. Synchronous updating, where each node v at each iteration updates its label based on the labels of its neighbors at the previous iteration, is usually avoided, since network subgraphs with bi-partite or nearly bi-partite structure can lead to oscillations of the labels.

The LPA algorithm has been proven to work well for static networks. For evolving networks, *Incremental LPA (ILPA)* is recommended that deals with the network changes incrementally [PCW09]. That is, when a new node (edge) joins or an old node (edge) leaves the network, the algorithm is executed locally instead of globally. To achieve this, the time domain is discretized into time intervals of the same length. At each iteration, only the nodes (edges) that changed at the previous iteration are considered. The algorithm is run locally and iteratively until no labels can be changed. The ILPA is described as follows:

1. For each edge added at time interval t, the two nodes v_1, v_2, incident to the edge are labeled as $c_{v_1}^{(t)}$ and $c_{v_2}^{(t)}$, respectively. These two nodes are recorded as the new labeled nodes.

2. All new labeled nodes and their neighbors are added to the local node calculation sequence $\mathcal{V}_l$ in a random order.

3. The original LPA is sequentially applied to each node in $\mathcal{V}_l$. Similarly, nodes whose labels are changed will be recorded as new labeled nodes. If two or more different labels have the same number of neighbors, the most recent label is selected (i.e., the label of the neighbor with greater t).

4. Iterate Steps 2 and 3 until no labels are changed.

Each iteration of the LPA algorithm has a near-linear time complexity $\mathcal{O}(m)$ in the number of edges. In the case of ILPA, when a new edge is added, the labels of the two vertices which are adjacent to the edge will be updated. The vertices of the new-added edges and their neighbor vertices are added into local random calculation sequence V_l and each iteration of the ILPA algorithm has a near-linear time complexity $\mathcal{O}(m_l)$ in the number of local edges connected to the local vertices [PCW09].

11.6.2 Incremental Graph Clustering in Dynamic Social Networks

Graph clustering aims to partition the graph into several densely connected components based on various criteria, such as vertex connectivity or neighborhood similarity [ZCX10]. That is, the existing graph clustering methods mainly focus on the topological structure of a graph based on various criteria, such as normalized cut [SM00], modularity [NG04], structural density [XYFS07], or stochastic flows [SP09], so that each partition achieves a cohesive internal structure. Contrary to these methods that ignore the vertex attributes, graph summarization ignores the intra-cluster topological structures by exploiting the attribute similarity so that nodes with the same attribute values are grouped into one partition [YHP08].

Structural and attribute similarities are combined in *SA-Cluster* algorithm through a unified distance measure, which is based on a neighborhood random walk model estimated on the *attribute augmented graph* [ZCY09]. The attribute augmented graph is constructed by adding to the original graph a set of attribute vertices and attribute edges that connect attribute vertices sharing a common attribute value. Clustering is performed by following the k-Medoids framework with the random walk distance as the vertex similarity measure. The edge weights are iteratively adjusted to balance the importance between structural and attribute similarities imposing the recalculation of the corresponding random walk distances. The efficiency and scalability of the SA-Cluster algorithm is improved by the incremental algorithm *Inc-Cluster* that incrementally updates the random walk distances given the edge weight increments [ZCX10]. Both algorithms, SA-Cluster and its incremental counterpart (Inc-Cluster), are described in detail, next.

Let $\mathcal{G} = (\mathcal{V}, \mathcal{E}, \mathcal{A})$ be an *attributed graph* with $\mathcal{V}$ set of vertices v, set of edges $\mathcal{E}$ and $\mathcal{A} = \{a_1, \ldots, a_m\}$ be the set of m attributes associated with the vertices. A vertex $v \in \mathcal{V}$ is associated with an attribute vector $[a_1(v), \ldots, a_m(v)]$, where $a_i(v)$ is the i-th attribute value of vertex v. A vertex $v \in \mathcal{V}$ is called a structure vertex and an edge $(v_i, v_j) \in \mathcal{E}$ is called a structure edge. The domain of attribute a_i is denoted as $Dom(a_i) = \{a_{i1}, \ldots, a_{in_i}\}$ having size $|Dom(a_i)| = n_i$. An *attribute augmented graph* is denoted as $\mathcal{G}_a = (\mathcal{V} \cup \mathcal{V}_a, \mathcal{E} \cup \mathcal{E}_a)$ where $\mathcal{V}_a = \{v_{ij}\}_{i=1,j=1}^{m, n_i}$ is the set of attribute vertices and $\mathcal{E}_a \subseteq \mathcal{V} \times \mathcal{V}_a$ is the set of attribute edges. An attribute vertex $v_{ij} \in \mathcal{V}_a$ represents that attribute a_i takes the j-th value. An attribute edge $(v_i, v_{jk}) \in \mathcal{E}_a$, iff $a_j(v_i) = a_{jk}$, i.e., vertex v_i takes the value of a_{jk} on attribute a_j. Examples of an attributed graph and an attribute augmented graph are illustrated in Figure 11.6.1.

The *unified neighborhood random walk distance matrix* $\mathbf{R}_A$ of length L is defined as:

$$\mathbf{R}_A = \sum_{l=1}^{L} c\,(1-c)^l\,\mathbf{P}_A^l, \tag{11.6.1}$$

where $c \in (0, 1)$ is the random walk restart probability and $\mathbf{P}_A$ is the $|V \cup V_a| \times |V \cup V_a|$ transition probability matrix of the attribute augmented graph $\mathcal{G}_a$. The first $|V|$ rows (columns) correspond to the structure vertices and the last $|V_a|$ rows (columns) correspond to the attribute vertices.

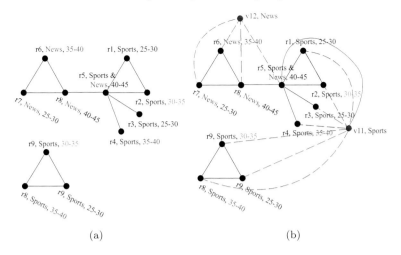

FIGURE 11.6.1: Graph examples for an article reader network with Two Attributes, "Topic" and "Age" [ZCX10]: a) Attributed graph and b) Attribute augmented graph: two attribute vertices v_{11} and v_{12} representing the topics "News" and "Sports" are added. Dashed lines connect authors with corresponding topics to the two vertices respectively. Attribute vertices and edges corresponding to the age attribute are omitted for the sake of clear presentation. **(See color insert.)**

Given the $|V| \times |V|$ matrix $\mathbf{P}_{V_1}$ holding the transition probabilities $p(v_i, v_j)$ between structure vertices v_i and v_j through a structure edge $(v_i, v_j) \in \mathcal{E}$, the $|V| \times |V_a|$ matrix $\mathbf{A}_1$ holding the transition probabilities $p(v_i, v_{jk})$ between structure and attribute vertices through an attribute edge $(v_i, v_{jk}) \in \mathcal{E}_a$, the $|V_a| \times |V|$ matrix $\mathbf{B}_1$ holding the transition probabilities between attribute and structure vertices through an attribute edge $(v_{ik}, v_j) \in \mathcal{E}_a$, and $\mathbf{O}$ the $|V_a| \times |V_a|$ zero matrix corresponding to the transition probabilities between attribute vertices $p(v_{ik}, v_{jq})$, $\mathbf{P}_A$ is partitioned as:

$$\mathbf{P}_A = \begin{bmatrix} \mathbf{P}_{V_1} & \mathbf{A}_1 \\ \mathbf{B}_1 & \mathbf{O} \end{bmatrix}. \tag{11.6.2}$$

Assuming, without loss of generality, that a structure edge has a fixed weight w_0 and the attribute edges corresponding to $a_1, \ldots, a_m$ have edge weights $w_1, \ldots, w_m$, respectively, the corresponding transition probabilities are given by:

$$p(v_i, v_j) = \begin{cases} \frac{w_0}{|N(v_i)|w_0 + w_1 + \ldots + w_m}, & \text{if } (v_i, v_j) \in \mathcal{E} \\ 0, & \text{otherwise} \end{cases} \tag{11.6.3}$$

$$p(v_i, v_{jk}) = \begin{cases} \frac{w_j}{|N(v_i)|w_0 + w_1 + \ldots + w_m}, & \text{if } (v_i, v_{jk}) \in \mathcal{E}_a \\ 0, & \text{otherwise,} \end{cases} \tag{11.6.4}$$

$$p(v_{ik}, v_j) = \begin{cases} \frac{1}{|N(v_{ik})|}, & \text{if } (v_{ik}, v_j) \in \mathcal{E}_a \\ 0, & \text{otherwise,} \end{cases} \tag{11.6.5}$$

where $N(\cdot)$ represents the set of structure vertices connected to the vertex within parenthesis. The transition probability between attribute vertices is 0, since there is no edge between attribute vertices, i.e.:

$$p(v_{ik}, v_{jq}) = 0, \quad \forall v_{ik}, v_{jq} \in \mathcal{V}_a. \tag{11.6.6}$$

The SA-Cluster algorithm is based on the k-Medoids clustering framework. At the beginning of the clustering process, the cluster centroids are initialized and the random walk distance is estimated. Then, the following four steps are repeated until convergence.

1. Assign the vertices to their closest centroids.

2. Update the cluster centroids.

3. Adjust attribute edge weights $\{w_1, \ldots, w_m\}$.

4. Re-calculate the random walk distance matrix $\mathbf{R}_A$.

Steps 3-4 are added on the top of the traditional k-Medoids (Steps 1-2). At each iteration, the attribute edge weights $\{w_1, \ldots, w_m\}$ are automatically adjusted to reflect the clustering tendencies of different attributes. Accordingly, the transition probability matrix $\mathbf{P}_A$ and the neighborhood random walk distance matrix $\mathbf{R}_A$ change. As a result, the random walk distance matrix has to be re-calculated at each iteration due to the edge weight changes.

The incremental approach aims to reduce the number of random walk distance estimates based on the observation that only the transition probabilities of the attribute edges and not those of the structure edges are affected by the attribute weight adjustments. This implies that many elements in the random walk distance matrix remain unchanged. Thus, given the original random walk distance matrix $\mathbf{R}_A$ and the weight increments $\{\Delta w_1, \ldots, \Delta w_m\}$, the incremental algorithm estimates the increment matrix $\Delta \mathbf{R}_A$, and then the updated random walk distance matrix $\mathbf{R}_{N,A} = \mathbf{R}_A + \Delta \mathbf{R}_A$ is obtained. Since $\mathbf{R}_A$ is the weighted sum of a series of matrices $\mathbf{P}_A^l$, where $\mathbf{P}_A^l$ is the l-th power of the transition probability matrix $\mathbf{P}_A$, $l = 1, \ldots, L$, to compute $\Delta \mathbf{R}_A$ one needs to estimate $\Delta \mathbf{P}_A^l$ for different l values. Therefore, the problem lies in computing the increment matrix $\Delta \mathbf{P}_A^l$ given the original matrix $\mathbf{P}_A^l$ and the edge weight increments $\{\Delta w_1, \ldots, \Delta w_m\}$.

Given that $\sum_{i=1}^m w_i = m$ and w_0 is fixed, from (11.6.3)–(11.6.6) it is seen that only the transition probabilities in the submatrix $\mathbf{A}_1$ are affected by the attribute weight increments, leaving the other three submatrices unchanged. Thus, the increment of the transition probability matrix $\Delta \mathbf{P}_A^1$ is denoted as:

$$\Delta \mathbf{P}_A^1 = \begin{bmatrix} \mathbf{0} & \Delta \mathbf{A}_1 \\ \mathbf{0} & \mathbf{0} \end{bmatrix}. \tag{11.6.7}$$

Let $w_j' = w_j + \Delta w_j$ be the new weight. The probability increment (11.6.4) is given by [ZCX10]:

$$\Delta p(v_i, v_{jk}) = \Delta w_j \, p(v_i, v_{jk}). \tag{11.6.8}$$

Denoting by $\mathbf{A}_{a_i}$ the $|\mathcal{V}| \times n_i$ matrix of the transition probabilities from structure vertices in $\mathcal{V}$ to attribute vertices corresponding to attribute a_i, then the transition probabilities submatrix can be written as $\mathbf{A}_1 = [\mathbf{A}_{a_1} | \ldots | \mathbf{A}_{a_n}]$. That is, the element $\mathbf{A}_i(p, q)$ represents the transition probability from the p-th structure vertex $v_p \in \mathcal{V}$ to the q-th attribute value a_{iq} of a_i. Thus, $\Delta \mathbf{A}_1$ is equal to:

$$\Delta \mathbf{A}_1 = [\Delta w_1 \, \mathbf{A}_{a_1} | \ldots | \Delta w_m \, \mathbf{A}_{a_m}] \tag{11.6.9}$$

and the new transition probability matrix $\mathbf{P}_{N,A}$ after the edge weight change is represented as:

$$\mathbf{P}_{N,A} = \begin{bmatrix} \mathbf{P}_{V_1} & \mathbf{A}_1 + \Delta \mathbf{A}_1 \\ \mathbf{B}_1 & \mathbf{0} \end{bmatrix} = \begin{bmatrix} \mathbf{P}_{V_1} & \mathbf{A}_{N,1} \\ \mathbf{B}_1 & \mathbf{0} \end{bmatrix}. \tag{11.6.10}$$

The computation of $\Delta \mathbf{P}_A^l$ is based on the representation of the original l-th power matrix $\mathbf{P}^l = \mathbf{P}^{l-1} \mathbf{P}_A$ and the new matrix $\mathbf{P}_{N,A}^l = \mathbf{P}_{N,A}^{l-1} \mathbf{P}_A$ given the weight increments

$\{\Delta w_1, \ldots, \Delta w_m\}$. Thus, the l-th power transition probability matrix increment $\Delta \mathbf{P}_A^l$ is given by [ZCX10]:

$$
\begin{aligned}
\Delta \mathbf{P}_A^l &= \mathbf{P}_{N,A}^l - \mathbf{P}_A^l = \begin{bmatrix} \Delta \mathbf{P}_{V_l} & \Delta \mathbf{A}_l \\ \Delta \mathbf{B}_l & \Delta \mathbf{C}_l \end{bmatrix} \\
&= \begin{bmatrix} \Delta \mathbf{P}_{V_{l-1}} \mathbf{P}_{V_1} + \Delta \mathbf{A}_{l-1} \mathbf{B}_1 & \mathbf{P}_{V_{l-1}} \Delta \mathbf{A}_1 + \Delta \mathbf{P}_{V_{l-1}} \mathbf{A}_{N,1} \\ \Delta \mathbf{P}_{B_{l-1}} \mathbf{P}_{V_1} + \Delta \mathbf{C}_{l-1} \mathbf{B}_1 & \mathbf{B}_{l-1} \Delta \mathbf{A}_1 + \Delta \mathbf{B}_{l-1} \mathbf{A}_{N,1} \end{bmatrix},
\end{aligned}
\tag{11.6.11}
$$

where $\Delta \mathbf{A}_1$ is given by (11.6.9).

The steps of the incremental algorithm Inc-Cluster are summarized in Algorithm 11.6.1.

Algorithm 11.6.1: Inc-Cluster algorithm [ZCX10]

Input: Original matrices $\mathbf{R}_A$, $\mathbf{P}_A$, $\mathbf{A}_l$, and the attribute edge weight increments $\{\Delta w_1, \ldots, \Delta w_m\}$.

Output: The new random walk distance matrix $\mathbf{R}_{N,A}$.

1. Estimate $\Delta \mathbf{P}_{A_1}$ according to Eq. (11.6.8).

2. Estimate the increment unified random walk distance matrix $\Delta \mathbf{R}_A = c\,(1-c)\,\Delta \mathbf{P}_A^l$.

3. For each $l = 2, \ldots, L$:

   a. Estimate $\Delta \mathbf{P}_A^l$ according to Eq. (11.6.11).

   b. Estimate $\Delta \mathbf{R}_A = \Delta \mathbf{R}_A + c\,(1-c)^l\,\Delta \mathbf{P}_A^l$.

4. Estimate $\mathbf{R}_{N,A} = \mathbf{R}_A + \Delta \mathbf{R}_A$.

11.7 Conclusions

In this chapter, state-of-the-art methods for big social network analysis have been surveyed and their corresponding dynamic alternatives for capturing the network dynamics have been studied. Moreover, their optimized counterparts that efficiently handle large volumes of data have also been outlined. These methods span latent model adaptation, tensor model adaptation, and spectral clustering. Some representative applications entailing social data analysis have also been described, namely incremental label propagation and incremental graph clustering on an attribute augmented graph.

Latent models have the ability to reveal the latent information inherent in the data. Three models have been studied, namely the LSA, the PLSA and the LDA. The incremental LSA methods build on the incremental approaches of the PSVD decomposition (i.e., SVD folding-in, SVD updating). PLSA methods include the PLSA folding-in, the incremental PLSA, the QB PLSA and MAP PLSA, and the oPLSA. The PLSA folding-in exploits an incremental variant of the EM algorithm. The incremental PLSA is based on the asymmetric and the symmetric formulations of PLSA folding-in. A Bayesian PLSA framework with a Dirichlet density kernel prior is assumed in the QB PLSA and the MAP PLSA methods. In the oPLSA, the PLSA model parameters are updated by deriving from first principles the corresponding probability estimates between successive EM steps. The incremental PLSA and the oPLSA are formulated on a moving window framework allowing both addition and deletion of new data (i.e., documents and words), while oPLSA additionally handles out-of-vocabulary words in a systematic way. The LDA updating methods vary depending on the optimization assumption or the inference assumption made in order to compute the posterior distribution of the LDA generative model. The online LDA, that resorts to Variational Bayesian Inference methods, is based on an online stochastic optimization that can also be applied on mini-batches of data. Considering Markov Chain Monte Carlo based methods, the o-LDA uses batch LDA (i.e., batch Gibbs sampler) to get initial estimates on an initial data subset and then uses collapsed Gibbs sampling on previous data. o-LSA may result in poor inference, when the initialization of LDA is not based on a representative subset (i.e., documents) of the full dataset. Skipping the batch initialization phase, incremental LDA produces more robust estimates by entailing resampling by means of the incremental Gibbs sampler.

Spectral clustering methods detect the structure of the data distributions by exploiting the information inherent in the data relationships (expressed in terms of similarity). In an evolving data framework, incremental spectral methods have been described that avoid re-computations by simulating the change of eigensystem or by extracting representative points to compress the dataset. An incremental spectral clustering method simulates the change of eigensystem and updates the eigenvectors of the generalized eigenproblem by finding the derivatives on the eigenvalues/vectors with respect to perturbations in all the quantities involved. A self-adaptive incremental spectral clustering method is also built on an iterative procedure that works on a few representative points for each cluster. This method provides the ability to increase the clusters number by applying a similarity threshold criterion between the cluster representative and any points in the cluster. In Incremental Approximate Spectral Clustering, the positive semi-definite shifted Laplacian is approximated with eigenupdate methods based on the SVD updating.

Tensor models can be efficiently used to represent the multi-faceted multi-dimensional nature of social data. For evolving social data, the incremental tensor analysis framework has been presented that entails dynamic, streaming, and window-based tensor analysis. In dynamic tensor and window-based analyses, an incremental update of the covariance matrices is used, while in streaming tensor analysis an incremental update of projection matrices is employed by means of the SPIRIT algorithm. In dynamic tensor and streaming tensor methods, a forgetting factor that fades away the variance matrices of earlier steps is employed. This is not the case in the window-based tensor methods, which entail a sliding window with the same weight to all timestamps in the window in order to handle time changes.

All the aforementioned methods (i.e., latent variable models, spectral clustering, and tensor models) and their incremental solutions can efficiently handle big data. However, approaches other than the incremental ones that resort to optimization techniques of the baseline algorithms have been formulated and implemented within a parallel distributed framework may be more suitable and efficient for processing big data, especially when online-processing is not the main issue. Optimization techniques have mainly been proposed for spectral clustering, including sparsification of the affinity matrix, approximate solutions for eigen-decomposition, and data dimensionality reduction by means of clustering or random projection. Latent variable models and tensors can efficiently handle big data by means of parallelization, where the baseline algorithms are formulated for paralllel programming models (i.e., MPI, OpenMP and MapReduce) that are widely used in distributed environments.

Finally, the incremental counterparts of the Label Propagation Algorithm (LPA), which is widely used to reveal communities, and of the SA-Cluster algorithm, which exploits the attribute augmented graph, have been outlined. Original LPA is a global algorithm that assigns to each node the label shared by most of its neighbors, while its incremental variant, Incremental LPA, treats the problem locally by discretizing the time domain into equal sized time intervals and taking into consideration only the nodes (edges) that change between two time intervals. The SA-Cluster algorithm combines both structural and attribute graph similarities by exploiting a neighborhood random walk model on the attribute augmented graph. Its variant, Inc-Cluster, incrementally updates the random walk distances given the edge weight increments by dividing the transition probability matrix into submatrices and incrementally updates each one.

Acknowledgment

This research has been co-financed by the European Union (European Social Fund - ESF) and Greek national funds through the Operation Program "Education and Lifelong Learning" of the National Strategic Reference Framework (NSRF) - Research Funding Program: THALIS-UOA-ERASITECHNIS MIS 375435.

Bibliography

[AMS01]   D. Achlioptas, F. McSherry, and B. Schölkopf. Sampling techniques for kernel methods. In *Proc. Advances in Neural Information Processing Systems*, pages 335–342, 2001.

[ASW08]   A. Asuncion, P. Smyth, and M. Welling. Asynchronous distributed learning of topic models. In *Proc. Advances in Neural Information Processing Systems*, pages 81–88, 2008.

[Att00]   H. Attias. A variational Bayesian framework for graphical models. In *Proc. Advances in Neural Information Processing Systems*, pages 209–215, 2000.

[AWST09]   A. Asuncion, M. Welling, P. Smyth, and Y.W. Teh. On smoothing and inference for topic models. In *Proc. 25th Conference on Uncertainty in Artificial Intelligence*, pages 27–34, 2009.

[BB07]   A. Banerjee and S. Basu. Topic models over text streams: a study of batch and online unsupervised learning. In *Proc. 7th SIAM International Conference on Data Mining*, pages 431–436, 2007.

[BB08]   L. Bottou and O. Bousquet. The tradeoffs of large scale learning. In *Proc. Advances in Neural Information Processing Systems*, pages 161–168, 2008.

[BDO94]   M. W. Berry, S. T. Dumais, and G. W. O'Brien. Using linear algebra for intelligent information retrieval. Technical Report UT-CS-94-270, 1994.

[BHK98]   J. S. Breese, D. Heckerman, and C. Kadie. Empirical analysis of predictive algorithms for collaborative filtering. In *Proc. 14th Conference on Uncertainty in Artificial Intelligence*, pages 43–52, 1998.

[BK11]   N. Bassiou and C. Kotropoulos. Long distance bigram models applied to word clustering. *Pattern Recognition*, 44(1):145–158, 2011.

[BK14]   N. Bassiou and C. Kotropoulos. Online PLSA: Batch updating techniques including out-of-vocabulary words. *IEEE Transactions on Neural Networks and Learning Systems*, 25(11):1953–1966, 2014.

[BNJ03]   D. M. Blei, A. Y. Ng, and M. I. Jordan. Latent Dirichlet allocation. *Journal of Machine Learning Research*, 3(5):993–1022, 2003.

[Bra05]   T. Brants. Test data likelihood for PLSA models. *Information Retrieval*, 8(2):181–196, 2005.

[BTHC06] T. H. Brants, I. Tsochantaridis, T. Hofmann, and F. R. Chen. Methods, apparatus, and program products for performing incremental probabilistic latent semantic analysis. Patent No. 20060112128, 2006.

[CC08]   T. C. Chou and M. C. Chen. Using incremental PLSI for threshold-resilient online event analysis. *IEEE Transactions on Knowledge and Data Engineering*, 20(3):289–299, 2008.

[CC11]   X. Chen and D. Cai. Large scale spectral clustering with landmark-based representation. In *Proc. 26th Conference on Artificial Intelligence*, pages 313–318, 2011.

[CDW13]   W. M. Campbell, C. K. Dagli, and C. J. Weinstein. Social network analysis with content and graphs. *Lincoln Laboratory Journal*, 20(1):62–81, 2013.

[CGL+06] B. Chen, B. Gao, T.-Y. Liu, Y.-F. Chen, and W.-Y. Ma. Fast spectral clustering of data using sequential matrix compression. In *Proc. 7th European Conf. Machine Learning (ECML'06)*, volume 4212 of *Lecture Notes in Computer Science*, pages 590–597. Springer, 2006.

[CH00]   D. Cohn and T. Hofmann. The missing link – a probabilistic model of document content and hypertext connectivity. In *Proc. Advances in Neural Information Processing Systems*, 2000.

[Chu97]   F. R. K. Chung. *Spectral Graph Theory*, volume 92. American Mathematical Society, 1997.

[CMD+00] R. Chandra, R. Menon, L. Dagum, D. Kohr, D. Maydan, and J. McDonald. *Parallel Programming in OpenMP*. Morgan Kaufmann, 2000.

[CSBL11] W. Y. Chen, Y. Song, H. Bai, and C. J. Lin. Parallel spectral clustering in distributed systems. *IEEE Transactions on Pattern Analysis and Machine Intelligence*, 33(3):568–586, 2011.

[CSG09]   K. Canini, L. Shi, and T. Griffiths. Online inference of topics with latent Dirichlet allocation. In *Proc. Internatinal Conference on Artificial Intelligence and Statistics*, volume 5, pages 65–72, 2009.

[CW08]   J. T. Chien and M. S. Wu. Adaptive Bayesian latent semantic analysis. *IEEE Transactions on Audio, Speech, and Language Processing*, 16(1):198–207, 2008.

[CY05]   H. Chang and D. Yeung. Robust path-based spectral clustering with application to image segmentation. In *Proc. International Conference on Computer Vision*, pages 278–285, 2005.

[CZG09]   J. Chen, O.R. Zaiane, and R. Goebel. A visual data mining approach to find overlapping communities in networks. In *Proc. International Conference on Advances in Social Networks Analysis and Mining*, pages 338–343, 2009.

[DDF+90]   S. Deerwester, S. T. Dumais, G. W. Furnas, T. K. Landauer, and R. Harshman. Indexing by latent semantic analysis. *Journal American Society Information Science*, 41(6):391–407, 1990.

[DG08]   J. Dean and S. Ghemawat. MapReduce: Simplified data processing on large clusters. *ACM Communications*, 51(1):107–113, 2008.

[DGC14]   C. Dhanjal, R. Gaudel, and S. Clémencon. Efficient eigen-updating for spectral graph clustering. *Neurocomputing*, 131:440–452, 2014.

[DGR07]   A. S. Das, M. Datarand A. Garg, and S. Rajaram. Google news personalization: Scalable online collaborative filtering. In *Proc. 16th International Conference on World Wide Web*, pages 271–280, 2007.

[Din04]   C. Ding. A tutorial on spectral clustering. In *Proc. International Conference Machine Learning*, 2004.

[DK04]   M. Deshpande and G. Karypis. Item-based top-n recommendation. *ACM Transactions on Information Systems*, 22(1):143–177, 2004.

[DLR77]   A. Dempster, N. Laird, and D. Rubin. Maximum likelihood from incomplete data via the EM algorithm (with discussion). *Journal Royal Statistical Society, Series B*, 39:1–38, 1977.

[FBCM04]   C. Fowlkes, S. Belongie, F. Chung, and J. Malik. Spectral grouping using the Nyström method. *IEEE Transactions on Pattern Analysis and Machine Intelligence*, 26(2):214–225, 2004.

[FbS15]   Facebook newsroom: Statistics. http://newsroom.fb.com/company-info/, 2015 (accessed March 27, 2015).

[FLG00]   G. W. Flake, S. Lawrence, and C. L. Giles. Efficient identification of web communities. In *Proc. 2000 ACM SIGKDD Conference on Knowledge Discovery and Data Mining*, pages 150–160, 2000.

[For10]   S. Fortunato. Community detection in graphs. *Physics Reports*, 486:75–174, 2010.

[GH99]   D. Gildea and T. Hofmann. Topic-based language models using EM. In *Proc. 6th European Conference on Speech Communication and Technology*, pages 2167–2170, 1999.

[GLMY11]   U. Gargi, W. Lu, V. Mirrokni, and S. Yoon. Large-scale community detection on youtube for topic discovery and exploration. In *Proc. 5th International AAAI Conference on Weblogs and Social Media*, pages 486–489, 2011.

[GN02]   M. Girvan and M. E. J. Newman. Community structure in social and biological networks. *Proc. National Academy of Sciences of the United States of America*, 99(12):7821–7826, 2002.

[GS04]   T. L. Griffiths and M. Steyvers. Finding scientific topics. *National Academy of Sciences of the USA*, 101(1):5228–5235, 2004.

[HBB10]   M. D. Hoffman, D. M. Blei, and F. Bach. Online learning for latent Dirichlet allocation. In *Proc. Advances in Neural Information Processing Systems*, volume 23, pages 856–864, 2010.

[HCZS08]   C. Hong, Y. Chen, W. Zheng, and J. Shan. Parallelization and characterization of probabilistic latent semantic analysis. In *Proc. 37th International Conference on Parallel Computing*, pages 628–635, 2008.

[HK92]   L. Hagen and A. B. Kahng. New spectral methods for ratio cut partitioning and clustering. *IEEE Transactions on Computer-Aided Design*, 11(9):1074–1085, 1992.

[Hof01]   T. Hofmann. Unsupervised learning by probabilistic latent semantic analysis. *Machine Learning*, 42(1-2):177–196, 2001.

[Hof04]   T. Hofmann. Latent semantic models for collaborative filtering. *ACM Transactions on Information Systems*, 22(1):89–115, 2004.

[HP98]   T. Hofmann and J. Puzicha. Unsupervised learning from dyadic data. Technical Report TR-98-042, International Computer Science Institute, 1998.

[IN14]   N. Ifada and R. Nayak. Tensor-based item recommendation using probabilistic ranking in social tagging systems. In *Proc. 23rd International Conference on World Wide Web companion*, pages 805–810, 2014.

[JB03]   Michael I. Jordan and Francis R. Bach. Learning spectral clustering. In *Proc. Advances in Neural Information Processing Systems 16*, 2003.

[JGJS99]   M. Jordan, Z. Ghahramani, T. Jaakkola, and L. Saul. Introduction to variational methods for graphical models. *Machine Learning*, 37:183–233, 1999.

[JGS+11]   Y. Jin, Y. Gao, Y. Shi, L. Shang, R. Wang, and Y. Yang. P^2LSA and P^2LSA+: Two paralleled probabilistic latent semantic analysis algorithms based on the MapReduce model. In *Proc. 12th International Conference on Intelligent Data Engineering and Automated Learning*, volume 6936, pages 385–393, 2011.

[JNCJ08]   H. Jiang, T.N. Nguyen, I. Chen, and H. Jaygarl. Incremental latent semantic indexing for automatic traceability link evolution management. In *Proc. IEEE/ACM International Conference on Automated Software Engineering*, pages 59–68, 2008.

[JX06]   X. Ji and W. Xu. Document clustering with prior knowledge. In *Proc. 29th International ACM SIGIR Conference on Research and Development in Information Retrieval*, pages 405–412, 2006.

[JZM04]   X. Jin, Y. Zhou, and B. Mobasher. Web usage mining based on probabilistic latent semantic analysis. In *Proc. 2004 ACM SIGKDD Conference Knowledge Discovery and Data Mining (KDD'04)*, pages 197–205, 2004.

[KBK05]   T. G. Kolda, B. W. Bader, and J. P. Kenny. Higher-order web link analysis using multi-linear algebra. In *Proc. 5th IEEE International Conference on Data Mining*, pages 242–249, 2005.

[KMF11]   U. Kang, B. Meeder, and C. Faloutsos. Spectral analysis for billion-scale graphs: Discoveries and implementation. In *Proc. 15th Pacific-Asia Conference*, volume 6635 of *Lecture Notes in Computer Science*, pages 13–25. Springer, 2011.

[KPHF12] U. Kang, E. Papalexakis, A. Harpale, and C. Faloutsos. Gigatensor: scaling tensor analysis up by 100 times - algorithms and discoveries. In *Proc. 18th ACM SIGKDD International Conference on Knowledge Discovery and Data Mining*, pages 316–324, 2012.

[KPSC10] I. A. Kovács, R. Palotai, M. S. Szalay, and P. Csermeley. Community landscapes: An integrative approach to determine overlapping network module hierarchy, identify key nodes and predict network dynamics. *PLoS ONE*, 5(9):e12528, 2010.

[KTS11] T. Kong, Y. Tian, and H. Shen. A fast incremental spectral clustering for large data sets. In *Proc. IEEE International Conference on Parallel and Distributed Computing, Applications, and Technologies (PDCAT)*, pages 1–5, 2011.

[LCSX11] Y.-R. Lin, K. S. Candan, H. Sundaram, and L .Xie. SCENT: Scalable compressed monitoring of evolving multi-relational social networks. *ACM Transactions on Multimedia Computing, Communications, and Applications*, 75(1):29:1–29:22, 2011.

[LDv12] M. Leginus, P. Dolog, and V. Žemaitis. Improving tensor based recommenders with clustering. In *User Modeling, Adaptation, and Personalization*, volume LNCS 7379, pages 151–163. Springer, Berlin / Heidelberg, 2012.

[LHLC09] I. X. Y. Leung, P. Hui, P. Lio, and J. Crowcroft. Towards real-time community detection in large networks. *Physical Review E*, (066107), 2009.

[LK09] P. Liang and D. Klein. Online EM for unsupervised models. In *Proc. of Human Language Technologies: The 2009 Annual Conference of the North American Chapter of the Association for Computational Linguistics*, pages 611–619, 2009.

[LPV13] H. Lu, K. Plataniotis, and A. Venetsanopoulos. *Multilinear Subspace Learning: Dimensionality Reduction of Multidimensional Data*. Taylor and Francis, 2013.

[LSY03] G. Linden, B. Smith, and J. York. Amazon.com recommendations: Item-to-item collaborative filtering. *IEEE Internet Computing*, pages 76–80, 2003.

[Lux07] U. Von Luxburg. A tutorial on spectral clustering. *Statistics and Computing*, 17(4):395–416, 2007.

[LV00] L. D. Lathauwer and J. Vandewalle. On the best rank-1 and rank-$(r_1, r_2, \ldots, r_n)$ approximation of higher-order tensors. *SIAM Journal on Matrix Analysis and Applications*, 21(4):1324–1342, 2000.

[LWDH13] J. Liu, C. Wang, M. Danilevsky, and J. Han. Large-scale spectral clustering on graphs. In *Proc. 23rd International Joint Conference on Artificial Intelligence*, 2013.

[LY08] N. N. Liu and Q. Yang. Eigenrank: a ranking-oriented approach to collaborative filtering. In *Proc. 31st ACM SIGIR Conference on Research and Development in Information Retrieval*, pages 83–90, 2008.

[LYZ+07] T.-Y. Liu, H.-Y. Yang, X. Zheng, T. Qin, and W.-Y. Ma. Fast large-scale spectral clustering by sequential shrinkage optimization. In *Proc. 29th European Conf. IR (ECIR07)*, volume 4425 of *Lecture Notes in Computer Science:*, pages 319–330. Springer, 2007.

[LZCS11] Z. Liu, Y. Zhang, E. Y. Chang, and M. Sun. PLDA+: Parallel latent dirichlet allocation with data placement and pipeline processing. *ACM Transactions on Intelligent Systems and Technology*, 2(3):26:1–26:18, 2011.

[MB88] G. McLachlan and K. E. Basford. *Mixture Models*, volume 84. Marcel Dekker Inc., 1988.

[MM07] D. Mimno and A. McCallum. Organizing the OCA: Learning faceted subjects from a library of digital books. In *Proc. ACM/IEEE Joint Conference on Digital Libraries*, pages 376–385, 2007.

[MS96] K. Maschhoff and D. Sorensen. A portable implementation of ARPACK for distributed memory parallel architectures. In *Proc. Copper Mountain Conference on Iterative Methods*, 1996.

[MTP10] A. Maronidis, A. Tefas, and I. Pitas. Frontal view recognition using spectral clustering and subspace learning methods. In *Proc. International Conference Artificial Neural Networks*, volume 6352 of *Lecture Notes in Computer Science: Proc. International Conference on Artificial Intelligence*, pages 460–469. Springer, 2010.

[MZ04] B. Marlin and R. Zemel. The multiple multiplicative factor model for collaborative filtering. In *Proc. 21st International Conference on Machine Learning*, volume 69, pages 576–583, 2004.

[MZL+11] H. Ma, D. Zhou, C. Liu, M. R. Lyu, and I. King. Recommender systems with social regularization. In *Proc. 4th ACM International Conference on Web Search and Data Mining*, pages 287–296, 2011.

[NASW07] D. Newman, A. Asuncion, P. Smyth, and M. Welling. Distributed inference for latent Dirichlet allocation. In *Proc. Advances in Neural Information Processing Systems*, pages 1081–1088, 2007.

[NASW09] R. Newman, A. Asuncion, P. Smyth, and M. Welling. Distributed algorithms for topic models. *Journal of Machine Learning Research*, 10:1801–1828, 2009.

[NCL07] R. Nallapati, W. Cohen, and J. Lafferty. Parallelized variational EM for latent dirichlet allocation: An experimental evaluation of speed and scalability. In *Proc. 7th IEEE International Conference on Data Mining Workshops*, pages 349–354, 2007.

[New07] M. E. J. Newman. Finding community structure in networks using the eigenvectors of matrices. *Physical Review E*, (036104), 2007.

[NG04] M. E. J. Newman and M. Girvan. Finding and evaluating community structure in networks. *Physical Review E*, (026113), 2004.

[NH98] R. M. Neal and G. E. Hinton. A view of the EM algorithm that justifies incremental, sparse, and other variants. In M. I. Jordan, editor, *Learning in Graphical Models*, pages 355–368. Kluwer Academic Publishers, 1998.

[NJW01] A. Ng, M. Jordan, and Y. Weiss. On spectral clustering: Analysis and an algorithm. In *Proc. Advances in Neural Information Processing Systems*, pages 849–856, 2001.

[NXC+07]   H. Ning, W. Xu, Y. Chi, Y. Gong, and T. Huang. Incremental spectral clustering with application to monitoring of evolving blog communities. In *Proc. SIAM International Conference on Data Mining*, pages 261–272, 2007.

[NXC+10]   H. Ning, W. Xu, Y. Chi, Y. Gong, and T. S. Huang. Incremental spectral clustering by efficiently updating the eigen-system. *Pattern Recognition*, 43(1):113–127, 2010.

[O'B94]   G. W. O'Brien. Information management tools for updating an SVD-encoded indexing scheme. Technical Report UT-CS-94-258, 1994.

[PCW09]   S. Pang, C. Chen, and T. Wei. A realtime clique detection algorithm: Time-based incremental label propagation. In *Proc. International Conference on Intelligent Information Technology Application*, volume 3, pages 459–462, 2009.

[PDFV05]   G. Palla, I. Derenyi, I. Farkas, and T. Vicsek. Uncovering the overlapping community structure of complex networks in nature and society. *Nature*, 435(7043):814–818, 2005.

[PKVS12]   S. Papadopoulos, Y. Kompatsiaris, A. Vakali, and P. Spyridonos. Community detection in social media performance and application considerations. *Springer Data Mining and Knowledge Discovery*, 24:515–554, 2012.

[PL05]   P. Pons and M. Latapy. Computing communities in large networks using random walks. In *Proc. Computer and Information Sciences*, volume LNCS 3733, pages 284–293, 2005.

[PM05]   W. Pentney and M. Meila. Spectral clustering of biological sequence data. In *Proc. National Conference on Artificial Intelligence (AAAI)*, volume 2, pages 845–850, 2005.

[PSF05]   S. Papadimitriou, J. Sun, and C. Faloutsos. Streaming pattern discovery in multiple time-series. In *Proc. 31st International Conference on Very Large Data Bases*, pages 697–708, 2005.

[PZK04]   J. Park, H. Zha, and R. Kasturi. Spectral clustering for robust motion segmentation. In *Proc. European Conference on Computer Vision*, pages 390–401, 2004.

[RAK07]   U. N. Raghavan, R. Albert, and S. Kumara. Near linear time algorithm to detect community structures in large-scale networks. *Physical Review E*, pages 036106+, 2007.

[RD13]   D. Rafailidis and P. Daras. The TFC model: Tensor factorization and tag clustering for item recommendation in social tagging systems. *IEEE Transactions on Systems, Man and Cybernetics, Part A: Systems and Humans*, 43(3):673–688, 2013.

[RIS+94]   P. Resnick, N. Iacovou, M. Suchak, P. Bergstrom, and J. Riedl. Grouplens: An open architecture for collaborative filtering of netnews. In *Proc. 1994 ACM Conference on Computer Supported Cooperative Work*, pages 175–186, 1994.

[RST10]   S. Rendle and L. Schmidt-Thieme. Pairwise interaction tensor factorization for personalized tag recommendation. In *Proc. 3rd ACM International Conference on Web Search and Data Mining*, pages 81–90, 2010.

[SBH02]   G. Shani, R. Brafman, and D. Heckerman. An MDP-based recommender system. In *Proc. 18th Conference on Uncertainty in Artificial Intelligence*, volume 6, pages 1265–1295, 2002.

[SFSZ05]   N. Speer, H. Fröhlich, C. Spieth, and A. Zell. Functional grouping of genes using spectral clustering and gene ontology. In *Proc. International Joint Conference on Neural Networks*, pages 298–303, 2005.

[SI09]   T. Sakai and A. Imiya. Fast spectral clustering with random projection and sampling. In *Proc. 3rd International Conference on Machine Learning and Data Mining*, volume 5632 of *Lecture Notes in Computer Science*, pages 372–384. Springer, 2009.

[SKKR00]   B. Sarwar, G. Karypis, J. Konstan, and J. Reidl. Application of dimensionality reduction in recommender systems – a case study. In *Proc. ACM-SIGKDD Conference on Knowledge Discovery in Databases*, pages 265–285, 2000.

[SKKR01]   B. Sarwar, G. Karypis, J. Konstan, and J. Reidl. Item-based collaborative filtering recommendation algorithms. In *Proc. 10th International Conference on World Wide Web*, 2001.

[SLTS05]   X. Song, C.-Y. Lin, B. L. Tseng, and M.-T. Sun. Modeling and predicting personal information dissemination behavior. In *Proc. 11th ACM SIGKDD International Conference on Knowledge Discovery and Data Mining*, pages 479–488, 2005.

[SM00]   J. Shi and J. Malik. Normalized cuts and image segmentation. *IEEE Transactions on Pattern Analysis and Machine Intelligence*, 22(8):888–905, 2000.

[SNM10]   P. Symeonidis, A. Nanopoulos, and Y. Manolopoulos. A unified framework for providing recommendations in social tagging systems based on ternary semantic analysis. *IEEE Transactions on Knowledge and Data Engineering*, 22(2):179–192, 2010.

[SP09]   V. Satuluri and S. Parthasarathy. Scalable graph clustering using stochastic flows: Applications to community discovery. In *Proc. 2009 ACM SIGKDD Conference on Knowledge Discovery and Data Mining*, pages 737–745, 2009.

[SS08]   H. Shinnou and M. Sasaki. Spectral clustering for a large data set by reducing the similarity matrix size. In *Proc. 6th International Conference on Language Resources and Evaluation*, pages 201–204, 2008.

[STF06]   J. Sun, D. Tao, and C. Faloutsos. Beyond streams and graphs: Dynamic tensor analysis. In *Proc. 12th ACM SIGKDD International Conference on Knowledge Discovery and Data Mining*, pages 374–383, 2006.

[STPF08]   J. Sun, D. Tao, S. Papadimitriou, and C. Faloutsos. Incremental tensor analysis: Theory and applications. *ACM Transactions on Knowledge Discovery from Data*, 2(3):11:1–11:37, 2008.

[TRG05]   R. Thakur, R. Rabenseinfer, and W. Gropp. Improving the performance of collective operations in MPICH. *International Journal of High Performance Computing Applications*, 19(1):49–66, 2005.

[TS07]   J. E. Tougas and R. J. Spiteri. Updating the partial singular value decomposition in latent semantic indexing. *Computational Statistics and Data Analysis*, 52:174–183, 2007.

[Tuc66]   L. R. Tucker. Some mathematical notes on three-mode factor analysis. *Psychometrika*, 31:279–311, 1966.

[Twi15]   Twitter usage. https://about.twitter.com/company, 2015 (accessed March 27, 2015).

[VDL07]   C. Valgren, T. Duckett, and A. Lilienthal. Incremental spectral clustering and its application to topological mapping. In *Proc. IEEE International Conference on Robotics and Automation*, pages 4283–4288, 2007.

[VM03]   D. Verma and M. Meila. A comparison of spectral clustering algorithms. Technical report, University of Washington, 2003.

[WAB12]   Y. Wang, E. Agichtein, and M. Benzi. TM-LDA: efficient online modeling of latent topic transitions in social media. In *Proc. ACM SIGKDD Conference on Knowledge Discovery and Data Mining*, pages 123–131, 2012.

[WAM09]   R. Wan, V. N. Ahn, and H. Mamitsuka. Efficient probabilistic latent semantic analysis through parallelization. In *Proc. 5th Asia Information Retrieval Symposium on Information Retrieval Technology*, pages 432–443, 2009.

[WBSC09]   Y. Wang, H. Bai, M. Stanton, and E. Chang. PLDA: Parallel latent dirichlet allocation for large-scale applications. In *Proc. International Conference on Algorithmic Aspects in Information and Management*, volume 5564 of *Lecture Notes in Computer Science*, pages 301–314. Springer, 2009.

[WC89]   Y. C. Wei and C. K. Cheng. Towards efficient hierarchical designs by ratio cut partitioning. In *Proc. International Conference on Computer Aided Design*, pages 298–301, 1989.

[Wei99]   Y. Weiss. Segmentation using eigenvectors: A unifying view. In *Proc. International Conference on Computer Vision*, volume 2, pages 975 – 982, 1999.

[WYL+09]   X. Wu, J. Yan, N. Liu, S. Yan, Y. Chen, and Z. Chen. Probabilistic latent semantic user segmentation for behavioral targeted advertising. In *Proc. 3rd International Workshop on Data Mining and Audience Intelligence for Advertising*, pages 10–17, 2009.

[WZWC08]   H. Wu, D. Zhang, Y. Wang, and X. Cheng. Incremental probabilistic latent semantic analysis for automatic question recommendation. In *Proc. ACM Conference on Recommender Systems*, pages 99–106, 2008.

[XYFS07]   X. Xu, N. Yuruk, Z. Feng, and T. A. J. Schweiger. Scan: a structural clustering algorithm for networks. In *Proc. 2007 ACM SIGKDD Conference on Knowledge Discovery and Data Mining*, pages 824–833, 2007.

[XYW+09]   J. Xu, G. Ye, Y. Wang, G. Herman, B. Zhang, and J. Yang. Incremental EM for probabilistic latent semantic analysis on human action recognition. In *Proc. IEEE International Conference on Advanced Video and Signal Based Surveillance*, pages 55–60, 2009.

[XYW+11] J. Xu, G. Ye, Y. Wang, W. Wang, and J. Yang. Online learning for PLSA-based visual recognition. In R. Kimmel, R. Klette, and A. Sugimoto, editors, *Computer Vision – ACCV 2010*, volume LNCS 6493, pages 95–108. Springer, 2011.

[XZMZ05] G. Xu, Y. Zhang, J. Ma, and X. Zhou. Discovering user access pattern based on probabilistic latent factor model. In *Proc. 16th Australasian Database Conference*, volume 39, pages 27–35, 2005.

[YHP08] Y.Tian, R. A. Hankins, and J. M. Patel. Efficient aggregation for graph summarization. In *Proc. 2008 ACM SIGMOD International Conference on Management of Data*, pages 567–580, 2008.

[YMM09] L. Yao, D. Mimno, and A. McCallum. Efficient methods for topic model inference on streaming document collections. In *Proc. 15th ACM SIGKDD International Conference on Knowledge Discovery and Data Mining*, pages 937–946, 2009.

[You15] Youtube press: Statistics. https://www.youtube.com/yt/press/statistics.html, 2015 (accessed March 27, 2015).

[YXQ09] F. Yan, N. Xu, and Y. Qi. Parallel inference for latent Dirichlet allocation on graphics processing units. In *Advances in Neural Information Processing Systems 22*, pages 2134–2142. 2009.

[ZCX10] Y. Zhou, H. Cheng, and J. Xu. Clustering large attributed graphs: An efficient incremental approach. In *Proc. IEEE International Conference on Data Mining*, pages 689–698, 2010.

[ZCY09] Y. Zhou, H. Cheng, and J. X. Yu. Graph clustering based on structural/attribute similarities. In *Proc. Very Large Data Bases Endowment*, pages 718–729, 2009.

[ZHD01] H. Zha, X. He, and C. H. Q. Ding. Spectral relaxation for k-means clustering. In *Proc. Advances in Neural Information Processing Systems*, pages 1057–1064, 2001.

[ZK07] Y. Zhang and J. Koren. Efficient Bayesian hierarchical user modeling for recommendation system. In *Proc. 30th ACM SIGIR Conference on Research and Development in Information Retrieval*, pages 47–54, 2007.

[ZS99] H. Zha and H. D. Simon. On updating problems in latent semantic indexing. *SIAM Journal Scientific Computing*, 21(2):782–791, 1999.

Chapter 12

Big Graph Storage, Processing and Visualization

Jaroslav Pokorny

Charles University, Czech Republic

Vaclav Snasel

VSB Technical University, Czech Republic

12.1 Introduction

Graphs are the most common abstract structure encountered in computer science and are widely used for abstract information representation. Any system that consists of discrete states (or sites) and connections between them can be modeled by a graph.

Graph data processing is an important topic of research in the database area. Graph-based logical models were studied in the context of object-oriented databases [Kim90] or graph-oriented logical database models in the 1980s [KV84]. The research of graph databases was popular in the 1990s with database models like GOOD [GPVdBVG94], GraphDB [Güt94] and graph query languages like GraphLog [CM90] and G [CMW87].

Graphs are ubiquitous in many areas of human activity. We can find graph data in areas like biology, software bug detection, information security, even in enterprise data, such as hierarchies of products or bill of material, financial data, and in a knowledge management is recent years. Other domains include the Web graph, social networks, and the Semantic Web.

With new Web applications a lot of graph data are generated and new requirements on their processing have emerged. For example, in the Web, Internet resources are represented as a graph of triples, subject-predicate-object, according to a standard model for data interchange on the Web RDF http://www.w3.org/RDF/ (retrieved on 14.7.2014). Compared to previous graph data, the approaches support of queries over structural properties of graphs is needed. In the last decade, the challenges in graph data management were influenced by the advent of large datasets called Big Data [SG14]. Large graphs (in the order of billions of nodes, edges, and attributes) have become increasingly important in the world of Big Data. Following the current terminology of Big Data, we will talk about Big Graphs in this chapter.

Big Data has been a challenge for the relational database management systems (DBMS) in the past. These graphs were stored in databases to allow for efficient queries using declarative query languages such as SQL. Traditional relational DBMSs (e.g., MySQL and PostgreSQL) have long been used for this purpose, but with the growing number of nodes and edges these DBMSs proved unsuitable for processing such data. This is in accordance with the fact that graph data are more and more processed by data mining algorithms [AW10]. Most data mining algorithms do not operate directly in the Big Data. They just get the data out, do whatever they need to do, and then store the results.

Commercial database products have to reflect these facts. Bigdata http://www.systap.com/ (retrieved on 14.7.2014) DBMS, handling very Big Graphs, scaled to 50 billion edges on a single machine and will scale to even larger graphs with its horizontally-scaled architecture.

Traditional graph algorithms assume the input graph fits in the memory or disks of a single machine. Because today's graphs are measured in terabytes and heading toward petabytes, with more than billions of nodes and edges, old assumptions about graphs have to be changed. This chapter is going to be focused on modern methods of Big Graph data storage and processing, its parallelization, compression and visualization. Special attention is devoted to graph DBMSs (see, e.g., [SP12] or [RWWE13]). Graph databases are well suited, e.g. for social networking analytics where relationships bind multiple entities across a messy pattern that is hard to break into structures of traditional relational DBMS.

First it is necessary to introduce the notion of a graph database, i.e., to specify its basic unit of information. Similarly as it is with XML or text databases, either one graph or a collection of graphs can be considered as a *graph database*. Then as it is expected, a *graph DBMS* is a DBMS optimized for managing graphs, i.e. highly-relational data. As it is usual today, we will use (imprecisely) the concept of a database also in the meaning of a DBMS itself in this chapter.

There are many types of graphs, and consequently, a lot of possibilities of how to set up graph databases. Obviously, graphs can be represented as relations (tables) in a relational DBMS and processed with SQL. This approach could be called *graph-enabled databases*. However, the general purpose relational DBMS allows only a small opportunity for graph specific optimizations, since it breaks down the graph structures into individual relations. A more interesting approach produces *native graph databases*, which have capabilities to query some structural properties of graphs. In a more dynamic environment, scalability, both in data size and the number of users is required. Recently, NoSQL databases [Cat11, Cel13], particularly their subcategory graph databases, have gained much attention due to their advantages in scalability.

Section 12.2 introduces some basic notions concerning various types of graphs and operations over them, especially to query graph data. In Section 12.3 we discuss possibilities for storing graph data. We start with DBMS architectures in general, mainly NoSQL ones, and those appropriate for storing Big Graph data. The rest of section is devoted to graph

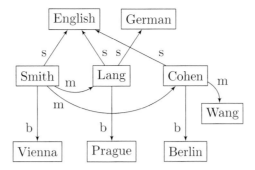

FIGURE 12.2.1: Labeled Digraph.

DBMSs and their data structures used for storing and indexing graphs. Section 12.4 deals with graph data processing, i.e. with possibilities to query graphs represented in different DBMSs. We consider relational DBMSs, Datalog implementations, and native graph databases. Section 12.5 describes graph data visualization. Conclusions summarize the chapter.

12.2   Basic Notions

Formally, a graph $G = (\mathcal{V}, \mathcal{E})$ is an ordered pair of a set of vertices $\mathcal{V} = \{v_i\}$ and a set of edges $\mathcal{E} \subset \mathcal{V} \times \mathcal{V}$. For more details see Chapter 2. Let v_i and v_j be two nodes from V; if $\{v_i, v_j\} \in \mathcal{E}$, then v_i and v_j are *adjacents*. In digraphs, for a node $u \in \mathcal{V}$, $v_1 \in \mathcal{V}$ is a *successor* of u if $(u, v_1) \in \mathcal{E}$. Similarly, $v_2 \in \mathcal{V}$ is a *predecessor* of u if $(v_2, u) \in \mathcal{E}$. Given a graph G, a *path* p is a sequence $v_0, v_1, \ldots, v_k$ of nodes that are connected by edges $(v_i, v_{i+1}), i = 1, \ldots, k - 1$. The node v_i is a *predecessor* of the node v_{i+1}. A *cycle* is a path p with $v_0 = v_k$ where $k > 1$. An *acyclic digraph* is a digraph with no cycles.

In many applications various properties associated with the edges and nodes are stored. Often *attributes* (*properties*) are allowed, i.e., both nodes and edges may have attributes, even a set of attributes. Attributes are expressed as couples (key, value). Sometimes both edges and nodes are labeled. In the simplest case, each directed edge is labeled with a symbol drawn from some finite alphabet Σ, hence $\mathcal{E} \subseteq \mathcal{V} \times \Sigma \times \mathcal{V}$. We talk about *property graphs* in this case. If edge labels are numbers, we can talk about *weighted graphs*.

Figure 12.2.1 shows a labeled digraph describing relationships of types *speaks* (s), *manages* (m), and *is_born_in* (b) in the domain containing entity types *Language*, *Person*, and *Town*. The digraph represents a fragment of a graph database. The graph representing *friendship* relationship on a social network like Facebook can be seen as an undirected graph. Weighting is useful in transit networks like roads and streets, or in flight networks. The weight of an edge can represent a distance or duration. Then, e.g., a weight of a path can be calculated as a sum of weights of their edges.

When multiple edges exist between the same nodes, we talk about *multigraphs*. An extension to the standard graph concept that allows an edge to point to more than two nodes is called *hypergraph*.

When a graph is considered as a database D, we have to define usual database operations enabling its manipulation, i.e. inserting, deleting, or changing graph parts, and mainly querying capabilities. Typical queries include *subgraph matching* (for a given graph find subgraphs of that match a given query pattern), *reachability* (can I get from u to v?), *shortest path* (find the quickest/shortest route from u to v), finding the *sum/min/max aggregations* over paths or subgraphs in a weighted graph. Other useful queries are finding the *immediate neighbors* of a node or its *out-neighbors* (*in-neighbors*) in digraphs.

In databases consisting of a collection of graphs, queries such as *supergraph matching* can be considered, e.g. queries like "Find graphs in D which are contained in the query graph Q" are possible. A *subgraph query* retrieves all those graphs in the database that are supergraphs of a Q.

Similarity subgraph matching is also useful. It deals with queries like "Find graphs in D which have some components of the query graph Q". Often similarity measure is useful in cases of noise and inconsistency in data. Well-known graph edit distance has three advantages: (1) it allows changes in both nodes and edges; (2) it reflects the topological information of graphs; and (3) it is sometimes a metric that can be applied to any type of graphs.

We will define formally some important query types on labeled graphs without association to real query languages appearing in today's graph databases. For formal query syntax we use a syntactic variant of Datalog rules (see, e.g. [Woo12]). The fact that there is the edge p between nodes u and v is expressed as (u, p, v). Let Σ be an alphabet.

A *conjunctive query over* Σ is an expression of the form:

$$Q(z_1, \ldots, z_n) \leftarrow (x_1, a_1, y_1), \ldots, (x_m, a_m, y_m), m \geq 1, \qquad (12.2.1)$$

where $m > 0$, x_i and y_i are node variables or constants ($1 \leq i \leq m$), $a_i \in \Sigma$ ($1 \leq i \leq m$), and z_i are either a x_j or y_j ($1 \leq i \leq n, 1 \leq j \leq m$), $Q(z_1, \ldots, z_n)$ is the *head* of the query; the expression on the right of the arrow is its *body*. The symbol "," between the triples denotes conjunction. Given a database D the *answer* $Q(D)$ *to* Q is a set of n-tuples of nodes satisfying its body.

If the head is $Q()$, then the query is of type *YES/NO*.

Example 1:

$$Q(x) \leftarrow (x, speaks, German), (x, speaks, English), (x, is\_born\_in, Prague). \qquad (12.2.2)$$

Using the labeled digraph from Figure 12.2.1, the associated database D contains facts (Lang, *speaks*, German), (Cohen, *speaks*, English), (Lang, *is_born_in*, Prague), etc. Then the answer to Q will be the set {Lang}.

A *regular path query over* Σ is an expression of the form:

$$Q(x, y) \leftarrow (x, r, y), \qquad (12.2.3)$$

where x and y are node variables, r is a regular expression over Σ. We use | for disjunction, · for concatenation, r^* for Kleene closure in regular expressions.

Given a database D the answer $Q(D)$ to Q is a set of couples of nodes, such that there is a path from x to y and the sequence of edge labels on this path satisfies r. When we restrict to matching only simple paths we talk about a *query with simple regular path over* Σ. A path is *simple*, if no node is repeated on it.

A *conjunctive regular path query over* Σ is an expression of the form:

$$Q(z_1, \ldots, z_n) \leftarrow (x_1, r_1, y_1), \ldots, (x_m, r_m, y_m), m \geq 1, \qquad (12.2.4)$$

where r_i is regular expression over Σ and z_i is a x_j or y_j.

Example 2:

We will add new relationship types *has_a_nationality*, *lives_in*, and *is_located_in* to our examples. They model various relationships between persons and towns, or between larger territorial units in the case of the transitive relation *lives_in*. The answer to

$$Q(x, y) \leftarrow (x, speaks, German), (x, speaks, English),$$
$$(x, has\_a\_nationality|((is\_born\_in|lives\_in) \cdot located\_in^*), y) \qquad (12.2.5)$$

contains pairs of persons x and places y, with x speaking German and English and being of nationality x or being born in or live in some place which is connected to y by a sequence of any number of *located_in* relationships.

More theoretical considerations about graph queries can be found, e.g., in [FG00] and [MW95].

12.3 Big Graph Data Storage

Today special attention is paid to storage and processing Big Data. In general, Big Data are most often characterized by several V's which also pose problems for their storage and processing:

- Volume - data scale in the range of TB to PB and more,

- Velocity - both how quickly data are being produced and how the data must be processed to meet demand,

- Variety - data are in many format types - structured, unstructured, semi-structured, text, media, etc.,

- Veracity - managing the reliability and predictability of inherently imprecise data.

In Big Graphs the first two dimensions seem to be the most relevant. This rather simplified and vague characterization does not consider explicitly structural relationship complexity of graphs. This complexity of graphs includes a necessity to connect and correlate relationships, hierarchies and multiple node relationships, which means in context of Big Graphs also new challenges for data analytics, i.e. a development of data mining algorithms. Thus, the volume means a lot of highly interconnected data in the case of Big Graphs. Consequently, graph data models applied in graph databases are used more and more to address the complexity problem, i.e. to provide analytical results from such complex data efficiently.

In Section 12.3.1 we describe some properties of relational and NoSQL DBMS architectures important for graph management. Section 12.3.2 is devoted to native graph DBMSs. Section 12.3.3 mentions some typical approaches to storing and indexing graph structures in these DBMSs.

12.3.1 DBMS Architectures

Database approaches to graph management can use two types of architectures, relational and NoSQL. Traditional relational DBMS is based on usage of SQL language and transactional properties guaranteeing ACID properties. ACID stands for atomicity, consistency, isolation, and durability and is fundamental to database transaction processing. One problem with the relational database model was that it was designed for character based data which could be modeled in terms of attributes and records translated into columns and rows in a table. With a growing volume of data and an increasing the number of users, today's applications often require more scalability and performance. And it is precisely these two aspects that are characteristic for a new category called NoSQL (Not Only SQL) databases. Their providers claim their products outperform and out-scale relational DBMS.

In the broad sense, NoSQL represent more categories. For example, they include also object-oriented and XML DBMSs. But only four approaches are considered as typical ones in the literature about NoSQL databases: key-value stores, column oriented and document stores, and obviously graph databases.

Since NoSQL databases are used in a non-reliable Internet environment, some important functionalities known, e.g. from traditional relational DBMSs, are modified in the world of NoSQL. For example,

- they have a simplified data model,

- database design is query driven,

- integrity constraints (IC) are not supported,

- there is no standard query language,

- unneeded complexity is reduced (simple API, simple get, put, and delete operations).

It is typical for distributed NoSQL databases that they are scalable [Pok13]. They use so called *horizontal scaling* (also *scale-out*) in which data are distributed horizontally in the network that means into groups of rows in the case of tabular data.

The high performance of NoSQL databases is often achieved by an in-memory processing approach, which means that the data are stored in a computer's memory to achieve faster access. Thanks to considerable technological advances during the last 30 years, *in-memory databases* have finally become available in commercial products. Various architectures are used for in-memory graph DBMSs. They have very different design characteristics. These databases often lack transactions, provide a single view of a graph, and only support durability through a snapshot of the graph to disk. For example, Trinity http://research.microsoft.com/en-us/projects/trinity/ (retrieved on 14.7.2014) is on a memory cloud, i.e., a globally addressable, in-memory key-value store over a cluster of machines.

Due to weakening ACID semantics, NoSQL provides little or no support for Online Transaction Processing (OLTP) as it is required for most enterprise applications. Indeed, CAP theorem [Bre00] has shown that a distributed computer system can only choose at most two out of three properties: Consistency, Availability, and tolerance to Partitions. Then, considering P in a network, NoSQL databases support A or C. In practice, A is the most preferred and the strict consistency is relaxed to so-called *eventual consistency*. At a high level, this means that if no new updates are made in the distributed storage system, eventually all accesses will return the last updated values.

However, most graph DBMSs are ACID-compliant and disk-backed. For example, Titan http://thinkaurelius.github.io/titan/ (retrieved on 14.7.2014) is a transactional DBMS that can support thousands of concurrent users executing complex graph traversals. Titan is a graph database layer which uses another database as a backend (NoSQL DBMS Cassandra http://cassandra.apache.org/ (retrieved on 14.7.2014) or HBase http://hbase.apache.org/ (retrievedon14.7.2014)), i.e. it affords the same transactional guarantees to the user as the underlying data store does. In the case of Cassandra, it is eventual consistency. Moreover, Titan can leverage both HBase and Cassandra's scalable architecture.

Graph-Based transactions can be found in GraphBase http://graphbase.net/ (retrieved on 14.7.2014). Sparksee http://sparsity-technologies.com/\#sparksee (retrieved on 14.7.2014) (formerly known as DEX) supports ACID with only partial isolation.

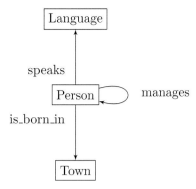

FIGURE 12.3.1: Graph database schema.

12.3.2 Graph DBMSs

By the graph DBMSs we will understand native graph databases in this section. These systems are mainly designed to provide efficient graph traversal functions, have a special implementation, often lack the support of declarative query interfaces, and do not use any query optimization strategies. As we will see, the diversity of these approaches, however, is huge. First, we mention data modeling used in native graph databases and then provide some examples of their implementations.

Data models used in native graph databases

As one expects, graph database models concern, as is usual in the database area, data structures for the database schema (*graph database schema*) and instances of this schema (graph database), i.e. a collection of graphs or one graph. In this case both types of data structures are modeled as graphs [AG08]. Figure 12.3.1 shows a graph database schema for the graph database in Figure 12.2.1. It contains the necessary entity types and relationship types.

An important part of each database schema is a set of ICs that enforce data consistency of associated databases. Examples of IC for graph databases are "labels are unique names," "graph does not contain a loop," etc. A graph database model usually describes simple relationships between two entities. In more complex approaches ISA-hierarchies known from classical conceptual modeling are possible. In practice, attributes of both entity types and relationship types are allowed. For example, a Person entity type can have the attributes *Id*, *Name*, and *Position*.

On the other hand, a graph database can be a collection of graphs or one graph without an a priori defined schema. Since we need to work with various graph types in applications, it is important to know precisely which types of graphs are allowed in a graph DBMS. For example, HyperGraphDB http://www.hypergraphdb.org/ (retrieved on 14.7.2014) serves for storing hypergraphs.

In practice, a graph database model is described rather intuitively without any formal fundamentals in most approaches. The terminology used is also very diverse and the difference between conceptual and database views of data is mostly blurred.

Several commercial and open source native graph databases

The well-maintained and structured Website http://nosql-database.org/ (retrieved on 14.7.2014) included 14 graph DBMSs in 2011. The most known are Neo4j http://www.neo4j.org/ (retrieved on 14.7.2014), InfiniteGraph http://www.objectivity.com/infinitegraph (retrieved on 14.7.2014), Sparksee, Titan, the Web graph database InfoGrid http://infogrid. org/trac/ (retrieved on 14.7.2014), HyperGraphDB, GraphBase, and Trinity.

Some products are intended for special graph applications. For example AllegroGraph http://franz.com/agraph/ (retrieved on 14.7.2014) works with RDF graphs and thus supports reasoning and ontology modeling. BrightStarDB http://brightstardb.com/ (retrieved on 14.7.2014), Bigdata and SparkleDB http://www.sparkledb.net/ (retrieved on 14.7.2014) (formerly known as Meronymy) serve similar purposes. Other software tools are DBMSs with restricted functionality. For example, WhiteDB http://whitedb.org/ (retrieved on 14.7.2014) is a lightweight NoSQL database library written in C, operating fully in memory. Another, rather a hybrid solution, is represented by Virtuoso Universal Server http://virtuoso.openlinksw.com/ (retrieved on 14.7.2014). Its functionality covers not only processing RDF data, but also relations, XML, text, and others.

Some of these projects are more mature than others, but each of them is trying to solve similar problems.

We describe some of graph DBMSs that are succesful in practice, in more detail.

Example 3: Neo4j

Neo4j is an open-source, highly scalable, robust (fully ACID compliant) native graph database that stores data in a graph. It is implemented in Java and belongs to the older NoSQL systems.

Neo4j stores data as nodes and relationships. Both nodes and relationships can hold properties in a key-value form. Attribute values can be either a primitive or an array of one primitive type. Nodes are often used to represent entities, but depending on the domain the relationships may be used for that purpose as well. Both the nodes and edges have internal unique identifiers that can be used for the data search. Nodes cannot refer to themselves directly. The semantics can be expressed by adding directed relationships between nodes.

Neo4j is a centralized system that lacks the computational power of a distributed, parallel system. It still suffers from scalability issues since it does not support partitioning still necessary in a distributed environment.

Processing of graphs in Neo4j entails mostly random data access which can be unsuitable for Big Graphs. Graphs that cannot fit into main memory may incur numerous disk accesses that significantly influences graph processing. Big Graphs similar to other Big Data collections must be partitioned over multiple machines to achieve scalable processing.

Example 4: Sparksee

In Sparksee a graph is a labeled directed attributed multigraph, where edges can be either directed or undirected. Both nodes and edges may have attributes. Sparksee also introduces the notion of a *virtual edge* that connects nodes having the same value for a given attribute. These edges are not materialized. A Sparksee graph is stored in a single file; values and identifiers are mapped by mapping functions into B+-trees. Bitmaps are used to store nodes and edges of a certain type.

The architecture of Sparksee includes the core, that manages and queries the graph structures, then an API layer to provide an application programming interface, and the higher layer applications, to extend the core capabilities and to visualize and browse the results. To speed up the different graph queries and other graph operations, Sparksee offers different types of indexing: i) attributes. ii) unique attributes, iii) edges to index their neighbors, and iv) indices on neighbors. Sparksee implements a number of graph algorithms, e.g. shortest-path, depth-first traversal, finding strong connected components of digraphs, etc.

12.3.3 Storing and indexing graph structures

We have mentioned in Section 12.3.1 the possibility to store graph data in a relational DBMS. However, graph databases are navigated mainly by the *following* relationships (e.g. *friendship* in social networks). This kind of storage and navigation is not possible in relational DBMSs due to the rigid relational structures and the inability to follow connections between the data in arbitrary way. On the other hand, there are attempts to provide internal support for graph data in relational DBMSs as well. Oracle's Version 12c, released in July 2013, includes the Oracle Spatial and Graph http://www.oracle.com/technetwork/database/options/spatialandgraph/overview/index.html (retrieved on 14.7.2014), which enables users to model and manipulate graph data in geographic information systems.

Similar to traditional relational DBMSs, some of the graph DBMS's products are distributed databases, e.g. InfiniteGraph. The distributed graph databases provide data persistence by storing graphs in distributed file systems. For example InfiniteGraph is built on a highly scalable, distributed database architecture where both data and processing are distributed across the network. A single graph database can be partitioned and distributed across multiple disk volumes and machines, with the ability to query data across machine boundaries. Rather than in-memory graphs, this system supports efficient traversal of graphs across distributed data stores. The same database client program can access the graph database locally or across a network in a native manner. Other graph databases, e.g. Neo4j, now start to support running in distributed mode on clusters; they are not designed to operate in a distribute environment.

The other example, Google's internal graph processing platform Pregel [MAB+10], is built on top of Hadoop http://hadoop.apache.org/(retrieved on 14.7.2014). Recall that Hadoop the software framework with its Hadoop Distributed File System (HDFS) and MapReduce framework provides batch jobs for processing the distributed nodes with message passing. The application developer is isolated from details of distribution that does the basic MapReduce programming model not optimal for graph processing because most graph algorithms are iterative and traverse the graph in some way. The Pregel library divides a graph into partitions, each consisting of a set of vertices and all of those vertices outgoing edges.

Distributed graph databases require a partitioning graph data which is a non-trivial problem. Optimal division of graphs requires finding the subgraphs of a graph. In practice, however, the number of edges is too large to efficiently compute an optimal partition; therefore most databases use random partitioning. In a dynamic environment, what looks like a good distribution one moment, may no longer be optimal a few seconds later. This is known to be an NP-complete problem in the general case.

Surfer [CWHY10] and GBASE [KTS+11] are examples of extensions for MapReduce that are proposed to help to process graphs more efficiently. Surfer offers a new primitive, *propagation*, which is an iterative computational pattern that transfers information along the edges from a node to its neighbors in the graph. GBASE uses a novel block compression for graph storage and a MapReduce algorithm to support incidence matrix based queries using the original adjacency matrix, without explicitly building the incidence matrix. More advanced solutions are now Giraph http://giraph.apache.org/ (retrieved on 14.7.2014), originated as the open-source counterpart to Pregel, and GraphLab http://graphlab.org/ (retrieved on 14.7.2014) projects. The GraphLab framework uses the Message Passing Interface http://en.wikipedia.org/wiki/Message_Passing_Interface (retrieved on 14.7.2014) model to scale and run complex algorithms using data in HDFS.

An interesting and effective technique for graph implementation is described in [BK14]. A flat description of the particular graph records uses a column-oriented storage model, bitmap columns enables fast access to parts of these graph records.

For graph databases it is important to implement a set of relationships between nodes. Native graph databases implement it with *index-free adjacency* (*adjacency lists*) where each node has explicit references to its adjacent nodes, and does not need to use an index to find them. The adjacency list format is simple and might be good for answering out-neighbors queries. In the case of only one relationship type, a digraph $G = (\mathcal{V}, \mathcal{E})$ can be represented by the *adjacency matrix* A, were $A_{ij} = 1$, if $(i, j) \in \mathcal{E}$ and $A_{ij} = 0$, otherwise. Unlike adjacency lists, an adjacency matrix is preferred if the graph is dense.

There is also an *incidence matrix* whose each row corresponds to an edge, and it has two non-zeros whose column Ids are the node Ids of the edge.

Since traversing neighbors (one of the most common operations) is too resource consuming, adjacency matrix is not usually used for Big Graph databases. GBASE uses matrix-vector multiplications on the adjacency and the incidence matrices for node-based and edge-based queries.

Another data structure is used for implementing graph databases, i.e., bitmaps. For example, in Sparksee a graph is represented through bitmap data structures and so-called *maps*. Node adjacencies are represented by bitmaps to minimize the space needed in memory. Bitmaps allow high compression rates. A map is an inverted index with key values associated to bitmaps or data values.

A lot of graph databases use an approach similar to the one used in Neo4j, i.e. nodes, relationships, relationship types, and attributes are stored in different store files. Neo4j actually stores adjacent relationships in doubly linked lists in the filesystem. Two separate caches serve as efficient block writes (a file system cache) and as efficient access reads (the object cache).

Graph querying is not an easy task. For example, to find graphs that contain a given graph pattern means to solve the subgraph isomorphism problem which has been proven to be NP-complete. Defining and computing the similarity between two graphs is also difficult.

Concerning query processing in Big Graphs a node indexing is necessary. Then candidate node sets obtained by index search serve to building subgraphs and constructing the query answer.

T-trees [LC86] seem to be widely used for in-memory databases. The latter results [LNT00] indicate that classical B-link trees outperform the T-tree if concurrency control is enforced. This condition is usually fulfilled in today's scalable systems.

Graph indexing is useful for graph pattern matching over a large collection of small graphs. Graph indexing distinguishes from indexing, e.g. relational databases. Because of queries on structural properties on the graph data some *structure indexes* are needed. Structure indexes can be based on reducing undirected graphs in two phases:

- performing depth-first traversal to obtain a tree,

- encoding the tree into a string.

Another line of graph indexing addresses reachability queries in large digraphs. Reachability queries correspond to recursive graph patterns, which are paths.

Often a *path index* is used in graph databases. GraphGrep index [SWG02] records all embeddings of paths (up to a maximal length) in the graph database. But a path is too simple and structural information hidden in the graph is lost. More sophisticated approaches index subgraphs [Sri11]. Authors of [YYH04] consider even a frequent structure-based approach considering only frequent "small" subgraphs (fragments). Their gIndex is of 10 times smaller size and achieves 3-10 times better performance in comparison with a typical path-based method. An efficient index, FG*-index, is proposed in [CKN09].

12.4   Graph Data Processing

One possibility to handle graph data is through constructing complex graph oriented programs based on direct calls to low level APIs. This style is very difficult to optimize. Similar to development of high-level languages for relational databases, there are approaches to define algebra with a set of query operations and an SQL-like language allowing a recursion with an engine that generates query plans. Therefore, these can be optimized and executed efficiently. An example of such a direction is the last development of the Sparksee graph database [MBDS14].

A lot of native graph databases have interfaces to some current open source software products http://www.tinkerpop.com/ (retrieved on 14.7.2014) from the graph space.

- *Blueprints* is a property graph model interface with provided implementations. Databases that implement the Blueprints interfaces automatically support Blueprints-enabled applications.

- *Frames* exposes the elements of a Blueprints graph as Java objects. Instead of writing software in terms of nodes and edges, with Frames, software is written in terms of domain objects and their relationships to each other.

- *Gremlin* is a domain specific language for traversing property graphs. This language has application in the areas of graph query, analysis, and manipulation.

- *Rexster* is a multi-faceted graph server that exposes any Blueprints graph through several mechanisms with a general focus on REST. Remember that Web API REST implements four basic methods denoted often as CRUD, i.e operation Create, Retrieve, Update, and Delete.

- *Furnace* is a property graph algorithms package. It provides implementations for standard graph analysis algorithms that can be applied to property graphs in a meaningful way.

- *Pipes* is a dataflow framework that enables the splitting, merging, filtering, and transformation of data from input to output. Computations are evaluated in a memory-efficient, lazy fashion.

For example, Titan DBMS provides native integration with the first four graph tools. Bigdata DBMS supports Blueprints and Sesam.

Usually two types of graph databases are studied in the context of graph querying. The first type consists of very large graphs, such as the Web graph and social networks. Typical querying tasks for such graph databases include finding the best connection between a given set of query nodes and matching subgraphs. The second type of graph database collects a large collection of small graphs, such as chemical compounds or those appearing in Customer Relationship Management software and Workflow Management Systems. Typical queries for this type of graph database include subgraph queries and similarity queries.

In Section 12.4.1 we describe possibilities of relational DBMSs for graph storage and querying. Section 12.4.2 explains a Datalog approach to graph data processing. Section 12.4.3 is devoted to query languages implemented in native graph DBMSs.

12.4.1 Querying graphs in relational DBMS

Since the 1999 version, the SQL standard has possibilities for graphs processing with "recursive union" used in WITH RECURSIVE clause. Graphs represented by tables in SQL database can be proceed by constructs of SQL. This was possible also earlier but even simple traversal algorithms required costly self joins on the table and programming in Embedded SQL. The WITH RECURSIVE clause allows formulating basic graph queries which require fixpoint computation, although not always in a simple way, but probably with a more effective implementation. The queries with WITH RECURSIVE work simply on acyclic graphs. To recognize cycling during query evaluation needs to construct rather tricky conditions in query expression.

Example 5:

Suppose a relational representation of a part of our graph database given by the entity type *persons* and relationship type *manages* is described by the relation schema Persons(Id, Name, Position, Manager_Id). Consider the query "Find all managers of Wang (including himself)." The table Managers will contain Wang's managers from all levels of the hierarchy given by relationships *manages* including the top manager.

```
WITH RECURSIVE Managers(Id, Name, Manager_Id) AS
(SELECT Id, Name, Manager_Id
    FROM Persons  -
    WHERE Name = 'Wang'
    UNION ALL
SELECT P.Id, P.Name, P.Manager_Id
    FROM Persons AS P
    INNER JOIN
    Managers AS M
    ON M.Manager_Id = P.Id)
SELECT * FROM Managers
```

Materialization of such a recursively defined view is given by the last SELECT. This query is an *ancestor_or_self* query applicable in acyclic digraphs. The following two tables represent relations Persons and Managers, respectively.

Persons	Id	Name	Position	Manager_Id
	1	Smith	director	NULL
	2	Lang	manager	1
	3	Cohen	manager	1
	4	Wang	assistant	3

Managers	Id	Name	Manager_Id
	1	Smith	NULL
	3	Cohen	1
	4	Wang	3

The last SELECT returns a table of persons (given by their attributes, Id and Name) who are the managers of Wang and Wang himself. Manager_Id will contain Ids of their direct managers.

However, expressiveness of WITH RECURSIVE clause is restricted in SQL. A recursive part of a query must not contain clauses SELECT DISTINCT, GROUP BY, HAVING, scalar aggregates, TOP, and OUTER JOIN.

12.4.2 Graph querying in Datalog

Datalog [AHV95] is an important rule-based language for inference using facts found in databases. In classical Datalog a *program P* is a finite set of rules of the form:

$$A \leftarrow A_1, \ldots, A_m \tag{12.4.1}$$

where A_i are predicates, or relation names with variables or constants as arguments. A is a relation name with variables as arguments. Datalog can be extended with negation and ICs. Datalog with negation usually allows a *stratified negation*, i.e., the conclusion of any rule and a negated hypothesis of any rule are not mutually recursive. Facts like (u, p, v) (see Section 12.2) correspond to $p(u, v)$ in Datalog. In fact, they are rules without body, i.e. $p(u, v) \leftarrow$. The main expressive advantage of Datalog is that it enables recursive queries. This is extremely important for a lot of graph queries. On the other hand, for graph data processing it is necessary to have an efficient implantation at one's disposal. The possibility to implement Datalog in a relational environment tends to algorithms based on variants of fixpoint computation that is not too feasible. However, there are methods that transform a Datalog program into an efficient specialized implementation that, given any set of facts, computes exactly the set of facts that can be inferred [LS03]. The authors of [LS06] used these methods for the development and implementation of a new graph-oriented language enabling them to formulate some variants of regular paths queries. Its optimized Datalog implementation using the method from [LS03] introduced in [TGL10]. An added value of such a language is that many queries can be written much more easily and clearly than directly in Datalog.

Example 6:

The ancestor_or_self query introduced in Example 5 can be expressed in Datalog as

$$
\begin{aligned}
Manages(i, n, m) &\leftarrow & Persons(i, n, p, m), n = Wang \\
Manages(i, n, m) &\leftarrow & Manages(i_1, n_1, i), Persons(i, n, p, m)
\end{aligned}
$$

Example 7:

Suppose a relation is_part_of containing facts that a location is a part of other location. Evaluation of a regular paths query (one conjunct of the query in Example 2)

$$Q(x, y) \leftarrow (x, has\_a\_nationality|((is\_born\_in|lives\_in) \cdot is\_located\_in^*), y)$$

can be represented by the following Datalog program:

$$
\begin{aligned}
asoc(x, y) &\leftarrow & is\_born\_in(x, y) \\
asoc(x, y) &\leftarrow & lives\_in(x, y) \\
is\_part\_of(x, y) &\leftarrow & is\_located\_in(x, y) \\
is\_part\_of(x, y) &\leftarrow & is\_located\_in(x, z), is\_part\_of(z, y) \\
Q(x, y) &\leftarrow & has\_a\_nationality(x, y) \\
Q(x, y) &\leftarrow & asoc(x, y) \\
Q(x, y) &\leftarrow & asoc(x, y), is\_part\_of(z, y)
\end{aligned}
$$

Compared to the WITH RECURSIVE clause in SQL, the capabilities of recursion in Datalog are more expressive. SQL enables only linear recursion, i.e. each FROM has at most one reference to a recursively defined relation.

12.4.3 Query languages in graph DBMS

Graph DBMSs offer a lot of graph languages (see, e.g., [AG08, Ang12]). Some of them have syntax similar to SQL and enable a set of basic graph queries extended by selection conditions (e.g., Cypher Query Language of Neo4j). Some of them provide a more advanced functionality like link analysis, social network analysis, pattern recognition, and keyword search (e.g., Sparksee). We will show some details of Cypher query language and mention briefly the graph traversal language Gremlin http://github.com/tinkerpop/gremlin/wiki (retrieved on 14.7.2014).

Example 8: Cypher Query Language

Cypher is a declarative graph query language that allows us to query and update the graphs. Being a declarative language, Cypher focuses on the clarity of expressing what to retrieve from a graph, not how to do it, in contrast to imperative languages like Java, and scripting languages like Gremlin and the JRuby, which can also be used in the Neo4j. Similar to relational DBMSs, a declarative approach to querying is not always an advantage. It restricts the ability to optimize the order of traversals in a query.

The syntax of Cypher statements can be compared with the classic SQL. For example, the SQL query

```
SELECT *
FROM Persons
WHERE Name = 'Lang'
```

has the following equivalent expression in Cypher:

```
START person=node:Person(Name = 'Lang')
RETURN person
```

The START clause specifies the starting point on the graph, from which the query is executed. Thus, the role of this phrase is something between the FROM and WHERE clauses of the SQL SELECT statement.

By the following expression we describe the query requiring the languages spoken by Lang:

```
START person=node:Person(Name = 'Lang')
MATCH person --> language
RETURN language.Name
```

where "-->" denotes directed edge. Other expression returns a hierarchy of places starting with the one where Lang is born.

```
START person=node:Person(Name = 'Lang')
MATCH person-[:is born_in]->[:located_in*]->place
RETURN place.Name
```

The notation principle used in Cypher reflects rewriting graph notation to linear ASCII notation with the help of regular expressions.

Cypher supports traversal and neighborhood queries. Cypher commands can embrace several parts, enabling

- data manipulations:
  CREATE: creates nodes and edges,
  DELETE: removes nodes, edges and attributes,
  SET: set values of the attributes,
  FOREACH: performs updating actions once per each element in a list.

- querying constructs:
  START: starting points in the graph, obtained via index lookups or by element Ids,
  MATCH: the graph pattern to match, bound to the starting points in START (it is equivalent to the SQL JOIN clause),
  WHERE: filtering criteria,
  RETURN: specifies the answer to query (it is equivalent to the SQL SELECT clause),
  ORDER BY: sorts the output,
  WITH: divides a query into multiple, distinct parts. The WITH clause is used to pipe the result from one query to the next one and to separate reading from updating of the graph.

Thus the Cypher expressions not only allow for data searching, but also their insertion, modification, or deletion. Therefore, it is not only a parallel to an SQL SELECT statement, but contains the UPDATE, INSERT, and DELETE statements as well.

In addition to Cypher, Neo4j also supports Gremlin and the Blueprints interface. Gremlin uses the concept of traversals to move from a starting location to (possibly multiple) ending locations in a graph. Gremlin is an open-source Turing-complete programming language based on the open source software Blueprints, that is based on traversals of a property graph with a syntax taken from object-oriented systems and the C programming language family. There is syntax for directed edges and more complex queries that looks more mathematical than SQL-like. Gremlin provides a command line prompt to insert/update/delete operations on the nodes and edges. As a query language, it can perform complex graph traversals compactly. To get the same query result, it uses much less code than using Java API.

12.5 Graph Data Visualization

Information Visualization (InfoVis) research focuses on the use of techniques to help people understand and analyze data. In particular, InfoVis considers how abstract data (i.e., without correspondence to the physical world) can best be visually represented. InfoVis, the study of transforming data, information, and knowledge into interactive visual representations, is very important to users because it provides mental models of information. The boom in Big Data analytics has triggered broad use of InfoVis in a variety of domains, ranging from science to art [SM14]. In [LCWL14], Liu et al. present a comprehensive survey and key insights into this fast-rising area.

In [SM14] the authors present an anthology of articles to foster the emerging convergence of arts, humanities, and complex networks. This book covers a kaleidoscope of different approaches, ranging from vigorous humanistic inquiry and pure natural science to free artistic expression.

Considering current exponential growth of InfoVis, this chapter cannot and does not present an exhaustive account of all relevant aspects of this field.

As an independent field, graph visualization arose in the 1990s with the Symposium on Graph Drawing, which was held in its 22nd edition in 2014. With increasing interest in information visualization, alternative visual representations of graphs have also been introduced, such as adjacency matrices.

Methods and tools for graph visualization are widely used in many applications, such as social contacts [HCL05], co-authorship network [KHS$^+$12], process mining [VDARS05], trajectories on maps [SHH11], and electronic communications [SFMB12]. According to

[VLKS$^+$11], graphs can be classified into two categories: static and dynamic, based on their time dependence.

Graph visualization methods compute a 2D/3D layout of the nodes and the edges, mainly based on node-link diagrams [Tut63, War13]. They play a fundamental role in graph visualization. Graph readability is affected by quantitative measurements called aesthetic criteria [EGHM10]. Thus, graph visualization generally deals with the ways of drawing graphs according to a set of predefined aesthetic criteria [Che06]. Different graph drawing algorithms may have their own aesthetic criteria to follow. As for aesthetic criteria, the widely accepted rules for drawing a comprehensible graph can be summed up as follows:

- Reflect the inherent symmetry

- Uniform edge length

- Minimized edge crossings

- Avoidance of sharp angles between edges

- Clutter reduction

- Even distribution of the vertices in the available space

Some of these criteria can be mutually exclusive, and problems which aim to optimize the criteria are often NP-hard. For example, a symmetrical graph may require a certain number of edge crossings, even if they might be avoided. And uniform edge lengths may not always produce the most appropriate results. Therefore, many graph drawing algorithms are heuristics or meta-heuristics [KKS$^+$14].

12.5.1 Static graph visualization

Graph visualization algorithms are categorized into the following approaches: force-directed layouts, the use of dimension reduction in graph layout, and computational improvements including multi-level techniques [GFV13]. Methods developed specifically for graph visualization often make use of node-attributes and are categorized based on whether the attributes are used to introduce constraints to the layout, provide a clustered view, or define an explicit representation in two-dimensional space.

Node-link diagrams have been the most used visual representation for graphs. Over the last decade graph drawing visualization has emerged as an exciting research area that addresses a significant problem: how to make sense of the ever increasing amounts of relational information that has become widely available. However, recent visualization work indicates that researchers have gradually shifted their attention from finding new layout algorithms [STT81, KK89, SCL$^+$09, KHKS12, LCWL14] to studying the usability in various applications.

E-learning log visualization and mining

E-learning is a method of education, which usually uses Learning Management Systems and the internet environment to ensure the maintenance of courses and to support the educational process. Moodle, one of such systems widely used, provides several statistical tools to analyze students' behavior in the system. However, none of these tools provides visualization of relations between students and their clustering into groups based on their similar behavior. In [SMDS14], the authors propose an approach for analysis of students' behavior in the system based on their profiles and on the students' profile similarity. The

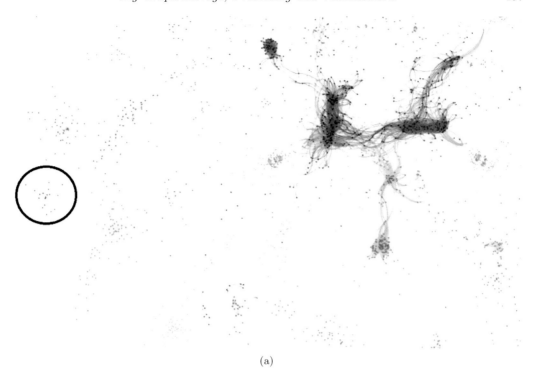

(a)

FIGURE 12.5.1: LMS Moodle log sequence visualization. (**See color insert.**)

approach uses process mining [VDARS05] for visualization of relations between students and groups of students Figure 12.5.1.

[MM08] proposes an approach to graph layout through the use of space filling curves which is very fast and guarantees that there will be no nodes that are colocated. The resulting layout is also aesthetic, and satisfies several criteria for graph layout effectiveness. Burch et al. [BKH$^+$11] conducted a user study to compare the readability of node-link diagrams and space-filling representations. They found that space-filling results are more space-efficient but more difficult to interpret.

In particular, orthogonal tree layouts significantly outperform radial tree layouts for some tasks, such as finding the least common ancestor of a set of marked leaf nodes. Yuan et al. [YCHZ12] argued that a good layout cannot be achieved simply by using automatic algorithms but need user inputs. Thus they proposed a framework that automatically stitches and maintains the layouts of individual subgraphs submitted by multiple users.

Another hot reseach topic with regard to improving usability is clutter reduction. Among all the solutions to reduce visual clutter, edge bundling is still the most popular one [HCL05, CZQ$^+$08, SHH11]. Recently, Selassie et al. [SHH11] proposed a bundling technique for directed graphs. At the same time, skeleton-based edge bundling was introduced in [EHP$^+$11]. They calculated the skeleton of edge distributions and used it to bundle the edges. Other ways to reduce clutter include density estimation, node aggregation, and level-of-detail rendering. Zinsmaier et al. [ZBDS12] presented an approach that combines these techniques and achieves a better time performance than other state-of-art methods, while generating appealing layouts.

The traditional matrix representation is suitable for visualizing dense graphs due to its non-overlapping visual encoding of edges. However, it may be ineffective for sparse graphs.

Recently, Dinkla et al. [DWvW12] designed compressed adjacency matrices, which aim to visualize sparse graphs, such as gene regulatory networks. Similar to matrix representations, PIWI [YLZ$^+$13] uses vertex plots that show nodes as colored dots without overlap, to display the neighborhood information of communities in a large graph.

Two-dimensional graph drawing, that is, graph drawing in the plane, has been widely studied. While this is not yet the case for graph drawing in 3D, there is nevertheless a growing body of research on this topic, motivated in part by advances in hardware for three-dimensional graphics, by experimental evidence suggesting that displaying a graph in three dimensions has some advantages over 2D displays [WF94, WF96, WM08], and by applications in information visualization [WF94, WM08], and software engineering [WHF93].

Not surprisingly, the mathematical literature is a source of results that can be regarded as early contributions to graph drawing. It is natural to generalize from drawing graphs in the plane to drawing graphs on other surfaces, such as the torus. Indeed, surface embeddings are the object of a vast amount of research in topological graph theory, with entire books devoted to the topic. We refer the interested reader to the book by Mohar and Thomassen [MT01] as an example. Numerous drawing styles or conventions for 3D drawings have been studied. These styles differ from one another in the way they represent nodes and edges.

12.5.2 Dynamic graph visualization

In a static network, the properties of nodes, links, and mapping functions remain unchanged over time. In a dynamic network, the number of nodes and links, the shape of the mapping function, and perhaps other properties of the graph change over time. Dynamic networks are time-varying networks. Visualization dynamics of the graph is very useful. It helps to understand dynamics features of the graph. Evolution of graph and communities over time can help us understand social mechanisms behind the graph. Time–varying changes leading to structural reorganization in a network, called *evolution* in some disciplines, is called *emergence* [Lew11].

To define a dynamic graph, we start from the definition of (static) graph (see section 12.2). Let $G_i = (\mathcal{V}, \mathcal{E})$ be a graph. Then, a *dynamic graph* is defined as a sequence $G = (G_1, G_2, \ldots, G_n)$ where $G_i = (\mathcal{V}_i, \mathcal{E}_i)$ are static graphs and indices refer to a sequence of time steps $t := (t_1, t_2, \ldots, t_n)$. In a dynamic graph, time can be modeled as discrete, ordinal or continuous. Dynamic graphs can be sampled based data on model. We also do not discern between instants and intervals: whether G_i is a snapshot at instant t_i or aggregates an interval around t_i.

While static graph visualizations are often divided into node-link and matrix representations, the representation of time was identified as the major distinguishing feature for dynamic graph visualizations: either graphs are represented as animated diagrams or as static charts based on a timeline. In [BBDW14], Beck et al. present a comprehensive survey of dynamic graph visualization.

Animation is a natural way to illustrate changes over time since it can effectively preserve a mental map [BdM06]. Visualization of the dynamics is not an easy task. There are several issues that have to be solved for correct visualization [YFDH01, BdM06, BBDW14].

The role of the mental map has been discussed since the first works on dynamic graph visualization and is probably their best evaluated aspect. While we briefly summarize results of related studies, Archambault and Purchase [AP13] review studies on the mental map in much greater detail. They have shown that preserving a mental map does not help much in gaining insights into animated dynamic graphs.

As a result, recent methods focus more on showing dynamic graphs statically [BKH$^+$11, TM12, LWW$^+$13]. To encode the time dimension in a static way, a timeline and small multiples are two popular choices.

Timeline-based approaches encode time as one axis and then draw and align the graph at each time point on the timeline. Abello et al. [AAK$^+$14] discuss the modeling and representation of time for dynamic graphs in greater detail.

Based on small multiples, Hadlak et al. [HSS11] proposed in-situ visualization, which allows users to interactively select multiple focused regions and choose suitable layouts for the selected data. They argued that a single visualization technique may not be enough, due to the complexity of large dynamic graphs. With their approach, a user can freely switch between different visualizations to adapt the analysis focus or the characteristics of regions of interest.

Examples: Dynamics in co-authorship network

Experiments presented use a weighted co-authors network based on the DBLP database [KHS$^+$12]. Paper [RKHS13] presented an approach to the visualization of weighted networks based on Sammon's projection and linear approximation. It introduces a method for visualizing dynamics of the social network. Results are illustrated by several 3D layout snapshots of the co-authorship graph extracted from the DBLP database.

Figures 12.5.2a and 12.5.2h depict the state of the network in two consecutive months 11/2011 and 12/2011, respectively. Figures (12.5.2b, . . . , 12.5.2g) show dynamic graph $G = (G_1, G_2, G_3, G_4, G_5, G_6)$. Vertices where distance is over a selected threshold are interpolated and these nodes are depicted in a different color (for illustrative purposes only). Other vertices in the interpolation layout are unchanged. Parts of the depicted layouts, highlighted manually by the circle and square, show how the communities are formed. Interpolation is very helpful in this case – the communities travel among the layout space and finally are connected to the new vertices.

12.6 Conclusions

In this chapter we have described some methods and tools for storage and processing graph data, and information visualization technologies particularly developed for this purpose. The development of Big Data processing has led to the widespread use of these technologies. Some of them are also appropriate for Big Graphs — a technology where data analysis is the main application requirement.

In other words, the technology provides a support for scalability, analytical algorithms integrated into query tools, etc. Transaction processing is probably not so actual in this case.

However, some experiences with processing graph data show that a graph database is not always the best choice; it depends on the exact types of queries that need to be performed. For example in [HSS11], the authors compared SQL processing of some queries in PostgreSQL DBMS and the same with Cypher in Neo4J. They reached better results with PostgreSQL by decomposing the complex SQL query into multiple simple SQL queries. By the way, this fact documents that query optimization in RDBMs has still reserves. Sometimes the differences between graph-enabled and native DBMS are huge. Architects of GraphBase DBMS argue that their product can be thousands of times faster than a relational solution.

Traditionally, however, graph database implementations also have problems that limit usefulness of some products:

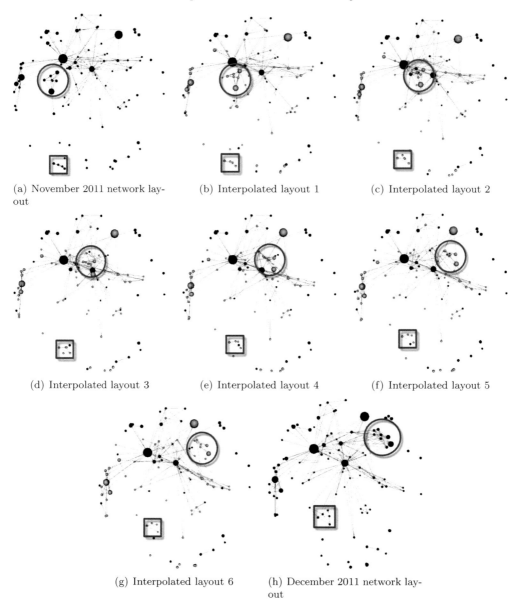

(a) November 2011 network layout

(b) Interpolated layout 1

(c) Interpolated layout 2

(d) Interpolated layout 3

(e) Interpolated layout 4

(f) Interpolated layout 5

(g) Interpolated layout 6

(h) December 2011 network layout

FIGURE 12.5.2: Layout linear interpolation between two consecutive months.

- Storing data in a graph can be slower than using data structures. This makes graph databases less-suited to high-throughput and transaction-processing tasks.

- Graph databases can impose unnecessary structural complexity. Sometimes simpler structures are easier to manage and provide better performance.

- They are difficult to query. The tools available for working with a graph DBMS are often poor.

We have seen that the diversity of graph data models and graph databases (or graph data stores) is huge. Unfortunately, no standard benchmarks are at our disposal. Hence, to compare these tools is difficult.

Some of the traditional graph data problems, particularly for Big Graphs, e.g., dynamic graph partitioning, workload balancing, integration of analytical methods with querying, still remain a challenge for a future.

Acknowledgment

This chapter has been partially supported by the grant from the Grant Agency of the Czech Republic – GACR No. P103/13/08195S.

Bibliography

[AAK+14]   D. Archambault, J. Abello, J. Kennedy, S. Kobourov, K.-L. Ma, S. Miksch, C. Muelder, and A. C. Telea. Temporal multivariate networks. In *Multivariate Network Visualization*, pages 151–174. Springer, 2014.

[AG08]   R. Angles and C. Gutierrez. Survey of graph database models. *ACM Computing Surveys*, 40(1):1, 2008.

[AHV95]   S. Abiteboul, R. Hull, and V. Vianu. *Foundations of databases*, volume 8. Addison-Wesley Reading, 1995.

[Ang12]   R. Angles. A comparison of current graph database models. In *Proc. IEEE 28th International Conference on Data Engineering Workshops*, pages 171–177, 2012.

[AP13]   D. Archambault and H. C. Purchase. The map in the mental map: Experimental results in dynamic graph drawing. *International Journal of Human-Computer Studies*, 71(11):1044–1055, 2013.

[AW10]   C. C. Aggarwal and H. Wang. *Managing and mining graph data*, volume 40. Springer, 2010.

[BBDW14]   F. Beck, M. Burch, S. Diehl, and D. Weiskopf. The state of the art in visualizing dynamic graphs. In *Proc. of the Eurographics Conference on Visualization*, 2014.

[BdM06]   S. Bender-deMoll and D. A. McFarland. The art and science of dynamic network visualization. *Journal of Social Structure*, 7(2):1–38, 2006.

[BK14]   D. Bleco and Y. Kotidis. Graph analytics on massive collections of small graphs. In *Proc. of the 17th International Conference on Extending Database Technology*, pages 523–534, 2014.

[BKH+11]   M. Burch, N. Konevtsova, J. Heinrich, M. Hoeferlin, and D. Weiskopf. Evaluation of traditional, orthogonal, and radial tree diagrams by an eye tracking study. *IEEE Transactions on Visualization and Computer Graphics*, 17(12):2440–2448, 2011.

[Bre00]   E. A. Brewer. Towards robust distributed systems. In *Proc. Principles of Distributed Computing*, page 7, 2000.

[Cat11]   R. Cattell. Scalable sql and nosql data stores. *ACM SIGMOD Record*, 39(4):12–27, 2011.

[Cel13]   J. Celko. *Joe Celkos Complete Guide to NoSQL: What Every SQL Professional Needs to Know about Non-Relational Databases*. Newnes, 2013.

[Che06]   C. Chen. *Information visualization: Beyond the horizon*. Springer Science & Business, 2006.

[CKN09]   J. Cheng, Y. Ke, and W. Ng. Efficient query processing on graph databases. *ACM Transactions on Database Systems*, 34(1):2, 2009.

[CM90]   M. P. Consens and A. O. Mendelzon. GraphLog: a visual formalism for real life recursion. In *Proc. of the 9th ACM SIGACT-SIGMOD-SIGART symposium on Principles of database systems*, pages 404–416, 1990.

[CMW87]   I. F. Cruz, A. O. Mendelzon, and P. T. Wood. A graphical query language supporting recursion. In *ACM SIGMOD Record*, volume 16, pages 323–330, 1987.

[CWHY10]   R. Chen, X. Weng, B. He, and M. Yang. Large graph processing in the cloud. In *Proc. of the 2010 ACM SIGMOD International Conference on Management of data*, pages 1123–1126, 2010.

[CZQ$^+$08] W. Cui, H. Zhou, H. Qu, P. C. Wong, and X. Li. Geometry-based edge clustering for graph visualization. *IEEE Transactions on Visualization and Computer Graphics*, 14(6):1277–1284, 2008.

[DWvW12]   K. Dinkla, M. A. Westenberg, and J. J. van Wijk. Compressed adjacency matrices: untangling gene regulatory networks. *IEEE Transactions on Visualization and Computer Graphics*, 18(12):2457–2466, 2012.

[EGHM10]   P. Eades, C. Gutwenger, S.-H. Hong, and P. Mutzel. Graph drawing algorithms. In *Algorithms and theory of computation handbook*, pages 6–6. Chapman & Hall/CRC, 2010.

[EHP$^+$11] O. Ersoy, C. Hurter, F. V. Paulovich, G. Cantareiro, and A. Telea. Skeleton-based edge bundling for graph visualization. *IEEE Transactions on Visualization and Computer Graphics*, 17(12):2364–2373, 2011.

[FG00]   S. Flesca and S. Greco. Querying graph databases. In *Advances in Database Technology*, Lecture Notes in Computer Science, pages 510–524. Springer, 2000.

[GFV13]   H. Gibson, J. Faith, and P. Vickers. A survey of two-dimensional graph layout techniques for information visualisation. *Information Visualization*, 12(3-4):324–357, 2013.

[GPVdBVG94] M. Gyssens, J. Paredaens, J. Van den Bussche, and D. Van Gucht. A graph-oriented object database model. *IEEE Transactions on Knowledge and Data Engineering*, 6(4):572–586, 1994.

[Güt94]   R. H. Güting. Graphdb: Modeling and querying graphs in databases. In *Proceedings of 20th International Conference on Very Large Data Bases*, pages 297–308, 1994.

[HCL05]   J. Heer, S. K. Card, and J. A. Landay. Prefuse: a toolkit for interactive information visualization. In *Proc. of the SIGCHI Conference on Human Factors in Computing Systems*, pages 421–430, 2005.

[HSS11]   S. Hadlak, H. Schulz, and H. Schumann. In situ exploration of large dynamic networks. *IEEE Transactions on Visualization and Computer Graphics*, 17(12):2334–2343, 2011.

[KHKS12]   M. Khoury, Y. Hu, S. Krishnan, and C. Scheidegger. Drawing large graphs by low-rank stress majorization. In *Computer Graphics Forum*, volume 31, pages 975–984, 2012.

[KHS+12]   M. Kudělka, Z. Horák, V. Snášel, P. Krömer, J. Platoš, and A. Abraham. Social and swarm aspects of co-authorship network. *Logic Journal of IGPL*, 20(3):634–643, 2012.

[Kim90]   W. Kim. *Introduction to object-oriented databases*, volume 90. MIT Press, 1990.

[KK89]   T. Kamada and S. Kawai. An algorithm for drawing general undirected graphs. *Information Processing Letters*, 31(1):7–15, 1989.

[KKS+14]   P. Kromer, M. Kudelka, V. Snael, M. Radvansky, and Z. Horák. Computing Sammon's projection of social networks by differential evolution. In *IEEE 28th International Conference on Advanced Information Networking and Applications*, pages 1001–1006, 2014.

[KTS+11]   U. Kang, H. Tong, J. Sun, C.-Y. Lin, and C. Faloutsos. Gbase: a scalable and general graph management system. In *Proc. of the 17th ACM SIGKDD International Conference on Knowledge Discovery and Data Mining*, pages 1091–1099, 2011.

[KV84]   G. M. Kuper and M. Y. Vardi. A new approach to database logic. In *Proc. of the 3rd ACM SIGACT-SIGMOD symposium on Principles of database systems*, pages 86–96, 1984.

[LC86]   T. J. Lehman and M. J. Carey. A study of index structures for main memory database management systems. In *Proc. of the 12th International Conference on Very Large Data Bases*, volume 294, pages 294–303, 1986.

[LCWL14]   S. Liu, W. Cui, Y. Wu, and M. Liu. A survey on information visualization: recent advances and challenges. *The Visual Computer*, pages 1–21, 2014.

[Lew11]   T. G. Lewis. *Network science: Theory and applications*. John Wiley & Sons, 2011.

[LNT00]   H. Lu, Y. Y. Ng, and Z. Tian. T-tree or b-tree: Main memory database index structure revisited. In *Proc. of 11th Australasian Database Conference*, pages 65–73, 2000.

[LS03]   Y. A. Liu and S. D. Stoller. From datalog rules to efficient programs with time and space guarantees. In *Proc. of the 5th ACM SIGPLAN International Conference on Principles and Practice of Declarative Programming*, pages 172–183, 2003.

[LS06]   Y. A. Liu and S. D. Stoller. Querying complex graphs. In *Practical Aspects of Declarative Languages*, pages 199–214. 2006.

[LWW+13]   S. Liu, Y. Wu, E. Wei, M. Liu, and Y. Liu. Storyflow: Tracking the evolution of stories. *IEEE Transactions on Visualization and Computer Graphics*, 19(12):2436–2445, 2013.

[MAB+10]   G. Malewicz, M. H. Austern, A. J. Bik, J. C. Dehnert, I. Horn, N. Leiser, and G. Czajkowski. Pregel: a system for large-scale graph processing. In *Proc. of the 2010 ACM SIGMOD International Conference on Management of Data*, pages 135–146, 2010.

[MBDS14]   N. Martinez-Bazan and D. Dominguez-Sal. Using semijoin programs to solve traversal queries in graph databases. In *Proc. of Workshop on Graph Data Management Experiences and Systems*, pages 1–6, 2014.

[MM08]   C. Muelder and K.-L. Ma. Rapid graph layout using space filling curves. *IEEE Transactions on Visualization and Computer Graphics*, 14(6):1301–1308, 2008.

[MT01]   B. Mohar and C. Thomassen. *Graphs on Surfaces*. Johns Hopkins University Press, 2001.

[MW95]   A. O. Mendelzon and P. T. Wood. Finding regular simple paths in graph databases. *SIAM Journal on Computing*, 24(6):1235–1258, 1995.

[Pok13]   J. Pokorny. NoSQL databases: a step to database scalability in web environment. *International Journal of Web Information Systems*, 9(1):69–82, 2013.

[RKHS13]   M. Radvansky, M. Kudelka, Z. Horak, and V. Snasel. Visualization of social network dynamics using Sammon's projection. In *Proc. of the IEEE 5th International Conference on Computational Aspects of Social Networks*, pages 56–61, 2013.

[RWWE13]   I. Robinson, J. Webber, J. Webber, and E. Eifrem. *Graph Databases*. O'Reilly Media, 2013.

[SCL+09]   L. Shi, N. Cao, S. Liu, W. Qian, L. Tan, G. Wang, J. Sun, and C.-Y. Lin. HiMap: Adaptive visualization of large-scale online social networks. In *Proc. Visualization Symposium, PacificVis*, pages 41–48, 2009.

[SFMB12]   M. Sedlmair, A. Frank, T. Munzner, and A. Butz. RelEx: Visualization for actively changing overlay network specifications. *IEEE Transactions on Visualization and Computer Graphics*, 18(12):2729–2738, 2012.

[SG14]   S. Sakr and M. Gaber. *Large Scale and Big Data: Processing and Management*. CRC Press, 2014.

[SHH11]   D. Selassie, B. Heller, and J. Heer. Divided edge bundling for directional network data. *IEEE Transactions on Visualization and Computer Graphics*, 17(12):2354–2363, 2011.

[SM14]   M. Schich and I. Meirelles. Arts, humanities and complex networks. *Leonardo*, 47(3):265–265, 2014.

[SMDS14]   K. Slaninová, J. Martinovič, P. Dráždilová, and V. Snašel. From Moodle log file to the students network. In *Proc. International Joint Conference SOCO13-CISIS13-ICEUTE13*, pages 641–650, 2014.

[SP12]   S. Sakr and E. Pardede. *Graph data management: techniques and applications*. Information Science Reference, 2012.

[Sri11]   S. Srinivasa. Data, storage and index models for graph databases. In *Graph Data Management: Techniques and Applications*, pages 47–70. IGI Global, 2011.

[STT81]   K. Sugiyama, S. Tagawa, and M. Toda. Methods for visual understanding of hierarchical system structures. *IEEE Transactions on Systems, Man and Cybernetics*, 11(2):109–125, 1981.

[SWG02]   D. Shasha, J. T. Wang, and R. Giugno. Algorithmics and applications of tree and graph searching. In *Proc. of the 21st ACM SIGMOD-SIGACT-SIGART Symposium on Principles of Database Systems*, pages 39–52, 2002.

[TGL10]   K. T. Tekle, M. Gorbovitski, and Y. A. Liu. Graph queries through datalog optimizations. In *Proc. of the 12th International ACM SIGPLAN Symposium on Principles and Practice of Declarative Programming*, pages 25–34, 2010.

[TM12]   Y. Tanahashi and K.-L. Ma. Design considerations for optimizing storyline visualizations. *IEEE Transactions on Visualization and Computer Graphics*, 18(12):2679–2688, 2012.

[Tut63]   W. T. Tutte. How to draw a graph. *Proceedings of the London Mathematical Society*, 13:743–768, 1963.

[VDARS05] W. M. Van Der Aalst, H. A. Reijers, and M. Song. Discovering social networks from event logs. *Computer Supported Cooperative Work*, 14(6):549–593, 2005.

[VLKS+11] T. Von Landesberger, A. Kuijper, T. Schreck, J. Kohlhammer, J. J. van Wijk, J.-D. Fekete, and D. W. Fellner. Visual analysis of large graphs: State-of-the-art and future research challenges. In *Proc. Computer Graphics Forum*, volume 30, pages 1719–1749, 2011.

[War13]   C. Ware. *Information visualization: perception for design*. Elsevier, 2013.

[WF94]   C. Ware and G. Franck. Viewing a graph in a virtual reality display is three times as good as a 2d diagram. In *Proc. IEEE Symposium on Visual Languages*, pages 182–183, 1994.

[WF96]   C. Ware and G. Franck. Evaluating stereo and motion cues for visualizing information nets in three dimensions. *ACM Transactions on Graphics*, 15(2):121–140, 1996.

[WHF93]   C. Ware, D. Hui, and G. Franck. Visualizing object oriented software in three dimensions. In *Proc. of the 1993 Conference of the Centre for Advanced Studies on Collaborative Research: Software Engineering-Volume 1*, pages 612–620, 1993.

[WM08]   C. Ware and P. Mitchell. Visualizing graphs in three dimensions. *ACM Transactions on Applied Perception*, 5(1):2, 2008.

[Woo12]   P. T. Wood. Query languages for graph databases. *ACM SIGMOD Record*, 41(1):50–60, 2012.

[YCHZ12]   X. Yuan, L. Che, Y. Hu, and X. Zhang. Intelligent graph layout using many users' input. *IEEE Transactions on Visualization and Computer Graphics*, 18(12):2699–2708, 2012.

[YFDH01]   K.-P. Yee, D. Fisher, R. Dhamija, and M. Hearst. Animated exploration of dynamic graphs with radial layout. In *Proc. IEEE Symposium on Information Visualization*, pages 43–43, 2001.

[YLZ+13]   J. Yang, Y. Liu, X. Zhang, X. Yuan, Y. Zhao, S. Barlowe, and S. Liu. Piwi: Visually exploring graphs based on their community structure. *IEEE Transactions on Visualization and Computer Graphics*, 19(6):1034–1047, 2013.

[YYH04]   X. Yan, P. S. Yu, and J. Han. Graph indexing: a frequent structure-based approach. In *Proc. of the 2004 ACM SIGMOD International Conference on Management of Data*, pages 335–346, 2004.

[ZBDS12]   M. Zinsmaier, U. Brandes, O. Deussen, and H. Strobelt. Interactive level-of-detail rendering of large graphs. *IEEE Transactions on Visualization and Computer Graphics*, 18(12):2486–2495, 2012.

Index

Printed and bound by CPI Group (UK) Ltd, Croydon, CR0 4YY

23/10/2024

01777708-0015